# Picture Fuzzy Logic and Its Applications in Decision Making Problems

# Advanced Studies in Complex Systems: Theory and Applications

**Series Editors:** Valentina Emilia Balas, Dumitru Baleanu and Hemen Dutta

# Picture Fuzzy Logic and Its Applications in Decision Making Problems

Chiranjibe Jana
Department of Applied Mathematics with Oceanology and Computer Programming
Vidyasagar University
Midnapore, West Bengal, India

Madhumangal Pal
Department of Applied Mathematics with Oceanology and Computer Programming
Vidyasagar University
Midnapore, West Bengal, India

Valentina Emilia Balas
Department of Automatics and Applied Software
Faculty of Engineering
Aurel Vlaicu University of Arad
Arad, Romania

Ronald R. Yager
Iona College
Machine Intelligence Institute
New Rochelle, NY, United States

Academic Press is an imprint of Elsevier
125 London Wall, London EC2Y 5AS, United Kingdom
525 B Street, Suite 1650, San Diego, CA 92101, United States
50 Hampshire Street, 5th Floor, Cambridge, MA 02139, United States
The Boulevard, Langford Lane, Kidlington, Oxford OX5 1GB, United Kingdom



**Notices**

Knowledge and best practice in this field are constantly changing. As new research and experience broaden our understanding, changes in research methods, professional practices, or medical treatment may become necessary.

Practitioners and researchers must always rely on their own experience and knowledge in evaluating and using any information, methods, compounds, or experiments described herein. In using such information or methods they should be mindful of their own safety and the safety of others, including parties for whom they have a professional responsibility.

To the fullest extent of the law, neither the Publisher nor the authors, contributors, or editors, assume any liability for any injury and/or damage to persons or property as a matter of products liability, negligence or otherwise, or from any use or operation of any methods, products, instructions, or ideas contained in the material herein.

ISBN: 978-0-443-22024-1

For information on all Academic Press publications
visit our website at https://www.elsevier.com/books-and-journals

*Publisher:* Mara Conner
*Acquisitions Editor:* Chris Katsaropoulos
*Editorial Project Manager:* Namrata Lama
*Production Project Manager:* Fahmida Sultana
*Cover Designer:* Greg Harris

Typeset by VTeX

# Contents

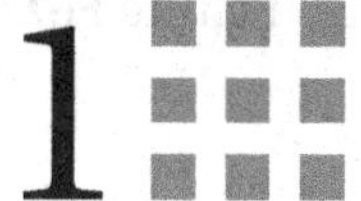

# Introduction to picture fuzzy sets and operators

## 1.1 Introduction

Fuzzy sets (FSs) are essential for describing uncertain, insufficient, or erroneous information. However, FSs are incompetent when they do not understand membership degrees. Then, in 1986, Atanassov [2] created intuitionistic fuzzy sets (IFSs), which are made up of an element's membership and nonmembership degrees. IFSs have received a lot of attention recently and are frequently used to solve problems involving multicriteria decision making (MCDM). These theories have received much attention from researchers in recent years and have been successfully applied to various real-world contexts, including decision making, pattern recognition, medical diagnosis, and clustering analysis [21,22,37]. Also, during the information-fusion process, weighted and ordered weighted aggregation operators [68,69] play a crucial role in aggregating all of the performance of the criteria for alternatives. For aggregating the various intuitionistic fuzzy numbers in that direction, Xu and Yager [65] offered a geometric aggregation operator, and Xu [66] presented a weighted averaging operator (IFNs). Moreover, Garg [23] suggested a number of interactive aggregation operations for IFNs. Using Einstein's $t$-norm and $t$-conorm operations, Garg [21] presented a generalized intuitionistic fuzzy interactive geometric aggregation operator. Wang and Liu [54] expanded these operators by employing Einstein norm operations in an IFS setting. To rank the IVIFS, Garg [24] presented a generalized enhanced score function. The intuitionistic fuzzy Einstein Choquet integral-based operators for decision-making issues were introduced by Xu et al. [67]. Wan et al. [60] provided a strategy for combining the various IVIF numbers with insufficient attribute weight in order to address the decision-making issues. Other than that, numerous MCDM approaches, including Elimination and Choice Expressing Reality (ELECTRE), Weighted Aggregated Sum Product Assessment (WASPAS) [72], Technique for Order Preference by Similarity to an Ideal Solution (TOPSIS) [29,34], TODIM [36], and VIKOR [20], have been proposed to handle MCDM problems.

The concept of neutrosophic (NS) sets developed by Smarandache [43,44] is a general platform that extends the concepts of the classical set and FS [71], IFS [2,15] and IVIFS [4]. In contrast to IFS and IVIFS [15,16], the indeterminacy is characterized explicitly in a neutrosophic set. A neutrosophic set has three basic components such as truth membership (T), indeterminacy membership (I), and falsity membership (F), which are defined independently of one another. However, a neutrosophic set will be more challenging to apply in real scientific and engineering fields. Therefore Wang et al. proposed the concept of a single-valued neutrosophic set (SVNS) and an interval neutrosophic set (INS), which are an

Picture Fuzzy Logic and Its Applications in Decision Making Problems. https://doi.org/10.1016/B978-0-44-322024-1.00005-4

instance of a neutrosophic set, and provide the set-theoretic operators and various properties of SVNSs and INSs. SVNSs present uncertainty, imprecise, inconsistent and incomplete information existing in the real world. Also, handling indeterminate and inconsistent information would be more suitable. Although SVNS and INS have been successfully applied in different areas, there are some real-life situations that SVNS or INS cannot represent. For instance, in the case of voting, human opinions involving more answers of the types: yes, abstain, no, refusal, cannot be accurately represented in a neutrosophic environment. Also, if an expert takes an opinion from a certain person about a certain object, then a person may say that 0.3 is the possibility that the statement is true, 0.4 says that the statement is false, and 0.2 says that he or she is unsure of it. This issue is also not handled by the neutrosophic environment. Thus handling this situation, Cuong [8,11,12] introduced picture fuzzy set (PFS) as a new concept of computational intelligence problems, which is characterized by three functions expressing the degree of membership, the degree of neutral membership, and the degree of the nonmembership. Some authors are currently working on specific issues that arise in the PFSs context. Singh [47] provided a correlation coefficient for the PFS. In order to overcome the clustering problem in an image-fuzzy environment, Son [50] introduced a generalized picture distance measure. Wei [61] provided a strategy for ranking the various options that were based on the picture fuzzy weighted crossentropy. Currently, studies on PFSs and their extensions mainly concentrate on the measures and aggregation operators and their application to MCDM problems and clustering analysis. Peng [40] has developed risk management-based multiattribute decision making (MADM) in PFSs. Later, a PFS analog decision-making measure was calculated by Son [51]. In the same environment, Thong [52] implemented picture composite cardinality and swarm optimization methods for automatic clustering. Selection of risk ranking for energy performance of a contracting project was based on a PFS and a novel MCDM model executed by Wang et al. [55]. In [56], Wang et al. proposed the MCDM method for calculating financial investment risk based on PFS Muirhead mean operators. Risk evaluation of construction project selection problems based on a normalized projection-based VIKOR model has been introduced by Wang et al. [57]. Wang et al. [58] have applied the PFS MCDM method to select a project building energy-efficiency retrofit. Later, the geometric aggregation-based-MADM method was introduced by Wang et al. [59]. Wei [62] has proposed PFS aggregation-based MADM. Again, Wei and others [64] focused on developing a projection-based PFS MADM model. Later, Zhang et al. [73] provided a distanced-based average solution for the MCGDM method under a picture 2-tuple linguistic environment. Picture 2-tuple linguistic aggregation operators based MCDM have been developed by Zhang et al. [74]. Tian et al. [53] studied the notion of a weighted PFS Choquet integral approach under a fuzzy Shapley measure and a power operator and developed these in connection with the MCDM problems. Luo and Zhang [35] utilized PFS in a new similarity measure. Then, Jana et al. [30] used Dombi operators to study MADM in a PFS environment. Singh and Kumar [46] proposed a new frame of the MCDM method for developing quality deployment under PFS. Haktanır and Kahraman [27] used CRITIC and REGIME methodology under PFS to the application of wearable health technology. Later, Haktanır and Kahraman [26] studied the defender–challenger problem for intelligent replacement analysis under PFS. Zhao et al. [76] developed the effective and failure model under the framework of flexible knowledge acquisition using the PFS argument. Jana and Pal [31] developed an enterprise-performance evaluation method using the PFS-Hamacher operator. Jan et al. [32] proposed to study generative adversarial network problems in complex PFS soft environments. Quality service transport-provider selection problems have been introduced by Gündoğdu et al. [25] under PFS-AHP and a linear assessment model. Singh and Ganie [49] used PFS similarity to apply the MADM pattern recognition and clustering model. Simic et al. [45] studied shredding facility location selection problems under the extended picture fuzzy CODAS MCDM method. Recently, Peng et al. [41] studied trust-relation-based social network problems in the PFS group decision-making method. A PFS distance and similarity measure has been developed for application with a complete lattice structure by Jin et al. [33].

## 1.2 Preliminaries

In this section, we annotate some essential ideas of PFSs of the universe. In a fuzzy set $\tilde{A}$, only the membership values of the members are considered. This membership value say $\mu$, $0 \le \mu \le 1$, indicates the belongingness of an element $x$ to the set $\tilde{A}$. It does not mean that $1-\mu$ represents the degree of nonbelongingness of the element $x$ to the set $\tilde{A}$. If it happens for all elements of $\tilde{A}$, then a fuzzy set is sufficient to explain all the nonrandom uncertainties. However, this is not possible for all types of nonrandom uncertainties. Hence, Atanassov [2] incorporated the concept of the nonmembership value of an element to the set $\tilde{A}$ and hence IFS is defined. In IFS, each element is associated with two real numbers, which lie between 0 and 1 and includes them. One is called a membership value, and the other is called a nonmembership value. The formal definition of IFS is given below.

**Definition 1.1.** [2] Let $X$ be the set of the universe. Then, an IFS $P$ over $X$ is defined as $P = \{(t, \mathcal{Y}_P(t), \mathcal{N}_P(t)) : t \in X\}$, where $\mathcal{Y}_P(t) \in [0,1]$ is the measure of membership and $\mathcal{N}_P(t) \in [0,1]$ is the measure of nonmembership of $t$ in $P$ with the condition $0 \leqslant \mathcal{Y}_P(t) + \mathcal{N}_P(t) \leqslant 1$ for all $t \in X$. The quantity $\pi(t) = 1 - (\mathcal{Y}_P(t) + \mathcal{N}_P(t))$ is called the hesitancy of $t$. This value lies between 0 and 1.

For more work on IFS see [3,5,6,13,14].

There are many extensions on IFS. For example, if the sum of squares of all components of all elements is less than or equal to 1, the set is called a Pythagorean fuzzy set (PyFS). If the sum of the cube of all components of an element is less than or equal to 1, then the set is called a Fermatean fuzzy set (FFS). If the sum of the $q$th ($q \ge 1$) powers of the components of an element is less than or equal to 1, the set is called a $q$-rung orthopair fuzzy (q-ROPF) set.

**Definition 1.2.** Let $X$ be the set of the universe. A q-ROPF set $P$ over $X$ is defined as $P = \{(t, \mathcal{Y}_P(t), \mathcal{N}_P(t)) : t \in X\}$, where $\mathcal{Y}_P(t) \in [0,1]$ is the measure of membership and $\mathcal{N}_P(t) \in$

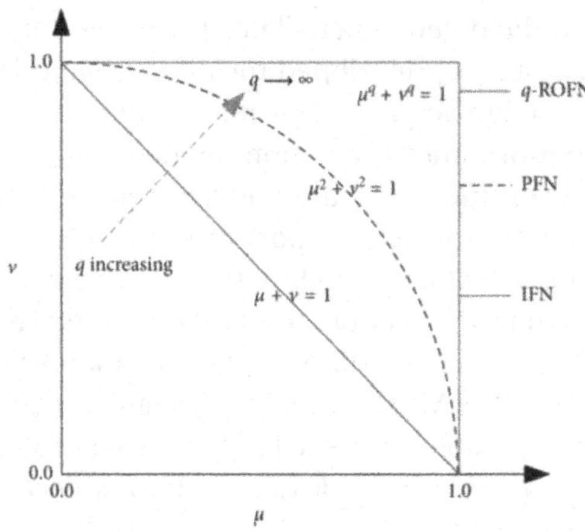


**FIGURE 1.1** Diagrammatic representation of IFS, PyFS, and q-ROPF sets.

[0, 1] is the measure of nonmembership of $t$ in $P$ with the condition $0 \leqslant \mathcal{Y}_P^q(t) + \mathcal{N}_P^q(t) \leqslant 1$ for all $t \in P$. The quantity $\pi(t) = (\mathcal{Y}_P^q(t) + \mathcal{N}_P^q(t) - \mathcal{Y}_P^q(t).\mathcal{N}_P^q(t))^{1/q}$ is called the indeterminacy of $t$.

A q-ROFS becomes IFS when $q = 1$, PyFS when $q = 2$, and FFS when $q = 3$. The diagrammatic representation of these sets is shown in Fig. 1.1.

The IFS is useful when membership and nonmembership cannot explain all possible cases with an element $x$. Suppose there is a proposition $\mathcal{P}$ and a set of people $X$. Assume that $c_1$ number of people from $X$ accept the proposition $\mathcal{P}$, $c_2$ number of people reject the proposition, and the remaining people, say $c_3$, either remain silent or do not give a clear response. In this situation, the degree of acceptance (membership value) is $c_1/|X|$, and the degree of nonacceptance (nonmembership value) is $c_2/|X|$. The value $1 - (c_1/|X| + c_2/|X|)$ (obviously $(c_1 + c_2 + c_3 = |X|)$ is called hesitancy. Note that the remaining $c_3$ number of people is of two types. Either they are neutral, or they give ambiguous responses. By considering the neutral case with IFS, Cuong et al. [8], defined a new type of fuzzy set called a picture fuzzy set. The mathematical definition of a PFS is provided below.

**Definition 1.3.** [8,11] A PFS $P$ over the fixed set $X$ is written as

$$P = \{(\mathcal{Y}_P(t), \mathcal{A}_P(t), \mathcal{N}_P(t)) : t \in X\},$$

$\mathcal{Y}_P(t) : X \to [0, 1]$, $\mathcal{A}_P(t) : X \to [0, 1]$ and $\mathcal{N}_P(t) : X \to [0, 1]$ are, respectively, presented positive membership degree, neutral membership degree and a nonmembership degree in a fuzzy set $P$, where $0 \leq \mathcal{Y}_P(t) + \mathcal{A}_P(t) + \mathcal{N}_P(t) \leq 1$ for $t \in X$. Also, the term $\pi_P(t) = 1 - (\mathcal{Y}_P(t) + \mathcal{A}_P(t) + \mathcal{N}_P(t))$ is called the refusal membership degree for $t$.

Let $P$ be a PFS. The height of $P$ is $h(P) = (h^+(P), h^0(P), h^-(P))$, where $h^+(P) = \sup\{\mathcal{Y}_P(t)\}$, $h^0(P) = \inf\{\mathcal{A}_P(t)\}$, $h^-(P) = \inf\{\mathcal{N}_P(t)\}$ for all $t \in X$.

A PFS $P$ is a picture fuzzy number (PFN) or a picture fuzzy value (PFV) if for at least one $t \in P$, $\sup\{\mathcal{Y}_P(t)\} = 1$, i.e., $h^+(P) = 1$. A PFN $P$ can be denoted by $(\mathcal{Y}_P, \mathcal{A}_P, \mathcal{N}_P)$.

Like FS and IFS, the $(c_1, c_2, c_3)$-cut of a PFS $P$ is a crisp set and it is denoted by $C_{(c_1,c_2,c_3)}(P)$ that is defined as

$$C_{(c_1,c_2,c_3)}(P) = \{(\mathcal{Y}_P(t), \mathcal{A}_P(t), \mathcal{N}_P(t)) : \mathcal{Y}_P(t) \geq c_1, \mathcal{A}_P(t) \leq c_2, \mathcal{N}_P(t) \leq c_3 : t \in X\},$$

where $c_1, c_2, c_3 \in [0, 1]$ and $c_1 + c_2 + c_3 \leq 1$.

The term $\mathcal{N}_P(x)$ is mentioned here as a nonmembership value. However, Cuong et al. and their followers referred to this quantity as a negative membership value. Unfortunately, those who referred to $\mathcal{N}_P(x)$ as a negative membership value considered a positive number as a 'negative membership value' during numerical computation, which violates the meaning of a negative number. The negative membership values are considered in the bipolar fuzzy set, and it has a different meaning, i.e., the negative membership value and the nonmembership value are different. Let us consider the following example to distinguish between nonmembership and negative membership.

Suppose a member of parliament (MP), say Mr. X received some money from the Government to develop his constituency. The following are the possibilities.

**(i)** Mr. X may spend the full amount he received;
**(ii)** Mr. X may spend most of the money on the development with the remaining part returned to the Government;
**(iii)** Mr. X spent a certain amount of money on the development, and the remaining part was spent on his personal work.

Case (i) is certain, and X, in this case, Mr. X will be a popular MP. Sometimes, this case may be treated using a fuzzy set. In case (ii), most people will accept Mr. X, but not all. The unspent money may be used for another constituency by the Government. This situation may be explained with IFS. However, case (iii) is dangerous, i.e., harmful to society, it has a negative impact. He spent the money on his own purposes without taking care of the development of society. Next time, the people will reject him as an MP. This situation can be formulated using a bipolar fuzzy set.

Let us consider another example where PFS is needed to explain the situation properly. Suppose U is a University and an agency is trying to grade the quality of four features of the university, say

$C_1$: Teaching and learning facility;
$C_2$: Placement opportunity;
$C_3$: Library and instrument facility;
$C_4$: Research quality.

Supposed to estimate these features, the agency collects information from a large number of academicians, say $M$, who are very familiar with the university. Note that all these four features cannot be measured as a numerical value, and different people will give different opinions. Hence, a data capture form (DCF) is designed in such a way that the people will respond against each feature as linguistic terms such as

**(a)** 'excellent', 'very very good', 'very good', 'good', 'satisfactory';
**(b)** 'not good', 'bad', 'very bad';
**(c)** 'no comments', 'don't know', 'can't say'.

The terms in (a) are positive responses with different degrees of membership (or acceptance), the terms in (b) are not good for the university, and these are considered as the degree of nonmembership (nonacceptance), and finally, the terms in (c) are considered as the degree of neutral responses.

Now, by counting all the terms of (a) from all responses, the membership value for a pacific feature can be determined by dividing total responses, i.e., by M. Similarly, the nonmembership value and neutral value of all features can be determined. Suppose the total count of the terms of (a) for $C_1$ is $m_1$, that of (b) is $m_3$ and that of (c) is $m_2$. Then, the picture for the representation for $C_1$ is $(m_1/M, m_2/M, m_3/M)$.

Note that $M$ is certain, but $m_1$, $m_2$, and $m_3$ are not certain because the classification of (a), (b), and (c) depends on the agency's mentality, knowledge, opinion, etc. Hence, this is an ideal example of PFS.

From the above examples, it is evident that FS and IFS can apply to any nonrandom uncertain problems, but other variants of FS, like PFS, neutrosophic fuzzy set, etc., do not apply to any nonrandom uncertain problems.

There is an extension of IFS known as the Pythagorean fuzzy set where some of the squares of membership and nonmembership values are less than or equal to 1. Similarly, the spherical fuzzy set (SFS) is defined from PFS where the sum of squares of all components of each element is less than or equal to 1. The formal definition is given below.

**Definition 1.4.** An SFS $P$ defined over the fixed set $X$ is written as

$$P = \{(\mathcal{Y}_P(t), \mathcal{A}_P(t), \mathcal{N}_P(t)) | t \in X\},$$

where $\mathcal{Y}_P(t) : X \to [0, 1]$, $\mathcal{A}_P(t) : X \to [0, 1]$ and $\mathcal{N}_P(t) : X \to [0, 1]$ represent, respectively, membership value, neutral value, and nonmembership value such that $0 \le \mathcal{Y}_P^2(t) + \mathcal{A}_P^2(t) + \mathcal{N}_P^2(t) \le 1$ for all $t \in X$.

The 3-dimensional representations of PFS and SFS are shown in Fig. 1.2.

If the sum of the $q$th ($q \ge 1$) powers of all components of an element of the PFS is less than or equal to 1, the set is known as $q$-rung PFS (q-RPFS), and it is defined below.

**Definition 1.5.** Let $X$ be the universe of discourse. For a given $q$, the q-RPFS $P$ defined over $X$ is

$$P = \{(\mathcal{Y}_P(t), \mathcal{A}_P(t), \mathcal{N}_P(t)) : t \in X\},$$

where $\mathcal{Y}_P(t) : X \to [0, 1]$, $\mathcal{A}_P(t) : X \to [0, 1]$ and $\mathcal{N}_P(t) : X \to [0, 1]$ represent, respectively, membership value, neutral value, and nonmembership value such that $0 \le \mathcal{Y}_P^q(t) + \mathcal{A}_P^q(t) + \mathcal{N}_P^q(t) \le 1$ for all $t \in X$.

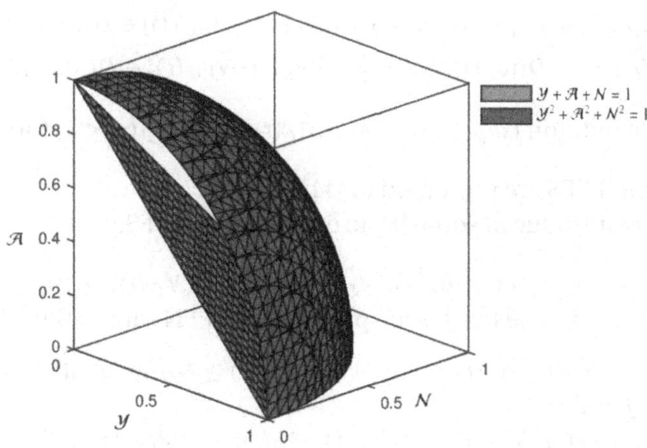


**FIGURE 1.2** 3D representations of PFS and SFS.

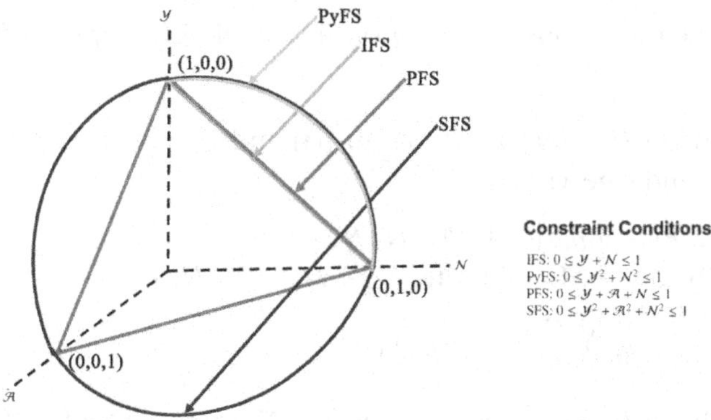


**FIGURE 1.3** 2D representations of IFS, PFS, PyFS, and SFS.

The quantity $\{1-(\mathcal{Y}_P^q(t)+\mathcal{A}_P^q(t)+N_P^q(t))^{1/q}\}$ is called the degree of refusal of $t \in X$.

All these sets are illustrated in Fig. 1.3.

Another extension of PFS is the interval-valued picture fuzzy set (IVPFS), where each component of an element of the PFS is a subinterval of $[0,1]$ instead of a fixed number within $[0,1]$. Its definition is given below. Let $D([0,1])$ be the set of all subintervals of $[0,1]$.

**Definition 1.6.** [11] An IVPFS $P$ on a universe of discourse $X$ is an object of the form

$$P=\{(t,\mathcal{Y}_P(t),\mathcal{A}_P(t),N_P(t)):t\in X\},$$

where

$$\mathcal{Y}_P:X\to D([0,1]),\mathcal{Y}_P(t)=[\mathcal{Y}_{PL}(t),\mathcal{Y}_{PU}(t)]\in D([0,1]),$$

$$\mathcal{A}_P : X \to D([0,1]), \mathcal{A}_P(t) = [\mathcal{A}_{PL}(t), \mathcal{A}_{PU}(t)] \in D([0,1]),$$
$$\mathcal{N}_P : X \to D([0,1]), \mathcal{N}_P(t) = [\mathcal{N}_{PL}(t), \mathcal{N}_{PU}(t)] \in D([0,1]),$$

which satisfies the condition $(sup\ \mathcal{Y}_P(t) + sup\ \mathcal{A}_P(t) + sup\ \mathcal{N}_P(t) \le 1)$ for all $t \in X$.

Several results on IFPFS are presented in [11].
Cuong et al. in [8] introduced some basic operations on PFS.

**Definition 1.7.** [8] Let $P = (\mathcal{Y}_P(t), \mathcal{A}_P(t), \mathcal{N}_P(t))$ and $Q = (\mathcal{Y}_Q(t), \mathcal{A}_Q(t), \mathcal{N}_Q(t))$ be any two PFNs over the set $X$, $t \in X$. The five basic operations on PFNs are defined below:

**(i)** $P \subseteq Q$, if $\mathcal{Y}_P(t) \le \mathcal{Y}_Q(t)$, $\mathcal{A}_P(t) \le \mathcal{A}_Q(t)$ and $\mathcal{N}_P(t) \ge \mathcal{N}_Q(t)$ for all $t \in X$;
**(ii)** $P = Q$ iff $P \subseteq Q$ and $Q \subseteq P$;
**(iii)** $P \cup Q = \{(t, \max\{\mathcal{Y}_P(t), \mathcal{Y}_Q(t)\}, \min\{\mathcal{A}_P(t), \mathcal{A}_Q(t)\}, \min\{\mathcal{N}_P(t), \mathcal{N}_Q\}) : t \in X\}$;
**(iv)** $P \cap Q = \{(t, \min\{\mathcal{Y}_P(t), \mathcal{Y}_Q(t)\}, \max\{\mathcal{A}_P(t), \mathcal{A}_Q(t)\}, \max\{\mathcal{N}_P(t), \mathcal{N}_Q(t)\}) : t \in X\}$;
**(v)** $\overline{P} = \{(t, \mathcal{N}_P(t), \mathcal{A}_P(t), \mathcal{Y}_P(t)) : t \in X\}$.

Depending on the operations of [65,66] on IFSs, Wei [62] proposed the following few operations on PFNs.

**Definition 1.8.** [62] Let $U = (\mathcal{N}_U(t), \mathcal{A}_U(t), \mathcal{Y}_U(t))$ and $V = (\mathcal{N}_V(t), \mathcal{A}_V(t), \mathcal{Y}_V(t))$ be two PFNs over the set $X$ and $\lambda$ be a scalar,

**(i)** $U \bigoplus V = \left(\mathcal{Y}_U + \mathcal{Y}_V - \mathcal{Y}_U\mathcal{Y}_V, \mathcal{A}_U\mathcal{A}_V, \mathcal{N}_U\mathcal{N}_V\right)$;
**(ii)** $U \bigotimes V = (\mathcal{Y}_U\mathcal{Y}_V, \mathcal{A}_U + \mathcal{A}_V - \mathcal{A}_U\mathcal{A}_V, \mathcal{N}_U + \mathcal{N}_V - \mathcal{N}_U\mathcal{N}_V)$;
**(iii)** $\lambda U = \left(1 - (1 - \mathcal{Y}_U)^\lambda, \mathcal{A}_U^\lambda, \mathcal{N}_U^\lambda\right)$;
**(iv)** $U^\lambda = \left(\mathcal{Y}_U^\lambda, 1 - (1 - \mathcal{A}_U)^\lambda, 1 - (1 - \mathcal{N}_U)^\lambda\right)$.

**Definition 1.9.** [30] Let $U = (\mathcal{Y}_U, \mathcal{A}_U, \mathcal{N}_U)$ be a PFN, then the score $\Lambda(U)$ and accuracy $\Phi(U)$ for PFN are defined as follows:

$$\Lambda(U) = \frac{1 + \mathcal{Y}_U - \mathcal{N}_U}{2}, \quad \Lambda(U) \in [0, 1],$$
$$\Phi(U) = \mathcal{Y}_U - \mathcal{N}_U, \quad \Phi(U) \in [-1, 1]. \tag{1.1}$$

Based on Definition 1.9, prioritized relations between two PFNs $U$ and $V$ are defined in the following ways.

**Definition 1.10.** [30] Let $U$ and $V$ be two PFNs. Then,

**(i)** If $\Lambda(U) < \Lambda(V)$, then $U \prec V$;
**(ii)** If $\Lambda(U) > \Lambda(V)$, then $U \succ V$;
**(iii)** If $\Lambda(U) = \Lambda(V)$, then
  **(a)** If $\Phi(U) < \Phi(V)$, then $U \prec V$;
  **(b)** If $\Phi(U) > \Phi(V)$, then $U \succ V$;
  **(c)** If $\Phi(U) = \Phi(V)$, then $U \sim V$.

Wei [62] derived the following operations.

**Definition 1.11.** Let us suppose that $U = (\mathcal{Y}_U, \mathcal{A}_U, \mathcal{N}_U)$ and $V = (\mathcal{Y}_V, \mathcal{A}_V, \mathcal{N}_V)$ are two PFNs on $X$ and $\lambda, \lambda_1, \lambda_2 > 0$, then

**(i)** $U \bigoplus V = V \bigoplus U$;
**(ii)** $U \bigotimes V = V \bigotimes U$;
**(iii)** $\lambda(U \bigoplus V) = \lambda U \bigoplus \lambda V$;
**(iv)** $(U \bigotimes V)^{\lambda} = U^{\lambda} \bigotimes V^{\lambda}$;
**(v)** $\lambda_1 U \bigoplus \lambda_2 U = (\lambda_1 + \lambda_2)U$;
**(vi)** $U^{\lambda_1} \bigotimes U^{\lambda_2} = U^{(\lambda_1+\lambda_2)}$;
**(vii)** $(U^{\lambda_1})^{\lambda_2} = U^{\lambda_1\lambda_2}$.

Some more operators on PFSs are defined below.

**Definition 1.12.** Let $U = \{(t, \mathcal{Y}_U(t), \mathcal{A}_U(t), \mathcal{N}_U(t)) : t \in X\}$ and $V = \{(t, \mathcal{Y}_V(t), \mathcal{A}_V(t), \mathcal{N}_V(t)) : t \in X\}$ be two PFSs defined over the universe $X$. The following operations are defined:

**(i)** $\boxplus U = \left\{\left(t, \frac{\mathcal{Y}_U(t)}{2}, \frac{\mathcal{A}_U(t)}{2}, \frac{1+\mathcal{N}_U(t)}{2}\right) : t \in X\right\}$;
**(ii)** $\diamond U = \{(t, 1 - \mathcal{A}_U(t) - \mathcal{N}_U(t), \mathcal{A}_U(t), \mathcal{N}_U(t)) : t \in X\}$;
**(iii)** $\Box U = \{(t, \mathcal{Y}_U(t), \mathcal{A}_U(t), 1 - \mathcal{Y}_U(t) - \mathcal{A}_U(t)) : t \in X\}$;
**(iv)** $U \cdot V = \{(t, \mathcal{Y}_U(t) \cdot \mathcal{Y}_V(t), \mathcal{A}_U(t) \cdot \mathcal{A}_V(t), \mathcal{N}_U(t) + \mathcal{N}_V(t) - \mathcal{N}_U(t) \cdot \mathcal{N}_V(t)) : t \in X\}$;
**(v)** $U \circ V = \{(t, \sqrt{\mathcal{Y}_U(t) \cdot \mathcal{Y}_V(t)}, \sqrt{\mathcal{A}_U(t) \cdot \mathcal{A}_V(t)}, \sqrt{\mathcal{N}_U(t) \cdot \mathcal{N}_V(t)}) : t \in X\}$;
**(vi)** $U * V = \left\{\left(t, \frac{\mathcal{Y}_U(t)+\mathcal{Y}_V(t)}{2}, \frac{\mathcal{A}_U(t)+\mathcal{A}_V(t)}{2}, \frac{\mathcal{N}_U(t)+\mathcal{N}_V(t)}{2}\right) : t \in X\right\}$;
**(vii)** $U \# V = \left\{\left(t, \frac{2\mathcal{Y}_U(t)\cdot\mathcal{Y}_V(t)}{\mathcal{Y}_U(t)+\mathcal{Y}_V(t)}, \frac{2\mathcal{A}_U(t)\cdot\mathcal{A}_V(t)}{\mathcal{A}_U(t)+\mathcal{A}_V(t)}, \frac{2\mathcal{N}_U(t)\cdot\mathcal{N}_V(t)}{\mathcal{N}_U(t)+\mathcal{N}_V(t)}\right) : t \in X\right\}$.

Some interesting properties related to the above operators are presented below.

**Proposition 1.1.** *Let* $U = \{(t, \mathcal{Y}_U(t), \mathcal{A}_U(t), \mathcal{N}_U(t)) : t \in X\}$ *and* $V = \{(t, \mathcal{Y}_V(t), \mathcal{A}_V(t), \mathcal{N}_V(t)) : t \in X\}$ *be two PFSs over the universe* $X$. *Then,*

**(i)** $\boxplus(U \cap V) = (\boxplus U) \cap (\boxplus V)$;
**(ii)** $\boxplus\Box U = \Box \boxplus U$;
**(iii)** $(\boxplus U) * (\boxplus V) = \boxplus(U * V)$;
**(iv)** $(\boxplus U) \circ (\boxplus V) \subseteq \boxplus(U \circ V)$;
**(v)** $\boxplus(U \cdot V) = U \cdot (\boxplus V)$;
**(vi)** $\overline{\Box\overline{U}} = \diamond U$;
**(vii)** $\overline{\diamond\overline{U}} = \Box U$;
**(viii)** $(\Box U) \cup (\diamond U) * ((\Box U) \cap (\diamond U) = (\Box U) * (\diamond U)$;
**(ix)** $\{(\Box U) * (\diamond U)\} \circ \{(\Box U)\#(\diamond U)\} = (\Box U) \circ (\diamond U)$.

*Proof.* (i)

$$\begin{aligned}&\boxplus(U \cap V)\\&= \boxplus\{(t, \mathcal{Y}_U(t) \wedge \mathcal{Y}_V(t), \mathcal{A}_U(t) \wedge \mathcal{A}_V(t), \mathcal{N}_U(t) \vee \mathcal{N}_V(t)) : t \in X\}\end{aligned}$$

$$= \left\{\left(t, \frac{\mathcal{Y}_U(t) \wedge \mathcal{Y}_V(t)}{2}, \frac{\mathcal{A}_U(t) \wedge \mathcal{A}_V(t)}{2}, \frac{1 + \mathcal{N}_U(t) \vee \mathcal{N}_V(t)}{2}\right) : t \in X\right\}$$
$$= \left\{\left(t, \frac{\mathcal{Y}_U(t) \wedge \mathcal{Y}_V(t)}{2}, \frac{\mathcal{A}_U(t) \wedge \mathcal{A}_V(t)}{2}, \frac{(1 + \mathcal{N}_U(t)) \vee (1 + \mathcal{N}_V(t))}{2}\right) : t \in X\right\}$$
$$= \left\{\left(t, \frac{\mathcal{Y}_U(t)}{2} \wedge \frac{\mathcal{Y}_V(t)}{2}, \frac{\mathcal{A}_U(t)}{2} \wedge \frac{\mathcal{A}_V(t)}{2}, \frac{(1 + \mathcal{N}_U(t))}{2} \vee \frac{(1 + \mathcal{N}_V(t))}{2}\right) : t \in X\right\}$$
$$= (\boxplus U) \cap (\boxplus V).$$

(ii) As per the definition

$$\boxplus \Box U = \boxplus\{(t, \mathcal{Y}_U(t), \mathcal{A}_U(t), 1 - \mathcal{Y}_U(t) - \mathcal{A}_U(t)) : t \in X\}$$
$$= \left\{\left(t, \frac{\mathcal{Y}_U(t)}{2}, \frac{\mathcal{A}_U(t)}{2}, \frac{1 + (1 - \mathcal{Y}_U(t) - \mathcal{A}_U(t))}{2}\right) : t \in X\right\}$$
$$= \left\{\left(t, \frac{\mathcal{Y}_U(t)}{2}, \frac{\mathcal{A}_U(t)}{2}, \frac{2 - \mathcal{Y}_U(t) - \mathcal{A}_U(t)}{2}\right) : t \in X\right\}$$
$$\text{and } \Box \boxplus U = \Box\left\{\left(t, \frac{\mathcal{Y}_U(t)}{2}, \frac{\mathcal{A}_U(t)}{2}, \frac{1 + \mathcal{N}_U(t)}{2}\right) : t \in X\right\}$$
$$= \left\{\left(t, \frac{\mathcal{Y}_U(t)}{2}, \frac{\mathcal{A}_U(t)}{2}, 1 - \frac{\mathcal{Y}_U(t)}{2} - \frac{\mathcal{A}_U(t)}{2}\right) : t \in X\right\}$$
$$= \left\{\left(t, \frac{\mathcal{Y}_U(t)}{2}, \frac{\mathcal{A}_U(t)}{2}, \frac{2 - \mathcal{Y}_U(t) - \mathcal{A}_U(t)}{2}\right) : t \in X\right\}.$$

Therefore $\boxplus \Box U = \Box \boxplus U$.

(iii) We see that

$$\boxplus U = \left\{\left(t, \frac{\mathcal{Y}_U(t)}{2}, \frac{\mathcal{A}_U(t)}{2}, \frac{1 + \mathcal{N}_U(t)}{2}\right) : t \in X\right\}$$
$$\text{and } \boxplus V = \left\{\left(t, \frac{\mathcal{Y}_V(t)}{2}, \frac{\mathcal{A}_V(t)}{2}, \frac{1 + \mathcal{N}_V(t)}{2}\right) : t \in X\right\}.$$

Now,

$$(\boxplus U) * (\boxplus V)$$
$$= \left\{\left(t, \frac{\mathcal{Y}_U(t) + \mathcal{Y}_V(t)}{4}, \frac{\mathcal{A}_U(t) + \mathcal{A}_V(t)}{4}, \frac{2 + \mathcal{N}_U(t) + \mathcal{N}_V(t)}{4}\right) : t \in X\right\}.$$

Also,

$$\boxplus (U * V)$$
$$= \boxplus\left\{\left(t, \frac{\mathcal{Y}_U(t) + \mathcal{Y}_V(t)}{2}, \frac{\mathcal{A}_U(t) + \mathcal{A}_V(t)}{2}, \frac{\mathcal{N}_U(t) + \mathcal{N}_V(t)}{2}\right) : t \in X\right\}$$
$$= \left\{\left(t, \frac{\mathcal{Y}_U(t) + \mathcal{Y}_V(t)}{4}, \frac{\mathcal{A}_U(t) + \mathcal{A}_V(t)}{4}, \frac{2 + \mathcal{N}_U(t) + \mathcal{N}_V(t)}{4}\right) : t \in X\right\}.$$

Therefore $(\boxplus U) * (\boxplus V) = \boxplus(U * V)$.
(iv) Here,

$$\begin{aligned}
&\boxplus (U \circ V)\\
&= \boxplus\{(t, \sqrt{\mathcal{Y}_U(t)\cdot\mathcal{Y}_V(t)}, \sqrt{\mathcal{A}_U(t)\cdot\mathcal{A}_V(t)}, \sqrt{\mathcal{N}_U(t)\cdot\mathcal{N}_V(t)}) : t \in X\}\\
&= \left\{\left(t, \frac{\sqrt{\mathcal{Y}_U(t)\cdot\mathcal{Y}_V(t)}}{2}, \frac{\sqrt{\mathcal{A}_U(t)\cdot\mathcal{A}_V(t)}}{2}, \frac{1+\sqrt{\mathcal{N}_U(t)\cdot\mathcal{N}_V(t)}}{2}\right) : t \in X\right\}
\end{aligned}$$

and

$$\begin{aligned}
&(\boxplus U) \circ (\boxplus V)\\
&= \left\{\left(t, \frac{\mathcal{Y}_U(t)}{2}, \frac{\mathcal{A}_U(t)}{2}, \frac{1+\mathcal{N}_U(t)}{2}\right) : t \in X\right\} \circ \left\{\left(t, \frac{\mathcal{Y}_V(t)}{2}, \frac{\mathcal{A}_V(t)}{2}, \frac{1+\mathcal{N}_V(t)}{2}\right) : t \in X\right\}\\
&= \left\{\left(t, \frac{\sqrt{\mathcal{Y}_U(t)\cdot\mathcal{Y}_V(t)}}{2}, \frac{\sqrt{\mathcal{A}_U(t)\cdot\mathcal{A}_V(t)}}{2}, \frac{\sqrt{(1+\mathcal{N}_U(t))\cdot(1+\mathcal{N}_V(t))}}{2}\right) : t \in X\right\}.
\end{aligned}$$

Let $\mathcal{N}_U(t) = p$ and $\mathcal{N}_V(t) = q$.
Our aim is to show $1 + \sqrt{\mathcal{N}_U(t)\cdot\mathcal{N}_V(t)} \leqslant \sqrt{(1+\mathcal{N}_U(t))(1+\mathcal{N}_V(t))}$

i.e., to show $1 + \sqrt{pq} \leqslant \sqrt{(1+p)\cdot(1+q)}$,

i.e., to show $(1+\sqrt{pq})^2 \leqslant (1+p)(1+q)$.

It is observed that $-(\sqrt{p}-\sqrt{q})^2 \leqslant 0$

i.e., $2\sqrt{pq} - p - q \leqslant 0$,

i.e., $1 + 2\sqrt{pq} + pq \leqslant 1 + p + q + pq$,

i.e., $(1+\sqrt{pq})^2 \leqslant (1+p)(1+q)$.

Thus $(\boxplus U) \circ (\boxplus V) \subseteq \boxplus(U \circ V)$.
(v) Here,

$$\begin{aligned}
&\boxplus (U \cdot V)\\
&= \boxplus\{(t, \mathcal{Y}_U(t)\cdot\mathcal{Y}_V(t), \mathcal{A}_U(t)\cdot\mathcal{A}_V(t), \mathcal{N}_U(t) + \mathcal{N}_V(t) - \mathcal{N}_U(t)\cdot\mathcal{N}_V(t)) : t \in X\}\\
&= \left\{\left(t, \frac{\mathcal{Y}_U(t)\cdot\mathcal{Y}_V(t)}{2}, \frac{\mathcal{A}_U(t)\cdot\mathcal{A}_V(t)}{2}, \frac{1+\mathcal{N}_U(t)+\mathcal{N}_V(t)-\mathcal{N}_U(t)\cdot\mathcal{N}_V(t)}{2}\right) : t \in X\right\}
\end{aligned}$$

and

$$\begin{aligned}
&U \cdot (\boxplus V)\\
&= \left\{\left(t, \frac{\mathcal{Y}_U(t)\cdot\mathcal{Y}_V(t)}{2}, \frac{\mathcal{A}_U(t)\cdot\mathcal{A}_V(t)}{2}, \mathcal{N}_U(t) + \frac{1+\mathcal{N}_V(t)}{2} - \mathcal{N}_U(t)\cdot\frac{1+\mathcal{N}_V(t)}{2}\right) : t \in X\right\}\\
&= \left\{\left(t, \frac{\mathcal{Y}_U(t)\cdot\mathcal{Y}_V(t)}{2}, \frac{\mathcal{A}_U(t)\cdot\mathcal{A}_V(t)}{2}, \frac{1+\mathcal{N}_U(t)+\mathcal{N}_V(t)-\mathcal{N}_U(t)\cdot\mathcal{N}_V(t)}{2}\right) : t \in X\right\}.
\end{aligned}$$

Thus $\boxplus(U \cdot V) = U \cdot (\boxplus V)$.
(vi) It is viewed that

$$\begin{aligned}\overline{\Box\overline{U}} &= \overline{\Box\{(t, \mathcal{N}_U(t), \mathcal{A}_U(t), \mathcal{Y}_U(t)) : t \in X\}} \\ &= \overline{\{(t, \mathcal{N}_U(t), \mathcal{A}_U(t), 1 - \mathcal{A}_U(t) - \mathcal{N}_U(t)) : t \in X\}} \\ &= \{(t, 1 - \mathcal{A}_U(t) - \mathcal{N}_U(t), \mathcal{A}_U(t), \mathcal{N}_U(t)) : t \in X\} \\ &= \diamond\{(t, \mathcal{Y}_U(t), \mathcal{A}_U(t), \mathcal{N}_U(t)) : t \in X\} \\ &= \diamond U.\end{aligned}$$

(vii) It is observed that

$$\begin{aligned}\overline{\diamond\overline{U}} &= \overline{\diamond\{(t, \mathcal{N}_U(t), \mathcal{A}_U(t), \mathcal{Y}_U(t)) : t \in X\}} \\ &= \overline{\{(t, 1 - \mathcal{Y}_U(t) - \mathcal{A}_U(t), \mathcal{A}_U(t), \mathcal{Y}_U(t)) : t \in X\}} \\ &= \{(t, \mathcal{Y}_U(t), \mathcal{A}_U(t), 1 - \mathcal{Y}_U(t) - \mathcal{A}_U(t)) : t \in X\} \\ &= \Box\{(t, \mathcal{Y}_U(t), \mathcal{A}_U(t), \mathcal{N}_U(t)) : t \in X\} \\ &= \Box U.\end{aligned}$$

(viii) We have,

$$\begin{aligned}\Box U &= \{(t, \mathcal{Y}_U(t), \mathcal{A}_U(t), 1 - \mathcal{Y}_U(t) - \mathcal{A}_U(t)) : t \in X\} \\ \text{and } \diamond U &= \{(t, 1 - \mathcal{A}_U(t) - \mathcal{N}_U(t), \mathcal{A}_U(t), \mathcal{N}_U(t)) : t \in X\}.\end{aligned}$$

It is clear that $\mathcal{Y}_U(t) \leqslant 1 - \mathcal{A}_U(t) - \mathcal{N}_U(t)$ because it will imply

$$\mathcal{Y}_U(t) + \mathcal{A}_U(t) + \mathcal{N}_U(t) \leqslant 1 \text{ and } \mathcal{N}_U(t) \leqslant 1 - \mathcal{Y}_U(t) - \mathcal{A}_U(t).$$

Therefore

$$\begin{aligned}\{(\Box U) \cup (\diamond U)\} &= \{(t, 1 - \mathcal{A}_U(t) - \mathcal{N}_U(t), \mathcal{A}_U(t), \mathcal{N}_U(t) : t \in X\} \\ \text{and } \{(\Box P) \cap (\diamond P)\} &= \{(t, \mathcal{Y}_U(t), \mathcal{A}_U(t), 1 - \mathcal{Y}_U(t) - \mathcal{A}_U(t) : t \in X\}.\end{aligned}$$

Thus

$$\begin{aligned}&\{(\Box U) \cup (\diamond U)\} * \{(\Box U) \cap (\diamond U)\} \\ &= \left\{\left(t, \frac{1 + \mathcal{Y}_U(t) - \mathcal{A}_U(t) - \mathcal{N}_U(t)}{2}, \mathcal{A}_U(t), \frac{1 + \mathcal{N}_U(t) - \mathcal{Y}_U(t) - \mathcal{A}_U(t)}{2}\right) : t \in X\right\}.\end{aligned}$$

Also,

$$\begin{aligned}&(\Box U) * (\diamond U) \\ &= \left\{\left(t, \frac{1 + \mathcal{Y}_U(t) - \mathcal{A}_U(t) - \mathcal{N}_U(t)}{2}, \mathcal{A}_U(t), \frac{1 + \mathcal{N}_U(t) - \mathcal{Y}_U(t) - \mathcal{A}_U(t)}{2}\right) : t \in X\right\}.\end{aligned}$$

Therefore $\{(\Box U) \cup (\diamond U)\} * \{(\Box U) \cap (\diamond U)\} = (\Box U) * (\diamond U)$.
(ix) We have,

$$\Box U = \{(t, \mathcal{Y}_U(t), \mathcal{A}_U(t), 1 - \mathcal{Y}_U(t) - \mathcal{A}_U(t)) : t \in X\}$$
$$\text{and } \diamond U = \{(t, 1 - \mathcal{A}_U(t) - \mathcal{N}_U(t), \mathcal{A}_U(t), \mathcal{N}_U(t)) : t \in X\}.$$

Therefore

$$\{(\Box U) * (\diamond U)\}$$
$$= \left\{\left(t, \frac{1 + \mathcal{Y}_U(t) - \mathcal{A}_U(t) - \mathcal{N}_U(t)}{2}, \mathcal{A}_U(t), \frac{1 + \mathcal{N}_U(t) - \mathcal{Y}_U(t) - \mathcal{A}_U(t)}{2}\right) : t \in X\right\}$$

and $\{(\Box U)\#(\diamond U)\}$

$$= \left\{\left(t, \frac{2\mathcal{Y}_U(t)(1 - \mathcal{A}_U(t) - \mathcal{N}_U(t))}{1 + \mathcal{Y}_U(t) - \mathcal{A}_U(t) - \mathcal{N}_U(t)}, \mathcal{A}_U(t), \frac{2\mathcal{N}_U(t)(1 - \mathcal{Y}_U(t) - \mathcal{A}_U(t))}{1 + \mathcal{N}_U(t) - \mathcal{Y}_U(t) - \mathcal{A}_U(t)}\right) : t \in X\right\}.$$

Thus

$$\{(\Box U) * (\diamond U)\} \circ \{(\Box U)\#(\diamond U)\}$$
$$= \{(t, \sqrt{\mathcal{Y}_U(t)(1 - \mathcal{A}_U(t) - \mathcal{N}_U(t))}, \mathcal{A}_U(t), \sqrt{\mathcal{N}_U(t)(1 - \mathcal{Y}_U(t) - \mathcal{A}_U(t)} : t \in X\}.$$

Also,

$$(\Box U) \circ (\diamond U)$$
$$= \{(t, \sqrt{\mathcal{Y}_U(t)(1 - \mathcal{A}_U(t) - \mathcal{N}_U(t))}, \mathcal{A}_U(t), \sqrt{\mathcal{N}_U(t)(1 - \mathcal{Y}_U(t) - \mathcal{A}_U(t))} : t \in X\}.$$

Therefore $\{(\Box U) * (\diamond U)\} \circ \{(\Box U)\#(\diamond U)\} = (\Box U) \circ (\diamond U)$. □

## 1.3 Relation on picture fuzzy set

Let $X_1$, $X_2$, and $X_3$ be three standard nonempty sets, i.e., a universe of discourses. The IF relation defines the picture fuzzy relation.

**Definition 1.13.** A picture fuzzy relation (PFR) $\rho$ is a picture fuzzy subset of $X_1 \times X_2$ that is defined as

$$\rho = \{((s, t), \mathcal{Y}_\rho(s, t), \mathcal{A}_\rho(s, t), \mathcal{N}_\rho(s, t)) : s \in X_1, t \in X_2\}, \tag{1.2}$$

where $\mathcal{Y}_\rho : X_1 \times X_2 \to [0, 1]$, $\mathcal{A}_\rho : X_1 \times X_2 \to [0, 1]$, $\mathcal{N}_\rho : X_1 \times X_2 \to [0, 1]$ satisfy the condition $\mathcal{Y}_\rho(s, t) + \mathcal{A}_\rho(s, t) + \mathcal{N}_\rho(s, t) \leq 1$ for all $(s, t) \in (X_1 \times X_2$.

Let $PFR(X_1 \times X_2)$ be the set of all picture fuzzy subsets on $X_1 \times X_2$.
The primary comparison operations on the PFRs are similar to PFSs.

**Definition 1.14.** Let $\rho_1$ and $\rho_2$ be two PFRs between $X_1$ and $X_2$. Then, for all $(s,t) \in X_1 \times X_2$,

**(i)** $\rho_1 \leq \rho_2 \Leftrightarrow (\mathcal{Y}_{\rho_1}(s,t) \leq \mathcal{Y}_{\rho_2}(s,t))$ and $(\mathcal{A}_{\rho_1}(s,t) \leq \mathcal{A}_{\rho_2}(s,t))$ and $(\mathcal{N}_{\rho_1}(s,t) \geq \mathcal{N}_{\rho_2}(s,t))$, (1.3)

**(ii)** $\rho_1 \wedge \rho_2 = \{((s,t), \mathcal{Y}_{\rho_1}(s,t) \wedge \mathcal{Y}_{\rho_2}(s,t), \mathcal{A}_{\rho_1}(s,t) \wedge \mathcal{A}_{\rho_2}(s,t), \mathcal{N}_{\rho_1}(s,t) \vee \mathcal{N}_{\rho_2}(s,t)) : s \in X_1, t \in X_2\}$, (1.4)

**(iii)** $\rho_1 \vee \rho_2 = \{((s,t), \mathcal{Y}_{\rho_1}(s,t) \vee \mathcal{Y}_{\rho_2}(s,t), \mathcal{A}_{\rho_1}(s,t) \wedge \mathcal{A}_{\rho_2}(s,t), \mathcal{N}_{\rho_1}(s,t) \wedge \mathcal{N}_{\rho_2}(s,t)) : s \in X_1, t \in X_2\}$, (1.5)

**(iv)** $\bar{\rho}_1 = \{((s,t), \mathcal{N}_{\rho_1}(s,t), \mathcal{A}_{\rho_1}(s,t), \mathcal{Y}_{\rho_1}(s,t)) : s \in X_1, t \in X_2\}$. (1.6)

**Proposition 1.2.** *[10] Let $\rho_1, \rho_2, \rho_3 \in PFR(X_1 \times X_2)$. Then,*

**(i)** $\rho_1 \wedge (\rho_2 \vee \rho_3) = (\rho_1 \wedge \rho_2) \vee (\rho_1 \wedge \rho_3)$;
**(ii)** $\rho_1 \vee (\rho_2 \wedge \rho_3) = (\rho_1 \vee \rho_2) \wedge (\rho_1 \vee \rho_3)$;
**(iii)** $\rho_1 \wedge \rho_2 \leq \rho_1$, $\rho_1 \wedge \rho_2 \leq \rho_2$;
**(iv)** *If* $(\rho_1 \geq \rho_2)$ *and* $(\rho_1 \geq \rho_3)$, *then* $(\rho_1 \geq \rho_2 \vee \rho_3)$;
**(v)** *If* $(\rho_1 \leq \rho_2)$ *and* $(\rho_1 \leq \rho_3)$, *then* $(\rho_1 \leq \rho_2 \wedge \rho_3)$.

The inverse relation is defined below.

**Definition 1.15.** [8] Let $\rho$ be a relation. The inverse relation, denoted as $\rho^{-1}$ between $X_2$ and $X_1$, is

$$\rho^{-1} = \{((t,s), \mathcal{Y}_{\rho^{-1}}(t,s), \mathcal{A}_{\rho^{-1}}(t,s), \mathcal{N}_{\rho^{-1}}(t,s)) : s \in X_1, t \in X_2\}, \tag{1.7}$$

where

$$\mathcal{Y}_{\rho^{-1}}(t,s) = \mathcal{Y}_{\rho}(s,t),$$
$$\mathcal{A}_{\rho^{-1}}(t,s) = \mathcal{A}_{\rho}(s,t),$$
$$\mathcal{N}_{\rho^{-1}}(t,s) = \mathcal{N}_{\rho}(s,t)$$

for all $(s,t) \in (X_1 \times X_2$.

A few results on inverse relations are given below.

**Proposition 1.3.** *Let $\rho_1, \rho_2, \rho_3$ be two relations on $X_1 \times X_2$. Then,*

**(i)** $((\rho_1)^{-1})^{-1} = \rho_1$;
**(ii)** $\rho_1 \leq \rho_2 \Rightarrow (\rho_1)^{-1} \leq (\rho_2)^{-1}$;
**(iii)** $(\rho_1 \vee \rho_2)^{-1} = (\rho_1)^{-1} \vee (\rho_2)^{-1}$;
**(iv)** $(\rho_1 \wedge \rho_2)^{-1} = (\rho_1)^{-1} \wedge (\rho_2)^{-1}$.

## 1.4 Picture fuzzy graph

In crisp graph theory, a graph is denoted as $G = (V, E)$, where $V$ is called the set of vertices and $E$ is the set of edges, and all the vertices and edges of crisp graphs are certain. However, in some real-life problems, it is seen that vertices and/or edges are uncertain, e.g., in social networks. To represent such a graph, the concept of the fuzzy graph (FG) and its variants/extensions are introduced by many graph theoreticians. An FG is denoted as $G = (V, \sigma, \mu)$, where $V$ is the set of vertices and $\sigma, \nu$ are two functions associated with a vertex and an edge, respectively. That is, $\sigma(s)$ represents the membership values of the vertex $s$ in $G$ and $\mu(s, t)$ represents the membership values of the edge $(s, t)$ in the graph $G$. In a crisp graph, a specific subset of $V \times V$ is a set of edges, while in FG, the set $V \times V$ is the set of edges. However, only the edges whose membership values are strictly greater than zero are considered as the set of edges, and this is denoted as $E$. Generally, in FG, the set of edges is not directly mentioned, but it is obvious there must be a set of edges. By incorporating the nonmembership values to each vertex and edge, the intuitionistic fuzzy graph (IFG) is defined. Like FG and IFG, the picture fuzzy graph (PFG) is also defined in a similar manner.

**Definition 1.16.** [77] Let $A = (\mathcal{Y}_A, \mathcal{A}_A, \mathcal{N}_A)$ be a PFS defined on a universe $V$ and $B = (\mathcal{Y}_B, \mathcal{A}_B, \mathcal{N}_B)$ be another PFS defined on $V \times V$. The tuple $G = (V, A, B)$ is called a PFG if for any $(s, t) \in V \times V$, such that

$$\begin{aligned} \mathcal{Y}_B(s,t) &\leq \max(\mathcal{Y}_A(s), \mathcal{Y}_A(t)), \\ \mathcal{A}_B(s,t) &\leq \min(\mathcal{A}_A(s), \mathcal{A}_A(t)), \\ \mathcal{N}_B(s,t) &\leq \min(\mathcal{N}_A(s), \mathcal{N}_A(t)). \end{aligned} \tag{1.8}$$

Also, $\mathcal{Y}_A(s) + \mathcal{A}_A(s) + \mathcal{N}_A(s) \leq 1$ for all $s \in V$ and $\mathcal{Y}_B(s,t) + \mathcal{A}_B(s,t) + \mathcal{N}_B(s,t) \leq 1$ for all $(s, t) \in V \times V$.

A PFG is denoted as $G = (V, A, B)$. The set of edges is denoted by $E$.

It was mentioned earlier that only the edges whose membership values are strictly greater than 0 are considered edges in FG. Now, what are the criteria for the existence of an edge in PFG? If at least one of the three quantities $(\mathcal{Y}_e, \mathcal{A}_e, \mathcal{N}_e)$ of the edge $e$ of a PFG $G$, is strictly greater than zero, then $e$ is an edge of the PFG $G$.

The PFGs are used to solve many real-life problems [1,7,42,77].

A PFG $G = (V, A, B)$ is called strong if $\mathcal{Y}_B(s,t) = \mathcal{Y}_A(s) \wedge \mathcal{Y}_A(t)$, $\mathcal{A}_B(s,t) = \mathcal{A}_A(s) \wedge \mathcal{A}_A(t)$, $\mathcal{N}_B(s,t) = \mathcal{N}_A(s) \vee \mathcal{N}_A(t)$ for all edges $(s, t) \in E$.

If we change the phrase "for all edges $(s, t) \in E$" to 'for all vertices $s, t \in V$', then the graph becomes complete PFG. Formally, A PFG $G = (V, A, B)$ is said to be complete, if $\mathcal{Y}_B(s,t) = \mathcal{Y}_A(s) \wedge \mathcal{Y}_A(t)$, $\mathcal{A}_B(s,t) = \mathcal{A}_A(s) \wedge \mathcal{A}_A(t)$, $\mathcal{N}_B(s,t) = \mathcal{N}_A(s) \vee \mathcal{N}_A(t)$ for all $s, t \in V$.

The complement of a PFG is defined by extending the same definition available for FG.

**Definition 1.17.** The complement of a PFG $G = (V, A, B)$ is denoted by $G^c = (V, A^c, B^c)$, where $A^c = (\mathcal{Y}_{A^c}, \mathcal{A}_{A^c}, \mathcal{N}_{A^c})$, $B^c = (\mathcal{Y}_{B^c}, \mathcal{A}_{B^c}, \mathcal{N}_{B^c})$, and $\mathcal{Y}_{A^c}(s) = \mathcal{Y}_A(s)$, $\mathcal{A}_{A^c}(s) = \mathcal{A}_A(s)$,

$\mathcal{N}_{A^c}(s) = \mathcal{N}_A(s)$,

$$\mathcal{Y}_{B^c}(s,t) = \mathcal{Y}_A(s) \wedge \mathcal{Y}_A(t) - \mathcal{Y}_B(s,t),$$
$$\mathcal{A}_{B^c}(s,t) = \mathcal{A}_A(s) \wedge \mathcal{A}_A(t) - \mathcal{A}_B(s,t),$$
$$\mathcal{N}_{B^c}(s,t) = \mathcal{N}_A(s) \vee \mathcal{N}_A(t) - \mathcal{N}_B(s,t)$$

for all $s, t \in V$.

It can be verified that for a PFG $G$, $(G^c)^c = G$. A PFG $G$ is said to be self-complement if $G^c = G$.

Like FGs and IFGs, several operations are defined in PFGs that are available in [77]. The same article introduces the busy and free nodes for PFGs.

The order and size of a PFG are defined as in FG or IFG as follows.

**Definition 1.18.** Let $G = (V, A, B)$ be a PFG, where $A = (\mathcal{Y}_A, \mathcal{A}_A, \mathcal{N}_A)$ and $B = (\mathcal{Y}_B, \mathcal{A}_B, \mathcal{N}_B)$. The order of $G$ is represented by $O(G) = (O_{\mathcal{Y}}(G), O_{\mathcal{A}}(G), O_{\mathcal{N}}(G))$, where

$$O_{\mathcal{Y}}(G) = \frac{1}{n}\sum_{s\in V} \mathcal{Y}_A(s), \quad O_{\mathcal{A}}(G) = \frac{1}{n}\sum_{s\in V} \mathcal{A}_A(s), \quad O_{\mathcal{N}}(G) = \frac{1}{n}\sum_{s\in V} \mathcal{N}_A(s).$$

Note that all these values lie between 0 and 1.

**Definition 1.19.** Let $G = (V, A, B)$ be a PFG, where $A = (\mathcal{Y}_A, \mathcal{A}_A, \mathcal{N}_A)$ and $B = (\mathcal{Y}_B, \mathcal{A}_B, \mathcal{N}_B)$.

The size of a PFG $G = (V, A, B)$ is denoted by $S(G) = (S_{\mathcal{Y}}(G), S_{\mathcal{A}}(G), S_{\mathcal{N}}(G))$, where

$$S_{\mathcal{Y}}(G) = \sum_{(s,t)\in E} \mathcal{Y}_B(s,t), \quad S_{\mathcal{A}}(G) = \sum_{(s,t)\in E} \mathcal{A}_B(s,t), \quad S_{\mathcal{N}}(G) = \sum_{(s,t)\in E} \mathcal{N}_B(s,t).$$

Sometimes in a crisp graph, the vertices and/or edges are associated with some weights. These graphs are known as weighted graphs. The same idea is used for all FGs. In an FG, all the vertices and edges are associated with membership values, i.e., some real numbers between 0 and 1. For some authors, these real numbers are considered as weights of the vertices and edges. However, actually, these numbers represent the membership values of the vertices and edges. Using this the concept of the weight $(w(G) = (w_{\mathcal{Y}}(G), w_{\mathcal{A}}(G), w_{\mathcal{N}}(G)))$ of a PFG $G$ is introduced, where

$$w_{\mathcal{Y}}(G) = \sum_{(s,t)\in E} \mathcal{Y}_A(s) \wedge \mathcal{Y}_A(t), \; w_{\mathcal{A}}(G) = \sum_{(s,t)\in E} \mathcal{A}_A(s) \wedge \mathcal{A}_A(t),$$
$$w_{\mathcal{N}}(G) = \sum_{(s,t)\in E} \mathcal{N}_A(s) \vee \mathcal{N}_A(t).$$

Using the size and weight of a PFG, Amanathulla et al. [1] defined the density of a PFG.

**Definition 1.20.** Let $G = (V, A, B)$ be a PFG. The density of $G$ is denoted by $\rho(G) = (\rho_{\mathcal{Y}}(G), \rho_{\mathcal{A}}(G), \rho_{N}(G))$, where

$$\rho_{\mathcal{Y}}(G) = \frac{S_{\mathcal{Y}}(G)}{w_{\mathcal{Y}}(G)}, \quad \rho_{\mathcal{Y}}(G) = \frac{S_{\mathcal{A}}(G)}{w_{\mathcal{A}}(G)}, \quad \rho_{N}(G) = \frac{S_{N}(G)}{w_{\mathcal{Y}}(G)}.$$

**Definition 1.21.** Let $G = (V, A, B)$ be a PFG and $H$ be a subgraph of $G$, i.e., the set of vertices (and edges) of $H$ is a subset of the set of vertices (and edges) of $G$.

**(i)** $H$ is called an intense picture fuzzy subgraph of $G$ if $\rho(S) \leq \rho(G)$ (i.e., $\rho_{\mathcal{Y}}(S) \leq \rho_{\mathcal{Y}}(G)$, $\rho_{\mathcal{A}}(S) \leq \rho_{\mathcal{A}}(G)$, $\rho_{N}(S) \leq \rho_{N}(G)$);

**(ii)** $H$ is said to be a feeble picture fuzzy subgraph of $G$ if $\rho(S) > \rho(G)$ (i.e., $\rho_{\mathcal{Y}}(S) > \rho_{\mathcal{Y}}(G)$, $\rho_{\mathcal{A}}(S) > \rho_{\mathcal{A}}(G)$, $\rho_{N}(S) > \rho_{N}(G)$).

**Example 1.1.** Let $G = (V, A, B)$ be a PFG where the set of vertices with membership values is

$$\{s_1(0.4, 0.4, 0.3), s_2(0.4, 0.2, 0.2), s_3(0.5, 0.3, 0.2), s_4(0.4, 0.3, 0.2), s_5(0.3, 0.2, 0.4)\}$$

and that the one for edges is

$$\begin{aligned}
&s_1s_2(0.14, 0.04, 0.15), s_1s_3(0.14, 0.04, 0.15), s_2s_3(0.28, 0.06, 0.10),\\
&s_2s_5(0.21, 0.06, 0.20), s_3s_4(0.25, 0.05, 0.08), s_3s_5(0.21, 0.06, 0.20), s_4s_5(0.24, 0.10, 0.22).
\end{aligned}$$

For this graph $\rho(G) = (0.7, 0.3, 0.5)$.
Let us consider four subgraphs $H_1$, $H_2$, $H_3$, $H_4$, where

$$\begin{aligned}
&H_1 = (V_1, A_1, B_1), \text{ where } V_1 = \{s_3, s_4, s_5\} \text{ and } E_1 = \{(s_3, s_5), (s_5, s_4), (s_3, s_4)\};\\
&H_2 = (V_2, A_2, B_2), \text{ where } V_2 = \{s_4, s_5\} \text{ and } E_2 = \{(s_5, s_4)\};\\
&H_3 = (V_3, A_3, B_3), \text{ where } V_3 = \{s_3, s_4\} \text{ and } E_3 = \{(s_3, s_4)\};\\
&H_4 = (V_4, A_4, B_4), \text{ where } V_4 = \{s_1, s_2, s_3\} \text{ and } E_4 = \{(s_1, s_2), (s_2, s_3), (s_1, s_3), (s_3, s_4)\}.
\end{aligned}$$

For these subgraphs, the densities are

$$\begin{aligned}
&\rho(H_1) = (0.7, 0.5, 0.55), \rho(H_2) = (0.8, 0.5, 0.55);\\
&\rho(H_3) = (0.625, 0.167, 0.040), \rho(H_4) = (0.675, 0.230, 0.480).
\end{aligned}$$

That is, the subgraphs $H_1$, $H_3$, and $H_4$ are intense picture fuzzy subgraphs, and $H_2$ is a feeble picture fuzzy subgraph.

The density of a PFG leads to another type of PFG called a balanced PFG. A PFG $G$ is called balanced if all its subgraphs are intense in $G$.
Several other results on PFGs are available in [1].

## 1.5 Arithmetics on picture fuzzy set

Based on Zadeh's extension principle, the arithmetic operations on PFS are defined in [19].

Let $U$ and $V$ be two PFSs. Then, $U * V$ (where * is any arithmetic operator such as $+, \times, /$) is

$$U * V = (u, \mathcal{Y}_{U*V}(u), \mathcal{A}_{U*V}(u), \mathcal{N}_{U*V}(u)),$$

where

$$\mathcal{Y}_{U*V}(u) = \bigvee_{s*t=u} [\mathcal{Y}_U(s) \wedge \mathcal{Y}_V(t)],$$

$$\mathcal{A}_{U*V}(u) = \bigwedge_{s*t=u} [\mathcal{A}_U(s) \vee \mathcal{A}_V(t)],$$

$$\mathcal{N}_{U*V}(u) = \bigwedge_{s*t=u} [\mathcal{N}_U(s) \vee \mathcal{N}_V(t)]$$

and $s * t = u$.

For illustration, let us consider two PFSs $U = \{(3, 0.4, 0.2, 0.2), (4, 0.3, 0.1, 0.6), (6, 0.4, 0.1, 0.3)\}$ and $V = \{(1, 0.2, 0.3, 0.2), (2, 0.5, 0.2, 0.3)\}$.

Then,

$$U + V = \{(u, \mathcal{Y}_{U+V}(u), \mathcal{A}_{U+V}(u), \mathcal{N}_{U+V}(u))\},$$

where

$$\mathcal{Y}_{U+V}(u) = \bigvee_{s+t=u} (\mathcal{Y}_U(s) \wedge \mathcal{Y}_U(t))$$

$$\mathcal{A}_{U+V}(u) = \bigwedge_{s+t=u} (\mathcal{A}_U(s) \vee \mathcal{A}_U(t))$$

$$\mathcal{N}_{U+V}(u) = \bigwedge_{s+t=u} (\mathcal{N}_U(s) \vee \mathcal{N}_U(t)).$$

Therefore for the above $U$ and $V$,

$$\begin{aligned} U + V =&\{(3+1, \min(0.4, 0.2), \max(0.2, 0.3), \max(0.2, 0.2)),\\ &(3+2, \min(0.4, 0.5), \max(0.2, 0.2), \max(0.2, 0.3)),\\ &(4+1, \min(0.3, 0.2), \max(0.1, 0.3), \max(0.6, 0.2)),\\ &(4+2, \min(0.3, 0.5), \max(0.1, 0.2), \max(0.6, 0.3)),\\ &(6+1, \min(0.4, 0.2), \max(0.1, 0.3), \max(0.3, 0.2)),\\ &(6+2, \min(0.4, 0.5), \max(0.1, 0.2), \max(0.3, 0.3))\}\\ =&\{(4, 0.2, 0.3, 0.2), (5, 0.4, 0.2, 0.3), (5, 0.2, 0.3, 0.6), (6, 0.3, 0.2, 0.6),\\ &(7, 0.2, 0.3, 0.3), (8, 0.4, 0.2, 0.3)\}\\ =&\{(4, 0.2, 0.3, 0.2), (5, \max(0.4, 0.2), \min(0.2, 0.3), \min(0.3, 0.6)), \end{aligned}$$

$$(6, 0.3, 0.2, 0.6), (7, 0.2, 0.3, 0.3), (8, 0.4, 0.2, 0.3)\}$$
$$= \{(4, 0.2, 0.3, 0.2), (5, 0.4, 0.2, 0.3), (6, 0.3, 0.2, 0.6), (7, 0.2, 0.3, 0.3), (8, 0.4, 0.2, 0.3)\}.$$

For PFSs, two types of Cartesian products are defined in [9]. Suppose $X_1$ and $X_2$ are two universes of discourse, and let $U = \{(s, \mathcal{Y}_U(s), \mathcal{A}_U(s), \mathcal{N}_U(s)) : s \in X_1\}$ and $V = \{(t, \mathcal{Y}_V(t), \mathcal{A}_V(t), \mathcal{N}_V(t)) : t \in X_2\}$ be two PFSs. The Cartesian products $\times_1$ and $\times_2$ on PFSs are defined below.

$$U \times_1 V = \{((s, t), \mathcal{Y}_U(s).\mathcal{Y}_V(t), \mathcal{A}_U(s).\mathcal{A}_V(t), \mathcal{N}_U(s).\mathcal{N}_V(t)) : s \in X_1, t \in X_2\}$$
$$U \times_2 V = \{((s, t), \mathcal{Y}_U(s) \wedge \mathcal{Y}_V(t), \mathcal{A}_U(s) \wedge \mathcal{A}_V(t), \mathcal{N}_U(s) \vee \mathcal{N}_V(t)) : s \in X_1, t \in X_2\}.$$

**Proposition 1.4.** *[9] Let $X_1$, $X_2$, $X_3$ be three universes of discourses and $U$, $V$ (defined on $X_1$), $W$ (defined on $X_2$), and $T$ (defined on $X_3$) be the four PFSs. Then, for $i = 1, 2$*

- **(i)** $U \times_i W = W \times_i U$;
- **(ii)** $(U \times_i W) \times_i T = U \times_i (W \times_i T)$;
- **(iii)** $(U \cup V) \times_i W = (U \times_i W) \cup (V \times_i W)$;
- **(iv)** $(U \cap V) \times_i W = (U \times_i W) \cap (V \times_i W)$.

## 1.6 Ordering of PFN

Like complex numbers, the PFNs are not ordered or compared. However, a particular case is defined for finding the maximum and minimum among two PFNs. If $(\mathcal{Y}_A, \mathcal{A}_A, \mathcal{N}_A)$ and $(\mathcal{Y}_B, \mathcal{A}_B, \mathcal{N}_B)$ are two PFNs.

- **(i)** If the conditions $\mathcal{Y}_A > \mathcal{Y}_B, \mathcal{A}_A > \mathcal{A}_B$, and $\mathcal{N}_A < \mathcal{N}_B$ hold together, then $(\mathcal{Y}_A, \mathcal{A}_A, \mathcal{N}_A) > (\mathcal{Y}_B, \mathcal{A}_B, \mathcal{N}_B)$.
- **(ii)** If the conditions $\mathcal{Y}_A < \mathcal{Y}_B, \mathcal{A}_A < \mathcal{A}_B$, and $\mathcal{N}_A > \mathcal{N}_B$ hold together, then $(\mathcal{Y}_A, \mathcal{A}_A, \mathcal{N}_A) < (\mathcal{Y}_B, \mathcal{A}_B, \mathcal{N}_B)$.

If none of the conditions hold, we cannot find larger or smaller PFNs.

**Example 1.2.** For two PFNs (0.5, 0.4, 0.1) and (0.45, 0.35, 0.2), we have $0.5 > 0.45, 0.4 > 0.35$, and $0.1 < 0.2$. Hence, $(0.5, 0.4, 0.1) > (0.45, 0.35, 0.2)$, i.e., (0.5. 0.4, 0.1) is larger than (0.45, 0.35, 0.2).

For two PFNs (0.4, 0.3, 0.1) and (0.5, 0.35, 0.15), we observe that

- **(i)** $0.4 \not> 0.5, 0.3 \not> 0.35$, but $0.1 < 0.15$;
- **(ii)** $0.4 < 0.5, 0.3 < 0.35$, but $0.1 \not> 0.15$.

Hence, neither $(\mathcal{Y}_A, \mathcal{A}_A, \mathcal{N}_A) > (\mathcal{Y}_B, \mathcal{A}_B, \mathcal{N}_B)$ nor $(\mathcal{Y}_A, \mathcal{A}_A, \mathcal{N}_A)) < (\mathcal{Y}_B, \mathcal{A}_B, \mathcal{N}_B)$. Hence, larger or smaller PFN cannot be found in this case.

If $(\mathcal{Y}_A, \mathcal{A}_A, \mathcal{N}_A) > (\mathcal{Y}_B, \mathcal{A}_B, \mathcal{N}_B)$, then

$$\max\{(\mathcal{Y}_A, \mathcal{A}_A, \mathcal{N}_A), (\mathcal{Y}_B, \mathcal{A}_B, \mathcal{N}_B)\} = (\mathcal{Y}_A, \mathcal{A}_A, \mathcal{N}_A)$$

and if $(\mathcal{Y}_A, \mathcal{A}_A, \mathcal{N}_A) < (\mathcal{Y}_B, \mathcal{A}_B, \mathcal{N}_B)$, then

$$\min\{(\mathcal{Y}_A, \mathcal{A}_A, \mathcal{N}_A), (\mathcal{Y}_B, \mathcal{A}_B, \mathcal{N}_B)\} = (\mathcal{Y}_A, \mathcal{A}_A, \mathcal{N}_A).$$

If a set of PFNs are comparable, then such a set of PFNs is called an order. The maximum and minimum values are only determined for this type of set. For other situations, maximum and minimum values cannot be determined. This is a natural limitation of PFNs. This is the conventional way to find the maximum and minimum.

We propose a defizified way for finding the maximum and minimum between two PFNs. Note that in a PFN, there are three components, and these components have different meanings or are important as per an application point of view. According to the importance of these components, one can impose the weights on the components. Let $(\mathcal{Y}_A, \mathcal{A}_A, \mathcal{N}_A)$ be a PFN and $(w_A^1, w_A^2, w_A^3)$ be a weight vector associated to this PFN, where $w_A^1, w_A^2, w_A^3 \geq 0$ and $w_A^1 + w_A^2 + w_A^3 = 1$. The defizified value of this PFN is

$$D_A = \frac{(\mathcal{Y}_A w_A^1 + \mathcal{A}_A w_A^2 + \mathcal{N}_A w_A^3)}{w_A^1 + w_A^2 + w_A^3} = (\mathcal{Y}_A w_A^1 + \mathcal{A}_A w_A^2 + \mathcal{N}_A w_A^3). \tag{1.9}$$

If the weights are equal then,

$$D_A = \frac{1}{3}(\mathcal{Y}_A + \mathcal{A}_A + \mathcal{N}_A).$$

By finding the defizified values of a set of PFNs, one can determine the maximum and minimum based on these defizified values. The weight vector may be the same for all PFNs or different for different PFNs.

## 1.7 Similarity measures between picture fuzzy sets

The similarity measures between any set type benefit pattern matching, decision making, and many other similar problems. Several types of similarity measures are available for different types of fuzzy sets. However, a specific definition must follow for any similarity measures, as given below.

**Definition 1.22.** Let $U = \{(t, \mathcal{Y}_U(t), \mathcal{A}_U(t), \mathcal{N}_U(t)) : t \in X\}$ and $V = \{(t, \mathcal{Y}_V(t), \mathcal{A}_V(t), \mathcal{N}_V(t)) : t \in X\}$ be two PFSs on the universe of discourse $X$. The similarity measure between $U$ and $V$ $(S(U, V))$ satisfies the following axioms:

**(S1)** $0 \leq S(U, V) \leq 1$;
**(S2)** $S(U, V) = 1$ if and only if $U = V$;
**(S3)** $S(U, V) = S(V, U)$;
**(S4)** Let $W$ be any PFS, if $U \subseteq V \subseteq W$, then $S(U, W) \leq S(U, V)$ and $S(U, W) \leq S(V, W)$.

For two PFSs $U$ and $V$, the refusal degrees of the element $t$ in $U$ and $V$ are, respectively, defined as $\pi_U(t) = 1 - \mathcal{Y}_U(t) - \mathcal{A}_U(t) - \mathcal{N}_U(t)$ and $\pi_V(t) = 1 - \mathcal{Y}_V(t) - \mathcal{A}_V(t) - \mathcal{N}_V(t)$. Let $X = \{t_1, t_2, \ldots, t_n\}$.

Several similarity measures between two PFSs are defined by [63], but some do not satisfy the above definition, justified by [75]. Only the correct similarity measures from [63] are given below:

$$S_w(U,V) = \frac{1}{n}\sum_{i=1}^{n}\cos\frac{\pi}{4}\left(|\mathcal{Y}_U(t_i) - \mathcal{Y}_V(t_i)| + |\mathcal{A}_U(t_i) - \mathcal{A}_V(t_i)|\right.$$
$$\left. + |\mathcal{N}_U(t_i) - \mathcal{N}_V(t_i)| + |\pi_U(t_i) - \pi_V(t_i)|\right). \quad (1.10)$$

Dinh and Thao [17] defined the following two similarity measures based on Hamming and Euclidean distances:

$$S_h(U,V) = 1 - \frac{1}{3n}\sum_{i=1}^{n}\left(|\mathcal{Y}_U(t_i) - \mathcal{Y}_V(t_i)| + |\mathcal{A}_U(t_i) - \mathcal{A}_V(t_i)| + |\mathcal{N}_U(t_i) - \mathcal{N}_V(t_i)|\right) \quad (1.11)$$

$$S_e(U,V)$$
$$= 1 - \left[\frac{1}{3n}\sum_{i=1}^{n}\left([\mathcal{Y}_U(t_i) - \mathcal{Y}_V(t_i)]^2 + [\mathcal{A}_U(t_i) - \mathcal{A}_V(t_i)]^2 + [\mathcal{N}_U(t_i) - \mathcal{N}_V(t_i)]^2\right)\right]^{1/2}. \quad (1.12)$$

Singh and Mishra's similarity measures [48] are

$$S_p(U,V) = 1 - \frac{1}{4n}\sum_{i=1}^{n}\left(|\mathcal{Y}_U(t_i) - \mathcal{Y}_V(t_i)| + |\mathcal{A}_U(t_i) - \mathcal{A}_V(t_i)|\right.$$
$$\left. + |\mathcal{N}_U(t_i) - \mathcal{N}_V(t_i)| + |\pi_U(t_i) - \pi_V(t_i)|\right). \quad (1.13)$$

A complicated similarity measure proposed by Luo and Zhang [35] is

$$S_l(U,V)$$
$$= 1 - \frac{1}{3n}\sum_{i=1}^{n}\left[2\sqrt{\mathcal{Y}_U(t_i)\mathcal{Y}_V(t_i)} + 2\sqrt{\mathcal{A}_U(t_i)\mathcal{A}_V(t_i)} + 2\sqrt{\mathcal{N}_U(t_i)\mathcal{N}_V(t_i)}\right.$$
$$+ \sqrt{(1 - \mathcal{Y}_U(t_i) - \mathcal{A}_U(t_i))(1 - \mathcal{Y}_V(t_i) - \mathcal{A}_V(t_i))}$$
$$+ \sqrt{(1 - \mathcal{Y}_U(t_i) - \mathcal{N}_U(t_i))(1 - \mathcal{Y}_V(t_i) - \mathcal{N}_V(t_i))}$$
$$\left. + \sqrt{(1 - \mathcal{A}_U(t_i) - \mathcal{N}_U(t_i))(1 - \mathcal{A}_V(t_i) - \mathcal{N}_V(t_i))}\right]. \quad (1.14)$$

A parametric similarity measure defined by Zhao et al. [75] is

$$S_m(U,V) = 1 - \left[\frac{1}{3n}\sum_{i=1}^{n}\{D^p_{UV1}(t_i) + D^p_{UV2}(t_i) + D^p_{UV3}(t_i)\}\right]^{1/p}, \quad (1.15)$$

where

$$D^p_{UV1}(t_i) = \frac{1}{w_1 + 1}\left|w_1(\mathcal{Y}_U(t_i) - \mathcal{Y}_V(t_i)) - (\mathcal{A}_U(t_i) - \mathcal{A}_V(t_i)) - (\mathcal{N}_U(t_i) - \mathcal{N}_V(t_i))\right|$$

$$D^p_{UV2}(t_i) = \frac{1}{2(w_2+1)}\left|w_2(\mathcal{A}_U(t_i) - \mathcal{A}_V(t_i)) - (\mathcal{Y}_U(t_i) - \mathcal{Y}_V(t_i)) + (\mathcal{N}_U(t_i) - \mathcal{N}_V(t_i))\right|$$

$$D^p_{UV1}(t_i) = \frac{1}{2(w_3+1)}\left|w_3(\mathcal{N}_U(t_i) - \mathcal{N}_V(t_i)) - (\mathcal{Y}_U(t_i) - \mathcal{Y}_V(t_i)) + (\mathcal{A}_U(t_i) - \mathcal{A}_V(t_i))\right|$$

$w_1, w_2, w_3 \in [0, \infty)$ and $\frac{1}{w_1+1} + \frac{1}{w_2+1} + \frac{1}{w_3+1} \in [0, 1]$ and $p$ is any integer.

The above definition is extended by incorporating the weights. The weighted similarity measure is defined as

$$S_m(U, V) = 1 - \left[\frac{1}{3n}\sum_{i=1}^{n} W_i\{D^p_{UV1}(t_i) + D^p_{UV2}(t_i) + D^p_{UV3}(t_i)\}\right]^{1/p}. \tag{1.16}$$

The weight $W_i$ is associated with each element $t_i \in X$, and the other terms are the same as in Eq. (1.15).

The other several similarity measures and indepth analyses are available in [75]. The similarity measure is also used to order a set of PFSs and solve decision-making problems.

The similarity measures find the closeness between two sets, while the distance between two sets measures the differences between such sets. Like similarity measures, the distance between the sets is defined.

Suppose $U$, $V$, and $W$ are three PFSs defined on a universe $X$. The distance between [38] two sets $U$ and $V$ is denoted by $D(U, V)$ and it satisfies the following properties.

**(D1)** $0 \le D(U, V) \le 1$;
**(D2)** $D(U, V) = D(V, U)$;
**(D3)** $D(U, V) = 0$ iff $U = V$;
**(D4)** If $U \subseteq V \subseteq W$ $D(U, V) \le D(U, W)$ and $D(V, W) \le D(U, W)$.

Like similarity measures, several distance functions are available for different fuzzy sets. Here, two valuable and simple distance functions are defined for PFSs.

The normalized Hamming distance $D_h(U, V)$ is

$$D_h(U, V) = \frac{1}{n}\sum_{i=1}^{n}\left[|\mathcal{Y}_U(t_i) - \mathcal{Y}_V(t_i)| + |\mathcal{A}_U(t_i) - \mathcal{A}_V(t_i)| + |\mathcal{N}_U(t_i) - \mathcal{N}_V(t_i)|\right]. \tag{1.17}$$

The normalized Euclidean distance $D_e(U, V)$ is

$$D_e(U, V) = \frac{1}{n}\sum_{i=1}^{n}\left[(\mathcal{Y}_U(t_i) - \mathcal{Y}_V(t_i))^2 + (\mathcal{A}_U(t_i) - \mathcal{A}_V(t_i))^2 + (\mathcal{N}_U(t_i) - \mathcal{N}_V(t_i))^2\right], \tag{1.18}$$

where $t_i \in X$ (the universal set).

A more general distance function is defined in [39]. Let $U$ and $V$ be two PFSs in $X$, then $D(U, V)$ is the distance measure between two PFSs $U$ and $V$ in $X = \{t_1, t_2, \ldots, t_n\}$:

$$D(U, V) = \left[\frac{1}{2n(a+1)^p}\sum_{i=1}^{n}(|(a+1-\lambda)(\mathcal{Y}^2_U(t_i) - \mathcal{A}^2_U(t_i)) - \lambda(\mathcal{N}^2_U(t_i) - \mathcal{N}^2_V(t_i))|^p\right.$$

$$+ |(a+1-\mu)(\mathcal{N}_U^2(t_i) - \mathcal{N}_U^2(t_i)) - \mu(\mathcal{Y}_U^2(t_i) - \mathcal{Y}_V^2(t_i))|^p)\Bigg]^{1/p},$$

where $a$, $\lambda$, and $\mu$ represent the level of uncertainty and $p$ is the $L_p$ norm, which satisfies the condition $\lambda + \mu \leq a+1, 0 < \lambda, \mu \leq a+1, a > 0$.

## 1.8 Convex combination of picture fuzzy sets

The convex combination of vectors is a vital tool in mathematics. Similarly, the same technique can be applied in PFS also. The convex combination of two PFSs is defined below.

Let $U$ and $V$ be two PFSs defined over the universe $X$, and $0 \leq \lambda_1 \leq 1, 0 \leq \lambda_2 \leq 1$ be two real numbers, then the convex combination of these two numbers is

$$C_\lambda(U, V) = \{(t, \mathcal{Y}_{C_\lambda}(t), \mathcal{A}_{C_\lambda}(t), \mathcal{N}_{C_\lambda}(t))\},$$

where

$$\mathcal{Y}_{C_\lambda}(t) = \lambda_1 \mathcal{Y}_U(t) + \lambda_2 \mathcal{Y}_V(t),$$
$$\mathcal{A}_{C_\lambda}(t) = \lambda_1 \mathcal{A}_U(t) + \lambda_2 \mathcal{A}_V(t),$$
$$\mathcal{N}_{C_\lambda}(t) = \lambda_1 \mathcal{N}_U(t) + \lambda_2 \mathcal{N}_V(t),$$

for all $t \in X$ and $\lambda = \{\lambda_1, \lambda_2\}$, $\lambda_1 + \lambda_2 = 1$.

This definition can be extended for any number of PFSs. Let $U_1, U_2, \ldots, U_n$ be a set of $n$ PFSs and $\lambda = \{\lambda_1, \lambda_2, \ldots, \lambda_n\}$ be a set of scalars such that $\lambda_1 + \lambda_2 + \ldots + \lambda_n = 1$. Then, the convex combination of these PFSs is

$$C_\lambda(U_1, U_2, \ldots, U_n) = \{(t, \mathcal{Y}_C(t), \mathcal{A}_C(t), \mathcal{N}_C(t))\},$$

where

$$\mathcal{Y}_{C_\lambda}(t) = \sum_{i=1}^{n} \lambda_i \mathcal{Y}_{U_i}(t),$$
$$\mathcal{A}_{C_\lambda}(t) = \sum_{i=1}^{n} \lambda_i \mathcal{A}_{U_i}(t),$$
$$\mathcal{N}_{C_\lambda}(t) = \sum_{i=1}^{n} \lambda_i \mathcal{N}_{U_i}(t),$$

for all $t \in X$.

In particular, let $\lambda_1 = \lambda$, then $\lambda_2 = 1 - \lambda$. In this case,

$$C_\lambda(U, V) = \{(t, \mathcal{Y}_{C_\lambda}(t), \mathcal{A}_{C_\lambda}(t), \mathcal{N}_{C_\lambda}(t))\},$$

where

$$\mathcal{Y}_{C_\lambda}(t) = \lambda \mathcal{Y}_U(t) + (1-\lambda)\mathcal{Y}_V(t),$$
$$\mathcal{A}_{C_\lambda}(t) = \lambda \mathcal{A}_U(t) + (1-\lambda)\mathcal{A}_V(t),$$
$$\mathcal{N}_{C_\lambda}(t) = \lambda \mathcal{N}_U(t) + (1-\lambda)\mathcal{N}_V(t),$$

for all $t \in X$ and $0 \leq \lambda \leq 1$.

When $\lambda = 0$, $C_\lambda(U, V) = V$ and when $\lambda = 1$, $C_\lambda(U, V) = U$. Thus if $U \subseteq V$, then $U \subseteq C_\lambda(U, V) \subseteq V$ for any $\lambda \in [0, 1]$. Also, if $U \subseteq V$ and $\lambda_1 \leq \lambda_2$, then $C_{\lambda_1}(U, V) \subseteq C_{\lambda_2}(U, V)$.

This convex combination is the weighted average among the PFSs and is also known as an aggregation of the PFSs. In particular, if $\lambda_i = \frac{1}{n}$ for all $i$, then the weighted average becomes the arithmetic mean of $n$ PFSs.

Another type of convex combination is also defined. In this new convex combination, the membership, neutral, and nonmembership values are combined for each PFS element instead of a combination of PFSs. This definition is given below.

**Definition 1.23.** Let $U = (\mathcal{Y}_U(t), \mathcal{A}_U(t), \mathcal{N}_U(t))$ be a PFS over a universe of discourse $X$. Then, an operator $\mathcal{D}$ on $U$ is defined as

$$\mathcal{D}_{\theta,\phi,\psi}(U) = \{(t, \mathcal{Y}_U(t) + \theta s_U(t), \mathcal{A}_U(t) + \phi s_U(t), \mathcal{N}_U(t) + \psi s_U(t)) : t \in X\},$$

where $s_U(t) = 1 - \mathcal{Y}_U(t) - \mathcal{A}_U(t) - \mathcal{N}_U(t)$, $0 \leq \theta, \phi, \psi \leq 1$ and $0 \leq \theta + \phi + \psi = 1$.

The set $\mathcal{D}_{\theta,\phi,\psi}(U)$ is a PFS because

$$\begin{aligned}&\mathcal{Y}_U(t) + \theta s_U(t) + \mathcal{A}_U(t) + \phi s_U(t) + \mathcal{N}_U(t) + \psi S_u(t)\\ &= (\mathcal{Y}_U(t) + \mathcal{A}_U(t) + \mathcal{N}_U(t)) + (\theta + \phi + \psi)s_U(t) \leq (\mathcal{Y}_U(t) + \mathcal{A}_U(t) + \mathcal{N}_U(t)) + s_U(t) \leq 1.\end{aligned}$$

Here, three quantities $\theta$, $\phi$, $\psi$ are independent within [0, 1].

If we choose $\psi = 1 - \theta - \phi$ in $\mathcal{D}_{\theta,\phi,\psi}(U)$ then a special type of operator is defined over a PFS.

**Definition 1.24.** Let $U = (\mathcal{Y}_U(t), \mathcal{A}_U(t), \mathcal{N}_U(t))$ be a PFS. Then, the operator $D$ on $U$ is defined as $D_{\theta,\phi}(U) = \{(t, \mathcal{Y}_U(t) + \theta s_U(t), \mathcal{A}_U(t) + \phi s_U(t), \mathcal{N}_U(t) + (1 - \theta - \phi)s_U(t)) : t \in X\}$, where $s_U(t) = 1 - \mathcal{Y}_U(t) - \mathcal{A}_U(t) - \mathcal{N}_U(t)$.

Note that, in $D_{\theta,\phi}$, only two parameters $\theta$, $\phi$ are independent and $\mathcal{Y}_U(t) + \theta s_U(t) + \mathcal{A}_U(t) + \phi s_U(t) + \mathcal{N}_U(t) + (1 - \theta - \phi)s_U(t) = 1$. Hence, $D_{\theta,\phi}$ is not, in general, a PFS, it is rather a IFS. In addition, if $\mathcal{A}_U(t) = 0$ for all $t \in X$, then $D_{\theta,\phi}$ becomes a FS.

Some special values for the operator $D_{\theta,\phi}$ are discussed below.

**Proposition 1.5.** *Let $U = (\mathcal{Y}_U, \mathcal{A}_U, \mathcal{N}_U)$ be a PFS over $X$. Then,*

**(i)** $D_{0,0}(U) = \Box U$;
**(ii)** $D_{1,0}(U) = \diamond U$;
**(iii)** $D_{\theta_2,\phi_2}(D_{\theta_1,\phi_1}(U)) = D_{\theta_1,\phi_1}(U)$.

*Proof.* (i) As per the definition

$$\begin{aligned} D_{0,0}(U) &= \{(t, \mathcal{Y}_U(t), \mathcal{A}_U(t), \mathcal{N}_U(t) + s_U(t)) : t \in X\} \\ &= \{(t, \mathcal{Y}_U(t), \mathcal{A}_U(t), \mathcal{N}_U(t) + 1 - \mathcal{Y}_U(t) - \mathcal{A}_U(t) - \mathcal{N}_U(t)) : t \in X\} \\ &= \{(t, \mathcal{Y}_U(t), \mathcal{A}_U(t), 1 - \mathcal{Y}_U(t) - \mathcal{A}_U(t)) : t \in X\} \\ &= \Box U. \end{aligned}$$

It is obvious that $D_{0,0}(U) = \Box U$ is not a PFS, as the sum of the components is equal to 1.
(ii) Now, $D_{1,0}$ is

$$\begin{aligned} D_{1,0} &= \{(t, \mathcal{Y}_U(t) + s_U(t), \mathcal{A}_U(t), \mathcal{N}_U(t)) : t \in X\} \\ &= \{(t, \mathcal{Y}_U(t) + 1 - \mathcal{Y}_U(t) - \mathcal{A}_U(t) - \mathcal{N}_U(t), \mathcal{A}_U(t), \mathcal{N}_U(t)) : t \in X\} \\ &= \{(t, 1 - \mathcal{A}_U(t) - \mathcal{N}_U(t), \mathcal{A}_U(t), \mathcal{N}_U(t)) : t \in X\} \\ &= \diamond U. \end{aligned}$$

(iii)

$$\begin{aligned} & D_{\theta_2,\phi_2}(D_{\theta_1,\phi_1})(U) \\ &= D_{\theta_2,\phi_2}\{(t, \mathcal{Y}_U(t) + \theta_1 s_U(t), \mathcal{A}_U(t) + \phi_1 s_U(t), \mathcal{N}_U(t) + (1 - \theta_1 - \phi_1) s_U(t)) : t \in U\} \\ &= \{(t, p, q, r) : t \in U\}, \end{aligned}$$

where

$$\begin{aligned} p &= \mathcal{Y}_U(t) + \theta_1 s_U(t) + \theta_2[1 - \mathcal{Y}_U(t) - \theta_1 s_U(t) - \mathcal{A}_U(t) - \phi_1 s_U(t) \\ &\quad - \mathcal{N}_U(t) - (1 - \theta_1 - \phi_1) s_U(t)] \\ &= \mathcal{Y}_U(t) + \theta_1 s_U(t), \\ q &= \mathcal{A}_U(t) + \phi_1 s_U(t) + \phi_2[1 - \mathcal{Y}_U(t) - \theta_1 s_U(t) - \mathcal{A}_U(t) - \phi_1 s_U(t) \\ &\quad - \mathcal{N}_U(t) - (1 - \theta_1 - \phi_1) s_U(t)] \\ &= \mathcal{N}_U(t) + \phi_1 s_U(t) \end{aligned}$$

and
$$\begin{aligned} r &= \mathcal{N}_U(t) + (1 - \theta_1 - \phi_1) s_U(t) + (1 - \theta_2 - \phi_2)[1 - \mathcal{Y}_U(t) - \theta_1 s_U(t) - \mathcal{A}_U(t) - \phi_1 s_U(t) \\ &\quad - \mathcal{N}_U(t) - (1 - \theta_1 - \phi_1) s_U(t)] \\ &= \mathcal{N}_U(t) + (1 - \theta_1 - \phi_1) s_U(t). \end{aligned}$$

Thus $D_{\theta_2,\phi_2}(D_{\theta_1,\phi_1})(P) = D_{\theta_1,\phi_1}(P)$. □

## 1.9 Picture fuzzy averaging operators

Several operators are defined on PFS, and they are used in decision-making problems. Some of them are discussed in detail in subsequent chapters. Here, introductory operators are defined.

**Definition 1.25.** Let $\{P_l = (\mathcal{Y}_{P_l}, \mathcal{A}_{P_l}, \mathcal{N}_{P_l}),\ l = 1, 2, ..., n\}$ be a set of $n$ PFNs. Also, let $\theta = (\theta_1, \theta_2, ..., \theta_n)^t$ ($\theta_l > 0$ for $l = 1, 2, .., n$ and $\sum_{l=1}^{n} \theta_l = 1$) be a weight vector. Then, the picture fuzzy weighted averaging operator (PFWAO) is a mapping $S^n \to S$, such that

$$PFWAO(P_1, P_2, ..., P_n) = \left(1 - \prod_{l=1}^{n}(1 - \mathcal{Y}_{P_l})^{\theta_l}, \prod_{l=1}^{n} \mathcal{A}_{P_l}^{\theta_l}, \prod_{l=1}^{n} \mathcal{N}_{P_l}^{\theta_l}\right), \tag{1.19}$$

where $S$ is a PFN.

This operator transferred $n$ PFNs into a single PFN.

**Example 1.3.** Let us suppose that the PFNs $P_1 = (\mathcal{Y}_{P_1}, \mathcal{A}_{P_1}, \mathcal{N}_{P_1}) = (0.3, 0.3, 0.3)$, $P_2 = (\mathcal{Y}_{P_2}, \mathcal{A}_{P_2}, \mathcal{N}_{P_2}) = (0.4, 0.2, 0.2)$, $P_3 = (\mathcal{Y}_{P_3}, \mathcal{A}_{P_3}, \mathcal{N}_{P_3}) = (0.35, 0.25, 0.2)$ and the corresponding weight vector $\theta = (\theta_1, \theta_2, \theta_3)^t = (0.2, 0.3, 0.5)$.

Then, PFWAO$(P_1, P_2, P_3) = (0.355, 0.242, 0.216)$.

The above operator is modified below.

**Definition 1.26.** Let $\{P_l = (\mathcal{Y}_{P_l}, \mathcal{A}_{P_l}, \mathcal{N}_{P_l}) : l = 1, 2, ..., n\}$ be a set of $n$ PFNs. Also, let $\theta = (\theta_1, \theta_2, ..., \theta_n)^t$ ($\theta_l > 0$ for $l = 1, 2, .., n$ and $\sum_{l=1}^{n} \theta_l = 1$) be the weight vector. Then, the picture fuzzy ordered weighted averaging operator (PFOWAO) is a mapping $S^n \to S$ such that

$$PFOWAO(P_1, P_2, ..., P_n) = \left(1 - \prod_{l=1}^{n}(1 - \mathcal{Y}_{P_{\delta(l)}})^{\theta_l}, \prod_{l=1}^{n} \mathcal{A}_{P_{\delta(l)}}^{\theta_l}, \prod_{l=1}^{n} \mathcal{N}_{P_{\delta(l)}}^{\theta_l}\right), \tag{1.20}$$

where $\delta$ is a permutation on $\{1, 2, 3, ..., n\}$ with $P_{\delta(l-1)} \geqslant P_{\delta(l)}$ for $l = 2, 3, ..., n$ and $S$ is a PFN.

**Example 1.4.** Let us suppose the PFNs and the corresponding weight vector are the same as in Example 1.3. Let $\delta(1) = 3$, $\delta(2) = 2$, and $\delta(3) = 1$. Then, PFOWAO$(p_1, p_2, p_3) = (0.341, 0.256, 0.244)$.

**Definition 1.27.** Let $\{P_l = (\mathcal{Y}_{P_l}, \mathcal{A}_{P_l}, \mathcal{N}_{P_l}) : l = 1, 2, ..., n\}$ be a collection on $n$ PFNs associated with a weight vector $\theta = (\theta_1, \theta_2, ..., \theta_n)^t$. Then, weighted PFNs are given by $\{\bar{P}_l = (1 - (1 - \mathcal{Y}_{P_l})^{n\theta_l}, \mathcal{A}_{P_l}^{n\theta_l}, \mathcal{N}_{P_l}^{n\theta_l})$ for $l = 1, 2, ..., n\}$.

**Definition 1.28.** Let $\{(\mathcal{Y}_l, \mathcal{A}_l, \mathcal{N}_l) : l = 1, 2, ..., n\}$ be a collection of $n$ PFNs. Let $(\theta_1, \theta_2, ..., \theta_n)^t$ ($\theta_l > 0$ for $l = 1, 2, .., n$ and $\sum_{l=1}^{n} \theta_l = 1$) be a weight vector. Then, the picture fuzzy hybrid weighted averaging operator (PFHWAO) is a mapping $S^n \to S$ such that

$$PFHWAO(P_1, P_2, ..., P_n) = \left(1 - \prod_{l=1}^{n}(1 - \mathcal{Y}_{\bar{P}_{\delta(l)}})^{\theta_l}, \prod_{l=1}^{n} \mathcal{A}_{\bar{P}_{\delta(l)}}^{\theta_l}, \prod_{l=1}^{n} \mathcal{N}_{\bar{P}_{\delta(l)}}^{\theta_l}\right), \tag{1.21}$$

where $\bar{P}_{\delta(l)}$ is the $l$th largest of the weighted PFNs $\bar{P}_l$ ($\bar{P}_l = n\theta_l P_l$ for $l = 1, 2, ..., n$) and $S$ is a PFN.

**Example 1.5.** Let us suppose the PFNs $P_1 = (0.4, 0.3, 0.2)$, $P_2 = (0.35, 0.2, 0.3)$, $P_3 = (0.5, 0.15, 0.1)$ and $P_4 = (0.35, 0.1, 0.1)$ with the weight vector $\theta = (\theta_1, \theta_2, \theta_3, \theta_4)^t = (0.2, 0.2, 0.3, 0.1)^t$.

Weighted PFNs are calculated as

$$\begin{aligned}
\bar{P}_1 &= \left(1-(1-0.4)^{(4\times0.2)}, (0.3)^{(4\times0.2)}, (0.2)^{(4\times0.2)}\right) = (0.335, 0.381, 0.275),\\
\bar{P}_2 &= \left(1-(1-0.35)^{(4\times0.2)}, (0.2)^{(4\times0.2)}, (0.3)^{(4\times0.2)}\right) = (0.291, 0.275, 0.381),\\
\bar{P}_3 &= \left(1-(1-0.5)^{(4\times0.3)}, (0.15)^{(4\times0.3)}, (0.1)^{(4\times0.3)}\right) = (0.564, 0.102, 0.063),\\
\bar{P}_4 &= \left(1-(1-0.35)^{(4\times0.1)}, (0.1)^{(4\times0.1)}, (0.1)^{(4\times0.1)}\right) = (0.158, 0.398, 0.398).
\end{aligned}$$

Then, the scores of these PFNs are calculated as

$$\begin{aligned}
s(\bar{P}_1) &= \frac{1+0.335+0.381-0.275}{3} = 0.480,\\
s(\bar{P}_2) &= \frac{1+0.291+0.275-0.381}{3} = 0.395,\\
s(\bar{P}_3) &= \frac{1+0.564+0.102-0.063}{3} = 0.534\\
\text{and } s(\bar{P}_4) &= \frac{1+0.158+0.398-0.398}{3} = 0.386.
\end{aligned}$$

Hence, $s(P_3) > s(P_1) > s(P_2) > s(P_4)$.

Thus

$$\begin{aligned}
\bar{P}_{\delta(1)} &= (0.564, 0.102, 0.063),\\
\bar{P}_{\delta(2)} &= (0.335, 0.381, 0.275),\\
\bar{P}_{\delta(3)} &= (0.291, 0.275, 0.381),\\
\bar{P}_{\delta(4)} &= (0.158, 0.398, 0.398).
\end{aligned}$$

Therefore $\text{PFHWAO}(P_1, P_2, P_3, P_4) = (0.307, 0.323, 0.303)$.

It is to be noted that when the components of the weight vector $\theta = (\theta_1, \theta_2, ..., \theta_n)^t$ are equal, i.e., $\theta_1 = \theta_2 = \cdots = \theta_n = \frac{1}{n}$, then $\bar{P}_l = n\theta_l P_l = P_l$. Hence, PFWAO and PFOWAO can be thought of as special cases of PFHWAO. Hence, PFHWAO is the generalization of PFWAO and PFOWAO.

## 1.10 Implication operator on picture fuzzy set

Like other classical and fuzzy sets, the implication operator plays a significant role in mathematical logic. Here, the picture fuzzy implication (PFI) operator is defined and presented with some properties.

**Definition 1.29.** Let $U = (\mathcal{Y}_U, \mathcal{A}_U, \mathcal{N}_U)$ and $V = (\mathcal{Y}_V, \mathcal{A}_V, \mathcal{N}_V)$ be two PFSs over the set of universe $X$. Then, the implication operation between $U$ and $V$ is defined as $U \mapsto V =$

$\{(t, \mathcal{Y}_V(t) \vee \mathcal{N}_U(t), \mathcal{A}_U(t) \wedge \mathcal{A}_V(t), \mathcal{N}_V(t) \wedge \mathcal{Y}_U(t)) : t \in X\}$. Here, $\mapsto$ is called the implication operator.

**Example 1.6.** Let us consider two PFSs $U$ and $V$ over the set of universe $X = \{s, t, u\}$ as follows:

$$U = \{(t, 0.55, 0.35, 0.1), (s, 0.45, 0.45, 0.1), (u, 0.4, 0.4, 0.2)\},$$
$$V = \{(t, 0.35, 0.35, 0.2), (s, 0.25, 0.35, 0.3), (u, 0.45, 0.4, 0.15)\}.$$

Then,

$$\begin{aligned}
&U \mapsto V \\
&= \{(t, 0.35 \vee 0.1, 0.35 \wedge 0.35, 0.2 \wedge 0.55), (s, 0.25 \vee 0.1, 0.45 \wedge 0.35, 0.3 \wedge 0.45), \\
&(u, 0.45 \vee 0.2, 0.4 \wedge 0.4, 0.15 \wedge 0.4)\} \\
&= \{(t, 0.35, 0.35, 0.2), (s, 0.25, 0.35, 0.3), (u, 0.45, 0.4, 0.15)\}.
\end{aligned}$$

**Proposition 1.6.** *If $U = (\mathcal{Y}_U, \mathcal{A}_U, \mathcal{N}_U)$ and $V = (\mathcal{Y}_V, \mathcal{A}_V, \mathcal{N}_V)$ are two PFSs over the set of universe $X$. Then,*

- **(i)** $U \mapsto \overline{U} = U$;
- **(ii)** $U \mapsto \overline{(1, 0, 0)} = (1, 0, 0)$;
- **(iii)** $U \mapsto \overline{(1, 0, 0)} \subseteq \overline{U}$;
- **(iv)** $(U \cap V) \mapsto W = (U \mapsto W) \cup (V \mapsto W)$;
- **(v)** $(U \cup V) \mapsto W = (U \mapsto W) \cap (V \mapsto W)$.

*Proof.* (i) We have,

$$\begin{aligned}
&U \mapsto \overline{U} \\
&= \{(t, \mathcal{Y}_U(t), \mathcal{A}_U(t), \mathcal{N}_U(t) : t \in X\} \mapsto \{(t, \mathcal{N}_U(t), \mathcal{A}_U(t), \mathcal{Y}_U(t) : t \in X\} \\
&= \{(t, \mathcal{N}_U(t) \vee \mathcal{N}_U(t), \mathcal{A}_U(t) \wedge \mathcal{A}_U(t), \mathcal{Y}_U(t) \wedge \mathcal{Y}_U(t) : t \in X\} \\
&= \{(t, \mathcal{N}_U(t), \mathcal{A}_U(t), \mathcal{Y}_U(t) : t \in X\}.
\end{aligned}$$

(ii) It is observed that

$$\begin{aligned}
&U \mapsto (1, 0, 0) \\
&= \{t, \mathcal{Y}_U(t), \mathcal{A}_U(t), \mathcal{N}_U(t) : t \in X\} \mapsto (1, 0, 0) \\
&= \{t, \mathcal{N}_U(t) \vee 1, \mathcal{A}_U(t) \wedge 0, \mathcal{Y}_U(t) \wedge 0 : t \in X\} \\
&= (1, 0, 0).
\end{aligned}$$

(iii) It is viewed that

$$\begin{aligned}
&U \mapsto \overline{(1, 0, 0)} \\
&= \{t, \mathcal{Y}_U(t), \mathcal{A}_U(t), \mathcal{N}_U(t) : t \in X\} \mapsto (0, 0, 1)
\end{aligned}$$

$$
\begin{aligned}
&= \{t, \mathcal{N}_U(t) \vee 0, \mathcal{A}_U(t) \wedge 0, \mathcal{Y}_U(t) \wedge 1 : t \in X\} \\
&= \{t, \mathcal{N}_U(t), 0, \mathcal{Y}_U(t) : t \in X\} \\
&\subseteq \{t, \mathcal{N}_U(t), \mathcal{A}_U(t), \mathcal{Y}_U(t) : t \in X\} \text{ [As } \mathcal{N}_U(t) \leqslant \mathcal{N}_U(t),\ 0 \leqslant \mathcal{A}_U(t) \text{ and } \mathcal{Y}_U(t) \geqslant \mathcal{Y}_U(t)] \\
&= \overline{U}.
\end{aligned}
$$

Thus $U \mapsto \overline{(1,0,0)} \subseteq \overline{U}$.

(iv) We see that

$$
\begin{aligned}
&(U \cap V) \mapsto W \\
&= [\{(t, \mathcal{Y}_U(t), \mathcal{A}_U(t), \mathcal{N}_U(t)) : t \in X\} \cap \{(t, \mathcal{Y}_V(t), \mathcal{A}_V, \mathcal{N}_V) : t \in X\}] \\
&\mapsto \{(t, \mathcal{Y}_W(t), \mathcal{A}_W(t), \mathcal{N}_W(t)) : t \in X\} \\
&= [\{(t, \mathcal{Y}_U(t) \wedge \mathcal{Y}_V(t), \mathcal{A}_U(t) \wedge \mathcal{A}_V(t), \mathcal{N}_U(t) \vee \mathcal{N}_V(t)) : t \in X\}] \\
&\mapsto \{(t, \mathcal{Y}_W(t), \mathcal{A}_W(t), \mathcal{N}_W(t)) : t \in X\} \\
&= \{(t, \mathcal{Y}_W(t) \vee (\mathcal{N}_U(t) \vee \mathcal{N}_V(t)), \mathcal{A}_W(t) \wedge (\mathcal{A}_U(t) \wedge \mathcal{A}_V(t)), \mathcal{N}_W(t) \wedge (\mathcal{Y}_U(t) \wedge \mathcal{Y}_V(t)) : t \in X\} \\
&= \{(t, (\mathcal{Y}_W(t) \vee \mathcal{N}_U(t)) \vee (\mathcal{Y}_W(t) \vee \mathcal{N}_V(t)), (\mathcal{A}_W(t) \wedge \mathcal{A}_U(t)) \wedge (\mathcal{A}_W(t) \wedge \mathcal{A}_V(t)), \\
&\quad (\mathcal{N}_W(t) \wedge \mathcal{Y}_U(t)) \wedge (\mathcal{N}_W(t) \wedge \mathcal{Y}_V(t)) : t \in X\} \\
&= (U \mapsto W) \cup (V \mapsto W)\}.
\end{aligned}
$$

Therefore $(U \cap V) \mapsto W = (U \mapsto W) \cup (V \mapsto W)$.

It is necessary to mention that $(U \cap V) \mapsto W$ may or may not be equal to $(U \mapsto W) \cap (V \mapsto W)$. When $U = V = W$ then clearly $(U \cap V) \mapsto W = (U \mapsto W) \cap (V \mapsto W)$. At the same time, below we consider an example that shows that $(U \cap V) \mapsto W \neq (U \mapsto W) \cap (V \mapsto W)$.

**Example 1.7.** Let us suppose three PFSs $U$, $V$, and $W$ over the set of universe $X = \{t, s\}$ are as follows:

$$
\begin{aligned}
U &= \{(t, 0.2, 0.3, 0.5), (s, 0.5, 0.3, 0.1)\}; \\
V &= \{(t, 0.8, 0.1, 0.1), (s, 0.5, 0.3, 0.2)\}; \\
\text{and } W &= \{(t, 0.1, 0.1, 0.1), (s, 0.7, 0.2, 0.1)\}.
\end{aligned}
$$

Therefore $U \cap V = \{(t, 0.2, 0.1, 0.5), (s, 0.5, 0.3, 0.2)\}$.

Now,

$$
\begin{aligned}
&(U \cap V) \mapsto W \\
&= \{(t, 0.2, 0.1, 0.5), (s, 0.5, 0.3, 0.2)\} \mapsto \{(t, 0.1, 0.1, 0.1), (s, 0.7, 0.2, 0.1)\} \\
&= \{(t, 0.5, 0.1, 0.1), (s, 0.7, 0.2, 0.1)\}.
\end{aligned}
$$

Also,

$$
\begin{aligned}
U \mapsto W &= \{(t, 0.5, 0.1, 0.1), (s, 0.7, 0.2, 0.1)\} \\
\text{and } V \mapsto W &= \{(t, 0.1, 0.1, 0.1), (s, 0.7, 0.2, 0.1)\}.
\end{aligned}
$$

Hence,

$$\begin{aligned}&(U \mapsto W) \cap (V \mapsto W)\\&= \{(t, 0.5, 0.1, 0.1), (s, 0.7, 0.2, 0.1)\} \cap \{(t, 0.1, 0.1, 0.1), (s, 0.7, 0.2, 0.1)\}\\&= \{(t, 0.1, 0.1, 0.1), (s, 0.7, 0.2, 0.1)\}.\end{aligned}$$

Thus in this case, $(U \cap V) \mapsto W = (U \mapsto W) \cap (V \mapsto W)$ does not hold.

(v) The proof is similar to the case of (iv). □

It is to be noted that the equality $(U \cup V) \mapsto W = (U \mapsto W) \cup (V \mapsto W)$ may or may not hold. When $U = V = W$, then the equality is true. An example is given below, showing that the equality is not valid.

**Example 1.8.** Let us consider three PFSs $U$, $V$, and $W$ over the set of universe $X = \{t, s\}$ as follows:

$$\begin{aligned}U &= \{(t, 0.3, 0.4, 0.1), (s, 0.7, 0.2, 0.1)\};\\V &= \{(t, 0.1, 0.5, 0.3), (s, 0.5, 0.3, 0.1)\};\\\text{and } W &= \{(t, 0.3, 0.3, 0.4), (s, 0.1, 0.5, 0.4)\}.\end{aligned}$$

Therefore $U \cup V = \{(t, 0.3, 0.4, 0.1), (s, 0.7, 0.2, 0.1)\}$.

Now, $(U \cup V) \mapsto W = \{(t, 0.3, 0.3, 0.3), (s, 0.1, 0.2, 0.4)\}$.

Also,

$$\begin{aligned}U \mapsto W &= \{(t, 0.3, 0.3, 0.3), (s, 0.1, 0.2, 0.4)\}\\\text{and } V \mapsto W &= \{(t, 0.3, 0.3, 0.1), (s, 0.1, 0.3, 0.4)\}.\end{aligned}$$

Hence,

$$\begin{aligned}&(U \mapsto W) \cup (V \mapsto W)\\&= \{(t, 0.3, 0.3, 0.3), (s, 0.1, 0.2, 0.4)\} \cap \{(t, 0.3, 0.3, 0.1), (s, 0.1, 0.3, 0.4)\}\\&= \{(t, 0.3, 0.3, 0.1), (s, 0.1, 0.2, 0.4)\}.\end{aligned}$$

Thus in this case, $(U \cup V) \mapsto W = (U \mapsto W) \cup (V \mapsto W)$ does not hold.

Now, we are looking to define two particular types of topological operators on PFSs.

## 1.11 Topological operators on picture fuzzy set

The current section introduces two special topological operators [12] on PFSs and presents some interesting results related to them.

**Definition 1.30.** Let $U=(\mathcal{Y}_U,\mathcal{A}_U,\mathcal{N}_U)$ be a PFS over $X$. Then, the topological operator of closure (TOC) $C$ and the topological operator of interior (TOI) $I$ on $U$ are defined as

$$C(U)=\{(s,\ \bigvee_{t\in X}\mathcal{Y}_U(t),\ \bigwedge_{t\in X}\mathcal{A}_U(t),\ \bigwedge_{t\in X}\mathcal{N}_U(t)):s\in X\}$$

$$\text{and } I(U)=\{(s,\ \bigwedge_{t\in X}\mathcal{Y}_U(t),\ \bigwedge_{t\in X}\mathcal{A}_U(t),\ \bigvee_{t\in X}\mathcal{N}_U(t)):s\in X\}.$$

**Example 1.9.** Let us consider the PFS, $U=\{(s,0.7,0.2,0.1),(t,0.5,0.1,0.2),(u,0.4,0.3,0.3)\}$ over the universe $X=\{s,t,u\}$.

Then,

$$\begin{aligned}
C(U)&=\{(s,0.7\vee 0.5\vee 0.4,0.2\wedge 0.1\wedge 0.3,0.1\wedge 0.2\wedge 0.3),\\
&\quad(t,0.7\vee 0.5\vee 0.4,0.2\wedge 0.1\wedge 0.3,0.1\wedge 0.2\wedge 0.3),\\
&\quad(u,0.7\vee 0.5\vee 0.4,0.2\wedge 0.1\wedge 0.3,0.1\wedge 0.2\wedge 0.3)\}\\
&=\{(s,0.7,0.1,0.1),(t,0.7,0.1,0.1),(u,0.7,0.1,0.1)\}\\
\text{and } I(U)&=\{(s,0.7\wedge 0.5\wedge 0.4,0.2\wedge 0.1\wedge 0.3,0.1\vee 0.2\vee 0.3),\\
&\quad(t,0.7\wedge 0.5\wedge 0.4,0.2\wedge 0.1\wedge 0.3,0.1\vee 0.2\vee 0.3),\\
&\quad(u,0.7\wedge 0.5\wedge 0.4,0.2\wedge 0.1\wedge 0.3,0.1\vee 0.2\vee 0.3)\}\\
&=\{(s,0.4,0.1,0.3),(t,0.4,0.1,0.3),(u,0.4,0.1,0.3)\}.
\end{aligned}$$

**Proposition 1.7.** *Let $U=(\mathcal{Y}_U,\mathcal{A}_U,\mathcal{N}_U)$ be a PFS on the universe $X$. Then, the following holds:*

**(i)** $\overline{I(\overline{U})}=C(U)$;

**(ii)** $C(U\cup V)=C(U)\cup C(V)$, *where $C$ and $I$ are TOC and TOI, respectively.*

*Proof.* (i) By the definition

$$\begin{aligned}
\overline{I(\overline{U})}&=\overline{I(\{(s,\mathcal{N}_U(s),\mathcal{A}_U(s),\mathcal{Y}_U(s)):s\in X\})}\\
&=\overline{\{(s,\ \bigwedge_{t\in X}\mathcal{N}_U(t),\ \bigwedge_{t\in X}\mathcal{A}_U(t),\ \bigvee_{t\in X}\mathcal{Y}_U(t)):s\in X\}}\\
&=\{(s,\ \bigvee_{t\in X}\mathcal{Y}_U(t),\ \bigwedge_{t\in X}\mathcal{A}_U(t),\ \bigwedge_{t\in X}\mathcal{N}_U(t)):s\in X\}\\
&=C(U).
\end{aligned}$$

(ii) Now,

$$\begin{aligned}
&C(U\cup V)\\
&=C(\{(s,\mathcal{Y}_U(s)\vee\mathcal{Y}_V(s),\mathcal{A}_U(s)\wedge\mathcal{A}_V(s),\mathcal{N}_U(s)\wedge\mathcal{N}_V(s)):s\in X\})\\
&=C(\{(s,\mathcal{Y}_U(s)\vee\mathcal{Y}_V(s),\mathcal{A}_U(s)\wedge\mathcal{A}_V(s),\mathcal{N}_U(s)\wedge\mathcal{N}_V(s)):s\in X\})\\
&=\{(s,(\bigvee_{t\in X}\mathcal{Y}_U(t))\vee(\bigvee_{t\in X}\mathcal{Y}_V(t)),(\bigwedge_{t\in X}\mathcal{A}_U(t))\wedge(\bigwedge_{t\in X}\mathcal{A}_V(t)),(\bigwedge_{t\in X}\mathcal{N}_U(t))\wedge(\bigwedge_{t\in X}\mathcal{N}_V(t))):s\in X\}
\end{aligned}$$

$$= \{(s, \underset{t\in X}{\vee} \mathcal{Y}_U(t), \underset{t\in X}{\wedge} \mathcal{A}_U(t), \underset{t\in X}{\wedge} N_U(t)) : s \in X\} \cup \{(s, \underset{t\in X}{\vee} \mathcal{Y}_V(t), \underset{t\in X}{\wedge} \mathcal{A}_V(t), \underset{t\in X}{\wedge} N_V(t)) : s \in X\}$$
$$= C(U) \cup C(V).$$

□

However, the property does not necessarily hold in the case of intersection, which the following two examples can prove. If $U$ and $V$ are two PFSs over the set of universe $X$ then Example 1.10 shows that $C(U \cap V) = C(U) \cap C(V)$, while Example 1.11 shows that $C(U \cap V) \neq C(U) \cap C(V)$.

**Example 1.10.** Let us consider two PFSs $U = \{(s, 0.5, 0.4, 0.2), (t, 0.5, 0.3, 0.15)\}$ and $V = \{(s, 0.3, 0.3, 0.3), (t, 0.4, 0.4, 0.2)\}$ on $X = \{s, t\}$.

Here,

$$\begin{aligned} C(U \cap V) &= C\{(s, 0.3, 0.3, 0.3), (t, 0.4, 0.3, 0.2)\} \\ &= \{(s, 0.4, 0.3, 0.2), (t, 0.4, 0.3, 0.2)\} \\ \text{and } C(U) \cap C(V) &= \{(s, 0.5, 0.3, 0.15), (t, 0.5, 0.3, 0.15)\} \cap \{(s, 0.4, 0.3, 0.2), (t, 0.4, 0.3, 0.2)\} \\ &= \{(s, 0.4, 0.3, 0.2), (t, 0.4, 0.3, 0.2)\}. \end{aligned}$$

Thus in this case, $C(U \cap V) = C(U) \cap C(V)$.

**Example 1.11.** Let us suppose two PFSs $U_1 = \{(s, 0.3, 0.35, 0.2), (t, 0.4, 0.25, 0.25)\}$ and $V_1 = \{(s, 0.5, 0.2, 0.3), (t, 0.3, 0.35, 0.3)\}$ on $X = \{s, t\}$.

Here,

$$\begin{aligned} C(U_1 \cap V_1) &= C\{(s, 0.3, 0.2, 0.3), (t, 0.3, 0.25, 0.3)\} \\ &= \{(s, 0.3, 0.2, 0.3), (t, 0.3, 0.2, 0.3)\}. \end{aligned}$$

However,

$$\begin{aligned} C(U_1) \cap C(V_1) &= \{(s, 0.4, 0.25, 0.2), (t, 0.4, 0.25, 0.2)\} \cap \{(s, 0.5, 0.2, 0.3), (t, 0.5, 0.2, 0.3)\} \\ &= \{(s, 0.4, 0.2, 0.3), (t, 0.4, 0.2, 0.3)\}. \end{aligned}$$

Thus in this case, $C(U_1 \cap V_1) \neq C(U_1) \cap C(V_1)$.

## 1.12 Dombi operations on PFNs

In fuzzy-set theory, the basic characteristics of the conjunctive and disjunctive operators, interpreted on fuzzy subsets, are:

1. the operator is associative;
2. the operator is commutative;
3. the correspondence principle is fulfilled, that is, if the fuzzy subset is crisp, then we obtain the conjunction and disjunction of the classical set theory.

Let $f(a)$ be the function assigned to the conjunctive operator $k(a,b)$, and $g(a)$ the one assigned to the disjunctive one $d(a,b)$. A series of operators can be constructed for the conjunctive (respectively, disjunctive) operators, the limit of which is the min (respectively, max) operator provided by Zadeh. Suppose $f(a)$, which is necessary for the production of the operators, with the help of the negation operator for every such function $f(a)$ a function $g(a)$ can be given from which a disjunctive can be constructed. For the general construction of the operator, we have a good connection between conjunctive, disjunctive, and negation operators, which leads to the necessary and sufficient condition for fulfilling the DeMorgan identity. Then, the third operator can be constructed with the help of any two operators. Based on construction, Hamacher's conditions belong to the DeMorgan class and Yager's operator system. For the fulfilment of the DeMorgan identity, necessary and sufficient conditions are given as $f_{\Re}(a)$ can be constructed for every $f(a)$, so that for the derived $K_{\Re}(a,b)$ and $d_{\Re}(a,b)$, $\lim_{\Re\to\infty} K_{\Re}(a,b)=\min(a,b)$ and $\lim_{\Re\to\infty} d_{\Re}(a,b)=\max(a,b)$. As Yager's operator is not reducible, for every $\Re$ there exist a $\beta$, for which, when $a<\beta$ and $b<\beta$, $K_{\Re}(a,b)=0$.

The measurement of fuzziness was derived from the general construction. It is noted that fuzzy operators can be utilized appropriately instead of arbitrarily applying different fuzzy measurements. When the fuzzy theory is applied in decision theory, the optimum is only defined, and its degree, i.e., it is not measured indeed. It will be helpful to construct a system in which one can conclude the sharpness of decision from the sharpness of the applied sets and operators.

Table 1.1 lists the characteristics of the operators proposed by Hamacher [28], Yager [70], and Dombi [18].

Here, the negation operator is $n(a)=1-a$. We see that

1. Yager's operator does not satisfy all the characteristics such as continuous, conjunctive, and disjunctive, but the Dombi operator satisfies these properties.
2. The form of the Hamacher operator can be obtained if the substitute parameter $\Re$ occurs in the table for $1/\Re$ in the case of the conjunctive operator, and for $1/(\Re'+1)$ in the case of the disjunctive operator. In the resulting transformed form, the condition of the fulfilment of the DeMorgan identity is $1/\Re=1/(\Re'+1)$, which is equivalent to the results of Hamacher when $n(a)=1-a$.
3. It can be shown that Yager's formula is equivalent to

   **(i)** $1-\min\left(1,\left((1-a)^{\Re}+(1-b)^{\Re}\right)^{1/\Re}\right)$, **(ii)** $\min\left(1,\left(a^{\Re}+b^{\Re}\right)^{1/\Re}\right)$.
4. In the case of all three operators if $\Re\le\Re'$, then $k_{\Re}(a,b)\le k_{\Re'}(a,b)$ and $d_{\Re}(a,b)\ge d_{\Re'}(a,b)$.
5. In the case of all three operators $\lim_{\Re\to 0} k_{\Re}(a,b)=0$ and $\lim_{\Re\to 0} d_{\Re}(a,b)=1$.

The above discussion shows that Dombi operators are more general than Hamacher and Yager operators. Wei [62,64] defined PFHWA, PFHOWA, PFHWA, PFHWG, PFHOWG, and PFHWG operators based on Hamacher operators. However, in this chapter we de-

**Table 1.1** Conjunction and Disjunction.

| Conjunction | | | | | |
|---|---|---|---|---|---|
| **Authors** | $f(a)=$ | $f^{-1}(a)=$ | $k_{\Re}(a,b)=$ | $k_1(a,b)=$ | $\lim_{\Re\to\infty} k_{\Re}(a,b)=$ |
| H. Hamacher | $\frac{e^{-a}}{\Re+(1-\Re)e^{-a}}$ | $-\ln\frac{\Re a}{1+(\Re-1)a}$ | $\frac{\Re ab}{1-(1-\Re)(a+b-ab)}$ | $ab$ | $\frac{ab}{a+b-ab}$ |
| R.R. Yager | $\begin{cases} 1-a^{1/\Re}, \\ \text{if } a<1 \\ 0, \\ \text{if } a\geq 1 \end{cases}$ | $\begin{cases} (1-a)^{\Re}, \\ \text{if } a<1 \\ 0, \\ \text{if } a\geq 1 \end{cases}$ | $\begin{cases} 1-\left((1-a)^{\Re}+(1-b)^{\Re}\right)^{1/\Re}, \\ \text{if } (1-a)^{\Re}+(1-b)^{\Re}<1 \\ 0, \\ \text{if } (1-a)^{\Re}+(1-b)^{\Re}\geq 1 \end{cases}$ | $\max(0,a+b-1)$ | $\min(x,y)$ |
| J. Dombi | $\frac{1}{1+a^{1/\Re}}$ | $(\frac{1}{a}-1)^{\Re}$ | $\frac{1}{1+\left((\frac{1}{a}-1)^{\Re}+(\frac{1}{b}-1)^{\Re}\right)^{1/\Re}}$ | $\frac{ab}{a+b-ab}$ if $a=0,\ b=0$ | $\min(a,b)$ |
| **Disjunction** | | | | | |
| | $g(a)=$ | $g^{-1}(a)=$ | $d_{\Re}(a,b)=$ | $d_1(a,b)=$ | $\lim_{\Re\to\infty} d_{\Re}(a,b)=$ |
| H. Hamacher | $\frac{\Re(1-e^{-a})}{\Re+(1-\Re)e^{-a}}$ | $-\ln\frac{\Re(1-a)}{\Re-(\Re-1)a}$ | $\frac{\Re(a+b)+ab(1-2\Re)}{\Re+ab(1-\Re)}$ | $a+b-ab$ | $\frac{a+b-2ab}{1-ab}$ |
| R.R. Yager | $\begin{cases} a^{1/\Re}, \\ \text{if } a<1 \\ 1, \\ \text{if } a\geq 1 \end{cases}$ | $\begin{cases} a^{\Re}, \\ \text{if } a<1 \\ 1, \\ \text{if } a\geq 1 \end{cases}$ | $\begin{cases} \left(a^{\Re}+b^{\Re}\right)^{1/\Re}, \\ \text{if } a^{\Re}+b^{\Re}<1 \\ 1, \\ \text{if } a^{\Re}+b^{\Re}\geq 1 \end{cases}$ | $\min(1,a+b)$ | $\max(x,y)$ |
| J. Dombi | $\frac{1}{1+a^{-1/\Re}}$ | $(\frac{1}{a}-1)^{-\Re}$ | $\frac{1}{1+\left((\frac{1}{a}-1)^{-\Re}+(\frac{1}{b}-1)^{-\Re}\right)^{-1/\Re}}$ | $\frac{a+b-2ab}{1-ab}$ if $a=1,\ b=1$ | $\max(a,b)$ |

fined PFDWA, PFDOWA, PFDHWA, PFDWG, PFDOWG, and PFDHWG operators based on Dombi operators, which are better than Wei's.

The Dombi norm (TN) and Dombi conorm (TCN) are defined below.

**Definition 1.31.** [18] Suppose that $f$ and $g$ are any two real numbers. Then, the Dombi norm and Dombi conorms can be defined as follows:

$$Dom(f,g)=\frac{1}{1+\{(\frac{1-f}{f})^{\varrho}+(\frac{1-g}{g})^{\varrho}\}^{1/\varrho}} \tag{1.22}$$

$$Dom^{c}(f,g)=1-\frac{1}{1+\{(\frac{f}{1-f})^{\varrho}+(\frac{g}{1-g})^{\varrho}\}^{1/\varrho}}, \tag{1.23}$$

where, $\varrho \geq 1$ and $(f,g)\in[0,1]\times[0,1]$.

## 1.13 Picture fuzzy Dombi aggregation operator

To explain Dombi operations on PFNS, let us first define Dombi $t$-norm and $t$-conorm operations as follows.

**Definition 1.32.** Let $p=(\mathcal{Y}_p,\mathcal{A}_p,\mathcal{N}_p)$ and $q=(\mathcal{Y}_q,\mathcal{A}_q,\mathcal{N}_q)$ be two PFNs with $s\geqslant 1$ and $k>0$. Then, Dombi $t$-norm and $t$-conorm operations on PFNs are defined as follows:

**(i)** $p\oplus_D q=\left(1-\frac{1}{1+\left[\left(\frac{\mathcal{Y}_p}{1-\mathcal{Y}_p}\right)^s+\left(\frac{\mathcal{Y}_q}{1-\mathcal{Y}_q}\right)^s\right]^{\frac{1}{s}}},\frac{1}{1+\left[\left(\frac{1-\mathcal{A}_p}{\mathcal{A}_p}\right)^s+\left(\frac{1-\mathcal{A}_q}{\mathcal{A}_q}\right)^s\right]^{\frac{1}{s}}},\frac{1}{1+\left[\left(\frac{1-\mathcal{N}_p}{\mathcal{N}_p}\right)^s+\left(\frac{1-\mathcal{N}_q}{\mathcal{N}_q}\right)^s\right]^{\frac{1}{s}}}\right);$

**(ii)** $p\otimes_D q=\left(\frac{1}{1+\left[\left(\frac{1-\mathcal{Y}_p}{\mathcal{Y}_p}\right)^s+\left(\frac{1-\mathcal{Y}_q}{\mathcal{Y}_q}\right)^s\right]^{\frac{1}{s}}},\frac{1}{1+\left[\left(\frac{1-\mathcal{A}_p}{\mathcal{A}_p}\right)^s+\left(\frac{1-\mathcal{A}_q}{\mathcal{A}_q}\right)^s\right]^{\frac{1}{s}}},1-\frac{1}{1+\left[\left(\frac{\mathcal{N}_p}{1-\mathcal{N}_p}\right)^s+\left(\frac{\mathcal{N}_q}{1-\mathcal{N}_q}\right)^s\right]^{\frac{1}{s}}}\right);$

**(iii)** $k\cdot p=\left(1-\frac{1}{1+\left\{k\left(\frac{\mathcal{Y}_p}{1-\mathcal{Y}_p}\right)^s\right\}^{\frac{1}{s}}},\frac{1}{1+\left\{k\left(\frac{1-\mathcal{A}_p}{\mathcal{A}_p}\right)^s\right\}^{\frac{1}{s}}},\frac{1}{1+\left\{k\left(\frac{1-\mathcal{N}_p}{\mathcal{N}_p}\right)^s\right\}^{\frac{1}{s}}}\right);$

**(iv)** $p^k=\left(\frac{1}{1+\left\{k\left(\frac{1-\mathcal{Y}_p}{\mathcal{Y}_p}\right)^s\right\}^{\frac{1}{s}}},\frac{1}{1+\left\{k\left(\frac{1-\mathcal{A}_p}{\mathcal{A}_p}\right)^s\right\}^{\frac{1}{s}}},1-\frac{1}{1+\left\{k\left(\frac{\mathcal{N}_p}{1-\mathcal{N}_p}\right)^s\right\}^{\frac{1}{s}}}\right).$

**Example 1.12.** Let us consider two PFNs $p=(0.6,0.1,0.03)$ and $q=(0.5,0.3,0.2)$ with $s=3$ and $k=2$. Then,

**(i)** $p\oplus_D q$

$$=\left(1-\frac{1}{1+\left[\left(\frac{0.6}{1-0.6}\right)^3+\left(\frac{0.5}{1-0.5}\right)^3\right]^{\frac{1}{3}}},\frac{1}{1+\left[\left(\frac{1-0.1}{0.1}\right)^3+\left(\frac{1-0.3}{0.3}\right)^3\right]^{\frac{1}{3}}},\frac{1}{1+\left[\left(\frac{1-0.03}{0.03}\right)^3+\left(\frac{1-0.2}{0.2}\right)^3\right]^{\frac{1}{3}}}\right)$$

$=(0.6205,0.099,0.0299);$

**(ii)** $p \otimes_D q$

$$= \left( \frac{1}{1+\left[\left(\frac{1-0.6}{0.6}\right)^3 + \left(\frac{1-0.5}{0.5}\right)^3\right]^{\frac{1}{3}}}, \frac{1}{1+\left[\left(\frac{1-0.1}{0.1}\right)^3 + \left(\frac{1-0.3}{0.3}\right)^3\right]^{\frac{1}{3}}}, 1 - \frac{1}{1+\left[\left(\frac{0.03}{1-0.03}\right)^3 + \left(\frac{0.2}{1-0.2}\right)^3\right]^{\frac{1}{3}}} \right)$$

$= (0.478, 0.099, 0.2001)$;

**(iii)** $2p$

$$= \left( 1 - \frac{1}{1+\{2(\frac{0.6}{1-0.6})^3\}^{\frac{1}{3}}}, \frac{1}{1+\{2(\frac{1-0.01}{0.1})^3\}^{\frac{1}{3}}}, \frac{1}{1+\{2(\frac{1-0.03}{0.03})^3\}^{\frac{1}{3}}} \right)$$

$= (0.653, 0.081, 0.023)$;

**(iv)** $p^2$

$$= \left( 1 - \frac{1}{1+\{2(\frac{1-0.6}{0.6})^3\}^{\frac{1}{3}}}, \frac{1}{1+\{2(\frac{1-0.01}{0.1})^3\}^{\frac{1}{3}}}, \frac{1}{1+\{2(\frac{0.03}{1-0.03})^3\}^{\frac{1}{3}}} \right)$$

$= (0.543, 0.081, 0.037)$.

Now, we define the picture fuzzy Dombi weighted averaging operator (PFDWAO), generalizing the concept of Dombi $t$-norm and $t$-conorm operations associated with some weight.

**Definition 1.33.** Let $\{p_l = (\mathcal{Y}_{p_l}, \mathcal{A}_{p_l}, \mathcal{N}_{p_l}) : l = 1, 2, ..., n\}$ be a set of PFNs. Then,

$$\text{PFDWAO}(p_1, p_2, ..., p_n)$$

$$= \left( 1 - \frac{1}{1+\left\{\sum_{i=1}^{n} \theta_l \left(\frac{\mathcal{Y}_{p_l}}{1-\mathcal{Y}_{p_l}}\right)^s\right\}^{\frac{1}{s}}}, \frac{1}{1+\sum_{i=1}^{n} \left\{\theta_l \left(\frac{1-\mathcal{A}_{p_l}}{\mathcal{A}_{p_l}}\right)\right\}^{s^{\frac{1}{s}}}}, \frac{1}{1+\sum_{i=1}^{n} \left\{\theta_l \left(\frac{1-\mathcal{N}_{p_l}}{\mathcal{N}_{p_l}}\right)^s\right\}^{\frac{1}{s}}} \right).$$

### 1.13.1 Properties of Dombi aggregation operators

The aggregation operators follow several mathematical properties.

The max and min among $n$ PFNs are defined below.

**Definition 1.34.** Let $\mathcal{Y}_U = \max\{\mathcal{Y}_{p_l} : l = 1, 2, \ldots, n\}$, $\mathcal{A}_U = \min\{\mathcal{A}_{p_l} : l = 1, 2, \ldots, n\}$ and $\mathcal{N}_U = \min\{\mathcal{N}_{p_l} : l = 1, 2, \ldots, n\}$ and $\mathcal{Y}_V = \min\{\mathcal{Y}_{p_l} : l = 1, 2, \ldots, n\}$, $\mathcal{A}_V = \min\{\mathcal{A}_{p_l} : l = 1, 2, \ldots, n\}$ and $\mathcal{N}_V = \max\{\mathcal{N}_{p_l} : l = 1, 2, \ldots, n\}$. Then,

$$P = \max\{p_l : l = 1, 2, \ldots, n\} = (\mathcal{Y}_P, \mathcal{A}_P, \mathcal{N}_P)$$
$$\text{and } Q = \min\{p_l : l = 1, 2, \ldots, n\} = (\mathcal{Y}_Q, \mathcal{A}_Q, \mathcal{N}_Q).$$

**Proposition 1.8.** *(Idempotency Property) Let* $\{p_l = (\mathcal{Y}_{p_l}, \mathcal{A}_{p_l}, \mathcal{N}_{p_l}) : l = 1, 2, ..., n\}$ *be a set of PFNs that are all identical, i.e.,* $p_l = p$ *for* $l = 1, 2, ..., n$*; then PFDWAO* $(p_1, p_2, ..., p_n) = p$.

*Proof.* From the definition,

$$\text{PFDWAO}(p_1, p_2, ..., p_n)$$
$$= \left(1 - \frac{1}{1 + \left\{\sum_{i=1}^{n} \theta_l \left(\frac{\mathcal{Y}_{p_l}}{1 - \mathcal{Y}_{p_l}}\right)^s\right\}^{\frac{1}{s}}}, \frac{1}{1 + \sum_{i=1}^{n} \left\{\theta_l \left(\frac{1 - \mathcal{A}_{p_l}}{\mathcal{A}_{p_l}}\right)^s\right\}^{\frac{1}{s}}}, \frac{1}{1 + \sum_{i=1}^{n} \left\{\theta_l \left(\frac{1 - \mathcal{N}_{p_l}}{\mathcal{N}_{p_l}}\right)^s\right\}^{\frac{1}{s}}}\right).$$

It is observed that

$$\begin{aligned}
&1 + \left\{\sum_{i=1}^{n} \theta_l \left(\frac{\mathcal{Y}_{p_l}}{1 - \mathcal{Y}_{p_l}}\right)^s\right\}^{\frac{1}{s}} \\
&= 1 + \left\{\sum_{i=1}^{n} \theta_l \left(\frac{\mathcal{Y}_{p}}{1 - \mathcal{Y}_{p}}\right)^s\right\}^{\frac{1}{s}} \\
&= 1 + \left\{\left(\frac{\mathcal{Y}_{p}}{1 - \mathcal{Y}_{p}}\right)^s (\theta_1 + \theta_2 + ... + \theta_n)\right\}^{\frac{1}{s}} \\
&= 1 + \frac{\mathcal{Y}_{p}}{1 - \mathcal{Y}_{p}} \; [\text{as } \theta_1 + \theta_2 + ... + \theta_n = 1] \\
&= \frac{1}{1 - \mathcal{Y}_{p}}.
\end{aligned}$$

Similarly, $1 + \left\{\sum_{i=1}^{n} \theta_l \left(\frac{1 - \mathcal{A}_{p_l}}{\mathcal{A}_{p_l}}\right)^s\right\}^{\frac{1}{s}} = \frac{1}{\mathcal{A}_p}$ and $1 + \left\{\sum_{i=1}^{n} \theta_l \left(\frac{1 - \mathcal{N}_{p_l}}{\mathcal{N}_{p_l}}\right)^s\right\}^{\frac{1}{s}} = \frac{1}{\mathcal{N}_p}$.
Thus PFDWAO $(p_1, p_2, ..., p_n) = (1 - \frac{1}{\frac{1}{1-\mathcal{Y}_p}}, \frac{1}{\frac{1}{\mathcal{A}_p}}, \frac{1}{\frac{1}{\mathcal{N}_p}}) = (\mathcal{Y}_p, \mathcal{A}_p, \mathcal{N}_p) = p$. □

**Proposition 1.9.** *(Boundedness Property) Let* $\{p_l = (\mathcal{Y}_{p_l}, \mathcal{A}_{p_l}, \mathcal{N}_{p_l}) : l = 1, 2, ..., n\}$ *be a set of PFNs. Then,* $Q \leqslant \text{PFDWAO}(p_1, p_2, ..., p_n) \leqslant P$*, where* $Q = \min\{p_l : l = 1, 2, ..., n\}$ *and* $P = \max\{p_l : l = 1, 2, ..., n\}$ *provided that* $\mathcal{A}_{p_1} = \mathcal{A}_{p_2} = \cdots = \mathcal{A}_{p_l}$.

*Proof.* Here,

$$\begin{aligned}
&\mathcal{Y}_Q \leqslant \mathcal{Y}_{p_l} \leqslant \mathcal{Y}_P \text{ for } l = 1, 2, ..., n \\
&\text{i.e., } 1 - \mathcal{Y}_P \leqslant 1 - \mathcal{Y}_{p_l} \leqslant 1 - \mathcal{Y}_Q \text{ for } l = 1, 2, ..., n, \\
&\text{i.e., } \frac{\mathcal{Y}_Q}{1 - \mathcal{Y}_Q} \leqslant \frac{\mathcal{Y}_{p_l}}{1 - \mathcal{Y}_{p_l}} \leqslant \frac{\mathcal{Y}_P}{1 - \mathcal{Y}_P} \text{ for } l = 1, 2, ..., n, \\
&\text{i.e., } 1 + \left\{\sum_{s=1}^{n} \theta_l \left(\frac{\mathcal{Y}_Q}{1 - \mathcal{Y}_Q}\right)^s\right\}^{\frac{1}{s}} \leqslant 1 + \left\{\sum_{s=1}^{n} \theta_l \left(\frac{\mathcal{Y}_{p_l}}{1 - \mathcal{Y}_{p_l}}\right)^s\right\}^{\frac{1}{s}} \leqslant 1 + \left\{\sum_{s=1}^{n} \theta_l \left(\frac{\mathcal{Y}_P}{1 - \mathcal{Y}_P}\right)^s\right\}^{\frac{1}{s}},
\end{aligned}$$

$$\text{i.e., } 1-\frac{1}{1+\left\{\sum_{s=1}^{n}\theta_l\left(\frac{\mathcal{Y}_Q}{1-\mathcal{Y}_Q}\right)^s\right\}^{\frac{1}{s}}}\leqslant 1-\frac{1}{1+\left\{\sum_{s=1}^{n}\theta_l\left(\frac{\mathcal{Y}_{p_l}}{1-\mathcal{Y}_{p_l}}\right)^s\right\}^{\frac{1}{s}}}\leqslant 1-\frac{1}{1+\left\{\sum_{s=1}^{n}\theta_l\left(\frac{\mathcal{Y}_P}{1-\mathcal{Y}_P}\right)^s\right\}^{\frac{1}{s}}}.$$

The left and rightmost terms are $\mathcal{Y}_Q$ and $\mathcal{Y}_P$, respectively.
Thus

$$\mathcal{Y}_Q\leqslant 1-\frac{1}{1+\left\{\sum_{s=1}^{n}\theta_l\left(\frac{\mathcal{Y}_{p_l}}{1-\mathcal{Y}_{p_l}}\right)^s\right\}^{\frac{1}{s}}}\leqslant \mathcal{Y}_P,$$

$\mathcal{A}_Q=\mathcal{A}_{p_l}=\mathcal{A}_P$ for $l=1,2,...,n$

$$\text{i.e., } \frac{1}{1+\left\{\sum_{s=1}^{n}\theta_l\left(\frac{1-\mathcal{A}_Q}{\mathcal{N}_Q}\right)^s\right\}^{\frac{1}{s}}}=\frac{1}{1+\left\{\sum_{s=1}^{n}\theta_l\left(\frac{1-\mathcal{A}_{p_l}}{\mathcal{A}_{p_l}}\right)^s\right\}^{\frac{1}{s}}}=\frac{1}{1+\left\{\sum_{s=1}^{n}\theta_l\left(\frac{1-\mathcal{N}_P}{\mathcal{N}_P}\right)^s\right\}^{\frac{1}{s}}},$$

$$\text{i.e., } \mathcal{A}_Q=\frac{1}{1+\left\{\sum_{s=1}^{n}\theta_l\left(\frac{1-\mathcal{A}_{p_l}}{\mathcal{A}_{p_l}}\right)^s\right\}^{\frac{1}{s}}}=\mathcal{A}_P$$

and $\mathcal{N}_P\leqslant\mathcal{N}_{p_l}\leqslant\mathcal{N}_Q$ for $l=1,2,...,n$

$$\text{i.e., } 1-\frac{1}{\mathcal{N}_P}\leqslant 1-\frac{1}{\mathcal{N}_{p_l}}\leqslant 1-\frac{1}{\mathcal{N}_Q}\ \text{for } l=1,2,...,n,$$

$$\text{i.e., } \frac{1-\mathcal{N}_Q}{\mathcal{N}_Q}\leqslant\frac{1-\mathcal{N}_{p_l}}{\mathcal{N}_{p_l}}\leqslant\frac{1-\mathcal{N}_P}{\mathcal{N}_P}\ \text{for } l=1,2,...,n,$$

$$\text{i.e., } 1+\left\{\sum_{s=1}^{n}\theta_l\left(\frac{1-\mathcal{N}_Q}{\mathcal{N}_Q}\right)^s\right\}^{\frac{1}{s}}\leqslant 1+\left\{\sum_{s=1}^{n}\theta_l\left(\frac{1-\mathcal{N}_{p_l}}{\mathcal{N}_{p_l}}\right)^s\right\}^{\frac{1}{s}}\leqslant 1+\left\{\sum_{s=1}^{n}\theta_l\left(\frac{1-\mathcal{N}_P}{\mathcal{N}_P}\right)^s\right\}^{\frac{1}{s}},$$

$$\text{i.e., } \frac{1}{1+\left\{\sum_{s=1}^{n}\theta_l\left(\frac{1-\mathcal{N}_P}{\mathcal{N}_P}\right)^s\right\}^{\frac{1}{s}}}\leqslant\frac{1}{1+\left\{\sum_{s=1}^{n}\theta_l\left(\frac{1-\mathcal{N}_{p_l}}{\mathcal{N}_{p_l}}\right)^s\right\}^{\frac{1}{s}}}\leqslant\frac{1}{1+\left\{\sum_{s=1}^{n}\theta_l\left(\frac{1-\mathcal{N}_Q}{\mathcal{N}_Q}\right)^s\right\}^{\frac{1}{s}}},$$

$$\text{i.e., } \mathcal{N}_P\leqslant\frac{1}{1+\left\{\sum_{s=1}^{n}\theta_l\left(\frac{1-\mathcal{N}_{p_l}}{\mathcal{N}_{p_l}}\right)^s\right\}^{\frac{1}{s}}}\leqslant\mathcal{N}_Q.$$

Thus finally, we obtain the inequality $Q\leqslant \mathrm{PFDWAO}(p_1,p_2,...,p_n)\leqslant P$. □

**Proposition 1.10.** *(Monotonicity Property) Let* $\{p_l = (\mathcal{Y}_{p_l}, \mathcal{A}_{p_l}, \mathcal{N}_{p_l}) : l = 1, 2, 3, ..., n\}$ *and* $\{p_l' = (\mathcal{Y}_{p_l'}, \mathcal{A}_{p_l'}, \mathcal{N}_{p_l'}) : l = 1, 2, 3, ..., n\}$ *be two sets of PFNs. Then, for* $l = 1, 2, ..., n$; $p_l \leqslant p_l' \Rightarrow$ *PFDWAO*$(p_1, p_2, ..., p_n) \leqslant$ *PFDWAO*$(p_1', p_2', ..., p_n')$.

*Proof.* Since $p_l \leqslant p_l'$ therefore $\mathcal{Y}_{p_l} \leqslant \mathcal{Y}_{p_l'}, \mathcal{A}_{p_l} \leqslant \mathcal{A}_{p_l'}$ and $\mathcal{N}_{p_l} \geqslant \mathcal{N}_{p_l'}$ for $l = 1, 2, 3, ..., n$.

Now,

$$\mathcal{Y}_{p_l} \leqslant \mathcal{Y}_{p_l'}.$$

This implies,

$$\frac{1}{1-\mathcal{Y}_{p_l}} - 1 \leqslant \frac{1}{1-\mathcal{Y}_{p_l'}} - 1.$$

This is,

$$1 + \left\{\sum_{s=1}^{n} \theta_l \left(\frac{\mathcal{Y}_{p_l}}{1-\mathcal{Y}_{p_l}}\right)^s\right\}^{\frac{1}{s}} \leqslant 1 + \left\{\sum_{s=1}^{n} \theta_l \left(\frac{\mathcal{Y}_{p_l}{}'}{1-\mathcal{Y}_{p_l'}}\right)^s\right\}^{\frac{1}{s}}.$$

Finally,

$$1 - \frac{1}{1 + \left\{\sum_{s=1}^{n} \theta_l \left(\frac{\mathcal{Y}_{p_l}}{1-\mathcal{Y}_{p_l}}\right)^s\right\}^{\frac{1}{s}}} \leqslant 1 - \frac{1}{1 + \left\{\sum_{s=1}^{n} \theta_l \left(\frac{\mathcal{Y}_{p_l'}}{1-\mathcal{Y}_{p_l'}}\right)^s\right\}^{\frac{1}{s}}}.$$

This implies,

$$\frac{1}{1 + \left\{\sum_{s=1}^{n} \theta_l \left(\frac{1-\mathcal{A}_{p_l}}{\mathcal{A}_{p_l}}\right)^s\right\}^{\frac{1}{s}}} \leqslant \frac{1}{1 + \left\{\sum_{s=1}^{n} \theta_l \left(\frac{1-\mathcal{A}_{p_l'}}{\mathcal{A}_{p_l'}}\right)^s\right\}^{\frac{1}{s}}}$$

and

$$\frac{1}{1 + \left\{\sum_{s=1}^{n} \theta_l \left(\frac{1-\mathcal{N}_{p_l}}{\mathcal{N}_{p_l}}\right)^s\right\}^{\frac{1}{s}}} \geqslant \frac{1}{1 + \left\{\sum_{s=1}^{n} \theta_l \left(\frac{1-\mathcal{N}_{p_l'}}{\mathcal{N}_{p_l'}}\right)^s\right\}^{\frac{1}{s}}}$$

whenever $\mathcal{N}_{p_l} \geqslant \mathcal{N}_{p_l'}$.

Thus PFDWAO$(p_1, p_2, ..., p_n) \leqslant$ PFDWAO$(p_1', p_2', ..., p_n')$ whenever $p_l \leqslant p_l'$ for $l = 1, 2, ..., n$. □

**Proposition 1.11.** *Let* $p = (\mathcal{Y}_p, \mathcal{A}_p, \mathcal{N}_p)$ *and* $q = (\mathcal{Y}_q, \mathcal{A}_q, \mathcal{N}_q)$ *be two PFNs. Then,*

**(i)** $p \oplus_D q = q \oplus_D p$;
**(ii)** $p \otimes_D q = q \otimes_D p$;
**(iii)** $k(p \oplus_D q) = kp \oplus_D kq$;

**(iv)** $(p \otimes_D q)^k = p^k \otimes_D q^k$;
**(v)** $(k_1 + k_2)p = k_1 p \oplus_D k_2 p$;
**(vi)** $p^{k_1} \otimes_D p^{k_2} = p^{(k_1+k_2)}$.

*Proof.* (i) The proofs of (i) and (ii) follow from the definitions.
(iii) We have,

$$
\begin{aligned}
& k(p \oplus_D q) \\
&= k\Big((\mathcal{Y}_p, \mathcal{A}_p, N_p) \oplus_D (\mathcal{Y}_q, \mathcal{A}_q, N_q)\Big) \\
&= k\Bigg(1 - \frac{1}{1+\left[\left(\frac{\mathcal{Y}_p}{1-\mathcal{Y}_p}\right)^s + \left(\frac{\mathcal{Y}_q}{1-\mathcal{Y}_q}\right)^s\right]^{\frac{1}{s}}}, \frac{1}{1+\left[\left(\frac{1-\mathcal{A}_p}{\mathcal{A}_p}\right)^s + \left(\frac{1-\mathcal{A}_q}{\mathcal{A}_q}\right)^s\right]^{\frac{1}{s}}}, \\
&\qquad \frac{1}{1+\left[\left(\frac{1-N_p}{N_p}\right)^s + \left(\frac{1-N_q}{N_q}\right)^s\right]^{\frac{1}{s}}}\Bigg) \\
&= k(t_1, t_2, t_3),
\end{aligned}
$$

where

$$
\begin{aligned}
t_1 &= 1 - \frac{1}{1+\left[\left(\frac{\mathcal{Y}_p}{1-\mathcal{Y}_p}\right)^s + \left(\frac{\mathcal{Y}_q}{1-\mathcal{Y}_q}\right)^s\right]^{\frac{1}{s}}}, \\
t_2 &= \frac{1}{1+\left[\left(\frac{1-\mathcal{A}_p}{\mathcal{A}_p}\right)^s + \left(\frac{1-\mathcal{A}_q}{\mathcal{A}_q}\right)^s\right]^{\frac{1}{s}}}, \\
t_3 &= \frac{1}{1+\left[\left(\frac{1-N_p}{N_p}\right)^s + \left(\frac{1-N_q}{N_q}\right)^s\right]^{\frac{1}{s}}}.
\end{aligned}
$$

Now,

$$
\begin{aligned}
& t_1 = 1 - \frac{1}{1+\left[\left(\frac{\mathcal{Y}_p}{1-\mathcal{Y}_p}\right)^s + \left(\frac{\mathcal{Y}_q}{1-\mathcal{Y}_q}\right)^s\right]^{\frac{1}{s}}} \\
& \text{i.e., } \left(\frac{1}{1-t_1} - 1\right)^s = \left(\frac{\mathcal{Y}_p}{1-\mathcal{Y}_p}\right)^s + \left(\frac{\mathcal{Y}_q}{1-\mathcal{Y}_q}\right)^s \\
& \text{i.e., } k\left(\frac{t_1}{1-t_1}\right)^s = k\left(\frac{\mathcal{Y}_p}{1-\mathcal{Y}_p}\right)^s + k\left(\frac{\mathcal{Y}_q}{1-\mathcal{Y}_q}\right)^s \\
& \text{i.e., } 1 + \left\{k\left(\frac{t_1}{1-t_1}\right)^s\right\}^{\frac{1}{s}} = 1 + \left\{k\left(\frac{\mathcal{Y}_p}{1-\mathcal{Y}_p}\right)^s + k\left(\frac{\mathcal{Y}_q}{1-\mathcal{Y}_q}\right)^s\right\}^{\frac{1}{s}}.
\end{aligned}
$$

Similarly,

$$\left\{k\left(\frac{1-t_2}{t_2}\right)^s\right\}^{\frac{1}{s}} = \left\{k\left(\frac{1-\mathcal{A}_p}{\mathcal{A}_p}\right)^s + k\left(\frac{1-\mathcal{A}_q}{\mathcal{A}_q}\right)^s\right\}^{\frac{1}{s}}$$

$$1+\left\{k\left(\frac{t_3}{1-t_3}\right)^s\right\}^{\frac{1}{s}} = 1+\left\{k\left(\frac{1-\mathcal{N}_p}{\mathcal{N}_p}\right)^s + k\left(\frac{1-\mathcal{N}_q}{\mathcal{N}_q}\right)^s\right\}^{\frac{1}{s}}.$$

Thus

$$k(p \oplus_D q)$$
$$= \left(1 - \frac{1}{1+\left\{k\left(\frac{\mathcal{Y}_p}{1-\mathcal{Y}_p}\right)^s + k\left(\frac{\mathcal{Y}_q}{1-\mathcal{Y}_q}\right)^s\right\}^{\frac{1}{s}}}, \frac{1}{1+\left\{k\left(\frac{1-\mathcal{A}_p}{\mathcal{A}_p}\right)^s + k\left(\frac{1-\mathcal{A}_q}{\mathcal{A}_q}\right)^s\right\}^{\frac{1}{s}}}, \frac{1}{1+\left\{k\left(\frac{1-\mathcal{N}_p}{\mathcal{N}_p}\right)^s + k\left(\frac{1-\mathcal{N}_q}{\mathcal{N}_q}\right)^s\right\}^{\frac{1}{s}}}\right).$$

Also,

$$kp \oplus_D kq$$
$$= \left(1 - \frac{1}{1+\left\{k\left(\frac{\mathcal{Y}_p}{1-\mathcal{Y}_p}\right)^s\right\}^{\frac{1}{s}}}, \frac{1}{1+\left\{k\left(\frac{1-\mathcal{A}_p}{\mathcal{A}_p}\right)^s\right\}^{\frac{1}{s}}}, \frac{1}{1+\left\{k\left(\frac{1-\mathcal{N}_p}{\mathcal{N}_p}\right)^s\right\}^{\frac{1}{s}}}\right)$$
$$\oplus_D \left(1 - \frac{1}{1+\left\{k\left(\frac{\mathcal{Y}_q}{1-\mathcal{Y}_q}\right)^s\right\}^{\frac{1}{s}}}, \frac{1}{1+\left\{k\left(\frac{1-\mathcal{A}_q}{\mathcal{A}_q}\right)^s\right\}^{\frac{1}{s}}}, \frac{1}{1+\left\{k\left(\frac{1-\mathcal{N}_q}{\mathcal{N}_q}\right)^s\right\}^{\frac{1}{s}}}\right)$$
$$= (m_1, n_1, r_1) \oplus_D (m_2, n_2, r_2) \text{ (say)}$$
$$= \left(1 - \frac{1}{1+\left[\left(\frac{m_1}{1-m_1}\right)^s + \left(\frac{m_2}{1-m_2}\right)^s\right]^{\frac{1}{s}}}, \frac{1}{1+\left[\left(\frac{1-n_1}{n_1}\right)^s + \left(\frac{1-n_2}{n_2}\right)^s\right]^{\frac{1}{s}}}, \frac{1}{1+\left[\left(\frac{1-r_1}{r_1}\right)^s + \left(\frac{1-r_2}{r_2}\right)^s\right]^{\frac{1}{s}}}\right).$$

Now,

$$m_1 = 1 - \frac{1}{1+\left\{k\left(\frac{\mathcal{Y}_p}{1-\mathcal{Y}_p}\right)^s\right\}^{\frac{1}{s}}},$$
$$\text{i.e., } \left(\frac{m_1}{1-m_1}\right)^s = k\left(\frac{\mathcal{Y}_p}{1-\mathcal{Y}_p}\right)^s.$$

Similarly,

$$\left(\frac{m_2}{1-m_2}\right)^s = k\left(\frac{\mathcal{Y}_q}{1-\mathcal{Y}_q}\right)^s.$$

Therefore,

$$1+\Big[\Big(\frac{m_1}{1-m_1}\Big)^s+\Big(\frac{m_2}{1-m_2}\Big)^s\Big]^{\frac{1}{s}}=1+\Big[k\Big(\frac{\mathcal{Y}_p}{1-\mathcal{Y}_p}\Big)^s+k\Big(\frac{\mathcal{Y}_q}{1-\mathcal{Y}_q}\Big)^s\Big]^{\frac{1}{s}}.$$

Thus

$$1+\Big[\Big(\frac{1-n_1}{n_1}\Big)^s+\Big(\frac{1-n_2}{n_2}\Big)^s\Big]^{\frac{1}{s}}=1+\Big[k\Big(\frac{1-\mathcal{A}_p}{\mathcal{A}_p}\Big)^s+k\Big(\frac{1-\mathcal{A}_q}{\mathcal{A}_q}\Big)^s\Big]^{\frac{1}{s}}.$$

Proceeding in a similar way it is obtained that

$$1+\Big[\Big(\frac{1-r_1}{r_1}\Big)^s+\Big(\frac{1-r_2}{r_2}\Big)^s\Big]^{\frac{1}{s}}=1+\Big[k\Big(\frac{1-\mathcal{N}_p}{\mathcal{N}_p}\Big)^s+k\Big(\frac{1-\mathcal{N}_q}{\mathcal{N}_q}\Big)^s\Big]^{\frac{1}{s}}.$$

Therefore

$$\begin{aligned} & kp \oplus_D kq \\ &= \Big(1-\frac{1}{1+\big[k\big(\frac{\mathcal{Y}_p}{1-\mathcal{Y}_p}\big)^s+k\big(\frac{\mathcal{Y}_q}{1-\mathcal{Y}_q}\big)^s\big]^{\frac{1}{s}}}, \frac{1}{1+\big[k\big(\frac{1-\mathcal{A}_p}{\mathcal{A}_p}\big)^s+k\big(\frac{1-\mathcal{A}_q}{\mathcal{A}_q}\big)^s\big]^{\frac{1}{s}}}, \\ & \qquad \frac{1}{1+\big[k\big(\frac{1-\mathcal{N}_p}{\mathcal{N}_p}\big)^s+k\big(\frac{1-\mathcal{N}_q}{\mathcal{N}_b}\big)^s\big]^{\frac{1}{s}}}\Big). \end{aligned}$$

Hence, $k(p \oplus_D q) = (kp \oplus_D kq)$.

Proofs of (iv), (v), and (vi) are similar. □

## 1.14 Conclusion

In this chapter, some algebraic operations on PFSs are defined and associated with some operators on PFSs. Some properties related to these are investigated. Two particular types of operators, namely TOC and TOI on PFSs are introduced here with suitable examples. Also, some corresponding results are established. A new kind of operator, namely an 'implication operator' is initiated here, and some important properties are studied. Different types of PFAO are defined with suitable examples. Also, the concept of PFDWAO and some related properties are studied. Our studies will likely motivate researchers to introduce some new operators in the perspective of other types of sets that can be utilized in different real-life applications.

## References


[1] S. Amanathulla, B. Bera, M. Pal, Balanced picture fuzzy graph with application, Artif. Intell. Rev. 54 (2021) 5255–5281.

[2] K.T. Atanassov, Intuitionistic fuzzy sets, Fuzzy Sets Syst. 20 (1986) 87–96.

[3] K.T. Atanassov, More on intuitionistic fuzzy sets, Fuzzy Sets Syst. 33 (1989) 37–45.
[4] K.T. Atanassov, Interval valued intuitionistic fuzzy set, Fuzzy Sets Syst. 31 (1989) 343–349.
[5] K.T. Atanassov, Some operators on intuitionistic fuzzy sets, in: First International Conference on IFS, Sofia, 1997.
[6] H. Bustince, P. Burillo, Structures on intuitionistic fuzzy relations, Fuzzy Sets Syst. 78 (1996) 293–303.
[7] P. Chellamani, D. Ajay, S. Broumi, et al., An approach to decision-making via picture fuzzy soft graphs, Granul. Comput. 7 (2022) 527–548.
[8] B.C. Cuong, V. Kreinovich, Picture fuzzy sets - a new concept for computational intelligence problems, in: Proceedings of the Third World Congress on Information and Communication Technologies WIICT, 2013.
[9] B.C. Cuong, Picture fuzzy sets – first results. Part 1, in: Seminar "Neuro-Fuzzy Systems with Applications", Preprint 03/2013, Institute of Mathematics, Hanoi, May 2013.
[10] B.C. Cuong, Picture fuzzy sets – first results. Part 2, in: Seminar "Neuro-Fuzzy Systems with Applications", Preprint 04/2013, Institute of Mathematics, Hanoi, June 2013.
[11] B.C. Cuong, Picture fuzzy sets, J. Comput. Sci. Cybern. 30 (2014) 409–420.
[12] B.C. Cuong, P.V. Hai, Some fuzzy logic operators for picture fuzzy sets, in: Seventh International Conference on Knowledge and Systems Engineering, 2015.
[13] M. Demirci, Axiomatic theory of intuitionistic fuzzy sets, Fuzzy Sets Syst. 110 (2000) 253–266.
[14] S.K. De, R. Biswas, A.R. Roy, Some operations on intuitionistic fuzzy set, Fuzzy Sets Syst. 114 (2000) 477–484.
[15] G. Deshrijver, C. Cornelis, E.E. Kerre, On the representation of intuitionistic fuzzy t-norms and t-conorms, IEEE Trans. Fuzzy Syst. 12 (2004) 45–61.
[16] G. Deshrijver, E.E. Kerre, Uninorms in L*-fuzzy set theory, Fuzzy Sets Syst. 148 (2004) 243–262.
[17] N.V. Dinh, N.X. Thao, Some measures of picture fuzzy sets and their application in multi-attribute decision making, Int. J. Math. Sci. Comput. 3 (2018) 23–41.
[18] J. Dombi, A general class of fuzzy operators, the demorgan class of fuzzy operators and fuzziness measures induced by fuzzy operators, Fuzzy Sets Syst. 8 (1982) 149–163.
[19] P. Dutta, S. Ganju, Some aspects of picture fuzzy set, Trans. A. Razmandze Math. Inst. 172 (2018) 164–175.
[20] Y. Du, P. Liu, Extended fuzzy VIKOR method with intuitionistic trapezoidal fuzzy numbers, Inf. Int. Interdiscip. J. 14 (8) (2011) 2575–2583.
[21] H. Garg, Generalized intuitionistic fuzzy interactive geometric interaction operators using Einstein t-norm and t-conorm and their application to decision making, Comput. Ind. Eng. 101 (2016) 53–69.
[22] H. Garg, Novel intuitionistic fuzzy decision making method based on an improved operation laws and its application, Eng. Appl. Artif. Intell. 60 (2017) 164–174.
[23] H. Garg, Some series of intuitionistic fuzzy interactive averaging aggregation operators, SpringerPlus 5 (1) (2016) 999, https://doi.org/10.1186/s40064-016-2591-9.
[24] H. Garg, A new generalized improved score function of interval-valued intuitionistic fuzzy sets and applications in expert systems, Appl. Soft Comput. 38 (2016) 988–999, https://doi.org/10.1016/j.asoc.2015.10.040.
[25] F.K. Gündoğdu, S. Duleba, S. Moslem, S. Aydın, Evaluating public transport service quality using picture fuzzy analytic hierarchy process and linear assignment model, Appl. Soft Comput. 100 (2021) 106920.
[26] E. Haktanır, C. Kahraman, Intelligent replacement analysis using picture fuzzy sets: defender-challenger comparison application, Eng. Appl. Artif. Intell. 121 (2023) 106018.
[27] E. Haktanır, C. Kahraman, A novel picture fuzzy CRITIC & REGIME methodology: wearable health technology application, Eng. Appl. Artif. Intell. 113 (2022) 104942.
[28] H. Hamacher, Uber Logische Agregationen Nicht Binar Explizierter Entscheidungskriterien, Rita G. Fischer Verlage, Frankfurt am Main, 1978.
[29] C.L. Hwang, K. Yoon, Multiple Attribute Decision Making Methods and Applications a State-of-the-Art Survey, Springer, Berlin, 1981.
[30] C. Jana, T. Senapati, M. Pal, R.R. Yager, Picture fuzzy Dombi aggregation operators: application to MADM process, Appl. Soft Comput. 74 (2019) 99–109.
[31] C. Jana, M. Pal, Assessment of enterprise performance based on picture fuzzy Hamacher aggregation operators, Symmetry 11 (1) (2019) 75.

[32] N. Jan, J. Gwak, D. Pamucar, Mathematical analysis of generative adversarial networks based on complex picture fuzzy soft information, Appl. Soft Comput. 137 (2023) 110088.
[33] J. Jin, H. Garg, T. You, Generalized picture fuzzy distance and similarity measures on the complete lattice and their applications, Expert Syst. Appl. 220 (2023) 119710.
[34] P. Liu, Multi-attribute decision making method research based on interval vague set and TOPSIS method, Technol. Econ. Dev. Econ. 15 (3) (2009) 453–463.
[35] M. Luo, Y. Zhang, A new similarity measure between picture fuzzy sets and its application, Eng. Appl. Artif. Intell. 96 (2020) 103956.
[36] R. Lourenzuttia, R.A. Krohlingb, A study of TODIM in a intuitionistic fuzzy and random environment, Expert Syst. Appl. 40 (16) (2013) 6459–6468.
[37] F.Y. Meng, Q. Zhang, H. Cheng, Approaches to multiple-criteria group decision making based on interval-valued intuitionistic fuzzy Choquet integral with respect to the generalized k-Shapley index, Knowl.-Based Syst. 37 (2013) 237–249.
[38] X. Peng, H. Yuan, Y. Yang, Pythagorean fuzzy information measures and their applications, J. Intell. Fuzzy Syst. 32 (2017) 991–1029.
[39] X. Peng, New similarity measure and distance measure for Pythagorean fuzzy set, Complex Intell. Syst. 5 (2019) 101–111.
[40] S.M. Peng, Study on enterprise risk management assessment based on picture fuzzy multiple attribute decision-making method, J. Intell. Fuzzy Syst. 33 (2017) 3451–3458.
[41] J.J. Peng, X.G. Chen, C. Tian, Z.Q. Zhang, H.Y. Song, F. Dong, Picture fuzzy large-scale group decision-making in a trust- relationship-based social network environment, Inf. Sci. 608 (2022) 1675–1701.
[42] X. Shi, S. Kosari, A.A. Talebi, et al., Investigation of the main energies of picture fuzzy graph and its applications, Int. J. Comput. Intell. Syst. 15 (2022) 31.
[43] F. Smarandache, A Unifying Field in Logics, Neutrosophy Neutrosophic Probability, and Logic, American Research Press, Rehoboth, 1999.
[44] F. Smarandache, A unifying field in logics neutrosophic logic, Mult. Valued Log. 8 (3) (2002) 385–438.
[45] V. Simic, S. Karagoz, M. Deveci, N. Aydin, Picture fuzzy extension of the CODAS method for multi-criteria vehicle shredding facility location, Expert Syst. Appl. 175 (2021) 114644.
[46] A. Singh, S. Kumar, Picture fuzzy set and quality function deployment approach based novel framework for multi-criteria group decision making method, Eng. Appl. Artif. Intell. 104 (2021) 104395.
[47] P. Singh, Correlation coefficients for picture fuzzy sets, J. Intell. Fuzzy Syst. 28 (2015) 591–604.
[48] P. Singh, N.K. Mishra, Risk analysis of flood disaster based on similarity measures in picture fuzzy environment, Afr. Mat. 29 (2018) 1019–1038.
[49] S. Sing, A.H. Ganie, Applications of picture fuzzy similarity measures in pattern recognition, clustering, and MADM, Expert Syst. Appl. 168 (2021) 114264.
[50] L.H. Son, Generalized picture distance measure and applications to picture fuzzy clustering, Appl. Soft Comput. 46 (2016) 284–295.
[51] L.H. Son, Measuring analogousness in picture fuzzy sets: from picture distance measures to picture association measures, Fuzzy Optim. Decis. Mak. 16 (2016) 1–20.
[52] P.H. Thong, A novel automatic picture fuzzy clustering method based on particle swarm optimization and picture composite cardinality, Knowl.-Based Syst. 109 (2016) 48–60.
[53] C. Tian, J.J. Peng, S. Zhang, W.Y. Zhang, J.Q. Wang, Weighted picture fuzzy aggregation operators and their applications to multi-criteria decision-making problems, Comput. Ind. Eng. 137 (2019) 106037.
[54] W. Wang, X. Liu, Intuitionistic fuzzy information aggregation using Einstein operations, IEEE Trans. Fuzzy Syst. 20 (5) (2012) 923–938.
[55] L. Wang, J.J. Peng, J.Q. Wang, A multi-criteria decision-making framework for risk ranking of energy performance contracting project under picture fuzzy environment, J. Clean. Prod. 191 (2018) 105–118.
[56] R. Wang, J. Wang, H. Gao, G. Wei, Methods for MADM with picture fuzzy Muirhead mean operators and their application for evaluating the financial investment risk, Symmetry 11 (1) (2019) 6.
[57] L. Wang, H.Y. Zhang, J.Q. Wang, L. Li, Picture fuzzy normalized projection-based VIKOR method for the risk evaluation of construction project, Appl. Soft Comput. 64 (2018) 216–226.
[58] L. Wang, H.Y. Zhang, J.Q. Wang, G.F. Wu, Picture fuzzy multi-criteria group decision-making method to hotel building energy efficiency retrofit project selection, RAIRO Oper. Res. 54 (1) (2020) 211–229.
[59] C. Wang, X. Zhou, H. Tu, S. Tao, Some geometric aggregation operators based on picture fuzzy sets and their application in multiple attribute decision making, Ital. J. Pure Appl. Math. 37 (2017) 477–492.

[60] S.P. Wan, G.L. Xu, F. Wang, J.Y. Dong, A new method for Atanassov's interval-valued intuitionistic fuzzy MAGDM with incomplete attribute weight information, Inf. Sci. 316 (2015) 329–347.
[61] G.W. Wei, Picture fuzzy cross-entropy for multiple attribute decision making problems, J. Bus. Econ. Manag. 17 (4) (2016) 491–502.
[62] G.W. Wei, Picture fuzzy aggregation operators and their application to multiple attribute decision making, J. Intell. Fuzzy Syst. 33 (2017) 713–724.
[63] G.W. Wei, Some cosine similarity measures for picture fuzzy sets and their applications to strategic decision making, Informatica 144 (2017) 547–564.
[64] G.W. Wei, F.E. Alsaadi, T. Hayat, A. Alsaed, Projection models for multiple attribute decision making with picture fuzzy information, Int. J. Mach. Learn. Cybern. 9 (2018) 713–719.
[65] Z.S. Xu, R.R. Yager, Some geometric aggregation operators based on intuitionistic fuzzy sets, Int. J. Gen. Syst. 35 (2006) 417–433.
[66] Z.S. Xu, Intuitionistic fuzzy aggregation operators, IEEE Trans. Fuzzy Syst. 15 (2007) 1179–1187.
[67] Y. Xu, H. Wang, J.M. Merigo, Intuitionistic fuzzy Einstein Choquet integral operators for multiple attribute decision making, Technol. Econ. Dev. Econ. 20 (2) (2014) 227–253.
[68] R.R. Yager, On ordered weighted averaging aggregation operators in multi-criteria decision making, IEEE Trans. Syst. Man Cybern. 18 (1) (1988) 183–190.
[69] R.R. Yager, J. Kacprzyk, The Ordered Weighted Averaging Operators: Theory and Applications, M.A., Kluwer, Boston, 1997.
[70] R.R. Yager, On a general class of fuzzy connectives, Technical report 78-18, Iona College, 1978.
[71] L.A. Zadeh, Fuzzy sets, Inf. Control 8 (1965) 338–353.
[72] E.K. Zavadskas, J. Antucheviciene, S.H.R. Hajiagha, S.S. Hashemi, Extension of weighted aggregated sum product assessment with interval-valued intuitionistic fuzzy numbers(WASPAS-IVIF), Appl. Soft Comput. 24 (2014) 1013–1021.
[73] S.Q. Zhang, H. Gao, G.W. Wei, Y. Wei, C. Wei, Evaluation based on distance from average solution method for multiple criteria group decision making under picture 2-tuple linguistic environment, Mathematics 7 (3) (2019) 243.
[74] X.Y. Zhang, J.Q. Wang, J.H. Hu, On novel operational laws and aggregation operators of picture 2-tuple linguistic information for MCDM problems, Int. J. Fuzzy Syst. 20 (3) (2018) 958–969.
[75] R.R. Zhao, M.X. Luo, S.G. Li, L.N. Ma, A parametric similarity measure between picture fuzzy sets and its applications in multi-attribute decision-making, Iran. J. Fuzzy Syst. 20 (1) (2023) 87–102.
[76] X.K. Zhao, X.M. Zhu, K.Y. Bai, R.T. Zhang, A novel failure mode and effect analysis method using a flexible knowledge acquisition framework based on picture fuzzy sets, Eng. Appl. Artif. Intell. 117 (Part A) (2023) 105625.
[77] C. Zuo, A. Pal, A. Dey, New concepts of picture fuzzy graphs with application, Mathematics 7 (2019) 470, https://doi.org/10.3390/math7050470.

# 2

# Picture fuzzy hybrid weighted operators and their application in the decision-making process

## 2.1 Introduction

Among the FSs [34], IFS [1,2,8], PFS takes into account sensible and valuable features. Cuong [3–5,7] and others initialized some features of PFSs and introduced a distance measure among the PFSs. Later, Cuong et al. [6] defined some fundamental operators: implications, disjunctions, conjunctions, and negations on PFS. Phong and coworkers [14] studied some contour soft PFS relations. Later, Wei [27–29] developed some techniques for similarity measures between PFSs. Due to its weight, many researchers have made attempts to enrich the concept of PFS in the real DMPs. An attempt was made to solve the problem by designing a proper mathematical model that aggregates [10] the various preferences of the DMs into a collective one. In this view, Wei [31] built up in the weighted form of picture fuzzy average (PFWA), order averaging (PFOWA), and hybrid averaging (PFHWA) operators. He also defined the weighted form of picture fuzzy geometric (PFWG), order geometric (PFOWG), and hybrid geometric (PFHWG) operators, respectively. Further, Wei [32] studied some weighted forms of picture fuzzy AOs based on Hamacher norms, and named picture fuzzy Hamacher aggregation, Hamacher correlated aggregation, induced Hamacher correlated aggregation, Hamacher prioritized aggregation, and Hamacher power aggregation operators. Khan et al. [11] focused on studying some weighted forms of Logarithm picture fuzzy AOs, and related Logarithm picture fuzzy averaging (LPFWA), order averaging (LPFOWA), hybrid averaging (LPFHWA) operators, respectively, and Logarithm picture fuzzy geometric (LPFWG), order geometric (LPFOWG) and hybrid geometric (LPFHWG) operators. Jana et al. [9] proposed some weighted forms of picture fuzzy Dombi norms, and they were considered as picture fuzzy Dombi averaging (PFDWA), Dombi order averaging (PFDOWA), Dombi hybrid averaging (PFDHWA) operators, and also, picture fuzzy Dombi geometric (PFDWG), Dombi ordered geometric (PFDOWG), Dombi hybrid geometric (PFDHWG) operators, respectively. At the same time, many researchers used various approach to enrich the decision-making process by using picture fuzzy information [13–28,30]. In all these AOs, data are required for evaluating the information in crisp numbers. However, in some cases, it is challenging to express DMs to give their rating in a single number rather than express them in interval numbers. In that direction, Khalil et al. [12] defined IVPFS, and also defined generalized IVPFS soft sets (IVPFSSs). Furthermore, they applied the proposed idea to solve DMPs. Actually, there are no aggregation-operator-based decision-making theories in the environment of IVPFSs.

Picture Fuzzy Logic and Its Applications in Decision Making Problems. https://doi.org/10.1016/B978-0-44-322024-1.00006-6

From the above results, it is clear that aggregation operators are a hot research topic in decision-making areas. Therefore aggregation is, more often than not, to include mathematical operators, which is not only an average but a general notion. The aggregated value obtained by the aggregation operator is helpful if the applying operator is not biased; it should never tend to one or some number whose weight is on the higher side and does not tend to maximum or minimum arguments.

After the detailed study of the existing AOs in the PFS environment, it is observed that if we used the existing PFWA or PFOWA operators under some specific cases, their accumulation might result in a maximum. Again, if we utilized PFWG or PFOWG operators, then the accumulated values follow a result with higher importance and hence removed the bias issue of some unique situations during the decision process. Thus the motivation comes from [33] in the IFS setting; we concentrate on extending these operators supporting the PFS and IVPFS environments. Thus the main intention of this chapter is to introduce new AOs with characteristics in both averaging and geometric notions. In that regard, we propose some hybrid AOs for the PFS setting. The proposed operators are calculated in the weighted forms of hybrid picture fuzzy averaging and hybrid geometric (H-PFWAG) and hybrid picture fuzzy order averaging, and order geometric (H-PFOWAG) operators. Further, we introduce the same type of hybrid AOs in the IVPFS environment. It is also observed that existing AOs are deduced from the proposed hybrid operators. Finally, an approach to MCDM problems has been constructed to demonstrate a case study for the efficiency of validating the issue.

## 2.2 Preliminaries

In this section, we introduced some AOs on PFNs.

**Definition 2.1.** [9] For a group of PFNs $\mathcal{R}_f = (\mathcal{Y}_f, \mathcal{A}_f, \mathcal{N}_f)$, $f = 1, 2, \ldots \varpi$. A picture fuzzy weighted averaging (PFWA) operator, weighted geometric (PFWG), ordered weighted averaging (PFOWA), order weighted geometric (PFOWG) are, respectively, functions of dimension $\varpi$ defined as follows: $\Theta^{\varpi} \rightarrow \Theta$ having weight vector $\Xi = (\Xi_1, \Xi_2, \ldots, \Xi_{\varpi})^T$ where $\Xi > 0$ and $\sum_{f=1}^{\varpi} \Xi_f = 1$, as

**(i)** $$PFWA_{\Xi}(\mathcal{R}_1, \mathcal{R}_2, \ldots, \mathcal{R}_{\varpi}) = \bigoplus_{f=1}^{\varpi}(\Xi_f \mathcal{R}_f) = \left(1 - \prod_{f=1}^{\varpi}(1 - \mathcal{Y}_f)^{\Xi_f}, \prod_{f=1}^{\varpi}\mathcal{A}_f{}^{\Xi_f}, \prod_{f=1}^{\varpi}\mathcal{N}_f{}^{\Xi_f}\right); \tag{2.1}$$

**(ii)** $$PFWG_{\Xi}(\mathcal{R}_1, \mathcal{R}_2, \ldots, \mathcal{R}_{\varpi}) = \bigoplus_{f=1}^{\varpi}(\Xi_f \mathcal{R}_f) = \left(\prod_{f=1}^{\varpi}\mathcal{Y}_f{}^{\Xi_f}, 1 - \prod_{f=1}^{\varpi}(1 - \mathcal{A}_f)^{\Xi_f}, 1 - \prod_{f=1}^{\varpi}(1 - \mathcal{N}_f)^{\Xi_f}\right); \tag{2.2}$$

**(iii)** $PFOWA_{\Xi}(\mathcal{R}_1, \mathcal{R}_2, \ldots, \mathcal{R}_{\varpi}) = \bigoplus_{f=1}^{\varpi} (\Xi_f \mathcal{R}_{\sigma(f)})$

$$= \left(1 - \prod_{f=1}^{\varpi} (1 - \mathcal{Y}_{\sigma(f)})^{\Xi_f}, \prod_{f=1}^{\varpi} \mathcal{A}_{\sigma(f)}{}^{\Xi_f}, \prod_{f=1}^{\varpi} \mathcal{N}_{\sigma(f)}{}^{\Xi_f}\right); \tag{2.3}$$

**(iv)** $PFOWG_{\Xi}(\mathcal{R}_1, \mathcal{R}_2, \ldots, \mathcal{R}_{\varpi}) = \bigoplus_{f=1}^{\varpi} (\mathcal{R}_{\sigma(f)})^{\Xi_f}$

$$= \left(\prod_{f=1}^{\varpi} \mathcal{Y}_{\sigma(f)}{}^{\Xi_f}, 1 - \prod_{f=1}^{\varpi} (1 - \mathcal{A}_{\sigma(f)})^{\Xi_f}, 1 - \prod_{f=1}^{\varpi} (1 - \mathcal{N}_{\sigma(f)}{}^{\Xi_f}\right). \tag{2.4}$$

## 2.3 Aggregation operators with PFN information

Let $\Theta$ be a set of all PFNs. In this section, we introduce some weighted forms of hybrid PFN AOs, namely, hybrid PFN averaging and geometric (H-PFWAG), and hybrid PFN order averaging and geometric (H-PFOWAG) operators to accumulate a collection of PFNs.

### 2.3.1 Hybrid PFN aggregation operators

**Definition 2.2.** For a group of PFNs $\mathcal{R}_f = (\mathcal{Y}_f, \mathcal{A}_f, \mathcal{N}_f)$, $f = 1, 2, \ldots, \varpi$, A H-PFWAG operator is a function $H - PFWAG : \Theta^{\varpi} \to \Theta$ defined below

$$H - PFWAG = (\mathcal{R}_1, \mathcal{R}_2, \ldots, \mathcal{R}_{\varpi}) = \left(\bigoplus_{f=1}^{\varpi} \Xi_f \mathcal{R}_f\right)^{\gamma} \bigotimes \left(\bigotimes_{f=1}^{\varpi} \mathcal{R}_b^{\Xi_f}\right)^{\gamma-1}, \tag{2.5}$$

where $\gamma$ is a real number $\gamma \in [0, 1]$ that presents the attitudinal natures of the DMs towards an aggregation process, so it gives the weighted operator to accumulate averaging and geometric AOs, where $\Xi > 0$ is the standardized weight of $\mathcal{R}_f$, $f = 1, 2, \ldots, \varpi$.

**Theorem 2.1.** *The accumulated value of PFNs applying the H-PFWAG operator for a group of PFNs* $\mathcal{R}_f = (\mathcal{Y}_f, \mathcal{A}_f, \mathcal{N}_f)$, $f = 1, 2, \ldots, \varpi$ *remains a PFN, and is given below*

$$H - PFWAG = (\mathcal{R}_1, \mathcal{R}_2, \ldots, \mathcal{R}_{\varpi})$$

$$= \left( (1 - \prod_{f=1}^{\varpi} (1 - \mathcal{Y}_f)^{\Xi_f})^{\gamma} \left(\prod_{f=1}^{\varpi} (\mathcal{Y}_f)^{\Xi_f}\right)^{1-\gamma}, 1 - \left(1 - \prod_{f=1}^{\varpi} \mathcal{A}_f^{\Xi_f}\right)^{\gamma} \left(\prod_{f=1}^{\varpi} (1 - \mathcal{A}_f)^{\Xi_f}\right)^{1-\gamma}, \right.$$
$$\left. 1 - \left(1 - \prod_{f=1}^{\varpi} \mathcal{N}_f^{\Xi_f}\right)^{\gamma} \left(\prod_{f=1}^{\varpi} (1 - \mathcal{N}_f)^{\Xi_f}\right)^{1-\gamma} \right). \tag{2.6}$$

*Proof.* From the operation rules for PFNs $\mathcal{R}_f$, $f = 1, 2, \ldots, \varpi$ and a real number $\gamma \in [0, 1]$, we obtain

$$\bigoplus_{f=1}^{\varpi} \Xi_f \mathcal{R}_f = \left(1 - \prod_{f=1}^{\varpi}(1 - \mathcal{Y}_f)^{\Xi_f}, \prod_{f=1}^{\varpi}(\mathcal{A}_f)^{\Xi_f}, \prod_{f=1}^{\varpi}(\mathcal{N}_f)^{\Xi_f}\right),$$

$$\text{and } \prod_{f=1}^{\varpi}(\mathcal{R}_f)^{\Xi_f} = \left(\prod_{f=1}^{\varpi}(\mathcal{Y}_f)^{\Xi_f}, 1 - \prod_{f=1}^{\varpi}(1 - \mathcal{A}_f)^{\Xi_f}, 1 - \prod_{f=1}^{\varpi}(1 - \mathcal{N}_f)^{\Xi_f}\right)$$

$$\left(\bigoplus_{f=1}^{\varpi} \Xi_f \mathcal{R}_f\right)^{\gamma}$$

$$= \left(\left(1 - \prod_{f=1}^{\varpi}(1 - \mathcal{Y}_f)^{\Xi_f}\right)^{\gamma}, \; 1 - \left(1 - \prod_{f=1}^{\varpi}(\mathcal{A}_f)^{\Xi_f}\right)^{\gamma}, \; 1 - \left(1 - \prod_{f=1}^{\varpi}(\mathcal{N}_f)^{\Xi_f}\right)^{\gamma}\right) \tag{2.7}$$

$$\left(\prod_{f=1}^{\varpi}(\mathcal{R}_f)^{\Xi_f}\right)^{\gamma-1}$$

$$= \left(\left(\prod_{f=1}^{\varpi}(\mathcal{Y}_f)^{\Xi_f}\right)^{1-\gamma}, \; 1 - \left(\prod_{f=1}^{\varpi}(1 - \mathcal{A}_f)^{\Xi_f}\right)^{1-\gamma}, \; 1 - \left(\prod_{f=1}^{\varpi}(1 - \mathcal{N}_f)^{\Xi_f}\right)^{1-\gamma}\right). \tag{2.8}$$

Thus by Definition 2.2, we obtain

$$\left(\left(1 - \prod_{f=1}^{\varpi}(1 - \mathcal{Y}_f)^{\Xi_f}\right)^{\gamma}, \; 1 - \left(1 - \prod_{f=1}^{\varpi}(\mathcal{A}_f)^{\Xi_f}\right)^{\gamma}, \; 1 - \left(1 - \prod_{f=1}^{\varpi}(\mathcal{N}_f)^{\Xi_f}\right)^{\gamma}\right)$$

$$\otimes \left(\left(\prod_{f=1}^{\varpi}(\mathcal{Y}_f)^{\Xi_f}\right)^{1-\gamma}, \; 1 - \left(\prod_{f=1}^{\varpi}(1 - \mathcal{A}_f)^{\Xi_f}\right)^{1-\gamma}, \; 1 - \left(\prod_{f=1}^{\varpi}(1 - \mathcal{N}_f)^{\Xi_f}\right)^{1-\gamma}\right)$$

$$= \left(\left(1 - \prod_{f=1}^{\varpi}(1 - \mathcal{Y}_f)^{\Xi_f}\right)^{\gamma}\left(\prod_{f=1}^{\varpi}(\mathcal{Y}_f)^{\Xi_f}\right)^{1-\gamma},\right.$$

$$\left\{1 - \left(1 - \prod_{f=1}^{\varpi}(\mathcal{A}_f)^{\Xi_f}\right)^{\gamma}\right\} + \left\{1 - \left(\prod_{f=1}^{\varpi}(1 - \mathcal{A}_f)^{\Xi_f}\right)^{1-\gamma}\right\},$$

$$\left.\left\{1 - \left(1 - \prod_{f=1}^{\varpi}(\mathcal{N}_f)^{\Xi_f}\right)^{\gamma}\right\} + \left\{1 - \left(\prod_{f=1}^{\varpi}(1 - \mathcal{N}_f)^{\Xi_f}\right)^{1-\gamma}\right\}\right)$$

$$= \left(\left(1 - \prod_{f=1}^{\varpi}(1 - \mathcal{Y}_f)^{\Xi_f}\right)^{\gamma}\left(\prod_{f=1}^{\varpi}(\mathcal{Y}_f)^{\Xi_f}\right)^{1-\gamma},\right.$$

$$1-\left(1-\prod_{f=1}^{\varpi}\mathcal{A}_f^{\Xi_f}\right)^{\gamma}\left(\prod_{f=1}^{\varpi}(1-\mathcal{A}_f)^{\Xi_f}\right)^{1-\gamma},$$

$$\left.1-\left(1-\prod_{f=1}^{\varpi}\mathcal{N}_f^{\Xi_f}\right)^{\gamma}\left(\prod_{f=1}^{\varpi}(1-\mathcal{N}_f)^{\Xi_f}\right)^{1-\gamma}\right).$$

Hence, the theorem is proved. □

**Example 2.1.** Let there be three PFNs $\mathcal{R}_1=(0.6,0.1,0.3)$, $\mathcal{R}_2=(0.5,0.05,0.2)$, and $\mathcal{R}_3=(0.09,0.2,0.6)$, and weight vector $\Xi=(0.5,0.3,0.2)$. Then, without loss of generality $\gamma=0.5$ is a real number. Then,

$$H-PFWAG=(\mathcal{R}_1,\mathcal{R}_2,\ldots,\mathcal{R}_{\varpi})$$

$$=\left(\left(1-\prod_{f=1}^{\varpi}(1-\mathcal{Y}_f)^{\Xi_f}\right)^{\gamma}\left(1-\prod_{f=1}^{\varpi}(\mathcal{Y}_f)^{\Xi_f}\right)^{1-\gamma},1-\left(1-\prod_{f=1}^{\varpi}\mathcal{A}_f^{\Xi_f}\right)^{\gamma}\left(\prod_{f=1}^{\varpi}(1-\mathcal{A}_f)^{\Xi_f}\right)^{1-\gamma},\right.$$

$$\left.1-\left(1-\prod_{f=1}^{\varpi}\mathcal{N}_f^{\Xi_f}\right)^{\gamma}\left(\prod_{f=1}^{\varpi}(1-\mathcal{N}_f)^{\Xi_f}\right)^{1-\gamma}\right)$$

$$=\left(\left(1-(1-0.6)^{0.5}(1-0.5)^{0.3}(1-0.09)^{0.2}\right)^{0.5}\left(0.6^{0.5}0.5^{0.3}0.09^{0.2}\right)^{1-0.5},\right.$$

$$1-\left(1-(0.1)^{0.5}(0.05)^{0.3}(0.2)^{0.2}\right)^{0.5}\left((1-0.1)^{0.5}(1-0.05)^{0.3}(1-0.2)^{0.2}\right)^{1-0.5},$$

$$\left.1-\left(1-(0.3)^{0.5}(0.2)^{0.3}(0.6)^{0.2}\right)^{0.5}\left((1-0.3)^{0.5}(1-0.2)^{0.3}(1-0.6)^{0.2}\right)^{1-0.5}\right)$$

$$=\left(0.4390,0.0452,0.1343\right).$$

**Remark 2.1.** It is observed that if we assigned some values of $\gamma$, then the proposed operator reduces to some existing AOs.

**(1)** if $\gamma=1$, then operator H-PFWAG reduces to the PFWA operator [31];
**(2)** if $\gamma=0$, then operator H-PFWAG reduces to the PFWG operator [31].

**Example 2.2.** For two PFS $\mathcal{R}_1=(0.4,0.2,0.3)$ and $\mathcal{R}_2=(0.6,0.05,0.2)$. In order to accumulate such numbers by considering weights $\Xi_1=0.7$ and $\Xi_2=0.3$ and to analyze the sensitivity of the parameter $\gamma$, the accumulated values for $\mathcal{R}_1$ and $\mathcal{R}_2$ having different values of parameter $\gamma$ ranges 0 to 1, and their scores are presented in Table 2.1. It has been observed that the corresponding aggregated value also increases with the increase of the parameter. When $\gamma=0.5$, it acts as neutrality; when $\gamma=1$, the proposed operator reduces to PFWA [31]. Again, when $\gamma=0$, then it reduces to PFWG [31]. From the table, it is seen that with the values of $\gamma$ from 0 to 1, then the scores of the alternatives increase. Therefore based on the different values of the parameter, the decision maker has scope to change the rank of the options. For example, if the DMs considered the accumulation of information of alternatives in an optimistic sense, then he or she takes the largest value of $\gamma$ so that the scores of the alternatives increase; consequently, the membership degree increases. On the other hand, contrary to optimistic DMs, pessimistic DMs choose the smaller value of $\gamma$. Thus based on the decision-makers' attitude, optimistic and pessimistic behavioral characteristics are according to their choice. Hence, the proposed hybrid AOs are more general than the different priorities of DMs.

**Table 2.1** Table for Example 2.2.

| $\gamma$ | **0.0** | **0.2** | **0.4** | **0.5** |
|---|---|---|---|---|
| *Aggregate value* | (0.5483,0.1577,0.2714) | (0.5194,0.0243,0.0465) | (0.5149,0.0496,0.0961) | (0.5069,0.0627,0.1221) |
| *Score value* | 0.6385 | 0.7365 | 0.7094 | 0.6927 |
| $\gamma$ | **0.6** | **0.8** | **1.0** | |
| *Aggregate value* | (0.4990,0.0760,0.1490) | (0.4836,0.1034,0.2054) | (0.4687,0.1320,0.2656) | |
| *Score value* | 0.6750 | 0.6391 | 0.6016 | |

**Theorem 2.2.** *For a group of PFNs $\mathcal{R}_f = (\mathcal{Y}_f, \mathcal{A}_f, \mathcal{N}_f)$, $f = 1, 2, \ldots, \varpi$, An H-PFWAG operator satisfies the following properties:*

**(1)** *(Idempotency) Let $\mathcal{R}_f = (\mathcal{Y}_f, \mathcal{A}_f, \mathcal{N}_f)$, $f = 1, 2, \ldots, \varpi$ PFNs and $\mathcal{R}_f = \mathcal{R}$ for all $f$, then*

$$H - PFWAG(\mathcal{R}_1, \mathcal{R}_2, \ldots, \mathcal{R}_\varpi) = \mathcal{R};$$

**(2)** *(Boundedness) For PFNs $\mathcal{R}_f$, $f = 1, 2, \ldots, \varpi$, we obtain*

$$\mathcal{R}^- \leq H - PFWAG(\mathcal{R}_1, \mathcal{R}_2, \ldots, \mathcal{R}_\varpi) \leq \mathcal{R}^+,$$

*where* $\mathcal{R}^- = (\min_f\{\mathcal{Y}_f\}, \max_f\{\mathcal{A}_f\}, \max_f\{\mathcal{N}_f\})$ *and* $\mathcal{R}^+ = (\max_f\{\mathcal{Y}_f\}, \min_f\{\mathcal{A}_f\}, \min_f\{\mathcal{N}_f\})$;

**(3)** *(Monotonicity) If $\mathcal{R}_f$ and $\mathcal{R}'_f$ be two PFNs such that $\mathcal{R}_f \leq \mathcal{R}'_f$ for all $f$, then*

$$H - PFWAG(\mathcal{R}_1, \mathcal{R}_2, \ldots, \mathcal{R}_\varpi) \leq H - PFWAG(\mathcal{R}'_1, \mathcal{R}'_2, \ldots, \mathcal{R}'_\varpi).$$

As these properties are satisfied by PFWA and PFWG operators, then by definition the H-PFWAG operator directly follows these properties.

**Theorem 2.3.** *For a group of PFNs $\mathcal{R}_f = (\mathcal{Y}_f, \mathcal{A}_f, \mathcal{N}_f)$, $f = 1, 2, \ldots, \varpi$, then H-PFWAG operator and existing operators PFWA and PFWG satisfy the following inequality:*

$$PFWG(\mathcal{R}_1, \mathcal{R}_2, \ldots, \mathcal{R}_\varpi) \leq H - PFWAG(\mathcal{R}_1, \mathcal{R}_2, \ldots, \mathcal{R}_\varpi) \leq PFWA(\mathcal{R}_1, \mathcal{R}_2, \ldots, \mathcal{R}_\varpi).$$

*Proof.* For PFNs $\mathcal{R}_f = (\mathcal{Y}_f, \mathcal{A}_f, \mathcal{N}_f)$, $f = 1, 2, \ldots, \varpi$, and their weight vector $\Xi = (\Xi_1, \Xi_2, \ldots, \Xi_\varpi)$, then we have

$$\prod_{f=1}^{\varpi} \mathcal{Y}_f^{\Xi_f} \leq 1 - \prod^{\varpi} \left(1 - \mathcal{Y}_f\right)^{\Xi_f}, \text{ and}$$

$$0 \leq \prod_{f=1}^{\varpi} \mathcal{Y}_f^{\Xi_f}, 1 - \prod^{\varpi}\left(1 - \mathcal{Y}_f\right)^{\Xi_f} \leq 1.$$

For a real number $\gamma \in [0, 1]$, we have

$$\left(\prod_{f=1}^{\varpi} \mathcal{Y}_f^{\Xi_f}\right)^{1-\gamma} \leq \left(1 - \prod^{\varpi}\left(1 - \mathcal{Y}_f\right)^{\Xi_f}\right)^{1-\gamma},$$

which indicates that

$$\left(1 - \prod^{\varpi}\left(1 - \mathcal{Y}_f\right)^{\Xi_f}\right)^{\gamma} \left(\prod_{f=1}^{\varpi} \mathcal{Y}_b^{\Xi_f}\right)^{1-\gamma} \leq 1 - \prod^{\varpi}\left(1 - \mathcal{Y}_f\right)^{\Xi_f}. \tag{2.9}$$

Similarly, we have

$$\prod_{f=1}^{\varpi} \mathcal{A}_f^{\Xi_f} \geq 1 - \prod^{\varpi}\left(1 - \mathcal{A}_f\right)^{\Xi_f}, \text{ and } \prod_{f=1}^{\varpi} \mathcal{N}_f^{\Xi_f} \geq 1 - \prod^{\varpi}\left(1 - \mathcal{N}_f\right)^{\Xi_f}.$$

Then, we have

$$\prod_{f=1}^{\varpi} \mathcal{A}_f^{\Xi_f} \leq 1 - \left(1 - \prod^{\varpi} \mathcal{A}_f^{\Xi_f}\right)^{\gamma} \left(\prod^{\varpi}\left(1 - \mathcal{A}_f\right)^{\Xi_f}\right)^{1-\gamma}. \tag{2.10}$$

$$\prod_{f=1}^{\varpi} \mathcal{N}_f^{\Xi_f} \leq 1 - \left(1 - \prod^{\varpi} \mathcal{N}_f^{\Xi_f}\right)^{\gamma} \left(\prod^{\varpi}\left(1 - \mathcal{N}_f\right)^{\Xi_f}\right)^{1-\gamma}. \tag{2.11}$$

Therefore the definition of the score function is obtained as

$$\begin{aligned}
& S(H - PFWAG(\mathcal{R}_1, \mathcal{R}_2, \ldots, \mathcal{R}_{\varpi})) \\
& = 1 - \prod^{\varpi}\left(1 - \mathcal{Y}_f\right)^{\Xi_f} - \prod_{f=1}^{\varpi} \mathcal{A}_f^{\Xi_f} - \prod_{f=1}^{\varpi} \mathcal{N}_f^{\Xi_f} \geq \left(1 - \prod^{\varpi}\left(1 - \mathcal{Y}_f\right)^{\Xi_f}\right)^{\gamma} \left(\prod_{f=1}^{\varpi} \mathcal{Y}_f^{\Xi_f}\right)^{1-\gamma} \\
& - 1 + \left(1 - \prod_{f=1}^{\varpi} \mathcal{A}_f^{\Xi_f}\right)^{\gamma} \left(\prod_{f=1}^{\varpi} (1 - \mathcal{A}_f)^{\Xi_f}\right)^{1-\gamma} \\
& - 1 + \left(1 - \prod_{f=1}^{\varpi} \mathcal{N}_f^{\Xi_f}\right)^{\gamma} \left(\prod_{f=1}^{\varpi} (1 - \mathcal{N}_f)^{\Xi_f}\right)^{1-\gamma} \\
& = S(H - PFWAG(\mathcal{R}_1, \mathcal{R}_2, \ldots, \mathcal{R}_{\varpi})).
\end{aligned}$$

Thus $H - PFWAG(\mathcal{R}_1, \mathcal{R}_2, \ldots, \mathcal{R}_{\varpi}) \leq PFWA(\mathcal{R}_1, \mathcal{R}_2, \ldots, \mathcal{R}_{\varpi})$.

Similarly, $PFWG(\mathcal{R}_1, \mathcal{R}_2, \ldots, \mathcal{R}_{\varpi}) \leq H - PFWAG(\mathcal{R}_1, \mathcal{R}_2, \ldots, \mathcal{R}_{\varpi})$. Hence, the proof is completed. □

**Example 2.3.** Let there be three collections of PFNs $\mathcal{R}_1 = (0.7, 0.05, 0.2), \mathcal{R}_1 = (0.09, 0.2, 0.3)$, and $\mathcal{R}_2 = (0.6, 0.1, 0.2)$, and let $\Xi = (0.3, 0.5, 0.2)$ be the weight vector of $\mathcal{R}_f$, $f = 1, 2, 3$. Then, the computed value of aggregation operators H-PFWAG, PFWA, PFWG are given as $(0.3297, 0.1263, 0.2483)$, $(0.4466, 0.1149, 0.2449)$, and $(0.2434, 0.1376, 0.2517)$, respectively. The score values of H-PFWAG, PFWA, and PFWG operators for $\mathcal{R}_1$, $\mathcal{R}_2$, and $\mathcal{R}_3$ are 0.5407, 0.6009, and 0.4959, respectively. It follows that

$$PFWG(\mathcal{R}_1, \mathcal{R}_2, \mathcal{R}_3) \leq H - PFWAG(\mathcal{R}_1, \mathcal{R}_2, \ldots, \mathcal{R}_3) \leq PFWA(\mathcal{R}_1, \mathcal{R}_2, \mathcal{R}_3).$$

## 2.3.2 Hybrid ordered PFN aggregation operators

The aggregation operators for PFN also have many interesting properties and applications.

**Definition 2.3.** For a group of PFNs $\mathcal{R}_f = (\mathcal{Y}_f, \mathcal{A}_f, \mathcal{N}_f)$, $f = 1, 2, \ldots, \varpi$, A H-PFOWAG operator is a function $H - PFOWAG : \Theta^{\varpi} \to \Theta$ defined below

$$H - PFOWAG(\mathcal{R}_1, \mathcal{R}_2, \ldots, \mathcal{R}_\varpi) = \Big(\bigoplus_{f=1}^{\varpi} \Xi_f p_{\sigma(f)}\Big)^{\gamma} \bigotimes \Big(\bigotimes_{f=1}^{\varpi} \mathcal{R}_{\sigma(f)}^{\Xi_f}\Big)^{\gamma-1}, \tag{2.12}$$

where $\gamma \in [0, 1]$ is a real number that presents the attitudinal natures of the DMs towards an aggregation method, hence it gives a weighted average for accumulating order averages and geometric AOs, $\Xi > 0$ is the standardized weight vector of $\mathcal{R}_f$, $f = 1, 2, \ldots, \varpi$.

**Theorem 2.4.** *The accumulated value of PFNs H-PFOWAG operator for a group of PFNs $\mathcal{R}_f = (\mathcal{Y}_f, \mathcal{A}_f, \mathcal{N}_f)$, $f = 1, 2, \ldots, \varpi$ remains a PFN, and is given below*

$$\begin{aligned} &H - PFOWAG(\mathcal{R}_1, \mathcal{R}_2, \ldots, \mathcal{R}_\varpi) \\ &= \Bigg(\Big(1 - \prod_{f=1}^{\varpi}(1 - \mathcal{Y}_{\sigma(f)})^{\Xi_f}\Big)^{\gamma} \Big(\prod_{f=1}^{\varpi}(\mathcal{Y}_{\sigma(f)})^{\Xi_f}\Big)^{1-\gamma}, \\ &1 - \Big(1 - \prod_{f=1}^{\varpi} \mathcal{A}_{\sigma(f)}^{\Xi_f}\Big)^{\gamma} \Big(\prod_{f=1}^{\varpi}(1 - \mathcal{A}_{\sigma(f)})^{\Xi_f}\Big)^{1-\gamma}, \\ &1 - \Big(1 - \prod_{f=1}^{\varpi} \mathcal{N}_{\sigma(f)}^{\Xi_f}\Big)^{\gamma} \Big(\prod_{f=1}^{\varpi}(1 - \mathcal{N}_{\sigma(f)})^{\Xi_f}\Big)^{1-\gamma}\Bigg). \end{aligned} \tag{2.13}$$

*Proof.* From the operation rules for PFNs $\mathcal{R}_f$, $f = 1, 2, \ldots, \varpi$ and a real number $\gamma \in [0, 1]$, we obtain

$$\bigoplus_{f=1}^{\varpi} \Xi_f \mathcal{R}_{\sigma(f)} = \Big(1 - \prod_{f=1}^{\varpi}(1 - \mathcal{Y}_{\sigma(f)})^{\Xi_f}, \prod_{f=1}^{\varpi}(\mathcal{A}_{\sigma(f)})^{\Xi_f}, \prod_{f=1}^{\varpi}(\mathcal{N}_{\sigma(f)})^{\Xi_f}\Big),$$

$$\text{and } \prod_{f=1}^{\varpi}(\mathcal{R}_f)^{\Xi_f} = \Big(\prod_{f=1}^{\varpi}(\mathcal{Y}_{\sigma(f)})^{\Xi_f}, 1 - \prod_{f=1}^{\varpi}(1 - \mathcal{A}_{\sigma(f)})^{\Xi_f}, 1 - \prod_{f=1}^{\varpi}(1 - \mathcal{N}_{\sigma(f)})^{\Xi_f}\Big)$$

$$\left(\bigoplus_{f=1}^{\varpi} \Xi_f \mathcal{R}_{\sigma(f)}\right)^{\gamma}$$

$$= \left(\left(1-\prod_{f=1}^{\varpi}(1-\mathcal{Y}_{\sigma(f)})^{\Xi_f}\right)^{\gamma}, 1-\left(1-\prod_{f=1}^{\varpi}(\mathcal{A}_{\sigma(f)})^{\Xi_f}\right)^{\gamma}, 1-\left(1-\prod_{f=1}^{\varpi}(\mathcal{N}_{\sigma(f)})^{\Xi_f}\right)^{\gamma}\right)$$

$$\left(\prod_{f=1}^{\varpi}(\mathcal{R}_{\sigma(f)})^{\Xi_f}\right)^{\gamma-1}$$

$$= \left(\left(\prod_{f=1}^{\varpi}(\mathcal{Y}_{\sigma(f)})^{\Xi_f}\right)^{1-\gamma}, 1-\left(\prod_{f=1}^{\varpi}(1-\mathcal{A}_{\sigma(f)})^{\Xi_f}\right)^{1-\gamma}, 1-\left(\prod_{f=1}^{\varpi}(1-\mathcal{N}_{\sigma(f)})^{\Xi_f}\right)^{1-\gamma}\right).$$

Thus by Definition 2.3, we obtain

$$\left(\left(1-\prod_{f=1}^{\varpi}(1-\mathcal{Y}_{\sigma(f)})^{\Xi_f}\right)^{\gamma}, 1-\left(1-\prod_{f=1}^{\varpi}(\mathcal{A}_{\sigma(f)})^{\Xi_f}\right)^{\gamma}, 1-\left(1-\prod_{f=1}^{\varpi}(\mathcal{N}_{\sigma(f)})^{\Xi_f}\right)^{\gamma}\right)$$

$$\bigotimes \left(\left(\prod_{f=1}^{\varpi}(\mathcal{Y}_{\sigma(f)})^{\Xi_f}\right)^{1-\gamma}, 1-\left(\prod_{f=1}^{\varpi}(1-\mathcal{A}_{\sigma(f)})^{\Xi_f}\right)^{1-\gamma}, 1-\left(\prod_{f=1}^{\varpi}(1-\mathcal{N}_{\sigma(f)})^{\Xi_f}\right)^{1-\gamma}\right)$$

$$= \left(\left(1-\prod_{f=1}^{\varpi}(1-\mathcal{Y}_{\sigma(f)})^{\Xi_f}\right)^{\gamma}\left(\prod_{f=1}^{\varpi}(\mathcal{Y}_{\sigma(f)})^{\Xi_f}\right)^{1-\gamma},\right.$$

$$\left\{1-\left(1-\prod_{f=1}^{\varpi}(\mathcal{A}_{\sigma(f)})^{\Xi_f}\right)^{\gamma}\right\}+\left\{1-\left(\prod_{f=1}^{\varpi}(1-\mathcal{A}_{\sigma(f)})^{\Xi_f}\right)^{1-\gamma}\right\},$$

$$\left.\left\{1-\left(1-\prod_{f=1}^{\varpi}(\mathcal{N}_{\sigma(f)})^{\Xi_f}\right)^{\gamma}\right\}+\left\{1-\left(\prod_{f=1}^{\varpi}(1-\mathcal{N}_{\sigma(f)})^{\Xi_f}\right)^{1-\gamma}\right\}\right)$$

$$= \left(\left(1-\prod_{f=1}^{\varpi}(1-\mathcal{Y}_{\sigma(f)})^{\Xi_f}\right)^{\gamma}\left(\prod_{f=1}^{\varpi}(\mathcal{Y}_{\sigma(f)})^{\Xi_f}\right)^{1-\gamma},\right.$$

$$1-\left(1-\prod_{f=1}^{\varpi}\mathcal{A}_{\sigma(f)}^{\Xi_f}\right)^{\gamma}\left(\prod_{f=1}^{\varpi}(1-\mathcal{A}_{\sigma(f)})^{\Xi_f}\right)^{1-\gamma},$$

$$\left.1-\left(1-\prod_{f=1}^{\varpi}\mathcal{N}_{\sigma(f)}^{\Xi_f}\right)^{\gamma}\left(\prod_{f=1}^{\varpi}(1-\mathcal{N}_{\sigma(f)})^{\Xi_f}\right)^{1-\gamma}\right).$$

Hence, the theorem is proved. □

**Example 2.4.** Let $\mathcal{R}_1 = (0.6, 0.1, 0.3)$, $\mathcal{R}_2 = (0.4, 0.05, 0.2)$, and $\mathcal{R}_3 = (0.09, 0.2, 0.6)$ be three PFNs and the weight vector be $\Xi = (0.5, 0.3, 0.2)$. Then, without loss of generality, let $\gamma =$

0.5 be a real number. To find the permutations of the three PFNs, we find scores $\Lambda(\mathcal{R}_1) = 0.65$, $\Lambda(\mathcal{R}_2) = 0.60$, and $\Lambda(\mathcal{R}_3) = 0.245$ that implies $\Lambda(\mathcal{R}_1) > \Lambda(\mathcal{R}_2) > \Lambda(\mathcal{R}_3)$. Here, $\mathcal{R}_{\sigma(1)} = \mathcal{R}_1$, $\mathcal{R}_{\sigma(2)} = \mathcal{R}_2$, and $\mathcal{R}_{\sigma(3)} = \mathcal{R}_3$. Without loss of generality, we have for $\gamma = 0.5$

$$
\begin{aligned}
&H-PFOWAG(\mathcal{R}_1,\mathcal{R}_2,\ldots,\mathcal{R}_\varpi)\\
&=\left(\left(1-\prod_{f=1}^{\varpi}(1-\mathcal{Y}_{\sigma(f)})^{\Xi_f}\right)^{\gamma}\left(1-\prod_{f=1}^{\varpi}(\mathcal{Y}_{\sigma(f)})^{\Xi_f}\right)^{1-\gamma},\right.\\
&\quad 1-\left(1-\prod_{f=1}^{\varpi}\mathcal{A}_{\sigma(f)}^{\Xi_f}\right)^{\gamma}\left(\prod_{f=1}^{\varpi}(1-\mathcal{A}_{\sigma(f)})^{\Xi_f}\right)^{1-\gamma},\\
&\quad \left.1-\left(1-\prod_{f=1}^{\varpi}N_{\sigma(f)}^{\Xi_f}\right)^{\gamma}\left(\prod_{f=1}^{\varpi}(1-N_{\sigma(f)})^{\Xi_f}\right)^{1-\gamma}\right)\\
&=\left(\left(1-(1-0.6)^{0.5}(1-0.4)^{0.3}(1-0.09)^{0.2}\right)^{0.5}\left(0.6^{0.5}0.4^{0.3}0.09^{0.2}\right)^{1-0.5},\right.\\
&\quad 1-\left(1-(0.1)^{0.5}(0.05)^{0.3}(0.2)^{0.2}\right)^{0.5}\left((1-0.1)^{0.5}(1-0.05)^{0.3}(1-0.2)^{0.2}\right)^{1-0.5},\\
&\quad \left.1-\left(1-(0.3)^{0.5}(0.2)^{0.3}(0.6)^{0.2}\right)^{0.5}\left((1-0.3)^{0.5}(1-0.2)^{0.3}(1-0.6)^{0.2}\right)^{1-0.5}\right)\\
&=\left(0.5196, 0.0452, 0.1343\right).
\end{aligned}
$$

**Theorem 2.5.** *For a group of PFNs $\mathcal{R}_f = (\mathcal{Y}_f, \mathcal{A}_f, N_f)$, $f = 1, 2, \ldots, \varpi$, an H-PFOWAG operator satisfies the following properties:*

**(1)** *(Idempotency) Let $\{\mathcal{R}_f = (\mathcal{Y}_f, \mathcal{A}_f, N_f), f = 1, 2, \ldots, \varpi\}$ be a set of PFNs and $\mathcal{R}_f = \mathcal{R}$ for all $f$, then*

$$H-PFOWAG(\mathcal{R}_1,\mathcal{R}_2,\ldots,\mathcal{R}_\varpi) = \mathcal{R};$$

**(2)** *(Boundedness) For PFNs $\mathcal{R}_f$, $f = 1, 2, \ldots, \varpi$, we obtain*

$$\mathcal{R}^- \leq H-PFOWAG(\mathcal{R}_1,\mathcal{R}_2,\ldots,\mathcal{R}_\varpi) \leq \mathcal{R}^+,$$

*where $\mathcal{R}^- = (\min_f\{\mathcal{Y}_f\}, \max_f\{\mathcal{A}_f\}, \max_f\{N_f\})$ and $\mathcal{R}^+ = (\max_f\{\mathcal{Y}_f\}, \min_f\{\mathcal{A}_f\}, \min_f\{N_f\})$;*

**(3)** *(Monotonicity) If $\mathcal{R}_f$ and $\mathcal{R}'_f$ be two PFNs such that $\mathcal{R}_f \leq \mathcal{R}'_f$ for all $f$, then*

$$H-PFOWAG(\mathcal{R}_1,\mathcal{R}_2,\ldots,\mathcal{R}_\varpi) \leq H-PFOWAG(\mathcal{R}'_1,\mathcal{R}'_2,\ldots,\mathcal{R}'_\varpi).$$

As these properties are satisfied by the PFOWA and PFOWG operators, the definition of the H-PFOWAG operator directly follows from these properties.

**Theorem 2.6.** *For a group of PFNs $\mathcal{R}_f = (\mathcal{Y}_f, \mathcal{A}_f, N_f)$, $f = 1, 2, \ldots, \varpi$, the H-PFOWAG operator and existing operators PFOWA and PFOWG satisfy the following inequality:*

$$PFOWG(\mathcal{R}_1, \mathcal{R}_2, \ldots, \mathcal{R}_\varpi) \leq H - PFOWAG(\mathcal{R}_1, \mathcal{R}_2, \ldots, \mathcal{R}_\varpi) \leq PFOWA(\mathcal{R}_1, \mathcal{R}_2, \ldots, \mathcal{R}_\varpi).$$

*Proof.* The proof of the theorem is the same as that for Theorem 2.3. □

## 2.4 Interval-valued picture fuzzy approach

This section extends the proposed hybrid operators to interval-valued PFS (IVPFS).

**Definition 2.4.** An interval-valued picture fuzzy sets (IVPFS) over the fixed set $X$

$$U = \{(\mathcal{Y}_U(x), \mathcal{A}_U(x), N_U(x)) | x \in X\},$$

$\mathcal{Y}_U(x) \in [0, 1]$, $\mathcal{A}_U(x) \in [0, 1]$ and $N_U(x) \in [0, 1]$ are interval-valued PFNs, where $0 \leq \sup(\mathcal{Y}_U(x)) + \sup(\mathcal{A}_U(x)) + \sup(N_U(x)) \leq 1$ for all $x \in X$. For simplicity, we used $\mathcal{Y}_U(x) = [\mathcal{Y}^l, \mathcal{Y}^r]$, $\mathcal{A}_U(x) = [\mathcal{A}^l, \mathcal{A}^r]$, and $N_U(x) = [N^l, N^r]$. The tuple is presented as $([\mathcal{Y}^l, \mathcal{Y}^r], [\mathcal{A}^l, \mathcal{A}^r], [N^l, N^r])$ and are named interval-valued picture fuzzy numbers (IVPFNs) or interval-valued picture fuzzy values (IVPFVs) such that $[\mathcal{Y}^l, \mathcal{Y}^r], [\mathcal{A}^l, \mathcal{A}^r], [N^l, N^r] \subseteq [0, 1]$ and $0 \leq \mathcal{Y}^r + \mathcal{A}^r + N^r \leq 1$.

We define score $\Lambda$ and accuracy $\Phi$ functions for IVPFNs $\mathcal{R}_f$, $f = 1, 2, \ldots, \varpi$ below:

**Definition 2.5.** For an IVPFNs $\mathcal{R} = ([\mathcal{Y}^l, \mathcal{Y}^r], [\mathcal{A}^l, \mathcal{A}^r], [N^l, N^r])$, a score function $\Lambda$ is defined as:

$$\Lambda(\mathcal{R}) = \frac{\mathcal{Y}^l + \mathcal{Y}^r - \mathcal{A}^l - \mathcal{A}^r - N^l - N^r}{2}, \Lambda(\mathcal{R}) \in [-1, 1] \tag{2.14}$$

and an accuracy function $\Phi$ is defined as

$$\Phi(\mathcal{R}) = \frac{\mathcal{Y}^l + \mathcal{Y}^r + \mathcal{A}^l + \mathcal{A}^r + N^l + N^r}{2}, \Phi(\mathcal{R}) \in [0, 1]. \tag{2.15}$$

**Definition 2.6.** For a group of IVPFNs $\mathcal{R}_f = ([\mathcal{Y}^l, \mathcal{Y}^r], [\mathcal{A}^l, \mathcal{A}^r], [N^l, N^r])$, $f = 1, 2, \ldots, \varpi$, a H-IVPFWAG operator is a function $H - IVPFWAG : \Theta^\varpi \rightarrow \Theta$ defined as:

$$H - IVPFWAG(\mathcal{R}_1, \mathcal{R}_2, \ldots, \mathcal{R}_\varpi) = \Big(\bigoplus_{f=1}^{\varpi} \Xi_f \mathcal{R}_f\Big)^\gamma \otimes \Big(\bigotimes_{f=1}^{\varpi} \mathcal{R}_f^{\Xi_f}\Big)^{\gamma-1}, \tag{2.16}$$

where $\gamma \in [0, 1]$ is a real number that presents the attitudinal natures of the DMs towards an aggregation method, hence it gives a weighted average to accumulate averaging and geometric AOs, $\Xi > 0$ is the standardized weight vector of $\mathcal{R}_f$, $f = 1, 2, \ldots, \varpi$.

**Theorem 2.7.** *The accumulated value of IVPFNs H-IVPFWAG operator for a group of IVPFNs*

$$\mathcal{R}_f = ([\mathcal{Y}^l_f, \mathcal{Y}^r_f], [\mathcal{A}^l_f, \mathcal{A}^r_f], [\mathcal{N}^l_f, \mathcal{N}^r_f]),\ f = 1, 2, \ldots, \varpi$$

*remains a IVPFN and is given below*

$$\begin{aligned}
&H - IVPFWAG(\mathcal{R}_1, \mathcal{R}_2, \ldots, \mathcal{R}_\varpi)\\
&= \Big(\bigoplus_{f=1}^{\varpi} \Xi_f \mathcal{R}_f\Big)^{\gamma} \otimes \Big(\bigotimes_{f=1}^{\varpi} \mathcal{R}_f^{\Xi_f}\Big)^{\gamma-1}\\
&= \Bigg(\Big(\Big[1 - \prod_{f=1}^{\varpi}(1-\mathcal{Y}^l_f)^{\Xi_f}\Big)^{\gamma}\Big(\prod_{f=1}^{\varpi}(\mathcal{Y}^l_f)^{\Xi_f}\Big)^{1-\gamma}, 1 - \prod_{f=1}^{\varpi}(1-\mathcal{Y}^r_f)^{\Xi_f}\Big)^{\gamma}\Big(\prod_{f=1}^{\varpi}(\mathcal{Y}^r_f)^{\Xi_f}\Big)^{1-\gamma}\Big],\\
&\Big[1 - \Big(1 - \prod_{f=1}^{\varpi}(\mathcal{A}^l_f)^{\Xi_f}\Big)^{\gamma}\Big(\prod_{f=1}^{\varpi}(1-\mathcal{A}^l_f)^{\Xi_f}\Big)^{1-\gamma}, 1 - \Big(1 - \prod_{f=1}^{\varpi}(\mathcal{A}^r_f)^{\Xi_f}\Big)^{\gamma}\Big(\prod_{f=1}^{\varpi}(1-\mathcal{A}^r_f)^{\Xi_f}\Big)^{1-\gamma}\Big],\\
&\Big[1 - \Big(1 - \prod_{f=1}^{\varpi}(\mathcal{N}^l_f)^{\Xi_f}\Big)^{\gamma}\Big(\prod_{f=1}^{\varpi}(1-\mathcal{N}^l_f)^{\Xi_f}\Big)^{1-\gamma}, 1 - \Big(1 - \prod_{f=1}^{\varpi}(\mathcal{N}^r_f)^{\Xi_f}\Big)^{\gamma}\Big(\prod_{f=1}^{\varpi}(1-\mathcal{N}^r_f)^{\Xi_f}\Big)^{1-\gamma}\Big]\Bigg),
\end{aligned} \tag{2.17}$$

*where* $(\sigma(1), \sigma(2), \ldots, \sigma(\varpi))$ *is a permutation of* $\{1, 2, \ldots, \varpi\}$ *such that* $\mathcal{R}_{\sigma(f-1)} \geq \mathcal{R}_{\sigma(f)}$, $f = 1, 2, \ldots, \varpi$, *and* $\gamma \in [0, 1]$ *is a real number to serve the attitudinal character of the DMs towards the aggregation function, and* $\Xi_f > 0$ *is the standard weight vector of* $\mathcal{R}_f$, $f = 1, 2, \ldots, \varpi$, *where* $\sum_{f=1}^{\varpi} \Xi_f = 1$.

*Proof.* The proof is the same as that for Theorem 2.3. □

**Definition 2.7.** For a group of IVPFNs $\mathcal{R}_f = ([\mathcal{Y}^l_f, \mathcal{Y}^r_f], [\mathcal{A}^l_f, \mathcal{A}^r_f], [\mathcal{N}^l_f, \mathcal{N}^r_f])$, $f = 1, 2, \ldots, \varpi$, a H-IVPFOWAG operator is a function $H - IVPFOWAG : \Theta^{\varpi} \to \Theta$ defined as:

$$H - IVPFOWAG(\mathcal{R}_1, \mathcal{R}_2, \ldots, \mathcal{R}_\varpi) = \Big(\bigoplus_{f=1}^{\varpi} \Xi_f \mathcal{R}_{\sigma(f)}\Big)^{\gamma} \otimes \Big(\bigotimes_{f=1}^{\varpi} \mathcal{R}_{\sigma(f)}^{\Xi_f}\Big)^{\gamma-1}, \tag{2.18}$$

where $\gamma \in [0, 1]$ is a real number that serves the attitudinal natures of the DMs towards an aggregation manner, hence it gives a weighted average to accumulate averaging and geometric AOs, $\Xi > 0$ is the standardized weight vector of $\mathcal{R}_f$, $f = 1, 2, \ldots, \varpi$.

**Theorem 2.8.** *The accumulated value of a PFNs H-IVPFOWAG operator for a group of IVPFNs* $\mathcal{R}_f = ([\mathcal{Y}^l, \mathcal{Y}^r], [\mathcal{A}^l, \mathcal{A}^r], [\mathcal{N}^l, \mathcal{N}^r])$, $f = 1, 2, \ldots, \varpi$ *remains an IVPFN and is given as:*

$$
\begin{aligned}
&H-IVPFOWAG(\mathcal{R}_1,\mathcal{R}_2,\ldots,\mathcal{R}_\varpi)\\
&=\Big(\bigoplus_{f=1}^{\varpi}\Xi_f\mathcal{R}_{\sigma(f)}\Big)^{\gamma}\bigotimes\Big(\bigotimes_{f=1}^{\varpi}\mathcal{R}_{\sigma(f)}^{\Xi_f}\Big)^{\gamma-1}\\
&=\Bigg(\Big[1-\Big(\prod_{f=1}^{\varpi}(1-\mathcal{Y}^l_{\sigma(f)})^{\Xi_f}\Big)^{\gamma}\Big(\prod_{f=1}^{\varpi}(\mathcal{Y}^l_{\sigma(f)})^{\Xi_f}\Big)^{1-\gamma},\\
&\qquad 1-\Big(\prod_{f=1}^{\varpi}(1-\mathcal{Y}^r_{\sigma(f)})^{\Xi_f}\Big)^{\gamma}\Big(\prod_{f=1}^{\varpi}(\mathcal{Y}^r_{\sigma(f)})^{\Xi_f}\Big)^{1-\gamma}\Big],\\
&\quad\Big[1-\Big(1-\prod_{f=1}^{\varpi}(\mathcal{A}^l)^{\Xi_f}_{\sigma(f)}\Big)^{\gamma}\Big(\prod_{f=1}^{\varpi}(1-\mathcal{A}^l_{\sigma(f)})^{\Xi_f}\Big)^{1-\gamma},\\
&\qquad 1-\Big(1-\prod_{f=1}^{\varpi}(\mathcal{A}^r)^{\Xi_f}_{\sigma(f)}\Big)^{\gamma}\Big(\prod_{f=1}^{\varpi}(1-\mathcal{A}^r_{\sigma(f)})^{\Xi_f}\Big)^{1-\gamma}\Big],\\
&\quad\Big[1-\Big(1-\prod_{f=1}^{\varpi}(\mathcal{N}^l)^{\Xi_f}_{\sigma(f)}\Big)^{\gamma}\Big(\prod_{f=1}^{\varpi}(1-\mathcal{N}^l_{\sigma(f)})^{\Xi_f}\Big)^{1-\gamma},\\
&\qquad 1-\Big(1-\prod_{f=1}^{\varpi}(\mathcal{N}^r)^{\Xi_f}_{\sigma(f)}\Big)^{\gamma}\Big(\prod_{f=1}^{\varpi}(1-\mathcal{N}^r_{\sigma(f)})^{\Xi_f}\Big)^{1-\gamma}\Big]\Bigg),
\end{aligned}
\tag{2.19}
$$

*where* $(\sigma(1),\sigma(2),\ldots,\sigma(\varpi))$ *is a permutation of* $\{1,2,\ldots,\varpi\}$ *for which* $\mathcal{R}_{\sigma(f-1)}\geq\mathcal{R}_{\sigma(f)}$, $f=1,2,\ldots,\varpi$, *and* $\gamma\in[0,1]$ *is a real number that represents the attitudinal features of the DMs towards the aggregation operator.*

**Theorem 2.9.** *For a group of IVPFNs,* $\mathcal{R}_f=([\mathcal{Y}^l_f,\mathcal{Y}^r_f],[\mathcal{A}^l_f,\mathcal{A}^r_f],[\mathcal{N}^l_f,\mathcal{N}^r_f])$, $f=1,2,\ldots,\varpi$, *a H-IVPFWAG operator and existing operators IVPFWA and IVPFWG satisfy the following inequality:*

$$
\begin{aligned}
IVPFWG(\mathcal{R}_1,\mathcal{R}_2,\ldots,\mathcal{R}_\varpi)&\leq H-IVPFWAG(\mathcal{R}_1,\mathcal{R}_2,\ldots,\mathcal{R}_\varpi)\\
&\leq IVPFWA(\mathcal{R}_1,\mathcal{R}_2,\ldots,\mathcal{R}_\varpi).
\end{aligned}
$$

*Proof.* The proof is the same as that for Theorem 2.3. □

**Theorem 2.10.** *For a group of IVPFNs,* $\mathcal{R}_f=([\mathcal{Y}^l_f,\mathcal{Y}^r_f],[\mathcal{A}^l_f,\mathcal{A}^r_f],[\mathcal{N}^l_f,\mathcal{N}^r_f])$, $f=(1,2,\ldots,\varpi)$, *a H-IVPFOWAG operator and the operators IVPFOWA and IVPFOWG satisfy the following inequality:*

$$
\begin{aligned}
IVPFOWG(\mathcal{R}_1,\mathcal{R}_2,\ldots,\mathcal{R}_\varpi)&\leq H-IVPFOWAG(\mathcal{R}_1,\mathcal{R}_2,\ldots,\mathcal{R}_\varpi)\\
&\leq IVPFOWA(\mathcal{R}_1,\mathcal{R}_2,\ldots,\mathcal{R}_\varpi).
\end{aligned}
$$

*Proof.* The proof is the same as that for Theorem 2.3. □

## 2.5 MCDM based on the proposed operators

In this section, the multicriteria decision-making approach is provided based on the proposed operators.

### 2.5.1 Proposed operators-based decision-making approach

For this decision-making approach, considered $\xi$ alternatives $\aleph_1, \aleph_2, \ldots, \aleph_\xi$ that are evaluated by a team of experts based on the $\varpi$ criteria $C_1, C_2, \ldots, C_\varpi$. Then, each criteria is performed with the importance of a weight vector as $\Xi = (\Xi_1, \Xi_2, \ldots, \Xi_\varpi)$, where $\Xi_f > 0$ with $\sum_{f=1}^{\varpi} \Xi_f = 1$. The evaluation of alternatives is executed by the expert team and their rating in terms of PFNs $\Upsilon_{\xi\varpi}$ by the processing of scores and presented data. All the information is summarized in the form of a decision matrix as

$$\mathcal{R} = [\aleph_{gf}^{\delta}]_{\xi\times\varpi} = \begin{array}{c} \\ \aleph_1 \\ \aleph_2 \\ \vdots \\ \aleph_\varpi \end{array} \begin{array}{c} \begin{array}{cccc} C_1 & C_2 & \cdots & C_\xi \end{array} \\ \left[\begin{array}{cccc} \left(\mathcal{Y}_{11}, \mathcal{A}_{11}, \mathcal{N}_{11}\right) & \left(\mathcal{Y}_{12}, \mathcal{A}_{12}, \mathcal{N}_{12}\right) & \cdots & \left(\mathcal{Y}_{1\varpi}, \mathcal{A}_{1\varpi}, \mathcal{N}_{1\varpi}\right) \\ \left(\mathcal{Y}_{21}, \mathcal{A}_{21}, \mathcal{N}_{21}\right) & \left(\mathcal{Y}_{22}, \mathcal{A}_{22}, \mathcal{N}_{22}\right) & \cdots & \left(\mathcal{Y}_{2\varpi}, \mathcal{A}_{2\varpi}, \mathcal{N}_{2\varpi}\right) \\ \vdots & \vdots & \ddots & \vdots \\ \left(\mathcal{Y}_{\xi 1}, \mathcal{A}_{\xi 1}, \mathcal{N}_{\xi 1}\right) & \left(\mathcal{Y}_{\xi 2}, \mathcal{A}_{\xi 2}, \mathcal{N}_{\xi 2}\right) & \cdots & \left(\mathcal{Y}_{\xi\varpi}, \mathcal{A}_{\xi\varpi}, \mathcal{N}_{\xi\varpi}\right) \end{array}\right] \end{array}. \tag{2.20}$$

In the following, we provided two MCDM methods under a PFNs and IVPFNs environment.

*Approach 1*

For this proposed approach, assume $\mathcal{R}_{gf}$ is PFNs given by $\mathcal{R}_{gf} = (\mathcal{Y}_{gf}, \mathcal{A}_{gf}, \mathcal{N}_{gf})$, $g = 1, 2, \ldots, \xi$, $f = 1, 2, \ldots, \varpi$, where $\mathcal{Y}_{gf}, \mathcal{A}_{gf}, \mathcal{N}_{gf}$, respectively, represents membership, neutral, and nonmembership values of each alternative $\aleph_g$ of each criteria $C_f$ such that $0 \leq \mathcal{Y}_{gf}, \mathcal{A}_{gf}, \mathcal{N}_{gf} \leq 1$ with $0 \leq \mathcal{Y}_{gf} + \mathcal{A}_{gf} + \mathcal{N}_{gf} \leq 1$, $g = 1, 2, \ldots, \xi$, $f = 1, 2, \ldots, \varpi$. An MCDM approach includes the following steps for H-PFWAG and H-PFOWAG operators.

**Step 1.** Introduce the information in the decision matrix $\mathcal{R} = (\mathcal{R}_{gf})_{(\xi\times\varpi)}$.
**Step 2.** Applying either the H-PFWAG operator

$$\begin{aligned} \Upsilon_g &= H - PFWAG(\mathcal{R}_{g1}, \mathcal{R}_{g2}, \ldots, \mathcal{R}_{g\varpi}) \\ &= \Bigg(\left(1 - \prod_{f=1}^{\varpi}(1-\mathcal{Y}_f)^{\Xi_f}\right)^{\gamma}\left(\prod_{f=1}^{\varpi}(\mathcal{Y}_f)^{\Xi_f}\right)^{1-\gamma}, 1 - \left(1 - \prod_{f=1}^{\varpi}\mathcal{A}_f^{\Xi_f}\right)^{\gamma}\left(\prod_{f=1}^{\varpi}(1-\mathcal{A}_f)^{\Xi_f}\right)^{1-\gamma}, \\ &\quad 1 - \left(1 - \prod_{f=1}^{\varpi}\mathcal{N}_f^{\Xi_f}\right)^{\gamma}\left(\prod_{f=1}^{\varpi}(1-\mathcal{N}_f)^{\Xi_f}\right)^{1-\gamma}\Bigg) \end{aligned}$$

or applying the H-PFOWAG operator

$$\Upsilon_g = H - PFOWAG(\mathcal{R}_{g1}, \mathcal{R}_{g2}, \ldots, \mathcal{R}_{g\varpi})$$
$$= \left( \left(1 - \prod_{f=1}^{\varpi} (1 - \mathcal{Y}_{\sigma(f)})^{\Xi_f}\right)^{\gamma} \left(\prod_{f=1}^{\varpi} (\mathcal{Y}_{\sigma(f)})^{\Xi_f}\right)^{1-\gamma}, \right.$$
$$1 - \left(1 - \prod_{f=1}^{\varpi} \mathcal{A}_{\sigma(f)}^{\Xi_f}\right)^{\gamma} \left(\prod_{f=1}^{\varpi} (1 - \mathcal{A}_{\sigma(f)})^{\Xi_f}\right)^{1-\gamma},$$
$$\left. 1 - \left(1 - \prod_{f=1}^{\varpi} \mathcal{N}_{\sigma(f)}^{\Xi_f}\right)^{\gamma} \left(\prod_{f=1}^{\varpi} (1 - \mathcal{N}_{\sigma(f)})^{\Xi_f}\right)^{1-\gamma} \right)$$

to aggregate individual PFNs $\mathcal{R}_{gf}$, $g = 1, 2, \ldots, \xi$, $f = 1, 2, \ldots, \varpi$ to accumulate into PFN $\Upsilon_g$, $g = 1, 2, \ldots, \xi$.

**Step 3.** Compute the score results $\Lambda(\Upsilon)$ of each aggregate number $\Upsilon_g$, $g = 1, 2, \ldots, \xi$ by using this definition. If there is no variation between two $\Lambda(\Upsilon_1)$ and $\Lambda(\Upsilon_2)$ for any two $\mathcal{R}_1$, $\mathcal{R}_2$, then we want to calculate the accuracy values of the alternatives as $\Phi(\mathcal{R}_1)$ and $\Phi(\mathcal{R}_2)$.

**Step 4.** Rank all the options $\aleph_g$, $g = 1, 2, \ldots, \xi$.

### *Approach 2*

In this proposed method, we assume that an expert team gives their decision in terms of interval numbers for each of the options $\aleph_g$ $(g = 1, 2, \ldots, \xi)$ is evaluated for all criteria as

$$\aleph_g = \left\{ C_f, \left[\inf \mathcal{Y}_Q(C_f), \sup \mathcal{Y}_Q(C_f)\right], \left[\inf \mathcal{A}_{\aleph}(C_f), \sup \mathcal{A}_{\aleph}(C_f)\right], \right.$$
$$\left. \left[\inf \mathcal{N}_{\aleph}(C_f), \sup \mathcal{N}_{\aleph}(C_f)\right] | C_f \in C \right\},$$

where $\left[\inf \mathcal{Y}_{\aleph}(C_f), \sup \mathcal{Y}_{\aleph}(C_f)\right] \subseteq [0, 1]$, $\left[\inf \mathcal{A}_{\aleph}(C_f), \sup \mathcal{A}_{\aleph}(C_f)\right] \subseteq [0, 1]$, and $\left[\inf \mathcal{N}_{\aleph}(C_f), \sup \mathcal{N}_{\aleph}(C_f)\right] \subseteq [0, 1]$ represents an interval of support, neutral, and opposition degrees, respectively, of each alternative $Q_a$ corresponding to the criteria $C_f$ for which $0 \leq \sup \mathcal{Y}_{\aleph}(C_f) + \sup \mathcal{A}_{\aleph}(C_f) + \sup \mathcal{N}_{\aleph}(C_f) \leq 1$. For convenience, the criteria value is denoted by $\mathcal{R}_{gf} = \left(\left[\mathcal{Y}^l_{gf}, \mathcal{Y}^r_{gf}\right], \left[\mathcal{A}^l_{gf}, \mathcal{A}^r_{gf}\right], \left[\mathcal{N}^l_{gf}, \mathcal{N}^r_{gf}\right]\right)$, where $\left[\mathcal{Y}^l_{gf}, \mathcal{Y}^r_{gf}\right] \subseteq [0, 1]$, $\left[\mathcal{A}^l_{gf}, \mathcal{A}^r_{gf}\right] \subseteq [0, 1]$, and $\left[\mathcal{N}^l_{gf}, \mathcal{N}^r_{gf}\right] \subseteq [0, 1]$ and $0 \leq \mathcal{Y}^r_{gf} + \mathcal{A}^r_{gf} + \mathcal{N}^r_{gf} \leq 1$. The following steps are proposed to aggregate IVPFNs utilizing the H-IVPFWAG and H-IVPFOWAG operators in this environment.

**Step 1.** Introduce the information in the decision matrix $\mathcal{R} = (\mathcal{R}_{gf})_{(\xi \times \varpi)}$.

**Step 2.** Applying either the H-IVPFWAG operator

$$\Upsilon_g = H - IVPFWAG(\mathcal{R}_{g1}, \mathcal{R}_{g2}, \ldots, \mathcal{R}_{g\varpi})$$

$$= \Bigg( \Big[1-\Big(\prod_{f=1}^{\varpi}(1-\mathcal{Y}^{l}_{f})^{\Xi_f}\Big)^{\gamma}\Big(\prod_{f=1}^{\varpi}(\mathcal{Y}^{l}_{f})^{\Xi_f}\Big)^{1-\gamma},$$

$$1-\Big(\prod_{f=1}^{\varpi}(1-\mathcal{Y}^{r}_{f})^{\Xi_f}\Big)^{\gamma}\Big(\prod_{f=1}^{\varpi}(\mathcal{Y}^{r}_{f})^{\Xi_f}\Big)^{1-\gamma}\Big],$$

$$\Big[1-\Big(1-\prod_{f=1}^{\varpi}(\mathcal{A}^{l}_{f})^{\Xi_f}\Big)^{\gamma}\Big(\prod_{f=1}^{\varpi}(1-\mathcal{A}^{l}_{f})^{\Xi_f}\Big)^{1-\gamma},$$

$$1-\Big(1-\prod_{f=1}^{\varpi}(\mathcal{A}^{r}_{f})^{\Xi_f}\Big)^{\gamma}\Big(\prod_{f=1}^{\varpi}(1-\mathcal{A}^{r}_{f})^{\Xi_f}\Big)^{1-\gamma}\Big],$$

$$\Big[1-\Big(1-\prod_{f=1}^{\varpi}(\mathcal{N}^{l}_{f})^{\Xi_f}\Big)^{\gamma}\Big(\prod_{f=1}^{\varpi}(1-\mathcal{N}^{l}_{f})^{\Xi_f}\Big)^{1-\gamma},$$

$$1-\Big(1-\prod_{f=1}^{\varpi}(\mathcal{N}^{r}_{f})^{\Xi_f}\Big)^{\gamma}\Big(\prod_{f=1}^{\varpi}(1-\mathcal{N}^{r}_{f})^{\Xi_f}\Big)^{1-\gamma}\Big]\Bigg)$$

or applying the H-IVPFOWAG operator

$$\Upsilon_g = H-IVPFOWAG(\mathcal{R}_{g1},\mathcal{R}_{g2},\ldots,\mathcal{R}_{g\varpi})$$

$$= \Bigg( \Big[1-\Big(\prod_{f=1}^{\varpi}(1-\mathcal{Y}^{l}_{\sigma(f)})^{\Xi_f}\Big)^{\gamma}\Big(\prod_{f=1}^{\varpi}(\mathcal{Y}^{l}_{\sigma(f)})^{\Xi_f}\Big)^{1-\gamma},$$

$$1-\Big(\prod_{f=1}^{\varpi}(1-\mathcal{Y}^{r}_{\sigma(f)})^{\Xi_f}\Big)^{\gamma}\Big(\prod_{f=1}^{\varpi}(\mathcal{Y}^{r}_{\sigma(f)})^{\Xi_f}\Big)^{1-\gamma}\Big],$$

$$\Big[1-\Big(1-\prod_{f=1}^{\varpi}(\mathcal{A}^{l})^{\Xi_f}_{\sigma(f)}\Big)^{\gamma}\Big(\prod_{f=1}^{\varpi}(1-\mathcal{A}^{l}_{\sigma(f)})^{\Xi_f}\Big)^{1-\gamma},$$

$$1-\Big(1-\prod_{f=1}^{\varpi}(\mathcal{A}^{r})^{\Xi_f}_{\sigma(f)}\Big)^{\gamma}\Big(\prod_{f=1}^{\varpi}(1-\mathcal{A}^{r}_{\sigma(f)})^{\Xi_f}\Big)^{1-\gamma}\Big],$$

$$\Big[1-\Big(1-\prod_{f=1}^{\varpi}(\mathcal{N}^{l})^{\Xi_f}_{\sigma(f)}\Big)^{\gamma}\Big(\prod_{f=1}^{\varpi}(1-\mathcal{N}^{l}_{\sigma(f)})^{\Xi_f}\Big)^{1-\gamma},$$

$$1-\Big(1-\prod_{f=1}^{\varpi}(\mathcal{N}^{r})^{\Xi_f}_{\sigma(f)}\Big)^{\gamma}\Big(\prod_{f=1}^{\varpi}(1-\mathcal{N}^{r}_{\sigma(f)})^{\Xi_f}\Big)^{1-\gamma}\Big]\Bigg)$$

to aggregate individual IVPFNs $\mathcal{R}_{gf}$, $g = 1,2,\ldots,\xi$, $f = 1,2,\ldots,\varpi$ to accumulate into IVPFN $\Upsilon_g$, $g = 1,2,\ldots,\xi$.

**Step 3.** Compute the score $\Lambda(\Upsilon)$ of each aggregate number $\Upsilon_g$ $(g = 1,2,\ldots,\xi)$ by using Definition 2.5. If there is no variation between the two $\Lambda(\Upsilon_1)$ and $\Lambda(\Upsilon_2)$ for any two

$\mathcal{R}_1$, $\mathcal{R}_2$, then we need to calculate the accuracy values of options as $\Phi(\mathcal{R}_1)$ and $\Phi(\mathcal{R}_2)$ by Definition 2.5.

**Step 4.** Rank all the options $\aleph_g$, $g = 1, 2, \ldots, \xi$.

### 2.5.2 Numerical MAGDM model for PFNs

At the current time, our civilization is now threatened due to the coronavirus outbreak throughout the whole world. India is not out of its danger.

On 24 March 2020 the Government of India announced a nationwide lockdown as a preventive measure against the coronavirus pandemic in India. India's informal and unorganized industry employs 94 per cent of the population, and they contribute 45 per cent to its overall output. Hence, the Indian economy now becomes threatened and may also drop the country's Gross Domestic Product (GDP). To overcome this situation, the Indian government called economists or decision makers who can handle this kind of situation and able to determine which sector (alternatives) of the Indian economy will be affected by the situation. In that view, decision makers considered five sectors on which the Indian economy depends and were given as:

($\aleph_1$): Agriculture;
($\aleph_2$): Real-Estate;
($\aleph_3$): Industrial;
($\aleph_4$): Information Technology;
($\aleph_5$): Education.

For this evaluation, the DMs consider the criterion in terms of how much the effect of the shutdown on the particular sector is as follows:

$C_1$: Very low effect;
$C_2$: Low effect;
$C_3$: Regular effect;
$C_4$: Higher effect;
$C_5$: Very high effect.

Here, each criteria weight $C_f$, $f = 1, 2, \ldots, \varpi$ is the vector that is provided as $\Xi = (0.2, 0.25, 0.15, 0.3, 0.1)^T$ and in order to make the decision pessimistic for the prospect goals, then manipulate the aggregation by using the OWA weight vector $\Xi = (0.1, 0.2, 0.2, 0.2, 0.3)^T$.

The succeeding steps are adopted for **Approach I**, the favorable alternative is evaluated as follows:

**Step 1:** The importance of each alternatives is given in Table 2.2 in terms of PFNs.

**Step 2:** By applying the proposed methods and some existing operators to accumulate the PFNs, aggregated values are shown in Table 2.4, which provides PFWA, PFWG, PFOWA, PFOWG, H-PFWAG, and H-PFOWAG operators.

**Table 2.2** Picture fuzzy decision matrix.

| | $C_1$ | $C_2$ | $C_3$ | $C_4$ | $C_5$ |
|---|---|---|---|---|---|
| $\aleph_1$ | $\left((0.4, 0.1, 0.2)\right)$ | $\left((0.3, 0.1, 0.2)\right)$ | $\left((0.3, 0.2, 0.1)\right)$ | $\left((0.4, 0.1, 0.2)\right)$ | $\left((0.5, 0.1, 0.1)\right)$ |
| $\aleph_2$ | $\left((0.3, 0.1, 0.2)\right)$ | $\left((0.5, 0.1, 0.1)\right)$ | $\left((0.4, 0.1, 0.2)\right)$ | $\left((0.3, 0.2, 0.1)\right)$ | $\left((0.5, 0.1, 0.1)\right)$ |
| $\aleph_3$ | $\left((0.4, 0.1, 0.3)\right)$ | $\left((0.3, 0.1, 0.2)\right)$ | $\left((0.3, 0.1, 0.1)\right)$ | $\left((0.3, 0.1, 0.2)\right)$ | $\left((0.3, 0.2, 0.1)\right)$ |
| $\aleph_4$ | $\left((0.2, 0.3, 0.1)\right)$ | $\left((0.3, 0.1, 0.2)\right)$ | $\left((0.5, 0.1, 0.1)\right)$ | $\left((0.4, 0.2, 0.1)\right)$ | $\left((0.5, 0.1, 0.1)\right)$ |
| $\aleph_5$ | $\left((0.3, 0.2, 0.1)\right)$ | $\left((0.4, 0.1, 0.1)\right)$ | $\left((0.3, 0.1, 0.2)\right)$ | $\left((0.4, 0.1, 0.2)\right)$ | $\left((0.5, 0.1, 0.1)\right)$ |

**Step 3:** The aggregated values of PFNs are shown in Table 2.3, and score values are summarized in Table 2.4.

**Table 2.3** Aggregated values of PFN operators.

| | H-PFWAG | H-PFOWAG | PFWA |
|---|---|---|---|
| $\aleph_1$ | $\left((0.3689, 0.1134, 0.1724)\right)$ | $\left((0.3384, 0.1179, 0.1469)\right)$ | $\left((0.3734, 0.1110, 0.1682)\right)$ |
| $\aleph_2$ | $\left((0.3832, 0.1272, 0.1319)\right)$ | $\left((0.4182, 0.1179, 0.1272)\right)$ | $\left((0.3920, 0.1231, 0.1275)\right)$ |
| $\aleph_3$ | $\left((0.3195, 0.1089, 0.1901)\right)$ | $\left((0.3097, 0.1272, 0.1550)\right)$ | $\left((0.3213, 0.1072, 0.1824)\right)$ |
| $\aleph_4$ | $\left((0.3555, 0.1637, 0.1225)\right)$ | $\left((0.4061, 0.1355, 0.1179)\right)$ | $\left((0.3689, 0.1534, 0.1189)\right)$ |
| $\aleph_5$ | $\left((0.3740, 0.1179, 0.1415)\right)$ | $\left((0.3986, 0.1089, 0.1367)\right)$ | $\left((0.3782, 0.1149, 0.1366)\right)$ |
| | **PFOWA** | **PFWG** | **PFOWG** |
| $\aleph_1$ | $\left((0.3958, 0.1149, 0.1414)\right)$ | $\left((0.3646, 0.1158, 0.1761)\right)$ | $\left((0.3812, 0.1210, 0.1515)\right)$ |
| $\aleph_2$ | $\left((0.4264, 0.1149, 0.1231)\right)$ | $\left((0.3745, 0.1312, 0.1363)\right)$ | $\left((0.4102, 0.1210, 0.1312)\right)$ |
| $\aleph_3$ | $\left((0.3107, 0.1231, 0.1473)\right)$ | $\left((0.3178, 0.1105, 0.1978)\right)$ | $\left((0.3088, 0.1312, 0.1627)\right)$ |
| $\aleph_4$ | $\left((0.4186, 0.1282, 0.1149)\right)$ | $\left((0.3426, 0.1738, 0.1261)\right)$ | $\left((0.3939, 0.1428, 0.1210)\right)$ |
| $\aleph_5$ | $\left((0.4050, 0.1072, 0.1320)\right)$ | $\left((0.3698, 0.1210, 0.1465)\right)$ | $\left((0.3923, 0.1105, 0.1414)\right)$ |

**Table 2.4** Score values of aggregated PFNs.

| Alternative | H-PFWAG | H-PFOWAG | PFWA | PFOWA | PFWG | PFOWG |
|---|---|---|---|---|---|---|
| $\aleph_1$ | 0.5984 | 0.5960 | 0.6026 | 0.6272 | 0.5943 | 0.6149 |
| $\aleph_2$ | 0.6257 | 0.6455 | 0.6323 | 0.6517 | 0.6395 | 0.6395 |
| $\aleph_3$ | 0.5647 | 0.5774 | 0.5695 | 0.5817 | 0.5600 | 0.5731 |
| $\aleph_4$ | 0.6165 | 0.6441 | 0.6250 | 0.6519 | 0.6083 | 0.6365 |
| $\aleph_5$ | 0.6163 | 0.6310 | 0. 6208 | 0.6365 | 0.6117 | 0.6255 |

**Step 4:** As per the results of the score values, the rank of the options is shown in Table 2.4; the symbol "$\succ$" means "preferred to". Table 2.5 concluded that $\aleph_2$ is the most favorable alternative for the proposed operators, whereas other existing operators provided different alternatives.

If we apply **Approach II** for the selection of the best alternatives, we follow the steps below:

**Table 2.5** Ranking order of the alternatives under PFNs.

| Some existing operators | Ranking orders | Proposed operators | Ranking orders |
|---|---|---|---|
| $PFWA$ | $\aleph_2 \succ \aleph_4 \succ \aleph_5 \succ \aleph_1 \succ \aleph_3$ | $H-PFWAG$ | $\aleph_2 \succ \aleph_4 \succ \aleph_5 \succ \aleph_1 \succ \aleph_3$ |
| $PFWG$ | $\aleph_4 \succ \aleph_2 \succ \aleph_5 \succ \aleph_1 \succ \aleph_3$ | $H-PFOWAG$ | $\aleph_2 \succ \aleph_4 \succ \aleph_5 \succ \aleph_1 \succ \aleph_3$ |
| $PFOWA$ | $\aleph_2 \succ \aleph_5 \succ \aleph_4 \succ \aleph_1 \succ \aleph_3$ | | |
| $PFOWG$ | $\aleph_2 \succ \aleph_4 \succ \aleph_5 \succ \aleph_1 \succ \aleph_3$ | | |

**Step 1:** The grade values of each option are measured in the form of IVPFNs summarized in Table 2.6.

**Table 2.6** Picture fuzzy decision matrix.

| | $C_1$ | $C_2$ | $C_3$ |
|---|---|---|---|
| $\aleph_1$ | $\left([0.4,0.5],[0.1,0.2],[0.2,0.3]\right)$ | $\left([0.3,0.4],[0.1,0.2],[0.2,0.3]\right)$ | $\left([0.3,0.4],[0.2,0.3],[0.1,0.2]\right)$ |
| $\aleph_2$ | $\left([0.3,0.4],[0.1,0.2],[0.2,0.3]\right)$ | $\left([0.5,0.6],[0.1,0.2],[0.1,0.2]\right)$ | $\left([0.3,0.5],[0.1,0.2],[0.1,0.3]\right)$ |
| $\aleph_3$ | $\left([0.3,0.4],[0.1,0.2],[0.3,0.4]\right)$ | $\left([0.3,0.4],[0.1,0.2],[0.2,0.3]\right)$ | $\left([0.3,0.4],[0.1,0.2],[0.1,0.2]\right)$ |
| $\aleph_4$ | $\left([0.2,0.3],[0.3,0.4],[0.1,0.2]\right)$ | $\left([0.3,0.4],[0.1,0.2],[0.2,0.3]\right)$ | $\left([0.4,0.5],[0.1,0.2],[0.2,0.3]\right)$ |
| $\aleph_5$ | $\left([0.3,0.4],[0.2,0.3],[0.1,0.2]\right)$ | $\left([0.4,0.5],[0.1,0.2],[0.1,0.2]\right)$ | $\left([0.3,0.4],[0.1,0.2],[0.2,0.3]\right)$ |
| | $C_4$ | $C_5$ | |
| $\aleph_1$ | $\left([0.4,0.5],[0.1,0.2],[0.1,0.2]\right)$ | $\left([0.4,0.6],[0.1,0.2],[0.1,0.2]\right)$ | |
| $\aleph_2$ | $\left([0.3,0.4],[0.2,0.3],[0.1,0.2]\right)$ | $(([0.5,0.6],[0.1,0.2],[0.1,0.2]))$ | |
| $\aleph_3$ | $\left([0.4,0.5],[0.1,0.2],[0.2,0.3]\right)$ | $\left([0.3,0.4],[0.2,0.3],[0.1,0.2]\right)$ | |
| $\aleph_4$ | $\left([0.4,0.5],[0.1,0.2,],[0.1,0.2]\right)$ | $\left([0.4,0.5],[0.1,0.2],[0.1,0.2]\right)$ | |
| $\aleph_5$ | $\left([0.4,0.5],[0.1,0.2],[0.2,0.3]\right)$ | $\left([0.5,0.6],[0.1,0.2],[0.1,0.2]\right)$ | |

**Step 2:** The accumulated values of each alternative for the proposed operators H-IVPFWAG and H-IVPFOWAG are given in Table 2.7.

**Table 2.7** Aggregated values of IVPFNs operators.

| | H-IVPFWAG | H-IVPFOWAG |
|---|---|---|
| $\aleph_1$ | $\left([0.3592,0.4699],[0.1134,0.2142],[0.1415,0.2434]\right)$ | $\left([0.3592,0.4900],[0.1179,0.2190],[0.1272,0.2286]\right)$ |
| $\aleph_2$ | $\left([0.3681,0.4850],[0.1272,0.2286],[0.1179,0.2335]\right)$ | $(([0.3977,0.5199],[0.1179,0.2190],[0.1089,0.2286]))$ |
| $\aleph_3$ | $\left([0.3293,0.4298],[0.1089,0.2094],[0.1901,0.2927]\right)$ | $\left([0.3195,0.4199],[0.2172,0.2286],[0.1550,0.2576]\right)$ |
| $\aleph_4$ | $\left([0.3317,0.4335],[0.1344,0.2373],[0.1367,0.2384]\right)$ | $\left([0.3577,0.4590],[0.1170,0.2185],[0.1367,0.2384]\right)$ |
| $\aleph_5$ | $\left([0.3740,0.4749],[0.1179,0.2190],[0.1415,0.2434]\right)$ | $\left([0.3986,0.5000],[0.1089,0.2094],[0.1367,0.2384]\right)$ |

**Step 3:** The score values are provided in Table 2.8.

**Step 4:** The rankings of the alternatives are shown in Table 2.9 and $\aleph_2$ is the attractive alternative for the two proposed operators.

**Table 2.8** Score values of aggregated IVPFNs.

| Alternative | H-IVPFWAG | H-IVPFOWAG |
|---|---|---|
| $\aleph_1$ | 0.1166 | 0.1565 |
| $\aleph_2$ | 0.1459 | 0.2432 |
| $\aleph_3$ | −0.0420 | −0.1190 |
| $\aleph_4$ | 0.0224 | 0.1061 |
| $\aleph_5$ | 0.1271 | 0.2052 |

**Table 2.9** Ranking order of the alternatives under IVPFNs.

| Proposed operators | Ranking orders |
|---|---|
| $H-IVPFWAG$ | $\aleph_2 \succ \aleph_5 \succ \aleph_1 \succ \aleph_4 \succ \aleph_3$ |
| $H-IVPFOWAG$ | $\aleph_2 \succ \aleph_5 \succ \aleph_1 \succ \aleph_4 \succ \aleph_3$ |

**Table 2.10** Effect of the parameter $\gamma$ on the H-PFWAG operator.

| $\gamma$ | $\Lambda(\Upsilon_1)$ | $\Lambda(\Upsilon_2)$ | $\Lambda(\Upsilon_3)$ | $\Lambda(\Upsilon_4)$ | $\Lambda(\Upsilon_5)$ | Ranking order |
|---|---|---|---|---|---|---|
| $\gamma=0$ | 0.5943 | 0.6191 | 0.5600 | 0.6083 | 0.5867 | $\aleph_2 \succ \aleph_4 \succ \aleph_1 \succ \aleph_5 \succ \aleph_3$ |
| $\gamma=0.2$ | 0.5959 | 0.6217 | 0.5619 | 0.6115 | 0.6135 | $\aleph_2 \succ \aleph_5 \succ \aleph_4 \succ \aleph_1 \succ \aleph_3$ |
| $\gamma=0.5$ | 0.5984 | 0.6257 | 0.5647 | 0.6165 | 0.6163 | $\aleph_2 \succ \aleph_4 \succ \aleph_5 \succ \aleph_1 \succ \aleph_3$ |
| $\gamma=0.7$ | 0.6001 | 0.6283 | 0.5666 | 0.6199 | 0.6181 | $\aleph_2 \succ \aleph_4 \succ \aleph_5 \succ \aleph_1 \succ \aleph_3$ |
| $\gamma=0.9$ | 0.6018 | 0.6310 | 0.5685 | 0.6233 | 0.6199 | $\aleph_2 \succ \aleph_4 \succ \aleph_5 \succ \aleph_1 \succ \aleph_3$ |
| $\gamma=1$ | 0.6026 | 0.6332 | 0.5695 | 0.6250 | 0.6208 | $\aleph_2 \succ \aleph_4 \succ \aleph_5 \succ \aleph_1 \succ \aleph_3$ |

**Table 2.11** Effect of the parameter $\gamma$ on the H-PFOWAG operator.

| $\gamma$ | $\Lambda(\Upsilon_1)$ | $\Lambda(\Upsilon_2)$ | $\Lambda(\Upsilon_3)$ | $\Lambda(\Upsilon_4)$ | $\Lambda(\Upsilon_5)$ | Ranking order |
|---|---|---|---|---|---|---|
| $\gamma=0$ | 0.6149 | 0.6395 | 0.5731 | 0.6365 | 0.6255 | $\aleph_2 \succ \aleph_4 \succ \aleph_5 \succ \aleph_1 \succ \aleph_3$ |
| $\gamma=0.2$ | 0.6173 | 0.6419 | 0.5747 | 0.6396 | 0.6277 | $\aleph_2 \succ \aleph_4 \succ \aleph_5 \succ \aleph_1 \succ \aleph_3$ |
| $\gamma=0.5$ | 0.5960 | 0.6455 | 0.5774 | 0.6441 | 0.6310 | $\aleph_2 \succ \aleph_4 \succ \aleph_5 \succ \aleph_1 \succ \aleph_3$ |
| $\gamma=0.7$ | 0.6235 | 0.6480 | 0.5791 | 0.6472 | 0.6332 | $\aleph_2 \succ \aleph_4 \succ \aleph_5 \succ \aleph_1 \succ \aleph_3$ |
| $\gamma=0.9$ | 0.6260 | 0.6504 | 0.5809 | 0.6502 | 0.6355 | $\aleph_2 \succ \aleph_4 \succ \aleph_5 \succ \aleph_1 \succ \aleph_3$ |
| $\gamma=1$ | 0.6272 | 0.6517 | 0.5817 | 0.6519 | 0.6365 | $\aleph_2 \succ \aleph_4 \succ \aleph_5 \succ \aleph_1 \succ \aleph_3$ |

## 2.6 Sensitivity analysis for the parameter $\gamma$

Here, the values of $\gamma$ vary from 0 to 1. The overall score calculated by using the operators H-PFWAG, H-PFOWAG, H-IVPFWAG, and H-IVPFOWAG is summarized, respectively, in Tables 2.10, 2.11, 2.12, and 2.13. Tables 2.10, 2.11, 2.12, and 2.13, indicate that with the increasing value of $\gamma$, the score values of the proposed operator are increased. The values of $\gamma$ impact decision makers as the proposed method is flexible as the DMs can choose values of $\gamma$ as per their preferences in the decision-making process. In that direction, if the DMs can decide as to the pessimistic nature; in that view, they choose a smaller value of $\gamma$ to accumulate to decrease the scores values. Contrary to this, when a

**Table 2.12** Effect of the parameter $\gamma$ on the H-IVPFWAG operator.

| $\gamma$ | $\Lambda(\Upsilon_1)$ | $\Lambda(\Upsilon_2)$ | $\Lambda(\Upsilon_3)$ | $\Lambda(\Upsilon_4)$ | $\Lambda(\Upsilon_5)$ | Ranking order |
|---|---|---|---|---|---|---|
| $\gamma=0$ | 0.5487 | 0.5578 | 0.4688 | 0.4896 | 0.5527 | $\aleph_2 \succ \aleph_5 \succ \aleph_1 \succ \aleph_4 \succ \aleph_3$ |
| $\gamma=0.2$ | 0.5526 | 0.5638 | 0.4729 | 0.4975 | 0.5572 | $\aleph_2 \succ \aleph_5 \succ \aleph_1 \succ \aleph_4 \succ \aleph_3$ |
| $\gamma=0.5$ | 0.5583 | 0.5730 | 0.4813 | 0.5092 | 0.5636 | $\aleph_2 \succ \aleph_5 \succ \aleph_1 \succ \aleph_4 \succ \aleph_3$ |
| $\gamma=0.7$ | 0.5621 | 0.5791 | 0.4831 | 0.5172 | 0.5678 | $\aleph_2 \succ \aleph_5 \succ \aleph_1 \succ \aleph_4 \succ_3$ |
| $\gamma=0.9$ | 0.5660 | 0.5853 | 0.4873 | 0.5256 | 0.5721 | $\aleph_2 \succ \aleph_5 \succ \aleph_1 \succ \aleph_4 \succ \aleph_3$ |
| $\gamma=1$ | 0.5679 | 0.5884 | 0.4893 | 0.5292 | 0.5743 | $\aleph_2 \succ \aleph_5 \succ \aleph_1 \succ \aleph_4 \succ \aleph_3$ |

**Table 2.13** Effect of the parameter $\gamma$ on the H-IVPFOWAG operator.

| $\gamma$ | $\Lambda(\Upsilon_1)$ | $\Lambda(\Upsilon_2)$ | $\Lambda(\Upsilon_3)$ | $\Lambda(\Upsilon_4)$ | $\Lambda(\Upsilon_5)$ | Ranking order |
|---|---|---|---|---|---|---|
| $\gamma=0$ | 0.5674 | 0.6078 | 0.4739 | 0.5394 | 0.5159 | $\aleph_2 \succ \aleph_1 \succ \aleph_4 \succ \aleph_5 \succ \aleph_3$ |
| $\gamma=0.2$ | 0.5718 | 0.6133 | 0.4785 | 0.5449 | 0.5204 | $\aleph_2 \succ \aleph_1 \succ \aleph_4 \succ \aleph_5 \succ \aleph_3$ |
| $\gamma=0.5$ | 0.5783 | 0.6216 | 0.4405 | 0.5531 | 0.6026 | $\aleph_2 \succ \aleph_5 \succ \aleph_4 \succ \aleph_1 \succ \aleph_3$ |
| $\gamma=0.7$ | 0.5825 | 0.6272 | 0.4902 | 0.5886 | 0.5316 | $\aleph_2 \succ \aleph_1 \succ \aleph_4 \succ \aleph_5 \succ \aleph_3$ |
| $\gamma=0.9$ | 0.5869 | 0.6328 | 0.4949 | 0.5641 | 0.5361 | $\aleph_2 \succ \aleph_1 \succ \aleph_4 \succ \aleph_5 \succ \aleph_3$ |
| $\gamma=1$ | 0.5890 | 0.6356 | 0.4972 | 0.5669 | 0.5384 | $\aleph_2 \succ \aleph_1 \succ \aleph_4 \succ \aleph_5 \succ \aleph_3$ |

higher value of $\gamma$ is chosen, the score will increase, providing an optimistic decision to the DMs. Although, with the variation of $\gamma$, the best option is the same, which influences whether outcomes are objective and cannot be changed. Therefore in this method, the ranking results are stable. Additionally, it is observed that when $\gamma = 0$, the proposed operator H-PFWAG/H-IVPFWAG reduces to PFWG/IVPFWG and H-PFOWAG/H-IVPFOWAG reduces to PFOWG/IVPFOWG. Also, when $\gamma = 1$, then H-PFWAG/H-IVPFWAG reduces to PFWA/IVPFWA and H-PFOWAG/ H-IVPFOWAG reduces to PFOWA/IVPFOWA.

## 2.7 Conclusion

The main objective of this proposed approach is to introduce some new picture fuzzy hybrid aggregation operators, which aggregate the averaging and geometric PFS information. Although it has been studied, the aggregated values obtained by averaging or geometric operators either tend towards the most powerful arguments or towards the greatest weight value. Hence, the proposed operators do not give unbiased aggregated values. On this restriction, some new picture fuzzy hybrid weighted average and geometric aggregation operators are proposed under the PFS, as well as an IVPFS environment, which provides some moderate values. Furthermore, based on the standardized decision matrix and the proposed operators, two MCDM methods have been demonstrated based on the PFS as well as IVPFS arguments. A numerical example has been illustrated using the proposed operators and their results are compared with some existing operators. This approach can be applied to other fuzzy uncertain environment.

## References


[1] K. Atanassov, Intuitionistic Fuzzy Sets: Theory and Applications, Studies in Fuzziness and Soft Computing, vol. 35, Physica-Verlag, Heidelberg, 1999.

[2] K.T. Atanassov, G. Gargov, Interval-valued intuitionistic fuzzy sets, Fuzzy Sets Syst. 31 (1989) 343–349.

[3] B.C. Cuong, Picture fuzzy sets -first results. Part 1, Seminar "Neuro-fuzzy systems with applications", Tech. Rep., Institute of Mathematics, Hanoi, 2013.

[4] B.C. Cuong, Picture fuzzy sets -first results. Part 2, Seminar "Neuro-fuzzy systems with applications", Tech. Rep., Institute of Mathematics, Hanoi, 2013.

[5] B.C. Cuong, Picture fuzzy sets, J. Comput. Sci. Cybern. 30 (2014) 409–420.

[6] B.C. Cuong, P.V. Hai, Some fuzzy logic operators for picture fuzzy sets, in: Seventh International Conference on Knowledge and Systems Engineering, 2015, pp. 132–137.

[7] B.C. Cuong, V. Kreinovich, R.T. Ngan, A classification of representable $t$-norm operators for picture fuzzy sets, Departmental Technical Reports (CS), Paper 1047, 2016.

[8] G. Deschrijver, C. Cornelis, E.E. Kerre, On the representation of intuitionistic fuzzy $t$-norms and $t$-conorms, IEEE Trans. Fuzzy Syst. 12 (2004) 45–61.

[9] C. Jana, T. Senapati, M. Pal, R.R. Yager, Picture fuzzy Dombi aggregation operators: application to MADM process, Appl. Soft Comput. 74 (1) (2019) 99–109.

[10] C. Jana, M. Pal, J.Q. Wang, Bipolar fuzzy Dombi aggregation operators and its application in multiple attribute decision making process, J. Ambient Intell. Humaniz. Comput. 10 (9) (2019) 3533–3549.

[11] S. Khan, S. Abdullah, L. Abduhhah, S. Asrhaf, Logarithmic aggregation operators of picture fuzzy numbers for multi-attribute decision making problems, Mathematics 7 (2019) 608, https://doi.org/10.3390/math7070608.

[12] A.M. Khalil, S.G. Li, H. Garg, H. Li, S. Ma, New operations on interval-valued picture fuzzy set, interval-valued picture fuzzy soft set and their applications, IEEE Access 7 (2019) 2169–3536.

[13] X. Peng, J. Dai, Algorithm for picture fuzzy multiple attribute decision making based on new distance measure, Int. J. Uncertain. Quantificat. 7 (2017) 177–187.

[14] P.H. Phong, D.T. Hieu, R.T.H. Ngan, P.T. Them, Some compositions of picture fuzzy relations, in: Proceedings of the 7th National Conference on Fundamental and Applied Information Technology Research, FAIR7, Thai Nguyen, 2014, pp. 19–20.

[15] P.T.M. Phuong, P.H. Thong, L.H. Son, Theoretical analysis of picture fuzzy clustering, J. Comput. Sci. Cybern. 34 (1) (2018) 17–31.

[16] P. Singh, Correlation coefficients for picture fuzzy sets, J. Intell. Fuzzy Syst. 27 (2014) 2857–2868.

[17] L.H. Son, DPFCM: a novel distributed picture fuzzy clustering method on picture fuzzy sets, Expert Syst. Appl. 2 (2015) 51–66.

[18] L.H. Son, Generalized picture distance measure and applications to picture fuzzy clustering, Appl. Soft Comput. 46 (2016) 284–295.

[19] L.H. Son, Measuring analogousness in picture fuzzy sets: from picture distance measures to picture association measures, Fuzzy Optim. Decis. Mak. 16 (3) (2017) 1–20.

[20] L.H. Son, P. Viet, P. Hai, Picture inference system: a new fuzzy inference system on picture fuzzy set, Appl. Intell. 46 (3) (2017) 652–669.

[21] P.H. Thong, L.H. Son, Picture fuzzy clustering for complex data, Eng. Appl. Artif. Intell. 56 (2016) 121–130.

[22] P.H. Thong, L.H. Son, A novel automatic picture fuzzy clustering method based on particle swarm optimization and picture composite cardinality, Knowl.-Based Syst. 109 (2016) 48–60.

[23] P.H. Thong, L.H. Son, A new approach to multi-variables fuzzy forecasting using picture fuzzy clustering and picture fuzzy rules interpolation method, in: 6th International Conference on *Knowledge and Systems Engineering*, Hanoi, Vietnam, 2015, pp. 679–690.

[24] P.H. Thong, L.H. Son, H. Fujita, Interpolative picture fuzzy rules: a novel forecast method for weather nowcasting, in: Fuzzy Systems (FUZZ-IEEE), 2016 IEEE International Conference on IEEE, Vancouver, Canada, July 24-29, 2016, pp. 86–93.

[25] N.T. Thong, L.H. Son, HIFCF: an effective hybrid model between picture fuzzy clustering and intuitionistic fuzzy recommender systems for medical diagnosis, Expert Syst. Appl. 42 (7) (2015) 3682–3701.

[26] P.V. Viet, H.T.M. Chau, L.H. Son, P.V. Hai, Some extensions of membership graphs for picture inference systems, in: Knowledge and Systems Engineering (KSE), 2015 Seventh International Conference on, Ho Chi Minh City, Vietnam, IEEE, October 8-10, 2015, pp. 192–197.
[27] G.W. Wei, F.E. Alsaadi, T. Hayat, A. Alsaedi, Projection models for multiple attribute decision making with picture fuzzy information, Int. J. Mach. Learn. Cybern. 9 (4) (2018) 713–719.
[28] G.W. Wei, H. Gao, The generalized Dice similarity measures for picture fuzzy sets and their applications, Informatica 29 (1) (2018) 1–18.
[29] G.W. Wei, Some similarity measures for picture fuzzy sets and their applications, Iran. J. Fuzzy Syst. 15 (1) (2018) 77–89.
[30] G.W. Wei, Picture fuzzy cross-entropy for multiple attribute decision making problems, J. Bus. Econ. Manag. 17 (4) (2016) 491–502.
[31] G.W. Wei, Picture fuzzy aggregation operators and their application to multiple attribute decision making, J. Intell. Fuzzy Syst. 33 (2017) 713–724.
[32] G.W. Wei, Picture fuzzy Hamacher aggregation operators and their application to multiple attribute decision making, Fundam. Inform. 157 (3) (2018) 271–320.
[33] J. Ye, Intuitionistic fuzzy hybrid arithmetic and geometric aggregation operators for the decision-making of mechanical design schemes, Appl. Intell. 47 (2017) 743–751.
[34] L.A. Zadeh, Fuzzy sets, Inf. Control 8 (1965) 338–353.

# 3

# Multicriteria group decision-making process based on a picture fuzzy soft parameterized environment

## 3.1 Introduction

Decision-making techniques are essential in the field of management, engineering, economics, etc. Our modern science and technology were rapidly developed with the progressive development of multicriteria group decision making (MCGDM) and multiattribute group decision making (MAGDM). Hence, decision makers (DMs) are faced with different types of social and economic uncertainties. To overcome the complexity of decision science and management domains, DMs may consider their judgment in the form of a crisp number, not considering the fuzziness or vagueness in the problem's domain. Presently, DM can only furnish their rating on an exact level with proper handling of uncertainties in decision science. A proper choice of data analysis is required to minimize the uncertainty level. To determine this, researchers applied fuzzy set [56] theory, and observing its successful implementation. They have extended it to intuitionistic fuzzy sets (IFS) [5,25,37] and interval-valued intuitionistic fuzzy sets (IVIFS) [3] as to minimize the uncertainties. After seeing the successful applications of this extension, many authors have used these sets in the decision-making areas. For instance, some traditional problems in decision science [7,8,18–23,51–53] have been developed based on weighted averaging and geometric aggregation operators in IFS and IVIFS environments. Wang and Liu [46,47] developed these aggregation operators using Einstein norms. Garg [14] established a generalized form of aggregation operators in the same areas. Using the IFS argument, Garg et al. [15] utilized entropy-based problems with unknown weight. Many MCDM problems can be handled in an IFS environment as follows: using the WASPAS method [57], VIKOR [13], TOPSIS [30], and TODIM [32]. Although IFS has been favorably implemented in various areas of application, researchers observed that some cases cannot be explained similarly. For example, to interpret the case of voting, human beings explained their opinion as of types: yes, no, abstain, and refusal; they cannot be presented in IFS. To measure these situations, Cuoung [11,12] has introduced an advanced extension of FS, namely PFS. After seeing its auspicious applications to minimize uncertainties in real-world problems, researchers are drawing attention to its implementation in different uncertain environments. For instance, Wei [49] developed averaging and geometric operators using the PFS argument and utilized them to construct the MADM process. Then, Wei [48] studied crossentropy technique-based decision problems in PFS. Later, Son [41] used PFS arguments to associate distance measures. Son et al. [42] studied inference systems-based clustering in the

**Picture Fuzzy Logic and Its Applications in Decision Making Problems. https://doi.org/10.1016/B978-0-44-322024-1.00007-8**

same environment. Peng and Dai [38] established an algorithm-based picture fuzzy decision making using the distance measurement method. Wei [50] later discussed Hamacher aggregation functions to construct PFS decision theory. Recently, Jana et al. [17] studied a novel MADM problem using Dombi norm operators to aggregate PFS information. Khan et al. [27] developed a weighted aggregation-based decision-making method using Einstein norms. For more information on the search decision-making process to aggregate picture fuzzy numbers using various aggregation operators we refer the readers to the following references [16,31,40,43–45].

Sine methods in the environment of IFS and IVIFS are constructed but have certain limitations of parameterizations. Hence, they cannot be considered as an effective tool to solve real-world phenomena. To cross the barrier, soft set theory [36] plays an eminent role of observation and has handled successfully typical factors. In that view, many researchers have shown interest in this topic [1,35]. Maji et al. [33,34] has combined a bridge connection between soft set (SS) and FS theory and built a new concept, namely, FSS and IFSS. The extended concepts of these theories are properly described the real-world issues with the help of their parameterizations. Hybrid models combined with soft set theory have drawn a great deal of attention in many disciplines: generalized IFSS [14], IVIFSS [24,54], fuzzy soft expert sets [2], and hesitant fuzzy soft sets [6]. Cagman et al. [9,10] studied parameterized factors based on soft set model and fuzzy soft expert sets [2]. Selvachandran and Peng [39] used soft theory in the TPOSIS model. Arora and Garg [4] aggregated soft aggregation information using parameterized keys in the IFSS environment. At that time, a new mathematical tool was constructed with a combination of a picture fuzzy set and a soft set, which can describe real-life phenomena containing parameterized factors. Khalil et al. [26] defined new operations of IVPFS and then constructed IVPFSS and its application. Khan et al. [29] studied GPFSS and then applied it in decision-support systems. Yang et al. [54] defined PFSS and established some relevant properties of them, and an adjustable soft discernibility approach based decision-making process. However, the existing papers in PFSS environment approaches are well under the restriction of parameters, and decision makers are at the same priority level. However, these results need to be relaxed for better decision analysis. Further, to date, research on PFSS and PFSS is only limited to their basic algebraic operations and applications. However, research must be done on the aggregation of information in soft environments. Hence, these ideas have a new interest for researchers in future development. Thus by considering that PFSS has powerful applications in decision science, this paper presented a series of aggregating operators: picture fuzzy soft-weighted average operators and soft-weighted geometric operators, labeled as PFSWA and PFSWG operators. See Table 3.1.

## 3.2 Basic concept of PFSS and PFSN

Let $E$, $U$, and $P(U)$ be used as a set of parameters, universe, and power set of $U$, respectively. Also, $A$ is a subset of $E$.

**Table 3.1** Contribution towards picture fuzzy numbers.

| Authors | Aggregation information | Parameter information |
|---|---|---|
| Harg [16] | Algebraic norms | No |
| Jana et al. [17] | Dombi norms | No |
| Khalil et al. [26] | No | Interval-valued picture fuzzy soft results |
| Khan et al. [27] | Einstein norms | No |
| Khalid et al. [29] | No | Generalized picture fuzzy soft results |
| Liu and Zhan [31] | Linguistic type | No |
| Peng and Dai [38] | Distance measures | No |
| Sin [40] | Correlation measures | No |
| Son [41] | Distance measures | No |
| Son, Viet and Hai [42] | Inference systems | No |
| Son [43] | Picture fuzzy clustering | No |
| Thong and Son [44] | Clustering complex data | No |
| Thong and Son [45] | Particle swarm optimization | No |
| Wei [48] | Crossentropy | No |
| Wei [49] | Algebraic norms | No |
| Wei [50] | Hamacher norms | No |
| Yang, Liang, Ji and Liu [55] | No | Adjustable soft discernibility approach |
| Proposed method | Yes | Yes |

Let $(F, E)$ be a soft set over $U$, where $F$ is a function $F : E \to P(U)$; otherwise, a parameterized family of subsets over $U$. For $\varepsilon \in A, \mathcal{F}(\varepsilon)$ used for the set of approximate element $\varepsilon$ of $(\mathcal{F}, A)$, or $\varepsilon$ as the approximate elements of $(F, E)$.

**Definition 3.1.** [33] Let the power set be $P(U)$ for $U$ and $A \subseteq E$, and the set of all fuzzy subsets be $P(U)$ over $U$, then $(F, A)$ is called a fuzzy soft set, having presentation as $F : A \to P(U)$.

**Example 3.1.** Let a set of four mobiles be $U = \{mob_1, mob_2, mob_3, mob_4\}$ and a parameter set be $E = \{costly(e_1),\ beautiful(e_2),\ battery\ backup(e_3)\ and\ apps(e_4)\}$, then FSS for describing the "attractiveness of the mobiles" is $(\mathcal{F}, A) = \{\mathcal{F}_{e_1}, \mathcal{F}_{e_2}, \mathcal{F}_{e_3}\}$, where $A = \{e_1, e_2, e_3\} \subseteq E$ and $(F, A)$ can be presented as:

$$\mathcal{F}_{e_1} = \{(mob_1, 0.4), (mob_2, 0.6), (mob_3, 0.5), (mob_4, 0.4)\};$$
$$\mathcal{F}_{e_2} = \{(mob_1, 0.6), (mob_2, 0.7), (mob_3, 0.4), (mob_4, 0.5)\};\ and$$
$$\mathcal{F}_{e_3} = \{(mob_1, 0.8), (mob_2, 0.6), (mob_3, 0.6), (mob_4, 0.7)\}.$$

**Definition 3.2.** Let the set of universe and parameters be $U$ and $E$, respectively. For $P \subset E$, let $\mathcal{P}(U)$ be called the subsets of PFSs over $U$. The term $(F_P, \mathcal{R})$ is called a picture fuzzy soft set (PFSS) of $U$, where $F_P$ follows as, $F_P : \mathcal{R} \to P(U)$.

**Example 3.2.** Let the set of four mobiles be $U = \{mob_1, mob_2, mob_3, mob_4\}$ and the set of parameters be $E = \{costly(e_1),\ beautiful(e_2),\ batterybackup(e_3)\ and\ apps(e_4)\}$ under

PFSS for describing the "attractiveness of the mobiles" is $(P, A) = \{\mathcal{F}_{e_1}, \mathcal{F}_{e_2}, \mathcal{F}_{e_3}\}$, where $A = \{e_1, e_2, e_3\} \subseteq E$ and $(P, A)$ can be considered as:

$$P_{e_1} = \{(mob_1, 0.67, 0.14, 0.20), (mob_2, 0.45, 0.15, 0.21), (mob_3, 0.56, 0.12, 0.23), (mob_4, 0.30, 0.16, 0.11)\};$$

$$P_{e_2} = \{(mob_1, 0.70, 0.10, 0.20), (mob_2, 0.60, 0.13, 0.19), (mob_3, 0.50, 0.13, 0.39), (mob_4, 0.49, 0.04, 0.11)\}; \text{ and}$$

$$P_{e_3} = \{(mob_1, 0.90, 0.01, 0.03), (mob_2, 0.55, 0.09, 0.15), (mob_3, 0.39, 0.05, 0.18), (mob_4, 0.60, 0.14, 0.24)\}.$$

For convenience, the pair $\mathcal{R}_{e_f}(x) = \{(\mathcal{Y}(x), \mathcal{A}(x), \mathcal{N}(x)) | x \in U\}$, i.e., for selecting the best alternatives, ranking of alternatives has been done based on the score value. For this, a score function of $\mathcal{R}_{e_{gf}}$ is provided as

$$\Lambda(\mathcal{R}_{e_{gf}}) = \mathcal{Y}_{gf} - \mathcal{N}_{gf}, \tag{3.1}$$

where, $\Lambda(\mathcal{R}_{e_{gf}}) \in [0, 1]$. It is understood that the larger the $\Lambda(\mathcal{R}_{e_{gf}})$, the larger is PFSN $\mathcal{R}_{e_{gf}}$ by the definition.

**Example 3.3.** Let $\mathcal{R}_{e_{11}} = (0.60, 0.20, 0.20)$ and $\mathcal{R}_{e_{12}} = (0.30, 0.15, 0.25)$ be two PFSNs, then by Eq. (3.1), we obtain $\Lambda(\mathcal{R}_{e_{11}}) = 0.40$ and $\Lambda(p_{e_{12}}) = 0.5$. As $\Lambda(\mathcal{R}_{e_{11}}) > \Lambda(\mathcal{R}_{e_{12}})$, this implies $\mathcal{R}_{e_{11}} > \mathcal{R}_{e_{12}}$.

For some instances, the above rule cannot be applied to compare PFSNs. For example, let $\mathcal{R}_{e_{11}} = (0.60, 0.20, 0.20)$ and $p_{e_{12}} = (0.50, 0.09, 0.10)$, then it is not easy to determine which of them is larger as $\Lambda(\mathcal{R}_{e_{11}}) = \Lambda(P_{e_{12}})$. Hence, we define the accuracy function of $\mathcal{R}_{e_{gf}}$ as follows:

$$\Phi(\mathcal{R}_{e_{gf}}) = \mathcal{Y}_{gf} + \mathcal{A}_{gf} + \mathcal{N}_{gf}, \tag{3.2}$$

where, $\Phi(\mathcal{R}_{e_{gf}}) \in [0, 1]$. Based on the score $\Lambda$ and accuracy $\Phi$ functions, then the order relation between two PFSNs $\mathcal{R}_{e_{gf}}$ and $\mathcal{Q}_{e_{gf}}$ is defined as:

**(i)** If $\Lambda(\mathcal{R}_{e_{gf}}) < \Lambda(\mathcal{Q}_{e_{gf}})$, imply $\mathcal{R}_{e_{gf}} \prec \mathcal{Q}_{e_{gf}}$;
**(ii)** If $\Lambda(\mathcal{R}_{e_{gf}}) > \Lambda(\mathcal{Q}_{e_{gf}})$, imply $\mathcal{R}_{e_{gf}} \succ \mathcal{Q}_{e_{gf}}$;
**(iii)** If $\Lambda(\mathcal{R}_{e_{gf}}) = \Lambda(\mathcal{Q}_{e_{gf}})$, imply
- **(1)** If $\Phi(\mathcal{R}_{e_{gf}}) < \Phi(\mathcal{Q}_{e_{gf}})$, imply $\mathcal{R}_{e_{gf}} \prec \mathcal{Q}_{e_{gf}}$;
- **(2)** If $\Phi(\mathcal{R}_{e_{gf}}) > \Phi(\mathcal{Q}_{e_{gf}})$, imply $\mathcal{R}_{e_{gf}} \succ \mathcal{Q}_{e_{gf}}$;
- **(3)** If $\Phi(\mathcal{R}_{e_{gf}}) = \Phi(\mathcal{Q}_{e_{gf}})$, imply $\mathcal{R}_{e_{gf}} \sim \mathcal{Q}_{e_{gf}}$.

## 3.3 Picture fuzzy soft weighted average operators

In this section, new aggregation functions on PFS, namely PFSWA and PFWGA operators, are defined and their related properties studied.

### 3.3.1 Operations for PFSNs

**Definition 3.3.** Let $\mathcal{R}_e = (\mathcal{Y}, \mathcal{A}, \mathcal{N})$, $\mathcal{R}_{e_{11}} = (\mathcal{Y}_{11}, \mathcal{A}_{11}, \mathcal{N}_{11}))$, and $\mathcal{R}_{e_{12}} = (\mathcal{Y}_{12}, \mathcal{A}_{12}, \mathcal{N}_{12}))$ be three PFSNs over $X$, defined by some operation in this regard:

**(i)** $\mathcal{R}_{e_{11}} \oplus \mathcal{R}_{e_{12}} = \left((\mathcal{Y}_{11} + \mathcal{Y}_{12} - \mathcal{Y}_{11}\mathcal{Y}_{12}, \mathcal{A}_{11}\mathcal{A}_{12}, \mathcal{N}_{11}\mathcal{N}_{12})\right)$;
**(ii)** $\mathcal{R}_{e_{11}} \otimes \mathcal{R}_{e_{12}} = \left((\mathcal{Y}_{11}\mathcal{Y}_{12}, \mathcal{A}_{11} + \mathcal{A}_{12} - \mathcal{A}_{11}\mathcal{A}_{12}, \mathcal{N}_{11} + \mathcal{N}_{12} - \mathcal{N}_{11}\mathcal{N}_{12})\right)$;
**(iii)** $\lambda\mathcal{R}_e = \left(1 - (1 - \mathcal{Y})^\lambda, \mathcal{A}^\lambda, \mathcal{N}^\lambda\right)$;
**(iv)** $\mathcal{R}_e^\lambda = \left(\mathcal{Y}^\lambda, 1 - (1 - \mathcal{A})^\lambda, 1 - (1 - \mathcal{N})^\lambda\right)$.

**Definition 3.4.** Let $\mathcal{R}_{e_{gf}} = (\mathcal{Y}_{gf}, \mathcal{A}_{gf}, \mathcal{N}_{gf})$ $(g = 1, 2, \ldots, \xi;\ f = 1, 2, \ldots, \varpi)$ be a set of PFSNs and $\Xi_g, \phi_f$ are the weight vectors of the experts $E_f$s and parameter $e_g$s, respectively, where $\Xi_g \in [0, 1]$, $\phi_f \in [0, 1]$ satisfying the conditions $\sum_{g=1}^{\xi} \Xi_g = 1$ and $\sum_{f=1}^{\varpi} \phi_f = 1$. The picture fuzzy soft weighted average (PFSWA) operator is the mapping $PFSWA : \Theta^\varpi \to \Theta$ such that

$$PFSWA(\mathcal{R}_{e_{11}}, \mathcal{R}_{e_{12}}, \ldots, \mathcal{R}_{e_{gf}}) = \bigoplus_{f=1}^{\varpi} \phi_f \left(\bigoplus_{g=1}^{\xi} \Xi_g \mathcal{R}_{e_{gf}}\right). \tag{3.3}$$

Based on this operation on PFSNs, the following theorem is established.

**Theorem 3.1.** *Let* $\mathcal{R}_{e_{gf}} = (\mathcal{Y}_{gf}, \mathcal{A}_{gf}, \mathcal{N}_{gf})$ $(g = 1, 2, \ldots, \xi;\ f = 1, 2, \ldots, \varpi)$ *be a set of PFSNs, then the aggregated value of PFSNs using the PFSWA operator is again PFSN, and further*

$$\begin{aligned} &PFSWA(\mathcal{R}_{e_{11}}, \mathcal{R}_{e_{12}}, \ldots, \mathcal{R}_{e_{gf}}) \\ &= \left(1 - \prod_{f=1}^{\varpi}\left(\prod_{g=1}^{\xi}(1 - \mathcal{Y}_{gf})^{\Xi_g}\right)^{\phi_f}, \prod_{f=1}^{\varpi}\left(\prod_{g=1}^{\xi}\left(\mathcal{A}_{gf}\right)^{\Xi_g}\right)^{\phi_f}, \prod_{f=1}^{\varpi}\left(\prod_{g=1}^{\xi}\left(\mathcal{N}_{gf}\right)^{\Xi_g}\right)^{\phi_f}\right). \end{aligned} \tag{3.4}$$

*Proof.* The proof of this theorem can be obtained by mathematical induction.

For $g = 1$, we obtain $\Xi_1 = 1$. Then, from Definition 3.4, we have

$$\begin{aligned} PFSWA(\mathcal{R}_{e_{11}}, \mathcal{R}_{e_{12}}, \ldots, \mathcal{R}_{e_{gf}}) &= \bigoplus_{f=1}^{\zeta} \varpi_f\left(\mathcal{R}_{e_{1f}}\right) \\ &= \left(1 - \prod_{f=1}^{\varpi}\left(1 - \mathcal{Y}_{1f}\right)^{\phi_f}, \prod_{g=1}^{\xi}(\mathcal{A}_{1f})^{\phi_f}, \prod_{f=1}^{\varpi}(\mathcal{N}_{1f})^{\phi_f}\right) \\ &= \left(1 - \prod_{f=1}^{\varpi}\left(\prod_{g=1}^{1}(1 - \mathcal{Y}_{gf})^{\Xi_g}\right)^{\phi_f}, \prod_{f=1}^{\varpi}\left(\prod_{g=1}^{1}(\mathcal{A}_{gf})^{\Xi_g}\right)^{\phi_f}, \prod_{f=1}^{\varpi}\left(\prod_{\beta=1}^{1}(\mathcal{N}_{gf})^{\Xi_g}\right)^{\phi_f}\right). \end{aligned}$$

Again, for $f = 1$ and $\phi_1 = 1$,

$$PFWA(\mathcal{R}_{e_{11}}, \mathcal{R}_{e_{12}}, \ldots, \mathcal{R}_{e_{gf}}) = \left(\bigoplus_{g=1}^{\xi} \mathcal{A}_{g1} p_{e_{1f}}\right)$$

$$= \left(1 - \prod_{f=1}^{1}\left(1 - \mathcal{Y}_{g1}\right)^{\Xi_g}, \prod_{g=1}^{\xi}(\mathcal{A}_{g1})^{\Xi_g}, \prod_{g=1}^{\xi}(\mathcal{N}_{a1})^{\Xi_g}\right)$$

$$= \left(1 - \prod_{f=1}^{1}\left(\prod_{g=1}^{\xi}(1 - \mathcal{Y}_{gf})^{\Xi_g}\right)^{\phi_f}, \prod_{f=1}^{1}\left(\prod_{g=1}^{\xi}(\mathcal{A}_{gf})^{\Xi_g}\right)^{\phi_f}, \prod_{f=1}^{1}\left(\prod_{g=1}^{\xi}(\mathcal{N}_{gf})^{\Xi_g}\right)^{\phi_f}\right).$$

Thus Eq. (3.4) is true for $g = 1$ and $f = 1$. Suppose Eq. (3.4) holds for $f = d_1 + 1$, $g = d_2$ and $f = d_1$, $g = d_2 + 1$, then

$$\bigoplus_{f=1}^{d_1+1} \varpi_f\left(\bigoplus_{g=1}^{d_2} \Xi_g \mathcal{R}_{e_{gf}}\right)$$

$$= \left(1 - \prod_{f=1}^{d_1+1}\left(\prod_{g=1}^{d_2}(1 - \mathcal{Y}_{gf})^{\Xi_g}\right)^{\phi_f}, \prod_{f=1}^{d_1+1}\left(\prod_{g=1}^{d_2}(\mathcal{A}_{gf})^{\Xi_g}\right)^{\phi_f}, \prod_{f=1}^{d_1+1}\left(\prod_{g=1}^{d_2}(\mathcal{N}_{gf})^{\Xi_g}\right)^{\phi_f}\right)$$

and

$$\bigoplus_{f=1}^{d_1} \varpi_f\left(\bigoplus_{g=1}^{d_2+1} \Xi_g \mathcal{R}_{e_{gf}}\right)$$

$$= \left(1 - \prod_{f=1}^{d_1}\left(\prod_{g=1}^{d_2+1}(1 - \mathcal{Y}_{gf})^{\Xi_g}\right)^{\phi_f}, \prod_{f=1}^{d_1}\left(\prod_{g=1}^{d_2+1}(\mathcal{A}_{gf})^{\Xi_g}\right)^{\phi_f}, \prod_{f=1}^{d_1}\left(\prod_{g=1}^{d_2+1}(\mathcal{N}_{gf})^{\Xi_g}\right)^{\phi_f}\right).$$

Now, for $f = d_1 + 1$ and $g = d_2 + 1$,

$$\bigoplus_{f=1}^{d_1+1} \phi_f\left(\bigoplus_{g=1}^{d_2+1} \Xi_g \mathcal{R}_{e_{gf}}\right) = \bigoplus_{f=1}^{d_1+1} \phi_f\left(\bigoplus_{g=1}^{d_2} \Xi_g \mathcal{R}_{e_{gf}} \oplus \Xi_{d_2+1}\mathcal{R}_{e_{(d_2+1)f}}\right)$$

$$= \bigoplus_{f=1}^{d_1+1}\bigoplus_{g=1}^{d_2} \phi_f \Xi_g \mathcal{R}_{e_{gf}} \bigoplus_{f=1}^{d_1+1} \phi_f \Xi_{d_2+1}\mathcal{R}_{e_{(d_2+1)f}}\Big)$$

$$= \left(1 - \prod_{f=1}^{d_1+1}\left(\prod_{g=1}^{d_2}(1 - \mathcal{Y}_{gf})^{\Xi_g}\right)^{\phi_f} \oplus 1 - \prod_{f=1}^{d_1+1}\left((1 - \mathcal{Y}_{(d_2+1)f})^{\Xi_{d_2+1}}\right)^{\phi_f},\right.$$
$$\prod_{f=1}^{d_1+1}\left(\prod_{g=1}^{d_2}(\mathcal{N}_{gf})^{\Xi_g}\right)^{\phi_f} \oplus \prod_{f=1}^{d_1+1}\left((\mathcal{A}_{(d_2+1)f})^{\Xi_{d_2+1}}\right)^{\phi_f},$$
$$\left.\prod_{f=1}^{d_1+1}\left(\prod_{g=1}^{d_2}(\mathcal{N}_{gf})^{\Xi_g}\right)^{\phi_f} \oplus \prod_{f=1}^{d_1+1}\left((\mathcal{N}_{(d_2+1)f})^{\Xi_{d_2+1}}\right)^{\phi_f}\right)$$

$$= \left(1 - \prod_{f=1}^{d_1+1}\left(\prod_{g=1}^{d_2+1}(1 - \mathcal{Y}_{gf})^{\Xi_g}\right)^{\phi_f}, \prod_{f=1}^{d_1+1}\left(\prod_{g=1}^{d_2+1}(\mathcal{N}_{gf})^{\Xi_g}\right)^{\phi_f}, \prod_{f=1}^{d_1+1}\left(\prod_{g=1}^{d_2+1}(\mathcal{N}_{gf})^{\Xi_g}\right)^{\phi_f}\right).$$

Thus Eq. (3.4) holds for $f = d_1 + 1, g = d_2 + 1$, by induction on $g, f \geq 1$.

Since, $0 \leq \mathcal{Y}_{gf} \leq 1 \Leftrightarrow 0 \leq \prod_{g=1}^{\xi}(1 - \mathcal{Y}_{gf})^{\Xi_g} \leq 1$ and thus $0 \leq 1 - \prod_{f=1}^{\varpi}(\prod_{g=1}^{\xi}(1 - \mathcal{Y}_{gf})^{\Xi_g})^{\phi_f} \leq 1$.

Also,

$$0 \leq \mathcal{A}_{gf} \leq 1 \Leftrightarrow 0 \leq \prod_{g=1}^{\xi}(\mathcal{A}_{gf})^{\Xi_g} \leq 1 \Leftrightarrow 0 \leq \prod_{f=1}^{\varpi}(\prod_{g=1}^{\xi}(\mathcal{A}_{gf})^{\Xi_g})^{\phi_f} \leq 1$$

$$\text{and } 0 \leq \mathcal{N}_{gf} \leq 1 \Leftrightarrow 0 \leq \prod_{g=1}^{\xi}(\mathcal{N}_{gf})^{\Xi_g} \leq 1 \Leftrightarrow 0 \leq \prod_{f=1}^{\varpi}(\prod_{g=1}^{\xi}(\mathcal{N}_{gf})^{\Xi_g})^{\phi_f} \leq 1.$$

Thus $0 \leq 1 - \prod_{f=1}^{\varpi}(\prod_{g=1}^{\xi}(1 - \mathcal{Y}_{gf})^{\Xi_g})^{\phi_f} + \prod_{f=1}^{\varpi}(\prod_{g=1}^{\xi}(\mathcal{A}_{gf})^{\Xi_g})^{\phi_f} + \prod_{f=1}^{\varpi}(\prod_{g=1}^{\xi}(\mathcal{N}_{gf})^{\Xi_g})^{\phi_f} \leq 1$.

Thus the PFSWA operator is a PFSN. □

**Corollary 3.1.** *For only one parameter $e_1$, i.e., $f = 1$, the PFSWA operator reduces to the PFWA operator:*

$$PFSWA(\mathcal{R}_{e_{11}}, \mathcal{R}_{e_{21}}, \ldots, \mathcal{R}_{e_{g1}}) = \left(1 - \prod_{g=1}^{\xi}(1 - \mathcal{Y}_g)^{\Xi_g}, \prod_{g=1}^{\xi}(\mathcal{A}_g)^{\Xi_g}, \prod_{g=1}^{\xi}(\mathcal{N}_g)^{\Xi_g}\right). \tag{3.5}$$

The PFSWA operator adopts the following properties.

**Theorem 3.2.** *(**Idempotency**) Let $\mathcal{R}_{e_{gf}} = (\mathcal{Y}_{gf}, \mathcal{A}_{gf}, \mathcal{N}_{gf})$ $(g = 1, 2, \ldots, \xi; f = 1, 2, \ldots, \varpi)$ be a set of PFSNs and $\mathcal{R}_{e_{gf}} = \mathcal{R}_e$ for all $g, f$, then*

$$PFSWA(\mathcal{R}_{e_{11}}, \mathcal{R}_{e_{12}}, \ldots, \mathcal{R}_{e_{gf}}) = \mathcal{R}_e. \tag{3.6}$$

*Proof.* Since $\mathcal{R}_{e_{gf}} = \mathcal{R}_e = (\mathcal{Y}, \mathcal{A}, \mathcal{N})$ for all $g, f$. Then,

$$\begin{aligned}
&PFSWA(\mathcal{R}_{e_{11}}, \mathcal{R}_{e_{12}}, \ldots, \mathcal{R}_{e_{gf}}) \\
&= \left(1 - \prod_{f=1}^{\varpi}\left(\prod_{g=1}^{\xi}(1 - \mathcal{Y})^{\Xi_g}\right)^{\phi_f}, \prod_{f=1}^{\varpi}\left(\prod_{g=1}^{\xi}(\mathcal{A})^{\Xi_g}\right)^{\phi_f}, \prod_{f=1}^{\varpi}\left(\prod_{g=1}^{\xi}(\mathcal{N})^{\Xi_g}\right)^{\phi_f}\right) \\
&= \left(1 - \left((1 - \mathcal{Y})^{\sum_{g=1}^{\xi}\Xi_g}\right)^{\sum_{f=1}^{\varpi}\phi_f}, \left((\mathcal{A})^{\sum_{g=1}^{\xi}\Xi_g}\right)^{\sum_{f=1}^{\varpi}\phi_f} \left((\mathcal{N})^{\sum_{g=1}^{\xi}\Xi_g}\right)^{\sum_{f=1}^{\varpi}\phi_f}\right) \\
&= \left(1 - (1 - \mathcal{Y}), \mathcal{A}, \mathcal{N}\right) \\
&= \left(\mathcal{Y}, \mathcal{A}, \mathcal{N}\right).
\end{aligned}$$

Hence, the proof is completed. □

**Theorem 3.3.** ***(Boundedness)*** *Let* $\mathcal{R}_{e_{gf}} = (\mathcal{Y}_{gf}, \mathcal{A}_{gf}, \mathcal{N}_{gf})$, $g = 1, 2, \ldots, \xi$; $f = 1, 2, \ldots, \varpi$ *be a set of PFSNs. Let* $\mathcal{N}_{e_{gf}} = (\min_f \min_g\{\mathcal{Y}_{gf}\}, \max_f \max_g\{\mathcal{A}_{gf}\}, \max_f \max_g\{\mathcal{N}_{gf}\})$ *and* $\mathcal{Y}_{gf} = (\max_f \max_g\{\mathcal{Y}_{gf}\}, \mathcal{A}_{gf} = \min_f \min_g\{\mathcal{A}_{gf}\}, \mathcal{N}_{gf} = \min_f \min_g\{\mathcal{N}_{gf}\})$.

*Proof.* Since $\mathcal{R}_{e_{gf}} = (\mathcal{Y}_{gf}, \mathcal{A}_{gf}, \mathcal{N}_{gf})$ is a PFSN and $\min_f \min_g\{\mathcal{Y}_{gf}\} \leq \mathcal{Y}_{gf} \leq \max_f \max_g\{\mathcal{Y}_{gf}\}$, this introduces

$$1 - \max_f \max_g\{\mathcal{Y}_{gf}\} \leq 1 - \mathcal{Y}_{gf} \leq 1 - \min_f \min_g\{\mathcal{Y}_{gf}\}$$

$$\Leftrightarrow (1 - \max_f \max_g\{\mathcal{Y}_{gf}\})^{\Xi_g} \leq (1 - \mathcal{Y}_{gf}) \leq (1 - \min_f \min_g\{\mathcal{Y}_{gf}\})^{\Xi_g}$$

$$\Leftrightarrow 1 - \max_f \max_g\{\mathcal{Y}_{gf}\} \leq \prod_{g=1}^{\xi}(1 - \mathcal{Y}_{gf})^{\Xi_g} \leq 1 - \min_f \min_g\{\mathcal{Y}_{gf}\}$$

$$\Leftrightarrow (1 - \max_f \max_g\{\mathcal{Y}_{gf}\})^{\sum_{f=1}^{\varpi}\phi_f} \leq \prod_{f=1}^{\varpi}(\prod_{g=1}^{\xi}(1 - \mathcal{Y}_{gf})^{\Xi_g})^{\sum_{f=1}^{\varpi}\phi_f} \leq (1 - \min_f \min_g\{\mathcal{Y}_{gf}\})^{\sum_{f=1}^{\varpi}\phi_f}$$

$$\Leftrightarrow 1 - \max_f \max_g\{\mathcal{Y}_{gf}\} \leq \prod_{f=1}^{\varpi}(\prod_{g=1}^{\xi}(1 - \mathcal{Y}_{gf})^{\Xi_g})^{\sum_{f=1}^{\varpi}\phi_f} \leq 1 - \min_f \min_g\{\mathcal{Y}_{gf}\}.$$

Therefore

$$\max_f \max_g\{\mathcal{Y}_{gf}\} \leq 1 - \prod_{f=1}^{\varpi}\left(\prod_{g=1}^{\xi}(1 - \mathcal{Y}_{gf})^{\Xi_g}\right)^{\sum_{f=1}^{\varpi}\phi_f} \leq \min_f \min_g\{\mathcal{Y}_{gf}\}. \tag{3.7}$$

Again,

$$\min_f \min_g\{\mathcal{A}_{gf}\} \leq \mathcal{A}_{gf} \leq \max_f \max_g\{\mathcal{Y}_{gf}\},$$

which finds

$$(\min_f \min_g\{\mathcal{A}_{gf}\})^{\sum_{g=1}^{\xi}\Xi_g} \leq \prod_{g=1}^{\xi}(\mathcal{A}_{gf})^{\Xi_g} \leq (\max_f \max_g\{\mathcal{A}_{gf}\})^{\sum_{g=1}^{\xi}\Xi_g}$$

$$\Leftrightarrow \min_f \min_g\{\mathcal{A}_{gf}\} \leq \prod_{g=1}^{\xi}(\mathcal{A}_{gf})^{\Xi_g} \leq \max_f \max_g\{\mathcal{A}_{gf}\}$$

$$\Leftrightarrow (\min_f \min_g\{\mathcal{A}_{gf}\})^{\phi_f} \leq (\prod_{g=1}^{\xi}(\mathcal{A}_{gf})^{\Xi_g})^{\phi_f} \leq (\max_f \max_g\{\mathcal{A}_{gf}\})^{\phi_f}$$

$$\Leftrightarrow (\min_f \min_g\{\mathcal{A}_{gf}\})^{\sum_{f=1}^{\varpi}\phi_f} \leq \prod_{f=1}^{\varpi}(\prod_{g=1}^{\xi}(\mathcal{A}_{gf})^{\Xi_g})^{\phi_f} \leq (\max_f \max_g\{\mathcal{A}_{gf}\})^{\sum_{f=1}^{\varpi}\phi_f}.$$

Hence,

$$\min_f \min_g\{\mathcal{A}_{gf}\} \le \prod_{f=1}^{\varpi}(\prod_{g=1}^{\xi}(\mathcal{A}_{gf})^{\Xi_g})^{\phi_f} \le \max_f \max_g\{\mathcal{A}_{gf}\} \tag{3.8}$$

and, $\min_f \min_g\{\mathcal{N}_{gf}\} \le \mathcal{N}_{gf} \le \max_f \max_g\{\mathcal{N}_{gf}\}$, which gives

$$(\min_f \min_g\{\mathcal{N}_{gf}\})^{\sum_{g=1}^{\xi}\Xi_g} \le \prod_{g=1}^{\xi}(\mathcal{N}_{gf})^{\Xi_g} \le (\max_f \max_g\{\mathcal{N}_{gf}\})^{\sum_{g=1}^{\xi}\mathcal{A}_g}$$

$$\Leftrightarrow \min_f \min_g\{\mathcal{N}_{gf}\} \le \prod_{g=1}^{\xi}(\mathcal{N}_{gf})^{\Xi_g} \le \max_f \max_g\{\mathcal{N}_{gf}\}$$

$$\Leftrightarrow (\min_f \min_g\{\mathcal{N}_{gf}\})^{\phi_f} \le (\prod_{g=1}^{\xi}(\mathcal{N}_{gf})^{\Xi_g})^{\phi_f} \le (\max_f \max_g\{\mathcal{N}_{gf}\})^{\phi_f}$$

$$\Leftrightarrow (\min_f \min_g\{\mathcal{N}_{gf}\})^{\sum_{f=1}^{\varpi}\phi_f} \le \prod_{f=1}^{\varpi}(\prod_{g=1}^{\xi}(\mathcal{N}_{gf})^{\Xi_g})^{\phi_f} \le (\max_f \max_g\{\mathcal{N}_{gf}\})^{\sum_{f=1}^{\varpi}\phi_f}.$$

Thus

$$\min_f \min_g\{\mathcal{N}_{gf}\} \le \prod_{f=1}^{\varpi}(\prod_{g=1}^{\xi}(\mathcal{N}_{gf})^{\Xi_g})^{\phi_f} \le \max_f \max_g\{\mathcal{N}_{gf}\}. \tag{3.9}$$

Let $\Theta \equiv PFSWA(\mathcal{R}_{e_{11}}, \mathcal{R}_{e_{12}}, \ldots, \mathcal{R}_{e_{gf}}) = (\mathcal{Y}_\Theta, \mathcal{A}_\Theta, \mathcal{N}_\Theta)$, then from Eqs. (3.7), (3.8), and (3.9), $\min_f \min_g\{\mathcal{Y}_{gf}\} \le \mathcal{Y}_\Theta \le \max_f \max_g\{\mathcal{Y}_{gf}\}$ and $\min_f \min_g\{\mathcal{A}_{gf}\} \le \mathcal{A}_\Theta \le \max_f \max_g\{\mathcal{A}_{gf}\}$, and $\min_f \min_g\{\mathcal{N}_{gf}\} \le \mathcal{N}_f \le \max_f \max_g\{\mathcal{N}_{gf}\}$.

Then, from the definition of the score function

$$\Lambda(\Theta) = \mathcal{Y}_\Theta - \mathcal{N}_\Theta \le \max_f \max_g\{\mathcal{Y}_{gf}\} - \min_f \min_g\{\mathcal{N}_{gf}\} = \Lambda(\mathcal{Y}_{gf})$$

$$\Lambda(\Theta) = \mathcal{Y}_\Theta - \mathcal{N}_\Theta \ge \min_f \min_g\{\mathcal{Y}_{gf}\} - \max_f \max_g\{\mathcal{N}_{gf}\} = \Lambda(\mathcal{N}_{gf}).$$

Now, three cases arise:

*Case 1.* If $\Lambda(\mathcal{R}_{e_{gf}}) < \Lambda(\mathcal{Y}_{gf})$ and $\Lambda(\mathcal{R}_{e_{gf}}) > \Lambda(\mathcal{N}_{gf})$, then for two PFSNs, we obtain

$$\mathcal{N}_{gf} \le PFSWA(\mathcal{R}_{e_{11}}, \mathcal{R}_{e_{12}}, \ldots, \mathcal{R}_{e_{gf}}) \le \mathcal{Y}_{gf}.$$

*Case 2.* If $\Lambda(\mathcal{R}_{e_{gf}}) = \Lambda(\mathcal{Y}_{gf})$, i.e., $\mathcal{Y}_\Theta + \mathcal{A}_\Theta + \mathcal{N}_\Theta = \max_f \max_g\{\mathcal{Y}_{gf}\} + \min_f \min_g\{\mathcal{A}_{gf}\} + \min_f \min_g\{\mathcal{N}_{gf}\}$, by the inequalities $\mathcal{Y}_\Theta = \max_f \max_g\{\mathcal{Y}_{gf}\}$ and $\mathcal{A}_\Theta = \min_f \min_g\{\mathcal{A}_{gf}\}$, and $\mathcal{N}_\Theta = \min_f \min_g\{\mathcal{N}_{gf}\}$.

Therefore

$$\Phi = \mathcal{Y}_\Theta + \mathcal{A}_\Theta + \mathcal{N}_\Theta = \max_f \max_g\{\mathcal{Y}_{gf}\} + \min_f \min_g\{\mathcal{A}_{gf}\} + \min_f \min_g\{\mathcal{N}_{gf}\} = \Phi(\mathcal{Y}_{gf}),$$

then by comparison of the two PFSNs, we have

$$PFSWA(\mathcal{R}_{e11}, \mathcal{R}_{e12}, \ldots, \mathcal{R}_{e_{gf}}) = \mathcal{Y}_{gf}.$$

*Case 3.* If $\Lambda(\mathcal{R}_{e_{gf}}) = \Lambda(\mathcal{N}_{gf})$, i.e., $\mathcal{Y}_\Theta + \mathcal{A}_\Theta + \mathcal{N}_\Theta = \min_f \min_g\{\mathcal{Y}_{gf}\} + \max_f \max_g\{\mathcal{A}_{gf}\} + \max_f \max_g\{\mathcal{N}_{gf}\}$, then by the above inequalities $\mathcal{Y}_\Theta = \min_f \min_g\{\mathcal{Y}_{gf}\}$, $\mathcal{A}_\Theta = \max_f \max_g\{\mathcal{A}_{gf}\}$, and $\mathcal{N}_\Theta = \max_f \max_g\{\mathcal{N}_{gf}\}$.

Hence,

$$\Phi = \mathcal{Y}_\Theta + \mathcal{A}_\Theta + \mathcal{N}_{gf} = \min_f \min_g\{\mathcal{Y}_{gf}\} + \max_f \max_g\{\mathcal{A}_{gf}\} + \max_f \max_g\{\mathcal{Y}_{gf}\} = \Phi(\mathcal{N}_{gf}).$$

Then, by comparison of the two PFSNs, we have

$$PFSWA(\mathcal{R}_{e11}, \mathcal{R}_{e12}, \ldots, \mathcal{R}_{e_{gf}}) = \mathcal{N}_{gf}.$$

Thus the proof is completed. □

**Theorem 3.4.** ***(Shift-invariance)*** *If $\mathcal{R}_e = (\mathcal{Y}, \mathcal{A}, \mathcal{N})$ is another PFSN, then*

$$PFSWA(\mathcal{R}_{e11} \oplus \mathcal{R}_e, \mathcal{R}_{e12} \oplus \mathcal{R}_e, \ldots, \mathcal{R}_{e_{gf}} \oplus \mathcal{R}_e) = PFSWA(\mathcal{R}_{e11}, \mathcal{R}_{e12}, \ldots, \mathcal{R}_{e_{gf}}) \oplus \mathcal{R}_e.$$

*Proof.* Since $\mathcal{R}_e$ and $\mathcal{R}_{e_{gf}}$ are PFSNs, we have

$$\mathcal{R}_e \oplus \mathcal{R}_{e_{gf}} = \Big(1 - (1 - \mathcal{Y})(1 - \mathcal{Y}_{gf}), \mathcal{A}\mathcal{A}_{gf}, \mathcal{N}\mathcal{N}_{gf}\Big).$$

Thus

$$\begin{aligned}
&PFSWA(\mathcal{R}_{e11} \oplus \mathcal{R}_e, \mathcal{R}_{e12} \oplus \mathcal{R}_e, \ldots, \mathcal{R}_{e_{gf}} \oplus \mathcal{R}_e) \\
&= \bigoplus_{f=1}^{\varpi} \phi_f \Big( \bigoplus_{g=1}^{\xi} \Xi_g (\mathcal{R}_{e_{gf}} \oplus \mathcal{R}_e) \Big) \\
&= \left( 1 - \prod_{f=1}^{\varpi} \Big( \prod_{g=1}^{\xi} (1 - \mathcal{Y}_{gf})^{\Xi_g} (1 - \mathcal{Y})^{\Xi_g} \Big)^{\phi_f}, \prod_{f=1}^{\varpi} \Big( \prod_{g=1}^{\xi} (\mathcal{A}_a)^{\Xi_g} (\mathcal{A})^{\Xi_g} \Big)^{\phi_f}, \right. \\
&\qquad \left. \prod_{f=1}^{\varpi} \Big( \prod_{g=1}^{\xi} (\mathcal{A}_{gf})^{\Xi_g} (\mathcal{A})^{\Xi_g} \Big)^{\phi_f} \right) \\
&= \left( 1 - (1 - \mathcal{Y}) \prod_{f=1}^{\varpi} \Big( \prod_{g=1}^{\xi} (1 - \mathcal{Y}_{gf})^{\Xi_g} \Big)^{\phi_f}, \mathcal{A} \prod_{f=1}^{\varpi} \Big( \prod_{g=1}^{\xi} (\mathcal{A}_{gf})^{\Xi_g} \Big)^{\phi_f}, \right. \\
&\qquad \left. \mathcal{N} \prod_{f=1}^{\varpi} \Big( \prod_{g=1}^{\xi} (\mathcal{N}_{gf})^{\Xi_g} \Big)^{\phi_f} \right)
\end{aligned}$$

$$= \left(1 - \prod_{f=1}^{\varpi}\left(\prod_{g=1}^{\xi}(1-\mathcal{Y}_{gf})^{\Xi_g}\right)^{\phi_f}, \prod_{f=1}^{\varpi}\left(\prod_{g=1}^{\xi}(\mathcal{N}_{gf})^{\Xi_g}\right)^{\phi_f}, \prod_{f=1}^{\varpi}\left(\prod_{g=1}^{\xi}(\mathcal{N}_{gf})^{\Xi_g}\right)^{\phi_f}\right)$$
$$\oplus \left(\mathcal{Y}, \mathcal{A}, \mathcal{N}\right)$$
$$= PFSWA(\mathcal{R}_{e11}, \mathcal{R}_{e12}, \ldots, \mathcal{R}_{e_{gf}}) \oplus \mathcal{R}_e.$$

Hence, the result is obtained. □

**Theorem 3.5.** ***(Homogeneity)*** *For any real number $\lambda > 0$,*

$$PFSWA(\lambda\mathcal{R}_{e11}, \lambda\mathcal{R}_{e12}, \ldots, \lambda\mathcal{R}_{e_{gf}}) = \lambda PFSWA(\mathcal{R}_{e11}, \mathcal{R}_{e12}, \ldots, \mathcal{R}_{e_{gf}}).$$

*Proof.* Let $\mathcal{R}_{e_{gf}} = (\mathcal{Y}_{gf}, \mathcal{A}_{gf}, \mathcal{N}_{gf})$ $(g = 1, 2, \ldots, \xi;\ f = 1, 2, \ldots, \varpi)$ be a set of PFSNs and $\lambda > 0$ be any real number.

Then, $\lambda\mathcal{R}_{e_{gf}} = \left(1 - (1-\mathcal{Y}_{gf})^{\lambda}, (\mathcal{A}_{gf})^{\lambda}, (\mathcal{N}_{gf})^{\lambda}\right)$.

Thus

$$PFSWA(\lambda\mathcal{R}_{e11}, \lambda\mathcal{R}_{e12}, \ldots, \lambda\mathcal{R}_{e_{gf}})$$
$$= \left(1 - \prod_{f=1}^{\varpi}\left(\prod_{g=1}^{\xi}(1-\mathcal{Y}_{gf})^{\lambda\Xi_g}\right)^{\phi_f}, \prod_{f=1}^{\varpi}\left(\prod_{g=1}^{\xi}(\mathcal{A}_{gf})^{\lambda\Xi_g}\right)^{\phi_f}, \prod_{f=1}^{\varpi}\left(\prod_{g=1}^{\xi}(\mathcal{N}_{gf})^{\lambda\Xi_g}\right)^{\phi_f}\right)$$
$$= \left(1 - \left(\prod_{f=1}^{\varpi}\left(\prod_{g=1}^{\xi}(1-\mathcal{Y}_{gf})^{\Xi_g}\right)^{\phi_f}\right)^{\lambda}, \left(\prod_{f=1}^{\varpi}\left(\prod_{g=1}^{\xi}(\mathcal{A}_{gf})^{\Xi_g}\right)^{\phi_f}\right)^{\lambda}, \left(\prod_{f=1}^{\varpi}\left(\prod_{g=1}^{\xi}(\mathcal{N}_{gf})^{\Xi_g}\right)^{\phi_f}\right)^{\lambda}\right)$$
$$= \lambda\ PFSWA(\mathcal{R}_{e11}, p_{e12}, \ldots, \mathcal{R}_{e_{gf}}).$$

Hence, the proof is completed. □

## 3.3.2 Picture fuzzy soft weighted geometric operator

In this subsection, the PFSWG operator and related properties are discussed.

**Definition 3.5.** Let $\mathcal{R}_{e_{gf}} = (\mathcal{Y}_{gf}, \mathcal{A}_{gf}, \mathcal{N}_{gf})$ $(g = 1, 2, \ldots, \xi;\ f = 1, 2, \ldots, \varpi)$ be a set of PFSNs and $\phi_f$, $\Xi_g$ are the weight vectors for $e_f$s and $e_g$s, respectively, satisfying $\phi_f \geq 0$, $\Xi_g \geq 0$ such that $\sum_{f=1}^{\varpi}\phi_f = 1$ and $\sum_{g=1}^{\xi}\Xi_g = 1$. Then, the picture fuzzy soft weighted geometric (PFSWG) operator is a function:

$$PFSWG : \Theta^{\varpi} \to \Theta$$

such that

$$PFSWG(\mathcal{R}_{e11}, \mathcal{R}_{e12}, \ldots, \mathcal{R}_{e_{gf}}) = \bigotimes_{f=1}^{\varpi}\left(\bigotimes_{g=1}^{\xi}\mathcal{R}_{e_{gf}}^{\Xi_g}\right)^{\phi_f}.$$

**Theorem 3.6.** *The picture fuzzy soft weighted geometric (PFSWG) operator is a function* $PFSWG : \Theta^{\varpi} \to \Theta$ *such that*

$$PFSWG(\mathcal{R}_{e_{11}}, \mathcal{R}_{e_{12}}, \ldots, \mathcal{R}_{e_{gf}})$$
$$= \left( \prod_{f=1}^{\varpi} \left( \prod_{g=1}^{\xi} (\mathcal{Y}_{gf})^{\Xi_g} \right)^{\phi_f}, 1 - \prod_{f=1}^{\varpi} \left( \prod_{g=1}^{\xi} (1 - \mathcal{A}_{gf})^{\Xi_g} \right)^{\phi_f}, 1 - \prod_{f=1}^{\varpi} \left( \prod_{g=1}^{\xi} (1 - \mathcal{N}_{gf})^{\Xi_g} \right)^{\phi_f} \right). \tag{3.10}$$

*Proof.* For $g = 1$ and $\Xi_1 = 1$ by Definition 3.3, we have

$$PFSWG(\mathcal{R}_{e_{11}}, \mathcal{R}_{e_{12}}, \ldots, \mathcal{R}_{e_{1f}}) = \bigotimes_{f=1}^{\varpi} \mathcal{R}_{e_{gf}}^{\phi_f}$$
$$= \left( \prod_{f=1}^{\varpi} (\mathcal{Y}_{1f})^{\phi_f}, 1 - \prod_{f=1}^{\varpi} (1 - \mathcal{A}_{1f})^{\phi_f}, 1 - \prod_{f=1}^{\varpi} (1 - \mathcal{N}_{1f})^{\phi_f} \right)$$
$$= \left( \prod_{f=1}^{\varpi} \left( \prod_{g=1}^{1} (\mathcal{Y}_{gf})^{\Xi_g} \right)^{\phi_f}, 1 - \prod_{f=1}^{\varpi} \left( \prod_{g=1}^{1} (1 - \mathcal{A}_{gf})^{\Xi_g} \right)^{\phi_f}, 1 - \prod_{f=1}^{\varpi} \left( \prod_{g=1}^{1} (1 - \mathcal{N}_{gf})^{\Xi_g} \right)^{\phi_f} \right).$$

For $f = 1$ and $\phi_1 = 1$, thus by Definition 3.5,

$$PFSWG(\mathcal{R}_{e_{11}}, \mathcal{R}_{e_{21}}, \ldots, \mathcal{R}_{e_{g1}}) = \bigotimes_{f=1}^{\varpi} \mathcal{R}_{e_{gf}}^{\Xi_g}$$
$$= \left( \prod_{g=1}^{\xi} (\mathcal{Y}_{g1})^{\Xi_g}, 1 - \prod_{g=1}^{\xi} (1 - \mathcal{A}_{g1})^{\Xi_g}, 1 - \prod_{g=1}^{\xi} (1 - \mathcal{N}_{g1})^{\Xi_g} \right)$$
$$= \left( \prod_{f=1}^{1} \left( \prod_{g=1}^{\xi} (\mathcal{Y}_{gf})^{\Xi_g} \right)^{\phi_f}, 1 - \prod_{f=1}^{1} \left( \prod_{g=1}^{\xi} (1 - \mathcal{A}_{gf})^{\Xi_g} \right)^{\phi_f}, 1 - \prod_{f=1}^{1} \left( \prod_{g=1}^{\xi} (1 - \mathcal{N}_{gf})^{\Xi_g} \right)^{\phi_f} \right).$$

Assume that Eq. (3.10) is true for $b = d_1 + 1, a = d_2$ and $b = d_1, a = d_2 + 1$, from which we obtain

$$\bigotimes_{f=1}^{d_1+1} \left( \bigotimes_{g=1}^{d_2} \mathcal{R}_{e_{gf}}^{\Xi_g} \right)^{\phi_f}$$
$$= \left( \prod_{f=1}^{d_1+1} \left( \prod_{g=1}^{d_2} (\mathcal{Y}_{gf})^{\Xi_g} \right)^{\phi_f}, 1 - \prod_{f=1}^{d_1+1} \left( \prod_{g=1}^{d_2} (1 - \mathcal{A}_{gf})^{\Xi_g} \right)^{\phi_f}, 1 - \prod_{f=1}^{d_1+1} \left( \prod_{g=1}^{d_2} (1 - \mathcal{N}_{gf})^{\Xi_g} \right)^{\phi_f} \right)$$

and

$$\bigotimes_{b=1}^{d_1}\Big(\bigotimes_{g=1}^{d_2+1}\mathcal{R}_{e_{gf}}^{\Xi_g}\Big)^{\phi_f}$$
$$=\left(\prod_{f=1}^{d_1}\Big(\prod_{g=1}^{d_2+1}(\mathcal{Y}_{gf})^{\Xi_g}\Big)^{\phi_f},\ 1-\prod_{f=1}^{d_1}\Big(\prod_{g=1}^{d_2+1}(1-\mathcal{A}_{gf})^{\Xi_g}\Big)^{\phi_f},\ 1-\prod_{f=1}^{d_1}\Big(\prod_{g=1}^{d_2+1}(1-\mathcal{N}_{gf})^{\Xi_g}\Big)^{\phi_f}\right).$$

Now, for $f=d_1+1$ and $g=d_2+1$, we obtain

$$\bigotimes_{f=1}^{d_1+1}\Big(\bigotimes_{g=1}^{d_2+1}\mathcal{R}_{e_{gf}}^{\Xi_g}\Big)^{\phi_f}$$
$$=\bigotimes_{f=1}^{d_1+1}\Big(\bigotimes_{a=1}^{d_2}\mathcal{R}_{e_{gf}}^{\Xi_g}\otimes\mathcal{R}_{e_{(d_2+1)f}}^{\psi_{d_2+1}}\Big)^{\phi_f}$$
$$=\bigotimes_{f=1}^{d_1+1}\Big(\bigotimes_{a=1}^{d_2}\mathcal{R}_{e_{gf}}^{\Xi_g}\Big)^{\phi_f}\bigotimes_{f=1}^{d_1+1}\Big(\Big(\mathcal{R}_{e_{(d_2+1)f}}^{\Xi_{d_2+1}}\Big)^{\phi_f}\Big)$$
$$=\left(\prod_{f=1}^{d_1+1}\Big(\prod_{g=1}^{d_2}(\mathcal{Y}_{gf})^{\Xi_g}\Big)^{\phi_f}\otimes\prod_{f=1}^{d_1+1}\Big((\mathcal{Y}_{(d_2+1)f})^{\psi_{d_2+1}}\Big)^{\phi_f},\right.$$
$$1-\prod_{f=1}^{d_1+1}\Big(\prod_{g=1}^{d_2}(1-\mathcal{A}_{gf})^{\Xi_g}\Big)^{\phi_f}\otimes1-\prod_{f=1}^{d_1+1}\Big((1-\mathcal{A}_{(d_2+1)f})^{\psi_{d_2+1}}\Big)^{\phi_f},$$
$$\left.1-\prod_{f=1}^{d_1+1}\Big(\prod_{g=1}^{d_2}(1-\mathcal{N}_{gf})^{\Xi_g}\Big)^{\phi_f}\otimes1-\prod_{f=1}^{d_1+1}\Big((1-\mathcal{N}_{(d_2+1)f})^{\psi_{d_2+1}}\Big)^{\phi_f}\right)$$
$$=\left(\prod_{f=1}^{d_1+1}\Big(\prod_{g=1}^{d_2+1}(\mathcal{Y}_{gf})^{\Xi_g}\Big)^{\phi_f},\ 1-\prod_{f=1}^{d_1+1}\Big(\prod_{g=1}^{d_2+1}(1-\mathcal{A}_{gf})^{\Xi_g}\Big)^{\phi_f},\ 1-\prod_{f=1}^{d_1+1}\Big(\prod_{g=1}^{d_2+1}(1-\mathcal{N}_{gf})^{\Xi_g}\Big)^{\phi_f}\right).$$

Therefore Eq. (3.10) holds for $f=d_1+1$, $g=d_2+1$, by induction the result holds for all $g,f\geq1$.

Since,

$$0\leq\mathcal{A}_{gf}\leq1\Leftrightarrow0\leq\prod_{g=1}^{\xi}(1-\mathcal{A}_{gf})^{\Xi_g}\leq1$$
$$\Leftrightarrow0\leq\prod_{f=1}^{\varpi}\Big(\prod_{g=1}^{\xi}(1-\mathcal{A}_{gf})^{\Xi_g}\Big)^{\phi_f}\leq1$$
$$\Leftrightarrow0\leq1-\prod_{f=1}^{\varpi}\Big(\prod_{g=1}^{\xi}(1-\mathcal{A}_{gf})^{\Xi_g}\Big)^{\phi_f}\leq1,\ \text{and}$$

$$0 \le \mathcal{N}_{gf} \le 1 \Leftrightarrow 0 \le \prod_{g=1}^{\xi}(1-\mathcal{N}_{gf})^{\Xi_g} \le 1$$
$$\Leftrightarrow 0 \le \prod_{f=1}^{\varpi}\left(\prod_{g=1}^{\xi}(1-\mathcal{N}_{gf})^{\Xi_g}\right)^{\phi_f} \le 1$$
$$\Leftrightarrow 0 \le 1-\prod_{f=1}^{\varpi}\left(\prod_{g=1}^{\xi}(1-\mathcal{N}_{gf})^{\Xi_g}\right)^{\phi_f} \le 1.$$

On the other hand,

$$0 \le \mathcal{Y}_{gf} \le 1$$
$$\Leftrightarrow 0 \le \prod_{g=1}^{\xi}(\mathcal{Y}_{gf})^{\Xi_g} \le 1$$
$$\Leftrightarrow 0 \le \prod_{f=1}^{\varpi}\left(\prod_{g=1}^{\xi}(\mathcal{Y}_{gf})^{\Xi_g}\right)^{\phi_f} \le 1.$$

Therefore $0 \le 1-\prod_{f=1}^{\varpi}\left(\prod_{g=1}^{\xi}(1-\mathcal{A}_{gf})^{\Xi_g}\right)^{\phi_f} + 1-\prod_{f=1}^{\varpi}\left(\prod_{g=1}^{\xi}(1-\mathcal{N}_{gf})^{\Xi_g}\right)^{\phi_f} + \prod_{f=1}^{\varpi}\left(\prod_{g=1}^{\xi}(\mathcal{Y}_{gf})^{\Xi_g}\right)^{\phi_f} \le 1.$

Thus the value obtained by the PFSWG aggregated operator is also a PFSN. □

The PFSWG operator satisfies the following properties:

**(1)** (**Idempotency**) If $\mathcal{R}_{e_{gf}} = \mathcal{R}_e = (\mathcal{Y}, \mathcal{A}, \mathcal{N})$ for all $g, f$, then

$$PFSWG(\mathcal{R}_{e11}, \mathcal{R}_{e12}, \ldots, \mathcal{R}_{e_{gf}}) = \mathcal{R}_e;$$

**(2)** (**Boundedness**) If $\mathcal{N}_{gf} = (\min_f \min_g\{\mathcal{Y}_{gf}\}, \max_f \max_g\{\mathcal{A}_{gf}\}, \max_f \max_g\{\mathcal{N}_{gf}\})$ and If $\mathcal{Y}_{gf} = (\max_f \max_g\{\mathcal{Y}_{gf}\}, \min_f \min_g\{\mathcal{Y}_{gf}\}, \min_b \min_a\{\mathcal{N}_{gf}\})$, then

$$\mathcal{Y}_{gf} \le PFSWG(\mathcal{R}_{e11}, \mathcal{R}_{e12}, \ldots, \mathcal{R}_{e_{gf}}) \le \mathcal{Y}_{gf};$$

**(3)** (**Shift-invariance**) Let $\mathcal{R}_e = (\mathcal{Y}, \mathcal{A}, \mathcal{N})$ be a PFSN then

$$PFSWG(\mathcal{R}_{e_{11}} \otimes \mathcal{R}_e, \mathcal{R}_{e_{12}} \otimes \mathcal{R}_e, \ldots, \mathcal{R}_{e_{gf}} \otimes \mathcal{R}_e) = PFSWG(\mathcal{R}_{e11}, \mathcal{R}_{e12}, \ldots, \mathcal{R}_{e_{gf}}) \otimes \mathcal{R}_e;$$

**(4)** (**Homogeneity**) For any real number $\lambda > 0$, then

$$PFSSWG(\mathcal{R}_{e11}^{\lambda}, \mathcal{R}_{e12}^{\lambda}, \ldots, \mathcal{R}_{e_{gf}}^{\lambda}) = \left(PFSWG(\mathcal{R}_{e11}, \mathcal{R}_{e12}, \ldots, \mathcal{R}_{e_{gf}})\right)^{\lambda}.$$

## 3.4 Model for MCGDM method using picture fuzzy soft information

In this section, we introduce the multicriteria group decision-making (MCGDM) method using PFSWA and PFSWG operators on a PFSS environment.

### 3.4.1 An approach based on proposed operators

Let a set of alternatives be $\aleph = \{\aleph_1, \aleph_2, \ldots, \aleph_L\}$ and evaluation is made by experts $\{m_1, m_2, \ldots, m_g\}$ under the $f$ parameters $E = \{e_1, e_2, \ldots, e_f\}$. Let $\Xi = (\Xi_1, \Xi_2, \ldots, \Xi_g)^T$ and $\phi = (\phi_1, \phi_2, \ldots, \phi_f)^T$ be weight vectors of the $g$ experts $x'_f s$ and $f$ parameters $e'_f s$ that $\Xi_g > 0$, $\Xi \in [0, 1]$ such that $\sum_{g=1}^{\xi} \Xi_g = 1$ and $\phi_f > 0$, $\phi \in [0, 1]$, where $\sum_{f=1}^{\varpi} \phi_f = 1$. To select the best ones based on values of $f$ experts in the form of PFSNs $\mathcal{R}_{e_{gf}} = \langle \mathcal{Y}_{gf}, \mathcal{G}_{gf}, \mathcal{N}_{gf} \rangle$, where $0 \leq \mathcal{Y}_{gf} + \mathcal{G}_{gf} + \mathcal{N}_{gf} \leq 1$ and the collective decision matrix is of the form $M = (\mathcal{R}_{e_{gf}})_{\xi \times \varpi}$. By these choice values of the experts, the accumulated PFS information $\mathcal{R}_{e_k}$ for the alternatives $\aleph_k$ $(k = 1, 2, \ldots, L)$ is $H_{e_k} = \langle \mathcal{Y}_k, \mathcal{G}_k, \mathcal{N}_k \rangle$ by applying average operators that are given in Eq. (1.5) and Eq. (1.8) and the corresponding order of the alternatives based on the score value of $\mathcal{R}_{e_k}$ $(k = 1, 2, \ldots, L)$.

We propose to solve MCGDM problems with picture fuzzy soft information using the PFSWA and PFSWG operators, as depicted in the algorithm below.

**Algorithm.**

**Step 1.** Collective information in the form of picture fuzzy soft matrix $P = \langle \mathcal{Y}_{gf}, \mathcal{G}_{gf}, \mathcal{N}_{gf} \rangle$ $(g = 1, 2, \ldots, \xi;\ f = 1, 2, \ldots, \varpi)$ for each alternatives under proposed parameters $e_k$ $(k = 1, 2, \ldots, L)$ as

$$\mathcal{R}_{\xi \times \varpi} = M = \begin{bmatrix} (\mathcal{Y}_{11}, \mathcal{G}_{11}, \mathcal{N}_{11}) & (\mathcal{Y}_{12}, \mathcal{G}_{12}, \mathcal{N}_{12}) & \cdots & (\mathcal{Y}_{1f}, \mathcal{G}_{1f}, \mathcal{N}_{1f}) \\ (\mathcal{Y}_{21}, \mathcal{G}_{21}, \mathcal{N}_{21}) & (\mathcal{Y}_{22}, \mathcal{G}_{22}, \mathcal{N}_{22}) & \cdots & (\mathcal{Y}_{2f}, \mathcal{G}_{2f}, \mathcal{N}_{2f}) \\ \vdots & \vdots & \ddots & \vdots \\ (\mathcal{Y}_{g1}, \mathcal{G}_{g1}, \mathcal{N}_{g1}) & (\mathcal{Y}_{g2}, \mathcal{G}_{g2}, \mathcal{N}_{g2}) & \cdots & (\mathcal{Y}_{gf}, \mathcal{G}_{gf}, \mathcal{N}_{gf}) \end{bmatrix}.$$

**Step 2.** To transform the values of cost type (B) into benefit (C) type in a normalized aggregated decision matrix by using the formula depicted in [53]:

$$N_{ij} = \begin{cases} \mathcal{R}^c_{e_{gf}}, & \text{if } e_f \in \tilde{B} \\ \mathcal{R}_{e_{gf}}, & \text{if } e_f \in \tilde{C}, \end{cases}$$

where $\mathcal{R}^c_{e_{gf}} = \langle 1 - \mathcal{N}_{gf}, \mathcal{G}_{gf}, \mathcal{Y}_{gf} \rangle$ is the complement of $\mathcal{R}_{e_{gf}} = \langle \mathcal{Y}_{gf}, \mathcal{G}_{gf}, \mathcal{N}_{gf} \rangle$.

**Step 3.** Aggregate the PFSNs $\mathcal{R}_{e_{gf}}$ $(g = 1, 2, \ldots, \xi;\ f = 1, 2, \ldots, \varpi)$ for each of the alternatives $\aleph_k$ $(k = 1, 2, \ldots, L)$ in a picture fuzzy soft decision matrix (PFSDM) using operators PFSWA or (PFSWG).

**Step 4.** Using Eq. (3.1) obtain the score value of $\Lambda_k$ $(k = 1, 2, \ldots, L)$ for each of the alternatives $\aleph_k$ $(k = 1, 2, \ldots, L)$.

**Step 5.** For the alternatives $\aleph_k$ $(k = 1, 2, \ldots, l)$ in order to choose the favorable alternatives according to $\Lambda_k$ $(k = 1, 2, \ldots, l)$.
**Step 6.** End.

## 3.5 Case study

In recent times, the Government of India looked forward to the development of the north-eastern region including many aspects: create new jobs, skill development, encouragement in small industries, transport structures, business facilities, and communication systems. The people of this region will be economically and socially stronger, and cannot feel themselves deprived compared to other parts of India. The people of this area cannot migrate to the big city to obtain more opportunities. For that purpose, the Government of India has taken the project to built the roads either to maintain what already exists or to engage in new roads. For successful completion of the proposed project, the Government of India appointed a team of five experts $m_1, m_2, m_3, m_4$, and $m_5$ whose weight vector is $\Xi_g = (0.2, 0.15, 0.2, 0.3, 0.15)^T$ and they will give the rating values of road contractors based on previous years' performance. In that view, the Indian Government has introduced global tender in the newspaper to select the contractors from the four road construction companies ($\aleph_1$) Jaihind Road Builders private (Pvt.) limited (Ltd.), ($\aleph_2$) J.K. Construction, ($\aleph_3$) Tata Infrastructure Ltd and ($\aleph_4$) Birla Pvt. Ltd that are the alternatives for these projects. The expert team will give their judgement based on the basis of performance of five road construction companies $\aleph_g$ $(g = 1, 2, \ldots, 4)$ under the four parameters (E):

$e_1$: Experience about contractor background;
$e_2$: Technical capability;
$e_3$: Tender price;
$e_4$: Completion time;

having weight vector $\phi_f = (0.2, 0.1, 0.3, 0.15, 0.25)^T$. Now, we applied the proposed approach to chose the favorable construction company.

### 3.5.1 Using the PFSWA operator

The steps of the proposed approach and their corresponding details are reviewed here:

**Step 1.** Each of the road construction companies is valuated by five experts to give their grades in terms of PFSNs that are given in Tables 3.2, 3.3, 3.4, and 3.5, their corresponding aggregated values are given in Table 3.6.

**Step 2.** Here, normalization of parameters is not required because all of them are benefit type.

**Step 3.** The opinions of the experts for each construction company $\aleph_k$ $(k = 1, 2, 3, 4)$ are accumulated by using Eq. (3.5) given as: $\Upsilon_1 = \langle 0.5845, 0.2627, 0.2999\rangle$, $\Upsilon_2 = \langle 0.6024, 0.1942, 0.2111\rangle$, $\Upsilon_3 = \langle 0.6172, 0.1688, 0.1909\rangle$, and $\Upsilon_4 = \langle 0.6216, 0.1578, 0.1994\rangle$.

**Table 3.2** PFS matrix for the road construction company $\aleph_1$.

| Experts | $e_1$ | $e_2$ | $e_3$ | $e_4$ | $e_5$ |
|---|---|---|---|---|---|
| $m_1$ | $\langle 0.7, 0.1, 0.2\rangle$ | $\langle 0.6, 0.2, 0.3\rangle$ | $\langle 0.5, 0.1, 0.3\rangle$ | $\langle 0.4, 0.2, 0.3\rangle$ | $\langle 0.6, 0.2, 0.4\rangle$ |
| $m_2$ | $\langle 0.5, 0.3, 0.4\rangle$ | $\langle 0.6, 0.2, 0.3\rangle$ | $\langle 0.6, 0.3, 0.4\rangle$ | $\langle 0.5, 0.1, 0.3\rangle$ | $\langle 0.5, 0.1, 0.2\rangle$ |
| $m_3$ | $\langle 0.6, 0.2, 0.3\rangle$ | $\langle 0.7, 0.3, 0.5\rangle$ | $\langle 0.4, 0.1, 0.2\rangle$ | $\langle 0.7, 0.2, 0.5\rangle$ | $\langle 0.7, 0.3, 0.4\rangle$ |
| $m_4$ | $\langle 0.5, 0.3, 0.4\rangle$ | $\langle 0.6, 0.4, 0.5\rangle$ | $\langle 0.7, 0.4, 0.5\rangle$ | $\langle 0.6, 0.2, 0.2\rangle$ | $\langle 0.4, 0.1, 0.1\rangle$ |
| $m_5$ | $\langle 0.6, 0.3, 0.3\rangle$ | $\langle 0.5, 0.2, 0.2\rangle$ | $\langle 0.7, 0.4, 0.5\rangle$ | $\langle 0.6, 0.3, 0.3\rangle$ | $\langle 0.6, 0.1, 0.2\rangle$ |

**Table 3.3** PFS matrix for the road construction company $\aleph_2$.

| Experts | $e_1$ | $e_2$ | $e_3$ | $e_4$ | $e_5$ |
|---|---|---|---|---|---|
| $m_1$ | $\langle 0.4, 0.1, 0.2\rangle$ | $\langle 0.7, 0.2, 0.3\rangle$ | $\langle 0.6, 0.2, 0.2\rangle$ | $\langle 0.5, 0.2, 0.3\rangle$ | $\langle 0.5, 0.2, 0.2\rangle$ |
| $m_2$ | $\langle 0.6, 0.2, 0.4\rangle$ | $\langle 0.4, 0.1, 0.1\rangle$ | $\langle 0.5, 0.2, 0.2\rangle$ | $\langle 0.7, 0.3, 0.3\rangle$ | $\langle 0.6, 0.4, 0.4\rangle$ |
| $m_3$ | $\langle 0.5, 0.1, 0.2\rangle$ | $\langle 0.6, 0.1, 0.3\rangle$ | $\langle 0.6, 0.2, 0.2\rangle$ | $\langle 0.6, 0.3, 0.3\rangle$ | $\langle 0.7, 0.3, 0.\rangle$ |
| $m_4$ | $\langle 0.7, 0.2, 0.2\rangle$ | $\langle 0.4, 0.1, 0.1\rangle$ | $\langle 0.7, 0.2, 0.2\rangle$ | $\langle 0.5, 0.2, 0.3\rangle$ | $\langle 0.7, 0.2, 0.3\rangle$ |
| $m_5$ | $\langle 0.4, 0.1, 0.2\rangle$ | $\langle 0.6, 0.2, 0.3\rangle$ | $\langle 0.4, 0.2, 0.1\rangle$ | $\langle 0.8, 0.3, 0.2\rangle$ | $\langle 0.7, 0.3, 0.2\rangle$ |

**Table 3.4** PFS matrix for the road construction company $\aleph_3$.

| Experts | $e_1$ | $e_2$ | $e_3$ | $e_4$ | $e_5$ |
|---|---|---|---|---|---|
| $m_1$ | $\langle 0.6, 0.2, 0.3\rangle$ | $\langle 0.8, 0.4, 0.3\rangle$ | $\langle 0.6, 0.1, 0.2\rangle$ | $\langle 0.5, 0.3, 0.2\rangle$ | $\langle 0.4, 0.1, 0.2\rangle$ |
| $m_2$ | $\langle 0.6, 0.3, 0.2\rangle$ | $\langle 0.5, 0.1, 0.2\rangle$ | $\langle 0.4, 0.1, 0.1\rangle$ | $\langle 0.5, 0.1, 0.2\rangle$ | $\langle 0.6, 0.2, 0.1\rangle$ |
| $m_3$ | $\langle 0.4, 0.2, 0.1\rangle$ | $\langle 0.7, 0.1, 0.1\rangle$ | $\langle 0.6, 0.4, 0.2\rangle$ | $\langle 0.7, 0.2, 0.4\rangle$ | $\langle 0.7, 0.2, 0.2\rangle$ |
| $m_4$ | $\langle 0.6, 0.1, 0.1\rangle$ | $\langle 0.6, 0.3, 0.3\rangle$ | $\langle 0.5, 0.2, 0.3\rangle$ | $\langle 0.7, 0.2, 0.2\rangle$ | $\langle 0.8, 0.1, 0.1\rangle$ |
| $m_5$ | $\langle 0.6, 0.1, 0.3\rangle$ | $\langle 0.4, 0.1, 0.3\rangle$ | $\langle 0.6, 0.2, 0.3\rangle$ | $\langle 0.7, 0.1, 0.4\rangle$ | $\langle 0.8, 0.4, 0.2\rangle$ |

**Table 3.5** PFS matrix for the road construction company $\aleph_4$.

| Experts | $e_1$ | $e_2$ | $e_3$ | $e_4$ | $e_5$ |
|---|---|---|---|---|---|
| $m_1$ | $\langle 0.6, 0.2, 0.3\rangle$ | $\langle 0.6, 0.1, 0.2\rangle$ | $\langle 0.4, 0.1, 0.2\rangle$ | $\langle 0.6, 0.2, 0.1\rangle$ | $\langle 0.5, 0.2, 0.1\rangle$ |
| $m_2$ | $\langle 0.4, 0.1, 0.1\rangle$ | $\langle 0.5, 0.1, 0.1\rangle$ | $\langle 0.8, 0.1, 0.1\rangle$ | $\langle 0.6, 0.3, 0.3\rangle$ | $\langle 0.7, 0.1, 0.2\rangle$ |
| $m_3$ | $\langle 0.6, 0.2, 0.2\rangle$ | $\langle 0.7, 0.2, 0.4\rangle$ | $\langle 0.6, 0.3, 0.3\rangle$ | $\langle 0.4, 0.1, 0.3\rangle$ | $\langle 0.7, 0.4, 0.4\rangle$ |
| $m_4$ | $\langle 0.8, 0.1, 0.2\rangle$ | $\langle 0.5, 0.1, 0.3\rangle$ | $\langle 0.7, 0.1, 0.1\rangle$ | $\langle 0.6, 0.2, 0.2\rangle$ | $\langle 0.5, 0.2, 0.3\rangle$ |
| $m_5$ | $\langle 0.7, 0.2, 0.2\rangle$ | $\langle 0.5, 0.1, 0.2\rangle$ | $\langle 0.5, 0.1, 0.2\rangle$ | $\langle 0.7, 0.2, 0.3\rangle$ | $\langle 0.7, 0.4, 0.3\rangle$ |

**Step 4.** The score value of each construction companies are: $\Lambda(\Upsilon_1) = 0.2846$, $\Lambda(\Upsilon_2) = 0.3913$, $\Lambda(\Upsilon_3) = 0.4263$, and $\Lambda(\Upsilon_4) = 0.4222$.

**Step 5.** Rank all the construction road companies $Q_k$ $(k = 1, 2, 3, 4)$ with the value of the score function $\Lambda(\Upsilon_k)$ $(k = 1, 2, 3, 4)$ based on the overall PFSNs as $\aleph_3 \succ \aleph_4 \succ \aleph_2 \succ \aleph_1$.

**Step 6.** Therefore $\aleph_3$ is taken as the most favorable road construction company.

## 3.5.2 Using the PFSWG operator

For applying the PFSWG operator on the same problem, we obtain:

**Step 3.** The accumulated values for each road construction company $\aleph_k$ $(k = 1, 2, 3, 4)$ using the operator PFSWG follows using Eq. (3.10): $\Upsilon_1 = \langle 0.5619, 0.2334, 0.3397\rangle$, $\Upsilon_2 = \langle 0.5753, 0.2094, 0.2273\rangle$, $\Upsilon_3 = \langle 0.5848, 0.1975, 0.2146\rangle$, and $\Upsilon_4 = \langle 0.5904, 0.1895, 0.2251\rangle$.

**Step 4.** The score values: $\Lambda(\Upsilon_1) = 0.2222$, $\Lambda(\Upsilon_2) = 0.3480$, $\Lambda(\Upsilon_3) = 0.3702$, and $\Lambda(\Upsilon_4) = 0.3653$.

**Step 5.** Rank all the road construction companies $Q_k$ $(k = 1, 2, 3, 4)$ based on score values $\Lambda(\Upsilon_k)$ $(k = 1, 2, 3, 4)$ of the overall PFS numbers as $\aleph_3 \succ \aleph_4 \succ \aleph_2 \succ \aleph_1$.

**Step 6.** Hence, $\aleph_3$ is the most attractive road construction company.

From the above analysis of the two methods, the overall rating based on two operators for the alternatives, but the order of the options are similar, and obtaining the most desirable option is $\aleph_3$.

## 3.6 Comparative studies

To compare the superiority of the proposed approach, this section contains a comparative study with the present approaches. It is distinguishable that PFSN can be used in analyzing the existing methods [17,27,49,50]. In that situation, the different parameters of the PFSS information are aggregated with an average operator by applying the weighted vector $(0.2, 0.1, 0.3, 0.15, 0.25)^T$ and then the obtained accumulated PFSDM for various road construction companies $Q_k$, $(k = 1, 2, 3, 4)$ is given in Table 3.6.

**Table 3.6** Aggregated value of PFS matrix for the road construction companies.

| Experts | $\aleph_1$ | $\aleph_2$ | $\aleph_3$ | $\aleph_4$ |
|---|---|---|---|---|
| $m_1$ | $\langle 0.5709, 0.1414, 0.2973\rangle$ | $\langle 0.5392, 0.1741, 0.2213\rangle$ | $\langle 0.5729, 0.1556, 0.2259\rangle$ | $\langle 0.5223, 0.1516, 0.1644\rangle$ |
| $m_2$ | $\langle 0.5427, 0.1856, 0.3130\rangle$ | $\langle 0.5734, 0.2358, 0.2709\rangle$ | $\langle 0.5223, 0.1481, 0.1366\rangle$ | $\langle 0.6647, 0.1179, 0.1402\rangle$ |
| $m_3$ | $\langle 0.6088, 0.1872, 0.3243\rangle$ | $\langle 0.6108, 0.1911, 0.1861\rangle$ | $\langle 0.6243, 0.2297, 0.1803\rangle$ | $\langle 0.6156, 0.2421, 0.3059\rangle$ |
| $m_4$ | $\langle 0.5754, 0.2407, 0.2787\rangle$ | $\langle 0.6529, 0.1866, 0.2195\rangle$ | $\langle 0.6555, 0.1525, 0.1722\rangle$ | $\langle 0.6546, 0.1320, 0.1872\rangle$ |
| $m_5$ | $\langle 0.6248, 0.2386, 0.3034\rangle$ | $\langle 0.5891, 0.2048, 0.1692\rangle$ | $\langle 0.6645, 0.1741, 0.2830\rangle$ | $\langle 0.6320, 0.1803, 0.2352\rangle$ |

**(1)** On applying the weighted form of the picture fuzzy Hamacher average operator (PFHWA) proposed by Wei [50] to the proposed data, then for $\gamma = 3$, the overall scores for each options $\aleph_k (k = 1, 2, 3, 4)$ are evaluated as $\Lambda(\Upsilon_1) = 0.2842$, $\Lambda(\Upsilon_2) = 0.3901$, $\Lambda(\Upsilon_3) = 0.4243$, and $\Lambda(\Upsilon_4) = 0.4200$. The corresponding ranking order shows $\aleph_3 \succ \aleph_4 \succ \aleph_2 \succ \aleph_1$, which predicts that $\aleph_3$ is the best choice.

**(2)** On utilizing the weighted form of the picture fuzzy Hamacher geometric operator (PFHWG) imposed by Wei [50] to the considered data, then for $\gamma = 3$, the score for each alternative $\aleph_k (k = 1, 2, 3, 4)$ are $\Lambda(\Upsilon_1) = 0.2831$, $\Lambda(\Upsilon_2) = 0.3865$, $\Lambda(\Upsilon_3) = 0.4177$, and $\Lambda(\Upsilon_4) = 0.4113$, then the order is as follows: $\aleph_3 \succ \aleph_4 \succ \aleph_2 \succ \aleph_1$, which gives $\aleph_3$ is the most favorable construction company.

**(3)** On applying the weighted form of the picture fuzzy Einstein averaging operator (PFEWA) applied by Khan et al. [27], the score values of the options are computed as: $\Lambda(\Upsilon_1) = 0.3173$, $\Lambda(\Upsilon_2) = 0.4088$, $\Lambda(\Upsilon_3) = 0.4408$, and $\Lambda(\Upsilon_4) = 0.4395$, which gives the corresponding ranking $\aleph_3 \succ \aleph_4 \succ \aleph_2 \succ \aleph_1$ from which we conclude that $\aleph_3$ is the optimal road construction company.

**(4)** On utilizing the weighted form of the picture fuzzy Einstein geometric operator (PFEWG) proposed by Khan et al. [27], the score values of the alternatives are $\Lambda(\Upsilon_1) = 0.2820$, $\Lambda(\Upsilon_2) = 0.3837$, $\Lambda(\Upsilon_3) = 0.4166$, and $\Lambda(\Upsilon_4) = 0.4099$, which gives the corresponding ranking order $\aleph_3 \succ \aleph_4 \succ \aleph_2 \succ \aleph_1$ from which the mentioned $\aleph_3$ is the optimal road construction company.

**(5)** For introducing the weighted form of Logarithmic picture fuzzy average (Log-PFWA) operator applied by Khan et al. [29], the computed scores are $\Lambda(\Upsilon_1) = 0.6117$, $\Lambda(\Upsilon_2) = 0.6761$, $\Lambda(\Upsilon_3) = 0.6974$, and $\Lambda(\Upsilon_4) = 0.6956$, which indicates the corresponding ordering of the alternatives as $\aleph_3 \succ \aleph_4 \succ \aleph_2 \succ \aleph_1$. Hence, $\aleph_3$ is the appropriate fit for the road construction company.

**(6)** On applying the weighted form of the picture fuzzy average operator (PFWA) considered by Wei [49], then the computed scores for each option are $\Lambda(\Upsilon_1) = 0.2846$, $\Lambda(\Upsilon_2) = 0.3913$, $\Lambda(\Upsilon_3) = 0.4263$, and $\Lambda(\Upsilon_4) = 0.4222$, the corresponding ordering of the road construction company $\aleph_3 \succ \aleph_4 \succ \aleph_2 \succ \aleph_1$, which shows that $\aleph_3$ is the required road construction company.

**(7)** On applying the weighted form of the picture fuzzy geometric operator (PFWG) proposed by Wei [49], the scores of each option are evaluated as $\Lambda(\Upsilon_1) = 0.2825$, $\Lambda(\Upsilon_2) = 0.3848$, $\Lambda(\Upsilon_3) = 0.4148$, and $\Lambda(\Upsilon_4) = 0.4076$, which gives the order $\aleph_3 \succ \aleph_4 \succ \aleph_2 \succ \aleph_1$, and again shows that $\aleph_3$ is the most suitable road construction company.

**(8)** Further, on applying the weighted form of the picture fuzzy Dombi averaging operator (PFDWA) proposed by Jana et al. [17], the score and ranking orders of the options are shown in Table 3.7.

**Table 3.7** Ranking order using the PFDWA operator.

| $\varrho$ | $\Lambda(\Upsilon_1)$ | $\Lambda(\Upsilon_2)$ | $\Lambda(\Upsilon_3)$ | $\Lambda(\Upsilon_4)$ | Ranking order |
|---|---|---|---|---|---|
| 1 | 0.2860 | 0.4182 | 0.4337 | 0.4314 | $\aleph_3 \succ \aleph_4 \succ \aleph_2 \succ \aleph_1$ |
| 2 | 0.2880 | 0.4131 | 0.4431 | 0.4395 | $\aleph_3 \succ \aleph_4 \succ \aleph_2 \succ \aleph_1$ |
| 3 | 0.2901 | 0.4076 | 0.4513 | 0.4509 | $\aleph_3 \succ \aleph_4 \succ \aleph_2 \succ \aleph_1$ |
| 4 | 0.2921 | 0.4018 | 0.4583 | 0.4581 | $\aleph_3 \succ \aleph_4 \succ \aleph_2 \succ \aleph_1$ |
| 5 | 0.3442 | 0.3956 | 0.4644 | 0.4640 | $\aleph_3 \succ \aleph_4 \succ \aleph_2 \succ \aleph_1$ |

From Table 3.7, it appears that no changes occur in the ranking orders of the alternatives for the range $1 \leq \varrho \leq 5$, but for all the cases $\aleph_3$ is the most desirable road construction company.

**(9)** Further, on the weighted form of the picture fuzzy Dombi geometric operator (PFDWG) considered by Jana et al. [17], then the score and ranking orders are as in Table 3.8.

From Table 3.8, it appears that there are some changes that occur in the ranking orders of the alternatives for the range $3 \leq \varrho \leq 5$, but in all the cases $\aleph_3$ is the most suitable road construction company for the PFDWG operator.

**Table 3.8** Ranking order using the PFDWAG operator.

| $\varrho$ | $\Lambda(\Upsilon_1)$ | $\Lambda(\Upsilon_2)$ | $\Lambda(\Upsilon_3)$ | $\Lambda(\Upsilon_4)$ | Ranking order |
|---|---|---|---|---|---|
| 1 | 0.2817 | 0.3827 | 0.4117 | 0.4030 | $\aleph_3 \succ \aleph_4 \succ \aleph_2 \succ \aleph_1$ |
| 2 | 0.2796 | 0.3763 | 0.3993 | 0.3862 | $\aleph_3 \succ \aleph_4 \succ \aleph_2 \succ \aleph_1$ |
| 3 | 0.2777 | 0.3698 | 0.3865 | 0.3689 | $\aleph_3 \succ \aleph_4 \succ \aleph_2 \succ \aleph_1$ |
| 4 | 0.2757 | 0.3636 | 0.3741 | 0.3525 | $\aleph_3 \succ \aleph_4 \succ \aleph_2 \succ \aleph_1$ |
| 5 | 0.2738 | 0.3576 | 0.3623 | 0.3375 | $\aleph_3 \succ \aleph_4 \succ \aleph_2 \succ \aleph_1$ |

Therefore from the above-executed comparative study with the present papers [17, 27,28,49,50] although rating grades for all options are different the ranking remains the same as $\aleph_3 \succ \aleph_4 \succ \aleph_2 \succ \aleph_1$ for the existing operators PFHWA (PFHWG), PFEWA (PFEWG), Log-PFWA, PFWA (PFWG), and PFDWA (PFDWG) operators with the proposed operators PFSWA (PFSWG). Hence, the proposed model is stable in the PFSS environment.

## 3.7 Advantages of the approach

It is seen from Section 3.6 that the considered method is stable compared with the present technique. The methods imply that $A_3$ is the best road construction company evaluated by experts. The proposed MCGDM in the PFS environment utilizing this advanced approach to compare the existing methods [17,27,28,49,50] where these decision-making processes only aggregated picture fuzzy arguments, but do not take into account its parameter information. Also, existing papers [26,27,55] under the PFS environment only considered algebraic operations on the PFS environment, but have not aggregating PFS arguments in a PFS environment. The advantages of the proposed approach facilitate the descriptions of decision information in real-world problems emphasizing their parameterizations property, which is shown in Table 3.9. Therefore the proposed MCGDM under a PFS environment can be utilized to explain the decision-making approach instead of other preexisting methods in the environment of PFSNs.

**Table 3.9** Characteristic comparisons with some of the existing methods.

| Methods | Whether it makes information aggregation easier | Provides aggregate parameter information |
|---|---|---|
| Jana et al. [17] | ✓ | × |
| Khan et al. [27] | ✓ | × |
| Khan et al. [28] | ✓ | × |
| Wei [49] | ✓ | × |
| Wei [50] | ✓ | × |
| Proposed method | ✓ | ✓ |

## 3.8 Conclusions

Here, we have studied MCGDM problems using PFSWA and PFSWG operators in a picture fuzzy soft environment. We have defined two new operators, namely, PFSWA and PFSWG operators to aggregate PFSNs. The properties of these operators was proved in detail. A road construction company selection problems has been developed using PFSWA and PFSWG operators in a PFS environment. In future, the proposed MCGDM model can be applied to risk analysis, re-grate theory based-decision making and intelligent diagnosis, management, business and economic decision making in picture fuzzy soft environments.

## References

[1] M. Ali, F. Feng, X. Liu, W. Min, M. Shabir, On some new operations in soft set theory, Comput. Math. Appl. 57 (9) (2009) 1547–1553.

[2] S. Alkhazaleh, A.R. Salleh, Fuzzy soft expert set and its application, Appl. Math. 5 (2014) 1349–1368.

[3] K.T. Atanassov, G. Gargov, Interval-valued intuitionistic fuzzy sets, Fuzzy Sets Syst. 31 (1989) 343–349.

[4] R. Arora, H. Garg, Prioritized averaging/geometric aggregation operators under the intuitionistic fuzzy soft set environment, Sci. Iran. E 25 (1) (2018) 466–482.

[5] K.T. Atanassov, On Intuitionistic Fuzzy Sets Theory, Studies in Fuzziness and Soft Computing, vol. 283, Springer-Verlag, Berlin Heidelberg, 2012.

[6] K.V. Babitha, S.J. John, Hesistant fuzzy soft sets, J. New Results Sci. 3 (2013) 98–107.

[7] G. Beliakov, A. Pradera, T. Calvo, Aggregation Functions: A Guide for Practitioners, Springer, Heidelberg, Berlin, New York, 2007.

[8] G. Beliakov, H. Bustince, D.P. Goswami, U.K. Mukherjee, N.R. Pal, On averaging operators for Atanassov's intuitionistic fuzzy sets, Inf. Sci. 181 (2011) 1116–1124.

[9] N. Cagman, F. Citak, S. Enginoglu, Fuzzy parameterized fuzzy soft set theory and its applications, Turk. J. Fuzzy Syst. 1 (1) (2001) 21–35.

[10] N. Cagman, I. Deli, Intuitionistic fuzzy parameterized soft set theory and its decision making, Appl. Soft Comput. 28 (2015) 109–113.

[11] B.C. Cuong, Picture fuzzy sets first results. Part 1, Seminar "Neuro-fuzzy systems with applications", Tech. Rep., Institute of Mathematics, Hanoi, 2013.

[12] B.C. Cuong, Picture fuzzy sets first results. Part 2, Seminar "Neuro-fuzzy systems with applications", Tech. Rep., Institute of Mathematics, Hanoi, 2013.

[13] Y. Du, P. Liu, Extended fuzzy VIKOR method with intuitionistic trapezoidal fuzzy numbers, Inf. Int. Interdiscip. J. 14 (8) (2011) 2575–2583.

[14] H. Garg, R. Arora, Generalized and group-based generalized intuitionistic fuzzy soft sets with applications in decision-making, Appl. Intell. 48 (2) (2018) 343–356.

[15] H. Garg, N. Agarwal, A. Tripathi, Entropy based multicriteria decision making method under fuzzy environment and unknown attribute weights, Glob. J. Technol. Optim. 6 (2015) 13–20.

[16] H. Garg, Some picture fuzzy aggregation operators and their applications to multicriteria decision-making, Arab. J. Sci. Eng. 42 (12) (2017) 5275–5290.

[17] C. Jana, T. Senapati, M. Pal, R.R. Yager, Picture fuzzy Dombi aggregation operators: application to MADM process, Appl. Soft Comput. 74 (2019) 99–109.

[18] C. Jana, M. Pal, J.Q. Wang, Bipolar fuzzy Dombi aggregation operators and its application in multiple attribute decision making process, J. Ambient Intell. Humaniz. Comput. 10 (9) (2019) 3533–3549.

[19] C. Jana, M. Pal, F. Karaaslan, J.Q. Wang, Trapezoidal neutrosophic aggregation operators and its application in multiple attribute decision -making process, Sci. Iran. E (2018), https://doi.org/10.24200/sci.2018.51136.2024.

[20] C. Jana, M. Pal, Assessment of enterprise performance based on picture fuzzy Hamacher aggregation operators, Symmetry 11 (1) (2019) 75, https://doi.org/10.3390/sym11010075.

[21] C. Jana, M. Pal, A robust single-valued neutrosophic soft aggregation operators in multi-criteria decision making, Symmetry 11 (1) (2019) 110, https://doi.org/10.3390/sym11010110.

[22] C. Jana, T. Senapati, M. Pal, Pythagorean fuzzy Dombi aggregation operators and its applications in multiple attribute decision-making, Int. J. Intell. Syst. 34 (2019) 2019–2038.

[23] C. Jana, M. Pal, J.Q. Wang, Bipolar fuzzy Dombi prioritized aggregation operators in multiple attribute decision making, Soft Comput. (2019), https://doi.org/10.1007/s00500-019-04130-z.

[24] Y. Jiang, Y. Tang, Q. Chen, H. Liu, J. Tang, Interval-valued intuitionistic fuzzy soft sets and their properties, Comput. Math. Appl. 60 (3) (2010) 906–918.

[25] A. Khalid, M. Abbas, Distance measures and operations in intuitionistic and interval- valued intuitionistic fuzzy soft set theory, Int. J. Fuzzy Syst. 17 (3) (2015) 490–497.

[26] A.M. Khalil, S.G. Li, H. Garg, H.X. Li, S.Q. Ma, New operations on interval-valued picture fuzzy set, interval-valued picture fuzzy soft set and their applications, IEEE Access 7 (1) (2019) 51236–51253.

[27] S. Khan, S. Abdullha, S. Ashraf, Picture fuzzy aggregation information based on Einstein operations and their application in decision making, Math. Sci. 13 (3) (2019) 213–229.

[28] S. Khan, S. Abdullha, L. Abdullha, S. Ashraf, Logarithmic aggregation operators of picture fuzzy numbers for multi-attribute decision making problems, Mathematics 7 (2019) 608, https://doi.org/10.3390/math7070608.

[29] M.J. Khan, P. Kumam, S. Ashraf, W. Kumam, Generalized picture fuzzy soft sets and their application in decision support systems, Symmetry 11 (2019) 415, https://doi.org/10.3390/sym11030415.

[30] P. Liu, Multi-attribute decision making method research based on interval vague set and TOPSIS method, Technol. Econ. Dev. Econ. 15 (3) (2009) 453–463.

[31] P. Liu, X. Zhang, A novel picture fuzzy linguistic aggregation operator and its application to group decision-making, Cogn. Comput. 10 (2) (2018) 242–259.

[32] R. Lourenzuttia, R.A. Krohlingb, A study of TODIM in a intuitionistic fuzzy and random environment, Expert Syst. Appl. 40 (16) (2013) 6459–6468.

[33] P.K. Maji, R. Biswas, A.R. Roy, Fuzzy soft sets, J. Fuzzy Math. 9 (3) (2001) 589–602.

[34] P.K. Maji, R. Biswas, A.R. Roy, Intuitionistic fuzzy soft sets, J. Fuzzy Math. 9 (3) (2001) 677–692.

[35] P.K. Maji, R. Biswas, A.R. Roy, Soft set theory, Comput. Math. Appl. 45 (4–5) (2003) 555–562.

[36] D. Molodtsov, Soft set theory-first results, Comput. Math. Appl. 27 (4–5) (1999) 19–31.

[37] J.J. Peng, J.Q. Wang, X.H. Wu, H.Y. Zhang, X.H. Chen, The fuzzy cross-entropy for intuitionistic hesitant fuzzy sets and its application in multi-criteria decision-making, Int. J. Syst. Sci. 46 (13) (2015) 2335–2350.

[38] X. Peng, J. Dai, Algorithm for picture fuzzy multiple attribute decision making based on new distance measure, Int. J. Uncertain. Quantificat. 7 (2017) 177–187.

[39] G. Selvachandran, X. Peng, A modified TOPSIS method based on vague parameterized vague soft sets and its application to supplier selection problems, Neural Comput. Appl. (2018), https://link.springer.com/article/10.1007/s00521-018-3409-1.

[40] P. Singh, Correlation coefficients for picture fuzzy sets, J. Intell. Fuzzy Syst. 27 (2014) 2857–2868.

[41] L.H. Son, Measuring analogousness in picture fuzzy sets: from picture distance measures to picture association measures, Fuzzy Optim. Decis. Mak. 16 (3) (2017) 1–20.

[42] L.H. Son, P. Viet, P. Hai, Picture inference system: a new fuzzy inference system on picture fuzzy set, Appl. Intell. 46 (3) (2017) 652–669.

[43] L.H. Son, DPFCM: a novel distributed picture fuzzy clustering method on picture fuzzy sets, Expert Syst. Appl. 2 (2015) 51–66.

[44] P.H. Thong, L.H. Son, Picture fuzzy clustering for complex data, Eng. Appl. Artif. Intell. 56 (2016) 121–130.

[45] P.H. Thong, L.H. Son, A novel automatic picture fuzzy clustering method based on particle swarm optimization and picture composite cardinality, Knowl.-Based Syst. 109 (2016) 48–60.

[46] W.Z. Wang, X.W. Liu, Intuitionistic fuzzy geometric aggregation operators based on Einstein operations, Int. J. Intell. Syst. 26 (2011) 1049–1075.

[47] W. Wang, X. Liu, Intuitionistic fuzzy information aggregation using Einstein operations, IEEE Trans. Fuzzy Syst. 20 (5) (2012) 923–938.

[48] G.W. Wei, Picture fuzzy cross-entropy for multiple attribute decision making problems, J. Bus. Econ. Manag. 17 (4) (2016) 491–502.

[49] G.W. Wei, Picture fuzzy aggregation operators and their application to multiple attribute decision making, J. Intell. Fuzzy Syst. 33 (2017) 713–724.

[50] G.W. Wei, Picture fuzzy Hamacher aggregation operators and their application to multiple attribute decision making, Fundam. Inform. 157 (3) (2018) 271–320.
[51] Z.S. Xu, R.R. Yager, Some geometric aggregation operators based on intuitionistic fuzzy sets, Int. J. Gen. Syst. 35 (2006) 417–433.
[52] Z.S. Xu, Intuitionistic fuzzy aggregation operators, IEEE Trans. Fuzzy Syst. 15 (2007) 1179–1187.
[53] Z.S. Xu, H. Hu, Projection models for intuitionistic fuzzy multiple attribute decision making, Int. J. Inf. Technol. Decis. Mak. 9 (2010) 267–280.
[54] X. Yang, T.Y. Lin, J. Yang, Y. Li, D. Yu, Combination of interval-valued fuzzy set and soft set, Comput. Math. Appl. 58 (3) (2009) 521–527.
[55] Y. Yang, C. Liang, S. Ji, T. Liu, Adjustable soft discernibility matrix based on picture fuzzy soft sets and its applications in decision making, J. Intell. Fuzzy Syst. 29 (2015) 1711–1722.
[56] L.A. Zadeh, Fuzzy sets, Inf. Control 8 (1965) 338–353.
[57] E.K. Zavadskas, J. Antucheviciene, S.H.R. Hajiagha, S.S. Hashemi, Extension of weighted aggregated sum product assessment with interval-valued intuitionistic fuzzy numbers (WASPAS-IVIF), Appl. Soft Comput. 24 (2014) 1013–1021.

# 4

# Picture fuzzy Dombi operators and their applications in multiattribute decision-making processes

## 4.1 Introduction

A generalization of the intuitionistic fuzzy set (IFS) [1], which is an extension of the fuzzy sets (FSs) [33], is the picture fuzzy set (PFS) [3,4]. Recently, some research was built into the PFSs environment: Sing [13] developed a PFS correlation metric before using it in a clustering technique. Based on a PFSs environment, Son and coworkers [14,20] offer time-series forecasting and weather forecasting based on numerous innovative fuzzy algorithms. The estimation that was provided was based on fuzzy computations in the PFS domain. In order to obtain a healthy conclusion and pay attention to supportive social frameworks, Thong et al. [22] examined a hybrid prototype technique in the PFS domain and IFS frameworks. The community models, generalized distance models, and illustrated separation were utilized in PFS networks [15,16]. A fuzzy derivation method inference system on PFS was used by Son et al. [17] to enhance the use of the conventional fuzzy reasoning methodology. Thong et al. [18,19] used particle swarm optimization along with the PFS clustering approach for complex data, and studied the generalized PFS distance measure and used it to resolve clustering analysis in the context of PFSs. Wei introduced AOs to collect PFS data and used them in MADM for ranking EPR models [28]. Wei [27] employed the crossentropy measurements of picture fuzzy preference in order to rank the possibilities. Garg [6] defined a novel method for compiling PFS data and created the MCDM method. An algorithm was created by Peng et al. [10] to rank the items in the decision-making process. Readers are directed to [11,21,23–26,32] for more information on PFS models.

Dombi [5] introduced a brand-new idea in 1982 known as Dombi norms, which have set a solid precedent for operating conditions. Liu et al. applied them to IFSs and proposed MAGDM difficulties in the context of Dombi–Bonferroni mean operators in that topic [9]. In the SVN environment, Chen and Ye [2] used Dombi norms to solve the MADM approach. Using Dombi norms, He [7] examined MADM challenges based on Typhoon disasters in the hesitant fuzzy environment. Dombi concepts have been extended to neutrosophic cubic sets by Shi and Ye [12], creating a problem with journey management.

These evaluations and examples lead us to believe that the PFS has enough fidelity to display the unbelievable and usable data that appear in real-world situations. As mentioned, decision-making problems (DMPs) in various fuzzy collecting environments under Dombi actions [7,9] and traditional AOs [30,31] motivate us to make significant improvements to the current study. The primary goal of this chapter is to make some AOs under PFS

Picture Fuzzy Logic and Its Applications in Decision Making Problems. https://doi.org/10.1016/B978-0-44-322024-1.00008-X

data known as PFD aggregations manifest for comparing the various priority of the options among the DMP. Regardless of the highly inventive prior methods used in this subject. We have worked hard and honestly, leaving no stone unturned to demonstrate the proposed solution in such a way that it would surpass all prior help to address the practical issue.

## 4.2 Preliminaries

In this section, we annotate some useful ideas of PFS of the universe X.

Wei [28] derived picture fuzzy AOs provided in the following definitions.

**Definition 4.1.** Let $\mathcal{R}_f = (\mathcal{Y}_f, \mathcal{A}_f, \mathcal{N}_f)$, $f = 1, 2, \ldots \varpi$ be a group of PFNs. Let PFWA be an operator of dimension $\varpi$ such that $\Theta^{\varpi} \to \Theta$ with weighting vector be $\Xi = (\Xi_1, \Xi_2, \ldots, \Xi_{\varpi})^T$, where $\Xi > 0$ and $\sum_{f=1}^{\varpi} \Xi_f = 1$, such as $PFWA_{\Xi}(\mathcal{R}_1, \tilde{\mathcal{R}}_2, \ldots, \mathcal{R}_{\varpi}) = \bigoplus_{f=1}^{\varpi}(\Xi_f \mathcal{R}_f) = \left(1 - \prod_{f=1}^{\varpi}(1-\mathcal{Y}_f)^{\Xi_f}, \prod_{f=1}^{\varpi} \mathcal{A}_b{}^{\Xi_f}, \prod_{f=1}^{\varpi} \mathcal{N}_f{}^{\Xi_f}\right)$.

**Definition 4.2.** Let $\mathcal{R}_f = (\mathcal{Y}_f, \mathcal{A}_f, \mathcal{N}_f)$ $f = 1, 2, \ldots \varpi$ be a group of PFNs. Let PFOWA be an operator of dimension $\varpi$, $\Theta^{\varpi} \to \Theta$ for the weighting vector $w = (\delta_1, \delta_2, \ldots, \delta_{\varpi})^T$, where $\delta_f > 0$ and $\sum_{f=1}^{\varpi} \delta_f = 1$. Furthermore,

$$PFOWA_w(\mathcal{R}_1, \mathcal{R}_2, \ldots, \mathcal{R}_{\varpi}) = \bigoplus_{f=1}^{\varpi}\left(\delta_f p_{\sigma(f)}\right) = \left(1 - \prod_{f=1}^{\varpi}\left(1 - \mathcal{Y}_{\sigma(f)}\right)^{\delta_f}, \prod_{f=1}^{\varpi} \mathcal{A}_{\sigma(f)}^{\delta_f}, \prod_{f=1}^{\varpi} \mathcal{N}_{\sigma(s)}^{\delta_f}\right),$$

where $(\sigma(1), \sigma(2), \ldots, \sigma(\varpi))$ is a permutation of $(1, 2, \ldots, \varpi)$, $p_{\sigma(f-1)} \geq p_{\sigma(f)}$ for all $f = 1, 2, \ldots, \varpi$.

**Definition 4.3.** Let $\mathcal{R}_f = (\mathcal{Y}_f, \mathcal{A}_f, \mathcal{N}_f)$, $f = 1, 2, \ldots, \varpi$ be a group of PFNs. Let PFHWA be a function of dimension $\varpi$, $\Theta^{\varpi} \to \Theta$ having associated weight vector $\Xi = (\Xi_1, \Xi_2, \ldots, \Xi_{\varpi})^T$, $\Xi > 0$ and $\sum_{f=1}^{\varpi} \Xi_f = 1$, as

$$PFHWA_w(\mathcal{R}_1, \mathcal{R}_2, \ldots, \mathcal{R}_{\varpi}) = \bigoplus_{f=1}^{\varpi}\left(\delta_f \mathcal{R}_{\sigma(f)}\right) = \left(1 - \prod_{f=1}^{\varpi}\left(1 - \dot{\mathcal{Y}}_{\sigma(f)}\right)^{\delta_f}, \prod_{f=1}^{\varpi} \dot{\mathcal{A}}_{\sigma(f)}^{\delta_f}, \prod_{f=1}^{\varpi} \dot{\mathcal{N}}_{\sigma(f)}^{\delta_f}\right),$$

and $\mathcal{R}_{\sigma(f)}$ is the $f$th largest weighted PFVs $\mathcal{R}_f$ ($\dot{\mathcal{R}}_f = \varpi \Xi_f \mathcal{R}_f$, $f = 1, 2, \ldots, \varpi$) and weighting vector $\Xi = (\Xi_1, \Xi_2, \ldots, \Xi_{\varpi})^T$ of $\mathcal{R}_f$, where $\Xi_f > 0$ and $\sum_{f=1}^{\varpi} \Xi_f = 1$, where $\varpi$ is the balancing coefficient.

**Definition 4.4.** [8] Let $\mathcal{R}_1 = (\mathcal{Y}_1, \mathcal{A}_1, \mathcal{N}_1)$ be PFNs, then score $\Lambda(\mathcal{R}_1)$ and accuracy $\Phi(\mathcal{R}_1)$ for PFN are defined as follows:

$$\Lambda(\mathcal{R}_1) = \frac{1 + \mathcal{Y}_1 - \mathcal{N}_1}{2}, \quad \Lambda(\mathcal{R}_1) \in [0, 1],$$
$$\Phi(\mathcal{R}_1) = \mathcal{Y}_1 - \mathcal{N}_1, \quad \Phi(\mathcal{R}_1) \in [-1, 1]. \tag{4.1}$$

Based on Definition 4.4, prioritized relations between two PFNs $\mathcal{R}_1$ and $\mathcal{R}_2$ are defined in the following ways.

**Definition 4.5.** **(i)** If $\Lambda(\mathcal{R}_1) < \Lambda(\mathcal{R}_2)$, indicates $\mathcal{R}_1 \prec \mathcal{R}_2$;
**(ii)** If $\Lambda(\mathcal{R}_1) > \Lambda(\mathcal{R}_2)$, indicates $\mathcal{R}_1 \succ \mathcal{R}_2$;
**(iii)** If $\Lambda(\mathcal{R}_1) = \Lambda(\mathcal{R}_2)$, then
 **(1)** If $\Phi(\mathcal{R}_1) < \Phi(\mathcal{R}_2)$, indicates $\mathcal{R}_1 \prec \mathcal{R}_2$;
 **(2)** If $\Phi(\mathcal{R}_1) > \Phi(\mathcal{R}_2)$, indicates $\mathcal{R}_1 \succ \mathcal{R}_2$;
 **(3)** If $\Phi(\mathcal{R}_1) = \Phi(\mathcal{R}_2)$, indicates $\mathcal{R}_1 \sim \mathcal{R}_2$.

## 4.3 Picture fuzzy Dombi weighted average operators

In this section, we describe Dombi arithmetic AOs in picture fuzzy environments such as PFDWA, PFDOWA, and PFDHWA operators.

**Theorem 4.1.** *Let $\mathcal{R} = (\mathcal{Y}, \mathcal{A}, \mathcal{N})$, $\mathcal{R}_1 = (\mathcal{Y}_1, \mathcal{A}_1, \mathcal{N}_1)$, $\mathcal{R}_2 = (\mathcal{Y}_2, \mathcal{A}_2, \mathcal{N}_2)$ be three PFNs, and $\lambda$, $\lambda_1$, $\lambda_2$ be the scalars. Then,*

**(i)** $\mathcal{R}_1 \oplus \mathcal{R}_2 = \mathcal{R}_2 \oplus \mathcal{R}_1$;
**(ii)** $\mathcal{R}_1 \otimes \mathcal{R}_2 = \mathcal{R}_2 \otimes \mathcal{R}_1$;
**(iii)** $\lambda(\mathcal{R}_1 \oplus \mathcal{R}_2) = \lambda\mathcal{R}_1 \oplus \lambda\mathcal{R}_2$, $\lambda > 0$;
**(iv)** $(\lambda_1 + \lambda_2)\mathcal{R} = \lambda_1\mathcal{R} \oplus \lambda_2\mathcal{R}$, $\lambda_1, \lambda_2 > 0$;
**(v)** $(\mathcal{R}_1 \otimes \mathcal{R}_2)^\lambda = \mathcal{R}_1^\lambda \otimes \mathcal{R}_2^\lambda$, $\lambda > 0$;
**(vi)** $\mathcal{R}^{\lambda_1} \otimes \mathcal{R}^{\lambda_2} = \mathcal{R}^{(\lambda_1+\lambda_2)}$, $\lambda_1, \lambda_2 > 0$.

*Proof.* Let three PFNs be $\mathcal{R}$, $\mathcal{R}_1$, and $\mathcal{R}_2$, and $\lambda, \lambda_1, \lambda_2 > 0$, we can obtain

**(i)**

$$
\begin{aligned}
&\mathcal{R}_1 \oplus \mathcal{R}_2 \\
&= \left(1 - \frac{1}{1+\left\{\left(\frac{\mathcal{Y}_1}{1-\mathcal{Y}_1}\right)^\varrho + \left(\frac{\mathcal{Y}_2}{1-\mathcal{Y}_2}\right)^\varrho\right\}^{1/\varrho}}, \frac{1}{1+\left\{\left(\frac{1-\mathcal{A}_1}{\mathcal{A}_1}\right)^\varrho + \left(\frac{1-\mathcal{A}_2}{\mathcal{A}_2}\right)^\varrho\right\}^{1/\varrho}}, \right. \\
&\qquad \left. \frac{1}{1+\left\{\left(\frac{1-\mathcal{N}_1}{\mathcal{N}_1}\right)^\varrho + \left(\frac{1-\mathcal{N}_2}{\mathcal{N}_2}\right)^\varrho\right\}^{1/\varrho}} \right) \\
&= \left(1 - \frac{1}{1+\left\{\left(\frac{\mathcal{Y}_2}{1-\mathcal{Y}_2}\right)^\varrho + \left(\frac{\mathcal{Y}_1}{1-\mathcal{Y}_1}\right)^\varrho\right\}^{1/\varrho}}, \frac{1}{1+\left\{\left(\frac{1-\mathcal{A}_2}{\mathcal{A}_2}\right)^\varrho + \left(\frac{1-\mathcal{A}_1}{\mathcal{A}1}\right)^\varrho\right\}^{1/\varrho}}, \right. \\
&\qquad \left. \frac{1}{1+\left\{\left(\frac{1-\mathcal{N}_2}{\mathcal{N}_2}\right)^\varrho + \left(\frac{1-\mathcal{N}_1}{\mathcal{N}_1}\right)^\varrho\right\}^{1/\varrho}} \right) \\
&= \mathcal{R}_2 \oplus \mathcal{R}_1.
\end{aligned}
$$

**(ii)** Similar to (i).

**(iii)** Let $t = 1 - \frac{1}{1+\left\{\left(\frac{\mathcal{Y}_1}{1-\mathcal{Y}_1}\right)^{\varrho}+\left(\frac{\mathcal{Y}_2}{1-\mathcal{Y}_2}\right)^{\varrho}\right\}^{1/\varrho}}$.

Then, $\frac{t}{1-t} = \left\{\left(\frac{\mathcal{Y}_1}{1-\mathcal{Y}_1}\right)^{\varrho}+\left(\frac{\mathcal{Y}_2}{1-\mathcal{Y}_2}\right)^{\varrho}\right\}^{1/\varrho}$.

Therefore $\left(\frac{t}{1-t}\right)^{\varrho} = \left(\frac{\mathcal{Y}_1}{1-\mathcal{Y}_1}\right)^{\varrho}+\left(\frac{\mathcal{Y}_2}{1-\mathcal{Y}_2}\right)^{\varrho}$.

Using the above discussion, we obtain

$$\begin{aligned}
&\lambda(\mathcal{R}_1 \oplus \mathcal{R}_2)\\
=&\lambda\left(1-\frac{1}{1+\left\{\left(\frac{\mathcal{Y}_1}{1-\mathcal{Y}_1}\right)^{\varrho}+\left(\frac{\mathcal{Y}_2}{1-\mathcal{Y}_2}\right)^{\varrho}\right\}^{1/\varrho}}, \frac{1}{1+\left\{\left(\frac{1-\mathcal{A}_1}{\mathcal{A}_1}\right)^{\varrho}+\left(\frac{1-\mathcal{A}_2}{\mathcal{A}_2}\right)^{\varrho}\right\}^{1/\varrho}},\right.\\
&\left.\frac{1}{1+\left\{\left(\frac{1-\mathcal{N}_1}{\mathcal{N}_1}\right)^{\varrho}+\left(\frac{1-\mathcal{N}_2}{\mathcal{N}_2}\right)^{\varrho}\right\}^{1/\varrho}}\right)\\
=&\left(1-\frac{1}{1+\left\{\lambda\left(\frac{\mathcal{Y}_1}{1-\mathcal{Y}_1}\right)^{\varrho}+\lambda\left(\frac{\mathcal{Y}_2}{1-\mathcal{Y}_2}\right)^{\varrho}\right\}^{1/\varrho}}, \frac{1}{1+\left\{\lambda\left(\frac{1-\mathcal{A}_1}{\mathcal{A}_1}\right)^{\varrho}+\lambda\left(\frac{1-\mathcal{A}_2}{\mathcal{A}_2}\right)^{\varrho}\right\}^{1/\varrho}},\right.\\
&\left.\frac{1}{1+\left\{\lambda\left(\frac{1-\mathcal{N}_1}{\mathcal{N}_1}\right)^{\varrho}+\lambda\left(\frac{1-\mathcal{N}_2}{\mathcal{N}_2}\right)^{\varrho}\right\}^{1/\varrho}}\right).
\end{aligned}$$

Now,

$$\begin{aligned}
&\lambda\mathcal{R}_1 \oplus \lambda\mathcal{R}_2\\
=&\left(1-\frac{1}{1+\left\{\lambda\left(\frac{\mathcal{Y}_1}{1-\mathcal{Y}_1}\right)^{\varrho}\right\}^{1/\varrho}}, \frac{1}{1+\left\{\lambda\left(\frac{1-\mathcal{A}_1}{\mathcal{A}_1}\right)^{\varrho}\right\}^{1/\varrho}}, \frac{1}{1+\left\{\lambda\left(\frac{1-\mathcal{N}_1}{\mathcal{N}_1}\right)^{\varrho}\right\}^{1/\varrho}}\right)\\
&\oplus\left(1-\frac{1}{1+\left\{\lambda\left(\frac{\mathcal{Y}_2}{1-\mathcal{Y}_2}\right)^{\varrho}\right\}^{1/\varrho}}, \frac{1}{1+\left\{\lambda\left(\frac{1-\mathcal{A}_2}{\mathcal{A}_2}\right)^{\varrho}\right\}^{1/\varrho}}, \frac{1}{1+\left\{\lambda\left(\frac{1-\mathcal{N}_2}{\mathcal{N}_2}\right)^{\varrho}\right\}^{1/\varrho}}\right)\\
=&\left(1-\frac{1}{1+\left\{\lambda\left(\frac{\mathcal{Y}_1}{1-\mathcal{Y}_1}\right)^{\varrho}+\lambda\left(\frac{\mathcal{Y}_2}{1-\mathcal{Y}_2}\right)^{\varrho}\right\}^{1/\varrho}}, \frac{1}{1+\left\{\lambda\left(\frac{1-\mathcal{A}_1}{\mathcal{A}_1}\right)^{\varrho}+\lambda\left(\frac{1-\mathcal{A}_2}{\mathcal{A}_2}\right)^{\varrho}\right\}^{1/\varrho}},\right.\\
&\left.\frac{1}{1+\left\{\lambda\left(\frac{1-\mathcal{N}_1}{\mathcal{N}_1}\right)^{\varrho}+\lambda\left(\frac{1-\mathcal{N}_2}{\mathcal{N}_2}\right)^{\varrho}\right\}^{1/\varrho}}\right)\\
=&\lambda(\mathcal{R}_1 \oplus \mathcal{R}_2).
\end{aligned}$$

**(iv)**

$$
\begin{aligned}
&\lambda_1\mathcal{R}\oplus\lambda_2\mathcal{R}\\
&=\left(1-\frac{1}{1+\left\{\lambda_1\left(\frac{\mathcal{Y}}{1-\mathcal{Y}}\right)^{\varrho}\right\}^{1/\varrho}},\frac{1}{1+\left\{\lambda_1\left(\frac{1-\mathcal{A}}{\mathcal{A}}\right)^{\varrho}\right\}^{1/\varrho}},\frac{1}{1+\left\{\lambda_1\left(\frac{1-\mathcal{N}}{\mathcal{N}}\right)^{\varrho}\right\}^{1/\varrho}}\right)\\
&\oplus\left(1-\frac{1}{1+\left\{\lambda_2\left(\frac{\mathcal{Y}}{1-\mathcal{Y}}\right)^{\varrho}\right\}^{1/\varrho}},\frac{1}{1+\left\{\lambda_2\left(\frac{1-\mathcal{A}}{\mathcal{A}}\right)^{\varrho}\right\}^{1/\varrho}},\frac{1}{1+\left\{\lambda_2\left(\frac{1-\mathcal{N}}{\mathcal{N}}\right)^{\varrho}\right\}^{1/\varrho}}\right)\\
&=\left(1-\frac{1}{1+\left\{(\lambda_1+\lambda_2)\left(\frac{\mathcal{Y}}{1-\mathcal{Y}}\right)^{\varrho}\right\}^{1/\varrho}},\frac{1}{1+\left\{(\lambda_1+\lambda_2)\left(\frac{1-\mathcal{A}}{\mathcal{A}}\right)^{\varrho}\right\}^{1/\varrho}},\right.\\
&\qquad\left.\frac{1}{1+\left\{(\lambda_1+\lambda_2)\left(\frac{1-\mathcal{N}}{\mathcal{N}}\right)^{\varrho}\right\}^{1/\varrho}}\right)\\
&=(\lambda_1+\lambda_2)\mathcal{R}.
\end{aligned}
$$

**(v)**

$$
\begin{aligned}
&(\mathcal{R}_1\otimes\mathcal{R}_2)^{\lambda}\\
&=\left(\frac{1}{1+\left\{\left(\frac{1-\mathcal{Y}_1}{\mathcal{Y}_1}\right)^{\varrho}+\left(\frac{1-\mathcal{Y}_2}{\mathcal{Y}_2}\right)^{\varrho}\right\}^{1/\varrho}},1-\frac{1}{1+\left\{\left(\frac{\mathcal{A}_1}{1-\mathcal{A}_1}\right)^{\varrho}+\left(\frac{\mathcal{A}_2}{1-\mathcal{A}_2}\right)^{\varrho}\right\}^{1/\varrho}},\right.\\
&\qquad\left.1-\frac{1}{1+\left\{\left(\frac{\mathcal{N}_1}{1-\mathcal{N}_1}\right)^{\varrho}+\left(\frac{\mathcal{N}_2}{1-\mathcal{N}_2}\right)^{\varrho}\right\}^{1/\varrho}}\right)^{\lambda}\\
&=\left(\frac{1}{1+\left\{\lambda\left(\frac{1-\mathcal{Y}_1}{\mathcal{Y}_1}\right)^{\varrho}+\lambda\left(\frac{1-\mathcal{Y}_2}{\mathcal{Y}_2}\right)^{\varrho}\right\}^{1/\varrho}},1-\frac{1}{1+\left\{\lambda\left(\frac{\mathcal{A}_1}{1-\mathcal{A}_1}\right)^{\varrho}+\lambda\left(\frac{\mathcal{A}_2}{1-\mathcal{A}_2}\right)^{\varrho}\right\}^{1/\varrho}},\right.\\
&\qquad\left.1-\frac{1}{1+\left\{\lambda\left(\frac{\mathcal{N}_1}{1-\mathcal{N}_1}\right)^{\varrho}+\lambda\left(\frac{\mathcal{N}_2}{1-\mathcal{N}_2}\right)^{\varrho}\right\}^{1/\varrho}}\right)\\
&=\left(\frac{1}{1+\left\{\lambda\left(\frac{1-\mathcal{Y}_1}{\mathcal{Y}_1}\right)^{\varrho}\right\}^{1/\varrho}},1-\frac{1}{1+\left\{\lambda\left(\frac{\mathcal{A}_1}{1-\mathcal{A}_1}\right)^{\varrho}\right\}^{1/\varrho}},1-\frac{1}{1+\left\{\lambda\left(\frac{\mathcal{N}_1}{1-\mathcal{N}_1}\right)^{\varrho}\right\}^{1/\varrho}}\right)\\
&\otimes\left(\frac{1}{1+\left\{\lambda\left(\frac{1-\mathcal{Y}_2}{\mathcal{Y}_2}\right)^{\varrho}\right\}^{1/\varrho}},1-\frac{1}{1+\left\{\lambda\left(\frac{\mathcal{A}_2}{1-\mathcal{A}_2}\right)^{\varrho}\right\}^{1/\varrho}},1-\frac{1}{1+\left\{\lambda\left(\frac{\mathcal{N}_2}{1-\mathcal{N}_2}\right)^{\varrho}\right\}^{1/\varrho}}\right)\\
&=\mathcal{R}_1^{\lambda}\otimes\mathcal{R}_2^{\lambda}.
\end{aligned}
$$

**(vi)**

$$
\begin{aligned}
&\mathcal{R}^{\lambda_1}\otimes\mathcal{R}^{\lambda_2}\\
&=\left(\frac{1}{1+\left\{\lambda_1\left(\frac{1-\mathcal{Y}}{\mathcal{Y}}\right)^{\varrho}\right\}^{1/\varrho}},1-\frac{1}{1+\left\{\lambda_1\left(\frac{\mathcal{A}}{1-\mathcal{A}}\right)^{\varrho}\right\}^{1/\varrho}},1-\frac{1}{1+\left\{\lambda_1\left(\frac{\mathcal{N}}{1-\mathcal{N}}\right)^{\varrho}\right\}^{1/\varrho}}\right)\\
&\otimes\left(\frac{1}{1+\left\{\lambda_2\left(\frac{1-\mathcal{Y}}{\mathcal{Y}}\right)^{\varrho}\right\}^{1/\varrho}},1-\frac{1}{1+\left\{\lambda_2\left(\frac{\mathcal{A}}{1-\mathcal{A}}\right)^{\varrho}\right\}^{1/\varrho}},1-\frac{1}{1+\left\{\lambda_2\left(\frac{\mathcal{N}}{1-\mathcal{N}}\right)^{\varrho}\right\}^{1/\varrho}}\right)\\
&=\left(\frac{1}{1+\left\{(\lambda_1+\lambda_2)\left(\frac{1-\mathcal{Y}}{\mathcal{Y}}\right)^{\varrho}\right\}^{1/\varrho}},1-\frac{1}{1+\left\{(\lambda_1+\lambda_2)\left(\frac{\mathcal{A}}{1-\mathcal{A}}\right)^{\varrho}\right\}^{1/\varrho}},\right.\\
&\qquad\left.1-\frac{1}{1+\left\{(\lambda_1+\lambda_2)\left(\frac{\mathcal{N}}{1-\mathcal{N}}\right)^{\varrho}\right\}^{1/\varrho}}\right)\\
&=\mathcal{R}^{(\lambda_1+\lambda_2)}.
\end{aligned}
$$

□

**Definition 4.6.** Let $\mathcal{R}_f=(\mathcal{Y}_f,\mathcal{A}_f,\mathcal{N}_f)$ $(f=1,2,\ldots,\varpi)$ be a group of PFNs. Let the PFDWA operator be a function $\Theta^{\varpi}\to\Theta$, where

$$PFDWA_{\Xi}(\mathcal{R}_1,\mathcal{R}_2,\ldots,\mathcal{R}_{\varpi})=\bigoplus_{f=1}^{\varpi}(\Xi_f\mathcal{R}_f)$$

and $\Xi=(\Xi_1,\Xi_2,\ldots,\Xi_{\varpi})^T$ be the weight of $\mathcal{R}_f$, $f=1,2,\ldots,\varpi$ such that $\Xi_f>0$ and $\sum_{f=1}^{\varpi}\Xi_f=1$.

The following theorem is based on Dombi norms.

**Theorem 4.2.** *Let $\mathcal{R}_f=(\mathcal{Y}_f,\mathcal{A}_f,\mathcal{N}_f)$ $(f=1,2,\ldots,\varpi)$ be a group of PFNs, then the accumulated result of them applying the PFDWA is again a PFN, and*

$$
\begin{aligned}
&PFDWA_{\Xi}(\mathcal{R}_1,\mathcal{R}_2,\ldots,\mathcal{R}_{\varpi})\\
&=\bigoplus_{f=1}^{\varpi}(\Xi_f\mathcal{R}_f)\\
&=\left(1-\frac{1}{1+\left\{\sum_{f=1}^{\varpi}\Xi_f\left(\frac{\mathcal{Y}_f}{1-\mathcal{Y}_f}\right)^{\varrho}\right\}^{1/\varrho}},\frac{1}{1+\left\{\sum_{f=1}^{\varpi}\Xi_f\left(\frac{1-\mathcal{A}_f}{\mathcal{A}_f}\right)^{\varrho}\right\}^{1/\varrho}},\right.\\
&\qquad\left.\frac{1}{1+\left\{\sum_{f=1}^{\varpi}\Xi_f\left(\frac{1-\mathcal{N}_f}{\mathcal{N}_f}\right)^{\varrho}\right\}^{1/\varrho}}\right),
\end{aligned}
\tag{4.2}
$$

*where weighting vector* $\Xi = (\Xi_1, \Xi_2, \ldots, \Xi_\varpi)$ *is of* $\mathcal{R}_f$, $f = 1, 2, \ldots, \varpi$, *where* $\Xi_f > 0$, *and* $\sum_{f=1}^{\varpi} \Xi_f = 1$.

*Proof.* This theorem can be proved by the induction method.

(i) When $\varpi = 2$, by Dombi rules on PFNs, we obtain

$$PFDWA_{\Xi}(\mathcal{R}_1, \mathcal{R}_2) = \mathcal{R}_1 \bigoplus \mathcal{R}_2 = (\mathcal{Y}_1, \mathcal{A}_1, \mathcal{N}_1) \bigoplus (\mathcal{Y}_2, \mathcal{A}_2, \mathcal{N}_2)$$

and for the right side of Eq. (4.2), we have

$$\left(1 - \frac{1}{1 + \left\{\Xi_1\left(\frac{\mathcal{Y}_1}{1-\mathcal{Y}_1}\right)^{\varrho} + \Xi_2\left(\frac{\mathcal{Y}_2}{1-\mathcal{Y}_2}\right)^{\varrho}\right\}^{1/\varrho}}, \frac{1}{1 + \left\{\Xi_1\left(\frac{1-\mathcal{A}_1}{\mathcal{A}_1}\right)^{\varrho} + \Xi_2\left(\frac{1-\mathcal{A}_2}{\mathcal{A}_2}\right)^{\varrho}\right\}^{1/\varrho}}, \frac{1}{1 + \left\{\Xi_1\left(\frac{1-\mathcal{N}_1}{\mathcal{N}_1}\right)^{\varrho} + \Xi_2\left(\frac{1-\mathcal{N}_2}{\mathcal{N}_2}\right)^{\varrho}\right\}^{1/\varrho}}\right)$$

$$= \left(1 - \frac{1}{1 + \left\{\sum_{s=1}^{2} \Xi_f\left(\frac{\mathcal{Y}_f}{1-\mathcal{Y}_f}\right)^{\varrho}\right\}^{1/\varrho}}, \frac{1}{1 + \left\{\sum_{f=1}^{2} \Xi_f\left(\frac{1-\mathcal{A}_f}{\mathcal{A}_f}\right)^{\varrho}\right\}^{1/\varrho}}, \frac{1}{1 + \left\{\sum_{f=1}^{2} \Xi_f\left(\frac{1-\mathcal{N}_f}{\mathcal{N}_f}\right)^{\varrho}\right\}^{1/\varrho}}\right).$$

Hence, Eq. (4.2) is valid for $\varpi = 2$.

(ii) Suppose Eq. (4.2) is valid for $\varpi = k$. Then, based on Eq. (4.2), we have

$$PFDWA_{\Xi}(\mathcal{R}_1, \mathcal{R}_2, \ldots, \mathcal{R}_k) = \bigoplus_{f=1}^{n} (\Xi_f \mathcal{R}_f)$$

$$= \left(1 - \frac{1}{1 + \left\{\sum_{f=1}^{k} \Xi_f\left(\frac{\mathcal{Y}_f}{1-\mathcal{Y}_f}\right)^{\varrho}\right\}^{1/\varrho}}, \frac{1}{1 + \left\{\sum_{f=1}^{k} \Xi_f\left(\frac{1-\mathcal{A}_f}{\mathcal{A}_f}\right)^{\varrho}\right\}^{1/\varrho}}, \frac{1}{1 + \left\{\sum_{f=1}^{k} \Xi_f\left(\frac{1-\mathcal{N}_f}{\mathcal{N}_f}\right)^{\varrho}\right\}^{1/\varrho}}\right).$$

If $\varpi = k + 1$, then

$$PFDWA_{\Xi}(\mathcal{R}_1, \mathcal{R}_2, \ldots, \mathcal{R}_k, \mathcal{R}_{k+1}) = \bigoplus_{f=1}^{k} (\Xi_f \mathcal{R}_f) \bigoplus (\Xi_{k+1} \mathcal{R}_{k+1})$$

$$=\left(1-\frac{1}{1+\left\{\sum_{f=1}^{k}\Xi_f\left(\frac{\mathcal{Y}_f}{1-\mathcal{Y}_f}\right)^{\varrho}\right\}^{1/\varrho}},\frac{1}{1+\left\{\sum_{s=1}^{k}\Xi_f\left(\frac{1-\mathcal{A}_f}{\mathcal{A}_f}\right)^{\varrho}\right\}^{1/\varrho}},\right.$$
$$\left.\frac{1}{1+\left\{\sum_{f=1}^{k}\Xi_f\left(\frac{1-\mathcal{N}_f}{\mathcal{N}_f}\right)^{\varrho}\right\}^{1/\varrho}}\right)$$
$$\oplus\left(1-\frac{1}{1+\left\{\Xi_{k+1}\left(\frac{\mathcal{Y}_{k+1}}{1-\mathcal{Y}_{k+1}}\right)^{\varrho}\right\}^{1/\varrho}},\frac{1}{1+\left\{\Xi_{k+1}\left(\frac{1-\mathcal{A}_{k+1}}{\mathcal{A}_{k+1}}\right)^{\varrho}\right\}^{1/\varrho}},\frac{1}{1+\left\{\Xi_{k+1}\left(\frac{1-\mathcal{N}_{k+1}}{\mathcal{N}_{k+1}}\right)^{\varrho}\right\}^{1/\varrho}}\right)$$
$$=\left(1-\frac{1}{1+\left\{\sum_{f=1}^{k+1}\Xi_f\left(\frac{\mathcal{Y}_f}{1-\mathcal{Y}_f}\right)^{\varrho}\right\}^{1/\varrho}},\frac{1}{1+\left\{\sum_{f=1}^{k+1}\Xi_f\left(\frac{1-\mathcal{A}_f}{\mathcal{A}_f}\right)^{\varrho}\right\}^{1/\varrho}},\right.$$
$$\left.\frac{1}{1+\left\{\sum_{f=1}^{k+1}\Xi_f\left(\frac{1-\mathcal{N}_f}{\mathcal{N}_f}\right)^{\varrho}\right\}^{1/\varrho}}\right).$$

Thus Eq. (4.2) is true for $\varpi=k+1$.
Therefore Eq. (4.2) holds for any $\varpi$. □

The PFWA operator exhibited the following properties.

**Theorem 4.3.** *(Idempotency property) If $\mathcal{R}_f=(\mathcal{Y}_f,\mathcal{A}_f,\mathcal{N}_f)$, $f=1,2,\ldots,\varpi$ is a group of all identical PFNs, i.e., $\mathcal{R}_f=\mathcal{R}$ for all $f$, then*

$$PFDWA_{\Xi}(\mathcal{R}_1,\mathcal{R}_2,\ldots,\mathcal{R}_{\varpi})=\mathcal{R}.$$

*Proof.* Since $\mathcal{R}_f=(\mathcal{Y}_f,\mathcal{A}_f,\mathcal{N}_f)=p$, $f=1,2,\ldots,\varpi$. Then, by Eq. (4.2),

$$PFDWA_{\Xi}(\mathcal{R}_1,\mathcal{R}_2,\ldots,\mathcal{R}_{\varpi})$$
$$=\bigoplus_{f=1}^{\varpi}(\Xi_f\mathcal{R}_f)$$
$$=\left(1-\frac{1}{1+\left\{\sum_{f=1}^{\varpi}\Xi_f\left(\frac{\mathcal{Y}_f}{1-\mathcal{Y}_f}\right)^{\varrho}\right\}^{1/\varrho}},\frac{1}{1+\left\{\sum_{f=1}^{\varpi}\Xi_f\left(\frac{1-\mathcal{A}_f}{\mathcal{A}_f}\right)^{\varrho}\right\}^{1/\varrho}},\right.$$
$$\left.\frac{1}{1+\left\{\sum_{f=1}^{\varpi}\Xi_f\left(\frac{1-\mathcal{N}_f}{\mathcal{N}_f}\right)^{\varrho}\right\}^{1/\varrho}}\right)$$
$$=\left(1-\frac{1}{1+\left\{\left(\frac{\mathcal{Y}}{1-\mathcal{Y}}\right)^{\varrho}\right\}^{1/\varrho}},\frac{1}{1+\left\{\left(\frac{1-\mathcal{A}}{\mathcal{A}}\right)^{\varrho}\right\}^{1/\varrho}},\frac{1}{1+\left\{\left(\frac{1-\mathcal{N}}{\mathcal{N}}\right)^{\varrho}\right\}^{1/\varrho}}\right)$$

$$=\left(1-\frac{1}{1+\frac{\mathcal{Y}}{1-\mathcal{Y}}},\frac{1}{1+\frac{1-\mathcal{A}}{\mathcal{A}}},\frac{1}{1+\frac{1-\mathcal{N}}{\mathcal{N}}}\right)=(\mathcal{Y},\mathcal{A},\mathcal{N})=\mathcal{R}.$$

Thus $PFDWA_{\Xi}(\mathcal{R}_1,\mathcal{R}_2,\ldots,\mathcal{R}_{\varpi})=\mathcal{R}$ holds. □

**Theorem 4.4.** *(Boundedness property) Let* $\mathcal{R}_f=(\mathcal{Y}_f,\mathcal{A}_f,\mathcal{N}_f)$, $f=1,2,\ldots,\varpi$ *be an ordered group of PFNs. Let*

$$\mathcal{R}^-=\min(\mathcal{R}_1,\mathcal{R}_2,\ldots,\mathcal{R}_{\varpi})\ and\ \mathcal{R}^+=\max(\mathcal{R}_1,\mathcal{R}_2,\ldots,\mathcal{R}_{\varpi}).$$

*Then,* $\mathcal{R}^-\le PFDWA_{\Xi}(\mathcal{R}_1,\mathcal{R}_2,\ldots,\mathcal{R}_{\varpi})\le\mathcal{R}^+$.

*Proof.* Let $\mathcal{R}_f=(\mathcal{Y}_f,\mathcal{A}_f,\mathcal{N}_f)$, $f=1,2,\ldots,\varpi$ be a group of PFNs. Let $\mathcal{R}^-=\min(\mathcal{R}_1,\mathcal{R}_2,\ldots,\mathcal{R}_{\varpi})=(\mathcal{Y}^-,\mathcal{A}^-,\mathcal{N}^-)$ and $\mathcal{R}^+=\max(\mathcal{R}_1,\mathcal{R}_2,\ldots,\mathcal{R}_{\varpi})=(\mathcal{Y}^+,\mathcal{A}^+,\mathcal{N}^+)$.

We have, $\mathcal{Y}^-=\min_f\{\mathcal{Y}_f\}$, $\mathcal{A}^-=\max_s\{\mathcal{A}_f\}$, $\mathcal{N}^-=\max_f\{\mathcal{N}_f\}$, $\mathcal{Y}^+=\max_f\{\mathcal{Y}_f\}$, $\mathcal{A}^+=\min_f\{\mathcal{A}_f\}$ and $\mathcal{N}^+=\min_f\{\mathcal{N}_f\}$.

There are inequalities:

$$1-\frac{1}{1+\left\{\sum_{f=1}^{\varpi}\Xi_f\left(\frac{\mathcal{Y}^-}{1-\mathcal{Y}^-}\right)^{\varrho}\right\}^{1/\varrho}}\le 1-\frac{1}{1+\left\{\sum_{f=1}^{\varpi}\Xi_f\left(\frac{\mathcal{Y}}{1-\mathcal{Y}}\right)^{\varrho}\right\}^{1/\varrho}}$$
$$\le 1-\frac{1}{1+\left\{\sum_{f=1}^{\varpi}\Xi_f\left(\frac{\mathcal{Y}^+}{1-\mathcal{Y}^+}\right)^{\varrho}\right\}^{1/\varrho}},$$
$$\frac{1}{1+\left\{\sum_{f=1}^{\varpi}\Xi_f\left(\frac{1-\mathcal{A}^+}{\mathcal{A}^+}\right)^{\varrho}\right\}^{1/\varrho}}\le\frac{1}{1+\left\{\sum_{f=1}^{\varpi}\Xi_f\left(\frac{1-\mathcal{A}}{\mathcal{A}}\right)^{\varrho}\right\}^{1/\varrho}}\le\frac{1}{1+\left\{\sum_{f=1}^{\varpi}\Xi_f\left(\frac{1-\mathcal{A}^-}{\mathcal{A}^-}\right)^{\varrho}\right\}^{1/\varrho}},$$
$$\frac{1}{1+\left\{\sum_{f=1}^{\varpi}\Xi_f\left(\frac{1-\mathcal{N}^+}{\mathcal{N}^+}\right)^{\varrho}\right\}^{1/\varrho}}\le\frac{1}{1+\left\{\sum_{f=1}^{\varpi}\Xi_f\left(\frac{1-\mathcal{N}}{\mathcal{N}}\right)^{\varrho}\right\}^{1/\varrho}}\le\frac{1}{1+\left\{\sum_{f=1}^{\varpi}\Xi_f\left(\frac{1-\mathcal{N}^-}{\mathcal{N}^-}\right)^{\varrho}\right\}^{1/\varrho}}.$$

Therefore $\mathcal{R}^-\le PFDWA_{\Xi}(\mathcal{R}_1,\mathcal{R}_2,\ldots,\mathcal{R}_{\varpi})\le\mathcal{R}^+$. □

**Theorem 4.5.** *(Monotonicity property) Let* $\mathcal{R}_f$ *and* $\mathcal{R}'_{\varpi}$, $f=1,2,\ldots,\varpi$ *be two groups of PFNs, such that* $\mathcal{R}_f\le\mathcal{R}'_{\varpi}$ *for all* $f$, *then*

$$PFDWA_{\Xi}(\mathcal{R}_1,\mathcal{R}_2,\ldots,\mathcal{R}_{\varpi})\le PFDWA_{\Xi}(\mathcal{R}'_1,\mathcal{R}'_2,\ldots,\mathcal{R}'_{\varpi}).$$

Here, we propose a PFDOWA operator.

**Definition 4.7.** Let $\mathcal{R}_f=(\mathcal{Y}_f,\mathcal{A}_f,\mathcal{N}_f)$, $f=1,2,\ldots,\varpi$ be a group of PFNs. The PFDOWA operator of dimension $\varpi$ is a function $PFDOWA:\Theta^{\varpi}\to\Theta$ with corresponding weighting vector $\delta=(\delta_1,\delta_2,\ldots,\delta_{\varpi})^T$ such that $\delta_f>0$, and $\sum_{f=1}^{\varpi}\delta_f=1$, as

$$PFDOWA_w(\mathcal{R}_1,\mathcal{R}_2,\ldots,\mathcal{R}_{\varpi})=\bigoplus_{f=1}^{\varpi}(\delta_f\mathcal{R}_{\sigma(f)}),\tag{4.3}$$

where $(\sigma(1), \sigma(2), \ldots, \sigma(\varpi))$ is the permutation of $(1, 2, \ldots, \varpi)$, for which $\mathcal{R}_{\sigma(f-1)} \geq \mathcal{R}_{\sigma(f)}$ for all $f = 1, 2, \ldots, \varpi$.

Definition 4.7 leads to the following theorem.

**Theorem 4.6.** *Let $\mathcal{R}_f = (\mathcal{Y}_f, \mathcal{A}_f, \mathcal{N}_f)$, $f = 1, 2, \ldots, \varpi$ be a group of PFNs. Let PFDOWA be an operator of dimension $\varpi$ such as $PFDOWA : \Theta^{\varpi} \to \Theta$ and weight vector be $\delta = (\delta_1, \delta_2, \ldots, \delta_{\varpi})^T$, where $\delta_f > 0$, and $\sum_{f=1}^{\varpi} \delta_f = 1$. Then,*

$$
\begin{aligned}
&PFDOWA_{\delta}(\mathcal{R}_1, \mathcal{R}_2, \ldots, \mathcal{R}_{\varpi}) \\
&= \bigoplus_{f=1}^{\varpi} (\delta_f \mathcal{R}_{\sigma(f)}) \\
&= \left( 1 - \frac{1}{1 + \left\{ \sum_{f=1}^{\varpi} \delta_f \left( \frac{\mathcal{Y}_{\sigma(f)}}{1 - \mathcal{Y}_{\sigma(f)}} \right)^{\varrho} \right\}^{1/\varrho}}, \frac{1}{1 + \left\{ \sum_{f=1}^{\varpi} \delta_f \left( \frac{1 - \mathcal{A}_{\sigma(f)}}{\mathcal{A}_{\sigma(f)}} \right)^{\varrho} \right\}^{1/\varrho}}, \right. \\
&\qquad \left. \frac{1}{1 + \left\{ \sum_{f=1}^{\varpi} \delta_f \left( \frac{1 - \mathcal{N}_{\sigma(f)}}{\mathcal{N}_{\sigma(f)}} \right)^{\varrho} \right\}^{1/\varrho}} \right),
\end{aligned} \tag{4.4}
$$

*where $(\sigma(1), \sigma(2), \ldots, \sigma(f))$ are the permutations of $(1, 2, \ldots, \varpi)$, for which $\mathcal{R}_{\sigma(f-1)} \geq \mathcal{R}_{\sigma(f)}$ for all $f = 1, 2, \ldots, \varpi$.*

The PFDOWA operator displayed the following properties.

**Theorem 4.7.** *(Idempotency property) If $\mathcal{R}_f$, $f = 1, 2, \ldots, \varpi$ are all identical PFNs, i.e., $\mathcal{R}_f = \mathcal{R}$ for all $f$, then*

$$PFDOWA_w(\mathcal{R}_1, \mathcal{R}_2, \ldots, \mathcal{R}_{\varpi}) = \mathcal{R}.$$

**Theorem 4.8.** *(Boundedness property) Let $\mathcal{R}_f$, $f = 1, 2, \ldots, \varpi$ be an ordered group of PFNs. Let $\mathcal{R}^- = \min_f \mathcal{R}_f$, $\mathcal{R}^+ = \max_f \mathcal{R}_f$. Then,*

$$\mathcal{R}^- \leq PFDOWA_w(\mathcal{R}_1, \mathcal{R}_2, \ldots, \mathcal{R}_{\varpi}) \leq \mathcal{R}^+.$$

**Theorem 4.9.** *(Monotonicity property) Let $\mathcal{R}_f$ and $\mathcal{R}'_{\varpi}$, $f = 1, 2, \ldots, \varpi$ be two groups of PFNs, such that $\mathcal{R}_f \leq \mathcal{R}'_{\varpi}$ for all $f$, then*

$$PFDOWA_w(\mathcal{R}_1, \mathcal{R}_2, \ldots, \mathcal{R}_{\varpi}) \leq PFDOWA_w(\mathcal{R}'_1, \mathcal{R}'_2, \ldots, \mathcal{R}'_{\varpi}).$$

**Theorem 4.10.** *(Commutativity property) Let $\mathcal{R}_f$ and $\mathcal{R}'_{\varpi}$, $f = 1, 2, \ldots, \varpi$ be two groups of PFNs, then*

$$PFDOWA_w(\mathcal{R}_1, \mathcal{R}_2, \ldots, \mathcal{R}_{\varpi}) = PFDOWA_w(\mathcal{R}'_1, \mathcal{R}'_2, \ldots, \mathcal{R}'_{\varpi}),$$

*where $\mathcal{R}'_{\varpi}$, $f = 1, 2, \ldots, \varpi$ is any permutation of $\mathcal{R}_f$, $f = 1, 2, \ldots, \varpi$.*

The PFDWA operator took the weights of the PFN into account in Definitions 4.6 and 4.7; once more, the PFDOWA weight denotes the ordered position of the PFNs rather than the weights of the PFNs themselves. As a result, distinct features of weights in both PFDWA and PFDOWA are followed; however, they gave it only a single thought. We provide the PFDHA operator with a way to circumvent this challenge.

**Definition 4.8.** Let PFDHA be an operator of dimension $\varpi$, where $PFDHA : \Theta^{\varpi} \to \Theta$, and the weighting vector be $\delta = (\delta_1, \delta_2, \ldots, \delta_{\varpi})$, and $\delta_f > 0$, and $\sum_{f=1}^{\varpi} \delta_f = 1$. Thus the PFDHWA operator further showed that

$$
\begin{aligned}
&PFDHWA_{\Xi,\delta}(\mathcal{R}_1, \mathcal{R}_2, \ldots, \mathcal{R}_{\varpi}) \\
&= \bigoplus_{f=1}^{\varpi} (\delta_f \dot{\delta}_{\sigma(f)}) \\
&= \left( 1 - \frac{1}{1 + \left\{ \sum_{f=1}^{\varpi} \delta_f \left( \frac{\dot{\mathcal{Y}}_{\sigma(f)}}{1 - \dot{\mathcal{Y}}_{\sigma(f)}} \right)^{\varrho} \right\}^{1/\varrho}}, \frac{1}{1 + \left\{ \sum_{f=1}^{\varpi} \delta_f \left( \frac{1 - \dot{\mathcal{A}}_{\sigma(f)}}{\dot{\mathcal{A}}_{\sigma(f)}} \right)^{\varrho} \right\}^{1/\varrho}}, \right. \\
&\qquad \left. \frac{1}{1 + \left\{ \sum_{f=1}^{\varpi} \delta_f \left( \frac{1 - \dot{\mathcal{N}}_{\sigma(f)}}{\dot{\mathcal{N}}_{\sigma(f)}} \right)^{\varrho} \right\}^{1/\varrho}} \right),
\end{aligned}
\tag{4.5}
$$

where $\dot{\mathcal{R}}_{\sigma(f)}$ is the $f$th largest PFN weighted value $\dot{\mathcal{R}}_f$ ($\dot{\mathcal{R}}_f = \varpi\, \Xi_f \mathcal{R}_f$, $f = 1, 2, \ldots, \varpi$), and $\Xi = (\Xi_1, \Xi_2, \ldots, \Xi_{\varpi})^T$ is the weight vector of $\dot{\mathcal{R}}_f$ with $\Xi_f > 0$ and $\sum_{f=1}^{\varpi} \Xi_f = 1$, where $\varpi$ is the balancing coefficient.

When $\delta = (1/\varpi, 1/\varpi, \ldots, 1/\varpi)$, the PFDHWA is a particular case of the PFDWA and PFDOWA operators, which reflects the degrees of the stated disagreements and their organized situations.

## 4.4 Picture fuzzy Dombi weighted geometric operators

Some critical geometric aggregation operators for PFN are discussed in this section.

**Definition 4.9.** For a group of PFNs, $\mathcal{R}_f = (\mathcal{Y}_f, \mathcal{A}_f, \mathcal{N}_f)$ $(f = 1, 2, \ldots, \varpi)$. Then, PFDWG is a function $\Theta^{\varpi} \to \Theta$ such that

$$
PFDWG_{\Xi}(\mathcal{R}_1, \mathcal{R}_2, \ldots, \mathcal{R}_{\varpi}) = \bigotimes_{f=1}^{\varpi} (\mathcal{R}_f)^{\Xi_f} \tag{4.6}
$$

and $\Xi = (\Xi_1, \Xi_2, \ldots, \Xi_{\varpi})^T$ is the weight vector of $\mathcal{R}_f$, $f = 1, 2, \ldots, \varpi$ such that $\Xi_f > 0$ and $\sum_{f=1}^{\varpi} \Xi_f = 1$.

The PFDWG operator leads to the following theorem.

**Theorem 4.11.** *Let $\mathcal{R}_f = (\mathcal{Y}_f, \mathcal{A}_f, \mathcal{N}_f)$ $(f = 1, 2, \ldots, \varpi)$ be a group of PFNs. The aggregation method by the PFDWG operator aggregated PFNs is likewise a PFN, and*

$$
\begin{aligned}
&PFDWG_{\Xi}(\mathcal{R}_1, \mathcal{R}_2, \ldots, \mathcal{R}_{\varpi}) \\
&= \bigotimes_{f=1}^{\varpi} (\mathcal{R}_f)^{\Xi_f} \\
&= \left( \frac{1}{1 + \left\{ \sum_{f=1}^{\varpi} \Xi_f \left( \frac{1-\mathcal{Y}_f}{\mathcal{Y}_f} \right)^{\varrho} \right\}^{1/\varrho}}, 1 - \frac{1}{1 + \left\{ \sum_{f=1}^{\varpi} \Xi_f \left( \frac{\mathcal{A}_f}{1-\mathcal{A}_f} \right)^{\varrho} \right\}^{1/\varrho}}, \right. \\
&\qquad \left. 1 - \frac{1}{1 + \left\{ \sum_{f=1}^{\varpi} \Xi_f \left( \frac{\mathcal{N}_f}{1-\mathcal{N}_f} \right)^{\varrho} \right\}^{1/\varrho}} \right),
\end{aligned} \tag{4.7}
$$

*where, $\Xi = (\Xi_1, \Xi_2, \ldots, \Xi_{\varpi})^T$ as a weight vector of $\mathcal{R}_f$, $f = 1, 2, \ldots, \varpi$, and $\Xi_f > 0$, and $\sum_{f=1}^{\varpi} \Xi_f = 1$.*

*Proof.* The proof of the theorem follows from Theorem 4.2. □

The PFDWG operator has validated the following properties.

**Theorem 4.12.** *(Idempotency property) If $\mathcal{R}_f = (\mathcal{Y}_f, \mathcal{A}_f, \mathcal{N}_f)$, $f = 1, 2, \ldots, \varpi$ are all the same, i.e., $\mathcal{R}_f = \mathcal{R}$ for all $f$, then*

$$PFDWG_{\Xi}(\mathcal{R}_1, \mathcal{R}_2, \ldots, \mathcal{R}_{\varpi}) = \mathcal{R}.$$

**Theorem 4.13.** *(Boundedness property) Let $\mathcal{R}_f = (\mathcal{Y}_f, \mathcal{A}_f, \mathcal{N}_f)$, $f = 1, 2, \ldots, \varpi$ be a group of ordered PFNs. Let $\mathcal{R}^- = \min_f \mathcal{R}_f$, $\mathcal{R}^+ = \max_f \mathcal{R}_f$. Then,*

$$\mathcal{R}^- \leq PFDWG_{\Xi}(\mathcal{R}_1, \mathcal{R}_2, \ldots, \mathcal{R}_{\varpi}) \leq \mathcal{R}^+.$$

**Theorem 4.14.** *(Monotonicity property) Let $\mathcal{R}_f = (\mathcal{Y}_f, \mathcal{A}_f, \mathcal{N}_f)$ and $\mathcal{R}'_{\varpi}$, $f = 1, 2, \ldots, \varpi$ be two groups of PFNs, such that $\mathcal{R}_f \leq \mathcal{R}'_{\varpi}$ for all b, then*

$$PFDWG_{\Xi}(\mathcal{R}_1, \mathcal{R}_2, \ldots, \mathcal{R}_{\varpi}) \leq PFDWG_{\Xi}(\mathcal{R}'_1, \mathcal{R}'_2, \ldots, \mathcal{R}'_{\varpi}).$$

Now, let us propose a PFDOWG operator.

**Definition 4.10.** Let $\mathcal{R}_f = (\mathcal{Y}_f, \mathcal{A}_f, \mathcal{N}_f)$, $f = 1, 2, \ldots, \varpi$ be a group of PFNs. Then, the PFDOWG operator is a function $PFDOWG : \Theta^{\varpi} \to \Theta$ with $\delta = (\delta_1, \delta_2, \ldots, \delta_{\varpi})^T$ the corresponding weight vector of $\delta_f > 0$, and $\sum_{f=1}^{\varpi} \delta_f = 1$. Therefore

$$PFDOWG_{\delta}(\mathcal{R}_1, \mathcal{R}_2, \ldots, \mathcal{R}_{\varpi}) = \bigotimes_{f=1}^{\varpi} (\mathcal{R}_{\sigma(f)})^{\delta_f}, \tag{4.8}$$

where, $(\sigma(1), \sigma(2), \ldots, \sigma(f))$ are the permutations of $(1, 2, \ldots, \varpi)$, for which $\mathcal{R}_{\sigma(f-1)} \geq \mathcal{R}_{\sigma(f)}$ for all $f = 1, 2, \ldots, \varpi$.

The PFDOWG operator provides the following theorem.

**Theorem 4.15.** *Let $\mathcal{R}_f = (\mathcal{Y}_f, \mathcal{A}_f, \mathcal{N}_f)$, $f = 1, 2, \ldots, \varpi$ be a group of PFNs. The PFDOWG operator is a function $PFDOWG : \Theta^{\varpi} \to \Theta$. Furthermore,*

$$PFDOWA_{\delta}(\mathcal{R}_1, \mathcal{R}_2, \ldots, \mathcal{R}_{\varpi}) = \bigotimes_{f=1}^{\varpi} (\mathcal{R}_f)^{\delta_f} = \left( \frac{1}{1+\left\{\sum_{f=1}^{\varpi} \delta_f \left(\frac{1-\mathcal{Y}_f}{\mathcal{Y}_f}\right)^{\varrho}\right\}^{1/\varrho}}, 1 - \frac{1}{1+\left\{\sum_{f=1}^{\varpi} \delta_f \left(\frac{\mathcal{A}_f}{1-\mathcal{A}_f}\right)^{\varrho}\right\}^{1/\varrho}}, 1 - \frac{1}{1+\left\{\sum_{f=1}^{\varpi} \delta_f \left(\frac{\mathcal{N}_f}{1-\mathcal{N}_f}\right)^{\varrho}\right\}^{1/\varrho}} \right), \tag{4.9}$$

*where $(\sigma(1), \sigma(2), \ldots, \sigma(f))$ is the permutation of $(1, 2, \ldots, \varpi)$, and $\mathcal{R}_{\sigma(f-1)} \geq \tilde{\mathcal{R}}_{\sigma(f)}$ for $f = 1, 2, \ldots, \varpi$, with associated weighting vector $\delta = (\delta_1, \delta_2, \ldots, \delta_{\varpi})^T$, $\delta_f > 0$, and $\sum_{f=1}^{\varpi} \delta_f = 1$.*

The PFDWG's weights for Definitions 4.9 and 4.10 took the PFNs into account. The PFDOWG weights again imply the ordered position of the PFNs of the weights of the PFN. As a result, several angles are taken to follow the weights between PFDWG and PFDOWG. However, they are only ever used once. We present the PFDHG operator to address such behavior.

**Definition 4.11.** The PFDHG operator is a function $PFDHG : \Theta^{\varpi} \to \Theta$ of the weight vector $\delta = (\delta_1, \delta_2, \ldots, \delta_{\varpi})$, where $\delta_f > 0$, and $\sum_{f=1}^{\varpi} \delta_f = 1$. Therefore PFDHWG is evaluated as

$$PFDHWG_{\delta,\Xi}(\mathcal{R}_1, \mathcal{R}_2, \ldots, \mathcal{R}_{\varpi}) = \bigotimes_{f=1}^{\varpi} (\dot{\mathcal{R}}_{\sigma(f)})^{\delta_f} = \left( \frac{1}{1+\left\{\sum_{f=1}^{\varpi} \delta_f \left(\frac{1-\dot{\mathcal{Y}}_{\sigma(f)}}{\dot{\mathcal{Y}}_{\sigma(f)}}\right)^{\varrho}\right\}^{1/\varrho}}, 1 - \frac{1}{1+\left\{\sum_{f=1}^{\varpi} \delta_f \left(\frac{\dot{\mathcal{A}}_{\sigma(f)}}{1-\dot{\mathcal{A}}_{\sigma(f)}}\right)^{\varrho}\right\}^{1/\varrho}}, 1 - \frac{1}{1+\left\{\sum_{f=1}^{\varpi} \delta_f \left(\frac{\dot{\mathcal{N}}_{\sigma(f)}}{1-\dot{\mathcal{N}}_{\sigma(f)}}\right)^{\varrho}\right\}^{1/\varrho}} \right), \tag{4.10}$$

where $\dot{\mathcal{R}}_{\sigma(f)}$ is the $f$th largest weighted PFNs $\dot{\mathcal{R}}_f$ and ($\dot{\mathcal{R}}_f = \varpi \Xi_f \mathcal{R}_f$, $f = 1, 2, \ldots, \varpi$), and for weight vector $\Xi = (\Xi_1, \Xi_2, \ldots, \Xi_\varpi)^T$ of $\dot{\mathcal{R}}_f$, where $\Xi_f > 0$ and $\sum_{f=1}^{\varpi} \Xi_f = 1$, where $\varpi$ is the balancing coefficient.

## 4.5 The MADM model based on PFN

In this section, the Dombi AOs are used for developing MADM problems, where the weights of the attributes are real and the values are in PFNs. Let a set of alternatives be $\aleph = \{\aleph_1, \aleph_2, \ldots, \aleph_a\}$ and attributes be $A = \{A_1, A_2, \ldots, A_f\}$. Let the weight vector be $\Xi = (\Xi_1, \Xi_2, \ldots, \Xi_\varpi)$ for the attribute $A_f$, where $\Xi_f > 0$ and $\sum_{f=1}^{\varpi} \Xi_f = 1$. Suppose that $R = (\mathcal{Y}_{gf}, \mathcal{A}_{gf}, \mathcal{N}_{gf})_{\xi \times \varpi}$ is the picture fuzzy decision matrix, where $\mathcal{Y}_{gf}$ is for PMD as $\aleph_g$ satisfies $A_f$ given by DMs, $\mathcal{A}_{gf}$ denotes for NMD for $\aleph_g$, which does not satisfy $A_f$, and $\mathcal{N}_{gf}$ provided for non-MD for $\aleph_g$ that does not fulfill $A_f$ by considering the DMs, where $\mathcal{Y}_{gf} \subset [0, 1]$, $\mathcal{A}_{gf} \subset [0, 1]$ and $\mathcal{N}_{gf} \subset [0, 1]$, and $0 \leq \mathcal{Y}_{gf} + \mathcal{A}_{gf} + \mathcal{N}_{gf} \leq 1$, $(g = 1, 2, \ldots, \xi)$.

We built up an algorithm for developing the MADM problem under PFDWA and PFDWG operators.

**Step 1.** Using data given in the picture fuzzy decision matrix $R$, and using the operator PFDWA

$$
\begin{aligned}
&\Upsilon_g \\
&= PFDWA(\mathcal{R}_{g1}, \mathcal{R}_{g2}, \ldots, \mathcal{R}_{g\varpi}) \\
&= \bigoplus_{f=1}^{\varpi} (\Xi_f \mathcal{R}_{gf}) \\
&= \left( 1 - \frac{1}{1 + \left\{ \sum_{f=1}^{\varpi} \Xi_f \left( \frac{\mathcal{Y}_f}{1-\mathcal{Y}_f} \right)^{\varrho} \right\}^{1/\varrho}}, \frac{1}{1 + \left\{ \sum_{f=1}^{\varpi} \Xi_f \left( \frac{1-\mathcal{A}_f}{\mathcal{A}_f} \right)^{\varrho} \right\}^{1/\varrho}}, \right. \\
&\qquad \left. \frac{1}{1 + \left\{ \sum_{f=1}^{\varpi} \Xi_f \left( \frac{1-\mathcal{N}_f}{\mathcal{N}_f} \right)^{\varrho} \right\}^{1/\varrho}} \right)
\end{aligned}
\tag{4.11}
$$

and

$$
\begin{aligned}
&\Upsilon_g \\
&= PFDWG(\mathcal{R}_{g1}, \mathcal{R}_{g2}, \ldots, \mathcal{R}_{g\varpi}) \\
&= \bigoplus_{f=1}^{\varpi} (\mathcal{R}_{gf})^{\Xi_f} \\
&= \left( \frac{1}{1 + \left\{ \sum_{f=1}^{\varpi} \Xi_f \left( \frac{1-\mathcal{Y}_f}{\mathcal{Y}_f} \right)^{\varrho} \right\}^{1/\varrho}}, 1 - \frac{1}{1 + \left\{ \sum_{f=1}^{\varpi} \Xi_f \left( \frac{\mathcal{A}_f}{1-\mathcal{A}_f} \right)^{\varrho} \right\}^{1/\varrho}}, \right.
\end{aligned}
$$

$$1 - \frac{1}{1+\left\{\sum_{f=1}^{\varpi} \Xi_f\left(\frac{N_f}{1-N_f}\right)^{\varrho}\right\}^{1/\varrho}}\Bigg) \tag{4.12}$$

to obtain the accumulated values of $\Upsilon_g$ $(g = 1, 2, \ldots, \xi)$ of the alternative $\aleph_g$.

**Step 2.** Evaluate the score value $\Lambda(\Upsilon_g)$ $(g = 1, 2, \ldots, \xi)$ based on the overall PFN $\Upsilon_g$ $(g = 1, 2, \ldots, \xi)$ to rank all $\aleph_g$ $(g = 1, 2, \ldots, \xi)$ for choosing the perfect $\aleph_g$. If there is no dissimilarity between $\Lambda(\Upsilon_g)$ and $\Lambda(\Upsilon_f)$, then we proceed to compute $\Phi(\Upsilon_g)$ and $\Phi(\Upsilon_f)$ based on the overall PFN of $\Upsilon_g$ and $\Upsilon_f$, and find the rank of $\aleph_g$ depending on $\Phi(\Upsilon_g)$ and $\Phi(\Upsilon_f)$.

**Step 3.** Rank all $\aleph_g$ $(g = 1, 2, \ldots, \xi)$ based on $\Lambda(\Upsilon_g)$ $(g = 1, 2, \ldots, \xi)$.

**Step 4.** Stop.

## 4.6 Numerical results

The recent advances in modern science and technology have made our society extremely complex, making it challenging to understand and predict human decision-making processes. The daily need for information technology has grown in today's society. Here, we deal with the challenge of evaluating technology commercialization to support the suggested approach. The panel $\aleph_g$, $g = 1, 2, \ldots, 5$ evaluates five emerging technology businesses (ETEs) as options. Based on four criteria, five ETE assessments have been made:

$A_1$: Technical advancement;
$A_2$: Market risk and potential market;
$A_3$: Human resources, industrialization framework, and financial investments;
$A_4$: The progress of science and technology and employment information.

In order to determine the dominance to each other, the DMs are chosen ETE based on their attribute's importance $\Xi = (0.2, 0.1, 0.3, 0.4)^T$. In this description, PFDM is supposed to be $\widetilde{R} = (\Upsilon_{gf})$, which is given in Table 4.1 with PFN data.

**Table 4.1** Picture fuzzy numbers.

| | $\aleph_1$ | $\aleph_2$ | $\aleph_3$ | $\aleph_4$ | $\aleph_5$ |
|---|---|---|---|---|---|
| $A_1$ | (0.55,0.32,0.10) | (0.69,0.12,0.09) | (0.87,0.08,0.04) | (0.81,0.08,0.07) | (0.84,0.08,0.07) |
| $A_2$ | (0.87,0.08,0.03) | (0.13,0.63,0.22) | (0.09,0.11,0.23) | (0.59,0.13,0.13) | (0.62,0.10,0.17) |
| $A_3$ | (0.42,0.33,0.18) | (0.07,0.79,0.12) | (0.08,0.75,0.07) | (0.06,0.83,0.07) | (0.07,0.81,0.05) |
| $A_4$ | (0.08,0.77,0.03) | (0.72,0.13,0.08) | (0.64,0.24,0.08) | (0.09,0.73,0.06) | (0.12,0.73,0.08) |

In order to find the most useful ETE $\aleph_g$, $g = 1, 2, \ldots, \xi$, we apply the operators PFDWA and PFDWG to solve a MADM method with PFN data given as follows:

- **Step 1.** Let $\varrho = 1$, we apply PFDWA to obtain the accumulate values $\Upsilon_g$ of ETE $\aleph_g$, $g = 1, 2, \ldots, 5$,

$$\Upsilon_1 = (0.5383, 0.3027, 0.0492), \; \Upsilon_2 = (0.6018, 0.1893, 0.0983),$$

$$\Upsilon_3 = (0.6759, 0.1826, 0.0679), \Upsilon_4 = (0.5134, 0.2393, 0.0686),$$
$$\Upsilon_5 = (0.5634, 0.2263, 0.0692).$$

- **Step 2.** Compute score values using Definition 4.4 as $\Lambda(\Upsilon_g)$, $g = 1, 2, \ldots, 5$ of the overall PFNs $\Upsilon_g$ as $\Lambda(\Upsilon_1) = 0.7446$, $\Lambda(\Upsilon_2) = 0.7518$, $\Lambda(\Upsilon_3) = 0.8040$, $\Lambda(\Upsilon_4) = 0.7224$, $\Lambda(\Upsilon_5) = 0.7471$.
- **Step 3.** Rank all the ETEs $\aleph_g$, $g = 1, 2, \ldots, 5$ in accordance with the score values $\Lambda(\Upsilon_a)$ $(a = 1, 2, \ldots, 5)$ of the overall PFNs as $\aleph_3 \succ \aleph_2 \succ \aleph_5 \succ \aleph_1 \succ \aleph_4$.
- **Step 4.** $\aleph_3$ is selected as the best ETE.

Also, the following results are obtained if operator PFDWG is applied instead of PFDWA.

- **Step 1.** For $\varrho = 1$, use the operator PFDWG to evaluate the overall values of $\Upsilon_g$ for ETEs $\aleph_g$, $g = 1, 2, \ldots, 5$, as follows:

$$\Upsilon_1 = (0.1615, 0.6139, 0.0938), \Upsilon_2 = (0.1695, 0.5809, 0.1101),$$
$$\Upsilon_3 = (0.1749, 0.5136, 0.0872), \Upsilon_4 = (0.1014, 0.7206, 0.0724),$$
$$\Upsilon_5 = (0.1247, 0.7049, 0.0793).$$

- **Step 2.** Applying Definition 4.4 to find the score values $\Lambda(\Upsilon_g)$, $g = 1, 2, \ldots, 5$ of PFNs $\Upsilon_g$, $g = 1, 2, \ldots, 5$ as $\Lambda(\Upsilon_1) = 0.5339$, $\Lambda(\Upsilon_2) = 0.5297$, $\Lambda(\Upsilon_3) = 0.5439$, $\Lambda(\Upsilon_4) = 0.5145$, $\Lambda(\Upsilon_5) = 0.5227$.
- **Step 3.** Rank all the ETEs $\aleph_g$, $g = 1, 2, \ldots, 5$ as per the values of the score functions $\Lambda(\Upsilon_g)$, $g = 1, 2, \ldots, 5$ of all PFNs as $\aleph_3 \succ \aleph_1 \succ \aleph_2 \succ \aleph_5 \succ \aleph_4$.
- **Step 4.** Return $\aleph_3$ as the desirable alternative for ETE.

According to the aforementioned computed findings, the alternatives' total values differ when the suggested operators are used, but the ordering of the alternatives is the same, and $\aleph_3$ is the most advantageous.

Tables 4.2 and 4.3 are presented to examine the impact of the parameter $\varrho \in [1, 10]$ for ranking the ETEs using the operators PFDWA and PFDWG.

**Table 4.2** The effect of $\varrho$ when using the PFDWA operator.

| $\varrho$ | $\Lambda(\Upsilon_1)$, $\Lambda(\Upsilon_2)$, $\Lambda(\Upsilon_3)$, $\Lambda(\Upsilon_4)$, $\Lambda(\Upsilon_5)$ | Ranking order |
|---|---|---|
| 1 | 0.7476, 0.7518, 0.8040, 0.7224, 0.7471 | $\aleph_3 \succ \aleph_2 \succ \aleph_5 \succ \aleph_1 \succ \aleph_4$ |
| 2 | 0.8243, 0.7806, 0.8495, 0.7974, 0.8201 | $\aleph_3 \succ \aleph_1 \succ \aleph_5 \succ \aleph_4 \succ \aleph_2$ |
| 3 | 0.8600, 0.7914, 0.8698, 0.8242, 0.8457 | $\aleph_3 \succ \aleph_1 \succ \aleph_5 \succ \aleph_4 \succ \aleph_2$ |
| 4 | 0.8774, 0.7971, 0.8810, 0.8373, 0.8583 | $\aleph_3 \succ \aleph_1 \succ \aleph_5 \succ \aleph_4 \succ \aleph_2$ |
| 5 | 0.8871, 0.8007, 0.8880, 0.8450, 0.8657 | $\aleph_3 \succ \aleph_1 \succ \aleph_5 \succ \aleph_4 \succ \aleph_2$ |
| 6 | 0.8933, 0.8032, 0.8928, 0.8501, 0.8707 | $\aleph_1 \succ \aleph_3 \succ \aleph_5 \succ \aleph_4 \succ \aleph_2$ |
| 7 | 0.8976, 0.8051, 0.8961, 0.8536, 0.8743 | $\aleph_1 \succ \aleph_3 \succ \aleph_5 \succ \aleph_4 \succ \aleph_2$ |
| 8 | 0.9006, 0.8066, 0.8986, 0.8563, 0.8770 | $\aleph_1 \succ \aleph_3 \succ \aleph_5 \succ \aleph_4 \succ \aleph_2$ |
| 9 | 0.9029, 0.8078, 0.9005, 0.8583, 0.8790 | $\aleph_1 \succ \aleph_3 \succ \aleph_5 \succ \aleph_4 \succ \aleph_2$ |
| 10 | 0.9048, 0.8088, 0.9021, 0.8599, 0.8807 | $\aleph_1 \succ \aleph_3 \succ \aleph_5 \succ \aleph_4 \succ \aleph_2$ |

**Table 4.3** The effect of $\varrho$ when using the PFDWG operator.

| $\varrho$ | $\Lambda(\Upsilon_1)$, $\Lambda(\Upsilon_2)$, $\Lambda(\Upsilon_3)$, $\Lambda(\Upsilon_4)$, $\Lambda(\Upsilon_5)$ | Ranking order |
|---|---|---|
| 1 | 0.5339, 0.5297, 0.5439, 0.5145, 0.5227 | $\aleph_3 \succ \aleph_1 \succ \aleph_2 \succ \aleph_5 \succ \aleph_4$ |
| 2 | 0.5019, 0.4984, 0.5091, 0.5049, 0.5082 | $\aleph_3 \succ \aleph_5 \succ \aleph_4 \succ \aleph_1 \succ \aleph_2$ |
| 3 | 0.4871, 0.4842, 0.4907, 0.4998, 0.4989 | $\aleph_5 \succ \aleph_4 \succ \aleph_3 \succ \aleph_1 \succ \aleph_2$ |
| 4 | 0.4788, 0.4744, 0.4779, 0.4956, 0.4914 | $\aleph_4 \succ \aleph_5 \succ \aleph_1 \succ \aleph_3 \succ \aleph_2$ |
| 5 | 0.4735, 0.4668, 0.4690, 0.4920, 0.4852 | $\aleph_4 \succ \aleph_5 \succ \aleph_1 \succ \aleph_3 \succ \aleph_2$ |
| 6 | 0.4698, 0.4610, 0.4624, 0.4888, 0.4804 | $\aleph_4 \succ \aleph_5 \succ \aleph_1 \succ \aleph_3 \succ \aleph_2$ |
| 7 | 0.4671, 0.4564, 0.4576, 0.4861, 0.4766 | $\aleph_4 \succ \aleph_5 \succ \aleph_1 \succ \aleph_3 \succ \aleph_2$ |
| 8 | 0.4650, 0.4528, 0.4538, 0.4839, 0.4736 | $\aleph_4 \succ \aleph_5 \succ \aleph_1 \succ \aleph_3 \succ \aleph_2$ |
| 9 | 0.4634, 0.4499, 0.4509, 0.4820, 0.4712 | $\aleph_4 \succ \aleph_5 \succ \aleph_1 \succ \aleph_3 \succ \aleph_2$ |
| 10 | 0.4621, 0.4476, 0.4484, 0.4805, 0.4692 | $\aleph_4 \succ \aleph_5 \succ \aleph_1 \succ \aleph_3 \succ \aleph_2$ |

## 4.7 Analysis on the effect of parameter $\varrho$ on decision making results

Here, we show how the parameter $\varrho$ affects the MADM method in the ranges $1 \le \varrho \le 10$ based on the operators PFDWA and PFDWG, which are shown in Tables 4.2, and 4.3, respectively. In Table 4.2, it is noted that for the variation of $\varrho$ for the operator PFDWA for rankings are nonidentical. When $1 \le \varrho \le 5$, we found $\aleph_3 \succ \aleph_2 \succ \aleph_5 \succ \aleph_1 \succ \aleph_4$ and $\aleph_3 \succ \aleph_1 \succ \aleph_5 \succ \aleph_4 \succ \aleph_2$, thus the favorable alternative is $\aleph_3$. When $5 \le \varrho \le 10$, then the corresponding ordering is $\aleph_1 \succ \aleph_3 \succ \aleph_5 \succ \aleph_4 \succ \aleph_2$, and $\aleph_1$ is desirable. In Table 4.3, for various values of $\varrho$ corresponding to the ranking for the operator PFDWG is different. Therefore in $1 \le \varrho \le 2$, then the orders are $\aleph_3 \succ \aleph_1 \succ \aleph_2 \succ \aleph_5 \succ \aleph_4$, $\aleph_3 \succ \aleph_5 \succ \aleph_4 \succ \aleph_1 \succ \aleph_2$ that reflects the alternative $\aleph_3$. When $\varrho = 3$, then the order is $\aleph_5 \succ \aleph_4 \succ \aleph_3 \succ \aleph_1 \succ \aleph_2$, and the best ETE is $\aleph_1$. When $4 \le \varrho \le 10$, then the corresponding order is $\aleph_4 \succ \aleph_5 \succ \aleph_1 \succ \aleph_3 \succ \aleph_2$, and the best ETE is $\aleph_4$.

For the different values of $\varrho$ that can modify the ordering of the alternatives for the PFDWG operator, which is less responsive to $\varrho$, this proposed MADM approach is based on the operators PFDWA and PFDWG. When the PFDWA operator, which is more sensitive to $\varrho$, is used, the ordering of the alternatives can be altered similarly for variations in $\varrho$.

## 4.8 Comparative analysis

The PFN is the subject of this application methodology. In real situations, the suggested MADM technique for the operators PFDWA and PFDWG demonstrates advanced reliability. For the purposes of comparison, Wei's methods [28,29] are employed here. Tables 4.4 and 4.5, present the corresponding results. In order to control picture fuzzy MADM difficulties, our suggested aggregation operators implement a new flexible measure for decision makers. Therefore our suggested methods are more inclusive and adaptable compared to other existing techniques for handling picture fuzzy MADM issues.

**Table 4.4** Comparative study with some of the preexisting methods.

| Methods | $\Lambda(\Upsilon_1)$ | $\Lambda(\Upsilon_2)$ | $\Lambda(\Upsilon_3)$ | $\Lambda(\Upsilon_4)$ | $\Lambda(\Upsilon_5)$ | Ranking order |
|---|---|---|---|---|---|---|
| Wei [28] | 0.6819 | 0.7195 | 0.7494 | 0.6549 | 0.6711 | $\aleph_3 \succ \aleph_2 \succ \aleph_1 \succ \aleph_5 \succ \aleph_4$ |
| Wei [29] | 0.6619 | 0.7027 | 0.7258 | 0.6221 | 0.6378 | $\aleph_3 \succ \aleph_2 \succ \aleph_1 \succ \aleph_5 \succ \aleph_4$ |
| Proposed method | 0.7446 | 0.7518 | 0.8040 | 0.7224 | 0.7471 | $\aleph_3 \succ \aleph_2 \succ \aleph_5 \succ \aleph_1 \succ \aleph_4$ |

**Table 4.5** Comparisons of the present method with some existing characteristic issues.

| Methods | Whether it describes fuzzy information easier | Whether it makes information aggregation more flexible by a parameter |
|---|---|---|
| Wei [28] | ✓ | × |
| Wei [29] | ✓ | × |
| Proposed method | ✓ | ✓ |

## 4.9 Conclusions

In this chapter, we looked at MADM issues with picture fuzzy information. From the inspiration of some Dombi operations in the picture fuzzy environment, we developed some Dombi aggregation operators, including the picture fuzzy Dombi weighted average (PFDWA), the picture fuzzy Dombi order weighted average (PFDOWA), the picture fuzzy Dombi hybrid weighted average (PFDHWA), the picture fuzzy Dombi weighted geometric (PFDWG), and the picture fuzzy Dombi order weighted geometric (PFDOWG) operators. The unique characteristics of those recommended operators are considered. To show the benefits of the proposed method and its applicability, we compared it to those already in use. Then, we expanded a few approaches to address multiattribute decision-making problems using those operators. Future applications of our suggested model include risk assessment, decision-making theory, and other fields in confusing contexts. Finally, a practical example of choosing an enterprise system for emerging technologies is given to help establish a plan and explain the value and efficacy of the suggested approach.

## References

[1] K. Atanassov, Intuitionistic Fuzzy Sets: Theory and Applications, Studies in Fuzziness and Soft Computing, vol. 35, Physica-Verlag, Heidelberg, 1999.
[2] J. Chen, J. Ye, Some single-valued neutrosophic Dombi weighted aggregation operators for multiple attribute decision-making, Symmetry 9 (82) (2017) 1–11.
[3] B.C. Cuong, Picture fuzzy sets - first results. Part 1, Seminar "Neuro-fuzzy systems with applications", Tech. Rep., Institute of Mathematics, Hanoi, 2013.
[4] B.C. Cuong, Picture fuzzy sets - first results. Part 2, Seminar "Neuro-fuzzy systems with applications", Tech. Rep., Institute of Mathematics, Hanoi, 2013.
[5] J. Dombi, A general class of fuzzy operators, the demorgan class of fuzzy operators and fuzziness measures induced by fuzzy operators, Fuzzy Sets Syst. 8 (1982) 149–163.
[6] H. Garg, Some picture fuzzy aggregation operators and their applications to multicriteria decision-making, Arab. J. Sci. Eng. 42 (12) (2017) 5275–5290.
[7] X. He, Typhoon disaster assessment based on Dombi hesitant fuzzy information aggregation operators, Nat. Hazards 90 (3) (2018) 1153–1175.

[8] C. Jana, T. Senapati, M. Pal, R.R. Yager, Picture fuzzy Dombi aggregation operators: application to MADM process, Appl. Soft Comput. 74 (1) (2019) 99–109.

[9] P. Liu, J. Liu, S.M. Chen, Some intuitionistic fuzzy Dombi Bonferroni mean operators and their application to multi-attribute group decision making, J. Oper. Res. Soc. 69 (1) (2018) 1–24.

[10] X. Peng, J. Dai, Algorithm for picture fuzzy multiple attribute decision making based on new distance measure, Int. J. Uncertain. Quantificat. 7 (2017) 177–187.

[11] P.T.M. Phuong, P.H. Thong, L.H. Son, Theoretical analysis of picture fuzzy clustering, J. Comput. Sci. Cybern. 34 (1) (2018) 17–31.

[12] L. Shi, J. Ye, Dombi aggregation operators of neutrosophic cubic sets for multiple attribute decision-making, Algorithms 11 (3) (2018), https://doi.org/10.3390/a11030029.

[13] P. Singh, Correlation coefficients for picture fuzzy sets, J. Intell. Fuzzy Syst. 27 (2014) 2857–2868.

[14] L.H. Son, DPFCM: a novel distributed picture fuzzy clustering method on picture fuzzy sets, Expert Syst. Appl. 2 (2015) 51–66.

[15] L.H. Son, Generalized picture distance measure and applications to picture fuzzy clustering, Appl. Soft Comput. 46 (2016) 284–295.

[16] L.H. Son, Measuring analogousness in picture fuzzy sets: from picture distance measures to picture association measures, Fuzzy Optim. Decis. Mak. 16 (3) (2017) 1–20.

[17] L.H. Son, P. Viet, P. Hai, Picture inference system: a new fuzzy inference system on picture fuzzy set, Appl. Intell. 46 (3) (2017) 652–669.

[18] P.H. Thong, L.H. Son, Picture fuzzy clustering for complex data, Eng. Appl. Artif. Intell. 56 (2016) 121–130.

[19] P.H. Thong, L.H. Son, A novel automatic picture fuzzy clustering method based on particle swarm optimization and picture composite cardinality, Knowl.-Based Syst. 109 (2016) 48–60.

[20] P.H. Thong, L.H. Son, A new approach to multi-variables fuzzy forecasting using picture fuzzy clustering and picture fuzzy rules interpolation method, in: 6th International Conference on *Knowledge and Systems Engineering*, Hanoi, Vietnam, 2015, pp. 679–690.

[21] P.H. Thong, L.H. Son, H. Fujita, Interpolative picture fuzzy rules: a novel forecast method for weather nowcasting, in: Fuzzy Systems (FUZZ-IEEE), 2016 IEEE International Conference on IEEE, Vancouver, Canada, July 24-29, 2016, pp. 86–93.

[22] N.T. Thong, L.H. Son, HIFCF: an effective hybrid model between picture fuzzy clustering and intuitionistic fuzzy recommender systems for medical diagnosis, Expert Syst. Appl. 42 (7) (2015) 3682–3701.

[23] P.V. Viet, H.T.M. Chau, L.H. Son, P.V. Hai, Some extensions of membership graphs for picture inference systems, in: Knowledge and Systems Engineering (KSE), 2015 Seventh International Conference on, Ho Chi Minh City, Vietnam, IEEE, October 8-10, 2015, pp. 192–197.

[24] G.W. Wei, F.E. Alsaadi, T. Hayat, A. Alsaedi, Projection models for multiple attribute decision making with picture fuzzy information, Int. J. Mach. Learn. Cybern. 9 (4) (2018) 713–719.

[25] G.W. Wei, H. Gao, The generalized Dice similarity measures for picture fuzzy sets and their applications, Informatica 29 (1) (2018) 1–18.

[26] G.W. Wei, Some similarity measures for picture fuzzy sets and their applications, Iran. J. Fuzzy Syst. 15 (1) (2018) 77–89.

[27] G.W. Wei, Picture fuzzy cross-entropy for multiple attribute decision making problems, J. Bus. Econ. Manag. 17 (4) (2016) 491–502.

[28] G.W. Wei, Picture fuzzy aggregation operators and their application to multiple attribute decision making, J. Intell. Fuzzy Syst. 33 (2017) 713–724.

[29] G.W. Wei, Picture fuzzy Hamacher aggregation operators and their application to multiple attribute decision making, Fundam. Inform. 157 (3) (2018) 271–320.

[30] Z.S. Xu, R.R. Yager, Some geometric aggregation operators based on intuitionistic fuzzy sets, Int. J. Gen. Syst. 35 (2006) 417–433.

[31] Z.S. Xu, Intuitionistic fuzzy aggregation operators, IEEE Trans. Fuzzy Syst. 15 (2007) 1179–1187.

[32] Y. Yang, C. Liang, S. Ji, T. Liu, Adjustable soft discernibility matrix based on picture fuzzy soft sets and its applications in decision making, J. Intell. Fuzzy Syst. 29 (4) (2015) 1711–1722.

[33] L.A. Zadeh, Fuzzy sets, Inf. Control 8 (1965) 338–353.

# 5 Picture fuzzy Dombi prioritized operators and their application in decision-making processes

## 5.1 Introduction

Cuong [3,4] presented a picture set (PFS), which is a generalization of intuitionistic fuzzy sets [1], which were generalized from fuzzy sets (FSs) [23]. Currently, researchers are motivated more to express PFSs in their studies: Sing [11] utilized picture fuzzy results to study the correlation coefficient and then applied it in clustering analysis. Son and others [12,13] proposed a measure of a generalized PF distance technique and integrated it into the hierarchical picture fuzzy clustering. Thong and Son [14] studied picture fuzzy clustering and then applied it in weather forecasting. Thong [15] utilized picture fuzzy recommenders information in medical diagnosis, which not only has theoretical aspects of fuzzy recommenders but also supports the application of healthcare management. Wei [20] exhibited picture fuzzy AOs method and used it to develop a MADM problem for ranking the EPR systems. Wei [19] researched a basic leadership technique in light of the PF-weighted crossentropy and used this to rank the choices. Wei [21] utilized Hamacher operations on picture fuzzy sets and constructed MADM problems.

In 1982, Dombi [5] proposed a new kind of $t$-norm (TN) and $t$-conorm (TCN) known as Dombi TN and Dombi TCN, which have the property of operational flexibility on parameters. With this supremacy, Liu et al. integrated Dombi operators into IFSs and developed MADM problems using a Dombi–Bonferroni mean operator in the environment of intuitionistic fuzzy (IF) information. Chen and Ye [2] attempted to consider the MADM approach using Dombi AOs in the single-valued neutrosophic environment. He [7] utilized Dombi operators in a hesitant fuzzy environment and applied this technique in typhoon disaster management. Lu and Ye [10] utilized cubic linguistic variable (LCV) to develop the linguistic cubic variable Dombi weighted arithmetic averaging (LCVDWAA) operator and linguistic cubic variable Dombi weighted geometric averaging (LCVDWGA) operator to aggregate LCV information. Wei and Wei [21] proposed to aggregate a SVN Dombi prioritized weighted averaging (SVNDPWA) operator and single-valued neutrosophic Dombi prioritized the weighted geometric (SVNDPWG) operator to handle aggregation SVNNs and investigated the interesting properties of these operators. Jana et al. [8] utilized the PF-Dombi weighted arithmetic PFDWA aggregation operator and PF-Dombi weighted geometric aggregation operators to aggregate PF-information. Existing studies of prioritized aggregation [6,16–18,24,25] and Dombi operation [5,8,9] still cannot be extended to aggregate PF-information. In this chapter, we develop some new AOs based on the combination of PF elements and Dombi operations to extend the prioritized AOs and Dombi operations of PFNs, considering the prioritized relationship of the PFNs. We propose some PF-Dombi prioritized weighted AOs for aggregating PF information and develop a multiattribute decision-making method based on PFDPW operators to solve MADM problems with picture fuzzy information. The fact is that the PF information has a worthwhile capacity to signify the uncertain data that appear in real-world problems. Solving the real-world issues applying under Dombi aggregation gives us enough motivation to construct our proposed model. The main object of this chapter is to exhibit some aggregation operators under PF data called PF-Dombi prioritized aggregations for the distinct priorities of the choices amid the decision-making process.

## 5.2 Preliminaries

In this section, we express some essential concepts related to PFSs over the universe of discourse $X$.

Jana et al. [8] utilized a picture fuzzy Dombi weighted average operator (PFDWA), and a picture fuzzy Dombi weighted geometric operator (PFDWG) to aggregate the picture fuzzy data.

**Definition 5.1.** [8] Let $\mathcal{R}_f = (\mathcal{Y}_f, \mathcal{A}_f, \mathcal{N}_f)$ $(f = 1, 2, \ldots, \varpi)$ be a group of PFNs. Then, the PFDWA operator is a mapping $\Theta^{\varpi} \to \Theta$, where

$$\begin{aligned}
&PFDWA_{\mho}(\mathcal{R}_1, \mathcal{R}_2, \ldots, \mathcal{R}_{\varpi})\\
&= \bigoplus_{f=1}^{\varpi} (\Xi_f \mathcal{R}_f)\\
&= \left(1 - \frac{1}{1+\left\{\sum_{f=1}^{\varpi} \Xi_f \left(\frac{\mathcal{Y}_f}{1-\mathcal{Y}_f}\right)^{\varrho}\right\}^{1/\varrho}}, \frac{1}{1+\left\{\sum_{f=1}^{\varpi} \Xi_f \left(\frac{1+\mathcal{A}_f}{\mathcal{A}_f}\right)^{\varrho}\right\}^{1/\varrho}},\right.\\
&\qquad \left.\frac{1}{1+\left\{\sum_{f=1}^{\varpi} \Xi_f \left(\frac{1+\mathcal{N}_f}{\mathcal{N}_f}\right)^{\varrho}\right\}^{1/\varrho}}\right),
\end{aligned} \tag{5.1}$$

where the weight vector of $\mathcal{R}_f$, $f = 1, 2, \ldots, \varpi$ is $\Xi = (\Xi_1, \Xi_2, \ldots, \Xi_{\varpi})^T$ with $\Xi_f > 0$ and $\sum_{f=1}^{\varpi} \Xi_f = 1$.

**Definition 5.2.** [8] Let $\mathcal{R}_f = (\mathcal{Y}_f, \mathcal{A}_f, \mathcal{N}_f)$ $(f = 1, 2, \ldots, \varpi)$ be a group of PFNs. Then, the PFDWG operator is a mapping $\Theta^{\varpi} \to \Theta$, where

$$\begin{aligned}
&PFDWG_{\Xi}(\mathcal{R}_1, \mathcal{R}_2, \ldots, \mathcal{R}_{\varpi})\\
&= \bigoplus_{f=1}^{\varpi} (\mathcal{R}_f)^{\Xi_f}
\end{aligned}$$

$$= \left( \frac{1}{1+\left\{\sum_{f=1}^{\varpi} \Xi_f \left(\frac{1-\mathcal{Y}_f}{\mathcal{Y}_f}\right)^{\varrho}\right\}^{1/\varrho}}, 1-\frac{1}{1+\left\{\sum_{f=1}^{\varpi} \Xi_f \left(\frac{\mathcal{A}_f}{1-\mathcal{A}_f}\right)^{\varrho}\right\}^{1/\varrho}}, \right.$$
$$\left. 1-\frac{1}{1+\left\{\sum_{f=1}^{\varpi} \Xi_f \left(\frac{\mathcal{N}_f}{1-\mathcal{N}_f}\right)^{\varrho}\right\}^{1/\varrho}} \right), \tag{5.2}$$

where the weight vector of $\mathcal{R}_f$, $f = 1, 2, \ldots, \varpi$ is $\Xi = (\Xi_1, \Xi_2, \ldots, \Xi_\varpi)^T$ with $\Xi_f > 0$ and $\sum_{f=1}^{\varpi} \Xi_f = 1$.

The prioritized average (PA) operator was originated by Yager [22], which is given in the following definition.

**Definition 5.3.** Let $D = \{D_1, D_2 \ldots, D_\varpi\}$ be a set of attributes that have a prioritized relation between the attributes by the linear ordering $D_1 \succ D_2 \succ D_3 \succ, \ldots, \succ D_r$, which implies that $D_g$ has a higher prioritized relation than $D_f$, if $g < f$. The value of $D_s(x)$ is the performance of any alternative $x$ under attribute $D_g$, which satisfies $D_g(x) \in [0, 1]$. If

$$PA(D_t(x)) = \sum_{s=1}^{\varpi} \Xi_s D_s(x), \tag{5.3}$$

where $\Xi_g = \frac{\hbar_g}{\sum_{g=1}^{\varpi} \hbar_g}$, $\hbar_f = \prod_{f=1}^{g-1} D_f(x)$, $f = 1, 2, \ldots, \varpi$, $\hbar = 1$, then PA is called the average operator.

The PA [22] generally used input arguments have exact values. In this chapter, we propose the PA operator [22] to the problem where input arguments are PFNs.

## 5.3 Picture fuzzy Dombi prioritized weighted arithmetic aggregation operators

**Definition 5.4.** Let $\mathcal{R}_f = (\mathcal{Y}_f, \mathcal{A}_f, \mathcal{N}_f)$, $f = 1, 2, \ldots, \varpi$ be a group of PFNs. Then, the PFDPA operator is given below:

$$PFDPA(\mathcal{R}_1, \mathcal{R}_2, \ldots, \mathcal{R}_\varpi) = \bigoplus_{f=1}^{\varpi} \left( \frac{\hbar_f}{\sum_{f=1}^{\varpi} \hbar_\theta} \mathcal{R}_f \right)$$
$$= \frac{\hbar_1}{\sum_{f=1}^{\varpi} \hbar_f} \mathcal{R}_1 \oplus \frac{\hbar_2}{\sum_{f=1}^{\varpi} \hbar_f} \mathcal{R}_2 \oplus \ldots \oplus \frac{\hbar_\varpi}{\sum_{f=1}^{\varpi} \hbar_f} \mathcal{R}_\varpi, \tag{5.4}$$

where $\hbar_f = \prod_{g=1}^{f-1} \Lambda(\mathcal{R}_g)$, $f = 2, 3, \ldots, \varpi$, $\hbar_1 = 1$ and $\Lambda(\mathcal{R}_f)$ is the score function implying $\Lambda(\mathcal{R}_f) = \frac{1+\mathcal{Y}_f - \mathcal{N}_f}{2}$.

The PFDPA operator is connected with the following theorem.

**Theorem 5.1.** *Let $\mathcal{R}_f = (\mathcal{Y}_f, \mathcal{A}_f, \mathcal{N}_f)$ $(f = 1, 2, \ldots, \varpi)$ be a group of PFNs. Then, the accumulated value of PFNs using PFDPA operation is also a PFN, and*

$$
\begin{aligned}
&PFDPA(\mathcal{R}_1, \mathcal{R}_2, \ldots, \mathcal{R}_\varpi)\\
&= \frac{\hbar_1}{\sum_{f=1}^{\varpi}\hbar_f}\mathcal{R}_1 \oplus \frac{\hbar_2}{\sum_{f=1}^{\varpi}\hbar_f}\mathcal{R}_2 \oplus \ldots \oplus \frac{\hbar_\varpi}{\sum_{f=1}^{\varpi}\hbar_f}\mathcal{R}_\varpi\\
&= \left(1 - \frac{1}{1+\left\{\sum_{f=1}^{\varpi}\frac{\hbar_f}{\sum_{f=1}^{\varpi}\hbar_f}\left(\frac{\mathcal{Y}_f}{1-\mathcal{Y}_f}\right)^{\varrho}\right\}^{1/\varrho}}, \frac{1}{1+\left\{\sum_{f=1}^{\varpi}\frac{\hbar_f}{\sum_{f=1}^{\varpi}\hbar_f}\left(\frac{1-\mathcal{A}_f}{\mathcal{A}_f}\right)^{\varrho}\right\}^{1/\varrho}},\right.\\
&\qquad\left.\frac{1}{1+\left\{\sum_{f=1}^{\varpi}\frac{\hbar_f}{\sum_{f=1}^{\varpi}\hbar_f}\left(\frac{1-\mathcal{N}_f}{\mathcal{N}_f}\right)^{\varrho}\right\}^{1/\varrho}}\right),
\end{aligned}
\tag{5.5}
$$

*where* $\hbar_f = \prod_{g=1}^{f-1}\Lambda(\mathcal{R}_g)$, $f = 2, 3, \ldots, \varpi$, $\hbar_1 = 1$, *and where the score value* $\Lambda(\mathcal{R}_f) = \frac{1+\mathcal{Y}_f-\mathcal{N}_f}{2}$.

The proof of Theorem 5.1 follows by mathematical induction.

*Proof.* (i) When $f = 2$, then by using Dombi norms on PFNs, we obtain the following results by mathematical induction:

$$PFDPA(\mathcal{R}_1, \mathcal{R}_2) = \mathcal{R}_1 \bigoplus \mathcal{R}_2 = (\mathcal{Y}_1, \mathcal{A}_1, \mathcal{N}_1) \bigoplus (\mathcal{Y}_2, \mathcal{A}_2, \mathcal{N}_2)$$

and the right side of Eq. (5.5) becomes

$$
\begin{aligned}
&\left(1 - \frac{1}{1+\left\{\frac{\hbar_1}{\sum_{f=1}^{\varpi}\hbar_f}\left(\frac{\mathcal{Y}_1}{1-\mathcal{Y}_1}\right)^{\varrho} + \frac{\hbar_2}{\sum_{f=1}^{\varpi}\hbar_f}\left(\frac{\mathcal{Y}_2}{1-\mathcal{Y}_2}\right)^{\varrho}\right\}^{1/\varrho}},\right.\\
&\qquad\frac{1}{1+\left\{\frac{\hbar_1}{\sum_{f=1}^{\varpi}\hbar_f}\left(\frac{1-\mathcal{A}_1}{\mathcal{A}_1}\right)^{\varrho} + \frac{\hbar_2}{\sum_{f=1}^{\varpi}\hbar_f}\left(\frac{1-\mathcal{A}_2}{\mathcal{A}_2}\right)^{\varrho}\right\}^{1/\varrho}},\\
&\qquad\left.\frac{1}{1+\left\{\frac{\hbar_1}{\sum_{f=1}^{\varpi}\hbar_f}\left(\frac{1-\mathcal{N}_1}{\mathcal{N}_1}\right)^{\varrho} + \frac{\hbar_2}{\sum_{f=1}^{\varpi}\hbar_f}\left(\frac{1-\mathcal{N}_2}{\mathcal{N}_2}\right)^{\varrho}\right\}^{1/\varrho}}\right)\\
&= \left(1 - \frac{1}{1+\left\{\sum_{f=1}^{2}\frac{\hbar_f}{\sum_{f=1}^{\varpi}\hbar_f}\left(\frac{\mathcal{Y}_f}{1-\mathcal{Y}_f}\right)^{\varrho}\right\}^{1/\varrho}}, \frac{1}{1+\left\{\sum_{f=1}^{2}\frac{\hbar_f}{\sum_{f=1}^{\varpi}\hbar_f}\left(\frac{1-\mathcal{A}_f}{\mathcal{A}_f}\right)^{\varrho}\right\}^{1/\varrho}},\right.\\
&\qquad\left.\frac{1}{1+\left\{\sum_{f=1}^{2}\frac{\hbar_f}{\sum_{f=1}^{\varpi}\hbar_f}\left(\frac{1-\mathcal{N}_f}{\mathcal{N}_f}\right)^{\varrho}\right\}^{1/\varrho}}\right).
\end{aligned}
$$

Thus Eq. (5.5) holds for $\varpi \geq 2$.

(ii) Suppose that Eq. (5.5) holds for $\varpi = r$, where $r$ is a positive integer, then from Eq. (5.5), we have

$$PFDPA(\mathcal{R}_1, \mathcal{R}_2, \ldots, \mathcal{R}_r) = \bigoplus_{f=1}^{r} \left( \frac{\hbar_f}{\sum_{f=1}^{\varpi} \hbar_f} \mathcal{R}_f \right)$$

$$= \left( 1 - \frac{1}{1 + \left\{ \sum_{f=1}^{r} \frac{\hbar_f}{\sum_{f=1}^{\varpi} \hbar_f} \left( \frac{\mathcal{Y}_f}{1-\mathcal{Y}_f} \right)^{\varrho} \right\}^{1/\varrho}}, \frac{1}{1 + \left\{ \sum_{f=1}^{r} \frac{\hbar_f}{\sum_{f=1}^{\varpi} \hbar_f} \left( \frac{1-\mathcal{A}_f}{\mathcal{A}_f} \right)^{\varrho} \right\}^{1/\varrho}}, \frac{1}{1 + \left\{ \sum_{f=1}^{r} \frac{\hbar_f}{\sum_{f=1}^{\varpi} \hbar_f} \left( \frac{1-\mathcal{N}_f}{\mathcal{N}_f} \right)^{\varrho} \right\}^{1/\varrho}} \right).$$

Now, for $\varpi = r + 1$, then

$$PFDPA(\mathcal{R}_1, \mathcal{R}_2, \ldots, \mathcal{R}_r, \mathcal{R}_{r+1}) = \bigoplus_{f=1}^{r} \left( \frac{\hbar_f}{\sum_{f=1}^{\varpi} \hbar_f} \mathcal{R}_f \right) \bigoplus \left( \frac{\hbar_{r+1}}{\sum_{f=1}^{\varpi} \hbar_f} \mathcal{R}_{r+1} \right)$$

$$= \left( 1 - \frac{1}{1 + \left\{ \sum_{f=1}^{r} \frac{\hbar_f}{\sum_{f=1}^{\varpi} \hbar_f} \left( \frac{\mathcal{Y}_f}{1-\mathcal{Y}_f} \right)^{\varrho} \right\}^{1/\varrho}}, \frac{1}{1 + \left\{ \sum_{f=1}^{r} \frac{\hbar_f}{\sum_{f=1}^{\varpi} \hbar_f} \left( \frac{1-\mathcal{A}_f}{\mathcal{A}_f} \right)^{\varrho} \right\}^{1/\varrho}}, \frac{1}{1 + \left\{ \sum_{f=1}^{r} \frac{\hbar_f}{\sum_{f=1}^{\varpi} \hbar_f} \left( \frac{1-\mathcal{N}_f}{\mathcal{N}_f} \right)^{\varrho} \right\}^{1/\varrho}} \right)$$

$$\bigoplus \left( 1 - \frac{1}{1 + \left\{ \frac{\hbar_{r+1}}{\sum_{f=1}^{\varpi} \hbar_f} \left( \frac{\mathcal{Y}_{r+1}}{1-\mathcal{Y}_{r+1}} \right)^{\varrho} \right\}^{1/\varrho}}, \frac{1}{1 + \left\{ \frac{\hbar_{r+1}}{\sum_{f=1}^{\varpi} \hbar_f} \left( \frac{1-\mathcal{A}_{r+1}}{\mathcal{A}_{r+1}} \right)^{\varrho} \right\}^{1/\varrho}}, \frac{1}{1 + \left\{ \frac{\hbar_{r+1}}{\sum_{f=1}^{\varpi} \hbar_f} \left( \frac{1-\mathcal{N}_{r+1}}{\mathcal{N}_{r+1}} \right)^{\varrho} \right\}^{1/\varrho}} \right)$$

$$= \left( 1 - \frac{1}{1 + \left\{ \sum_{f=1}^{r+1} \frac{\hbar_f}{\sum_{f=1}^{\varpi} \hbar_f} \left( \frac{\mathcal{Y}_f}{1-\mathcal{Y}_f} \right)^{\varrho} \right\}^{1/\varrho}}, \frac{1}{1 + \left\{ \sum_{f=1}^{r+1} \frac{\hbar_f}{\sum_{f=1}^{\varpi} \hbar_f} \left( \frac{1-\mathcal{A}_f}{\mathcal{A}_f} \right)^{\varrho} \right\}^{1/\varrho}}, \frac{1}{1 + \left\{ \sum_{f=1}^{r+1} \frac{\hbar_f}{\sum_{f=1}^{\varpi} \hbar_f} \left( \frac{1-\mathcal{N}_f}{\mathcal{N}_f} \right)^{\varrho} \right\}^{1/\varrho}} \right).$$

Hence, Eq. (5.5) is true for $\varpi = r + 1$.

Hence, we conclude that Eq. (5.5) is true for any positive integer. □

**Example 5.1.** Let $\mathcal{R}_1 = (0.60, 0.24, 0.10)$, $\mathcal{R}_2 = (0.50, 0.20, 0.30)$, $\mathcal{R}_3 = (0.30, 0.10, 0.40)$, and $\mathcal{R}_4 = (0.40, 0.30, 0.10)$ be the four PFNs. Now, from Definition 5.4, $\frac{\hbar_1}{\sum_{f=1}^{4}\hbar_f} = 0.4162$, $\frac{\hbar_2}{\sum_{f=1}^{4}\hbar_f} = 0.3122$, $\frac{\hbar_3}{\sum_{f=1}^{4}\hbar_f} = 0.1873$, and $\frac{\hbar_4}{\sum_{f=1}^{4}\hbar_f} = 0.0843$. Then, by Theorem 5.1, for $\varrho = 3$

$$
\begin{aligned}
&PFDPA(\mathcal{R}_1, \mathcal{R}_2, \ldots, \mathcal{R}_\varpi)\\
&= \left(1 - \frac{1}{1+\left\{\sum_{f=1}^{\varpi}\frac{\hbar_f}{\sum_{f=1}^{\varpi}\hbar_f}\left(\frac{\mathcal{Y}_f}{1-\mathcal{Y}_f}\right)^{\varrho}\right\}^{1/\varrho}}, \frac{1}{1+\left\{\sum_{f=1}^{\varpi}\frac{\hbar_f}{\sum_{f=1}^{\varpi}\hbar_f}\left(\frac{1-\mathcal{A}_f}{\mathcal{A}_f}\right)^{\varrho}\right\}^{1/\varrho}},\right.\\
&\quad\left.\frac{1}{1+\left\{\sum_{f=1}^{\varpi}\frac{\hbar_f}{\sum_{f=1}^{\varpi}\hbar_f}\left(\frac{1-\mathcal{N}_f}{\mathcal{N}_f}\right)^{\varrho}\right\}^{1/\varrho}}\right)\\
&= \left(1 - \frac{1}{1+\left\{0.4162\left(\frac{0.60}{1-0.60}\right)^3+0.3122\left(\frac{0.50}{1-0.50}\right)^3+0.1873\left(\frac{0.30}{1-0.30}\right)^3+0.0843\left(\frac{0.40}{1-0.40}\right)^3\right\}^{1/3}},\right.\\
&\quad\frac{1}{1+\left\{0.4162\left(\frac{1-0.24}{0.24}\right)^3+0.3122\left(\frac{1-0.20}{0.20}\right)^3+0.1873\left(\frac{1-0.10}{0.10}\right)^3+0.0843\left(\frac{1-0.30}{0.30}\right)^3\right\}^{1/3}},\\
&\quad\left.\frac{1}{1+\left\{0.4162\left(\frac{1-0.10}{0.10}\right)^3+0.3122\left(\frac{1-0.30}{0.30}\right)^3+0.1873\left(\frac{1-0.40}{0.40}\right)^3+0.0843\left(\frac{1-0.10}{0.10}\right)^3\right\}^{1/3}}\right)\\
&= \left(0.5468, 0.1527, 0.1223\right).
\end{aligned}
$$

The PFDPA operator executed the following theorems.

**Theorem 5.2.** *(Idempotency property) Let $\mathcal{R}_f = (\mathcal{Y}_f, \mathcal{A}_f, \mathcal{N}_f)$ $(f = 1, 2, \ldots, \varpi)$ be a group of PFNs, where $\hbar_f = \prod_{g=1}^{f-1}\Lambda(\mathcal{R}_g)$, $g = 2, 3, \ldots, \varpi$, $\hbar_1 = 1$, and $\Lambda(\mathcal{R}_g)$ shows the score function for $\mathcal{R}_f$. If all $\mathcal{R}_f$ are equal, i.e., $\mathcal{R}_f = \mathcal{R}$ for all $f$, then*

$$PFDPA(\mathcal{R}_1, \mathcal{R}_2, \ldots, \mathcal{R}_\varpi) = \mathcal{R}. \tag{5.6}$$

*Proof.* Since $\mathcal{R}_f = (\mathcal{Y}_f, \mathcal{A}_f, \mathcal{N}_f) = \mathcal{R}$, $f = 1, 2, \ldots, \varpi$, then from Theorem 5.1,

$$PFDPA(\mathcal{R}_1, \mathcal{R}_2, \ldots, \mathcal{R}_\varpi) = \bigoplus_{f=1}^{\varpi}\left(\frac{\hbar_f}{\sum_{f=1}^{\varpi}\hbar_f}\mathcal{R}_f\right)$$

$$= \left(1 - \frac{1}{1+\left\{\sum_{f=1}^{\varpi}\frac{\hbar_f}{\sum_{f=1}^{\varpi}\hbar_f}\left(\frac{\mathcal{Y}_f}{1-\mathcal{Y}_f}\right)^{\varrho}\right\}^{1/\varrho}}, \frac{1}{1+\left\{\sum_{f=1}^{\varpi}\frac{\hbar_f}{\sum_{f=1}^{\varpi}\hbar_f}\left(\frac{1-\mathcal{A}_f}{\mathcal{A}_f}\right)^{\varrho}\right\}^{1/\varrho}},\right.$$

$$\left.\frac{1}{1+\left\{\sum_{f=1}^{\varpi}\frac{\hbar_f}{\sum_{f=1}^{\varpi}\hbar_f}\left(\frac{1-\mathcal{N}_f}{\mathcal{N}_f}\right)^{\varrho}\right\}^{1/\varrho}}\right)$$

$$=\left(1-\frac{1}{1+\left\{\left(\frac{\mathcal{Y}}{1-\mathcal{Y}}\right)^{\varrho}\right\}^{1/\varrho}},\frac{1}{1+\left\{\left(\frac{1-\mathcal{A}}{\mathcal{A}}\right)^{\varrho}\right\}^{1/\varrho}},\frac{1}{1+\left\{\left(\frac{1-\mathcal{N}}{\mathcal{N}}\right)^{\varrho}\right\}^{1/\varrho}}\right)$$

$$=\left(1-\frac{1}{1+\frac{\mathcal{Y}}{1-\mathcal{Y}}},\frac{1}{1+\frac{1-\mathcal{A}}{\mathcal{A}}},\frac{1}{1+\frac{1-\mathcal{N}}{\mathcal{N}}}\right)=(\mathcal{Y},\mathcal{A},\mathcal{N})=\mathcal{R}.$$

Thus $PFDPA(\mathcal{R}_1,\mathcal{R}_2,\ldots,\mathcal{R}_\varpi)=\mathcal{R}$ holds. □

**Theorem 5.3.** *(Boundedness property) Let $\mathcal{R}_f=(\mathcal{Y}_f,\mathcal{A}_f,\mathcal{N}_f)$ $(f=1,2,\ldots,\varpi)$ be a group of ordered PFNs. Let*

$$\mathcal{R}^-=\min(\mathcal{R}_1,\mathcal{R}_2,\ldots,\mathcal{R}_\varpi)\ \textit{and}\ \mathcal{R}^+=\max(\mathcal{R}_1,\mathcal{R}_2,\ldots,\mathcal{R}_\varpi).$$

*Then, $\mathcal{R}^-\leq PFDPA(\mathcal{R}_1,\mathcal{R}_2,\ldots,\mathcal{R}_\varpi)\leq\mathcal{R}^+$.*

*Proof.* Let $\mathcal{R}_f=(\mathcal{Y}_f,\mathcal{A}_f,\mathcal{N}_f)$, $f=1,2,\ldots,\varpi$ be a group of PFNs.

$$\mathcal{R}^-=\min(\mathcal{R}_1,\mathcal{R}_2,\ldots,\mathcal{R}_\varpi)=(\mathcal{Y}'^-,\mathcal{A}'^-,\mathcal{N}'^-)\text{ and}$$
$$\mathcal{R}^+=\max(\mathcal{R}_1,\mathcal{R}_2,\ldots,\mathcal{R}_\varpi)=(\mathcal{Y}'^+,\mathcal{A}'^+,\mathcal{N}'^+),$$

where, $\mathcal{Y}'^-=\min_f\{\mathcal{Y}_f\}$, $\mathcal{A}'^-=\max_f\{\mathcal{A}_f\}$, $\mathcal{N}'^-=\max_f\{\mathcal{N}_f\}$, $\mathcal{Y}'^+=\max_f\{\mathcal{Y}_f\}$, $\mathcal{A}'^+=\min_f\{\mathcal{A}_f\}$ and $\mathcal{N}'^+=\min_f\{\mathcal{N}_f\}$. From this follows the inequalities:

$$1-\frac{1}{1+\left\{\sum_{f=1}^{\varpi}\frac{\hbar_f}{\sum_{f=1}^{\varpi}\hbar_f}\left(\frac{\mathcal{Y}'^-}{1-\mathcal{Y}'^-}\right)^{\varrho}\right\}^{1/\varrho}}\leq 1-\frac{1}{1+\left\{\sum_{f=1}^{\varpi}\frac{\hbar_f}{\sum_{f=1}^{\varpi}\hbar_f}\left(\frac{\mathcal{Y}}{1-\mathcal{Y}}\right)^{\varrho}\right\}^{1/\varrho}}$$
$$\leq 1-\frac{1}{1+\left\{\sum_{f=1}^{\varpi}\frac{\hbar_f}{\sum_{f=1}^{\varpi}\hbar_f}\left(\frac{\mathcal{Y}'^+}{1-\mathcal{Y}'^+}\right)^{\varrho}\right\}^{1/\varrho}}$$

$$\frac{1}{1+\left\{\sum_{f=1}^{\varpi}\frac{\hbar_f}{\sum_{f=1}^{\varpi}\hbar_f}\left(\frac{1-\mathcal{A}'^-}{\mathcal{A}'^-}\right)^{\varrho}\right\}^{1/\varrho}}\leq\frac{1}{1+\left\{\sum_{f=1}^{\varpi}\frac{\hbar_f}{\sum_{f=1}^{\varpi}\hbar_f}\left(\frac{1-\mathcal{A}}{\mathcal{A}}\right)^{\varrho}\right\}^{1/\varrho}}$$
$$\leq\frac{1}{1+\left\{\sum_{f=1}^{\varpi}\frac{\hbar_f}{\sum_{f=1}^{\varpi}\hbar_f}\left(\frac{1-\hat{\mathcal{A}}'^+}{\mathcal{A}'^+}\right)^{\varrho}\right\}^{1/\varrho}}$$

and

$$\frac{1}{1+\left\{\sum_{f=1}^{\varpi}\frac{\hbar_f}{\sum_{f=1}^{\varpi}\hbar_f}\left(\frac{1-\mathcal{N}'^-}{\mathcal{N}'^-}\right)^{\varrho}\right\}^{1/\varrho}}\leq\frac{1}{1+\left\{\sum_{f=1}^{\varpi}\frac{\hbar_f}{\sum_{f=1}^{\varpi}\hbar_f}\left(\frac{1-\mathcal{N}}{\mathcal{N}}\right)^{\varrho}\right\}^{1/\varrho}}$$

$$\leq \frac{1}{1+\left\{\sum_{f=1}^{\varpi} \frac{\hbar_f}{\sum_{f=1}^{\varpi} \hbar_f}\left(\frac{1-\mathcal{N}'^{+}}{\mathcal{N}'^{+}}\right)^{\varrho}\right\}^{1/\varrho}}.$$

Therefore

$$\mathcal{R}^{-} \leq PFDPA(\mathcal{R}_1, \mathcal{R}_2, \ldots, \mathcal{R}_{\varpi}) \leq \mathcal{R}^{+}.$$

□

**Theorem 5.4.** *(Monotonicity property) Let $\mathcal{R}_f$ and $\mathcal{R}'_f$, $f = 1, 2, \ldots, \varpi$ be two groups of PFNs, if $\mathcal{R}_f \leq \mathcal{R}'_f$ for all $f$, where $\hbar_f = \prod_{g=1}^{f-1} \Lambda(\mathcal{R}_g)$, $\hbar'_f = \prod_{g=1}^{f-1} \Lambda(\mathcal{R}'_g)$, $f = 2, 3, \ldots, \varpi$, $\hbar_1 = \hbar'_1 = 1$ and $\Lambda(\mathcal{R}_f)$ is the score value for $\mathcal{R}_f$, $f = 2, \ldots, \varpi$, $\Lambda(\mathcal{R}'_f)$ is the score value for $\mathcal{R}'_f$, $f = 1, 2, \ldots, \varpi$. Then,*

$$PFDPA(\mathcal{R}_1, \mathcal{R}_2, \ldots, \mathcal{R}_{\varpi}) \leq PFDPA(\mathcal{R}'_1, \mathcal{R}'_2, \ldots, \mathcal{R}'_{\varpi}).$$

*Proof.* Let $\mathcal{R}_f$, $f = 1, 2, \ldots, \varpi$ and $\mathcal{R}'_f$, $f = 1, 2, \ldots, \varpi$ be two groups of PFNs. If $\mathcal{R}'_1, \mathcal{R}'_2, \ldots, \mathcal{R}'_{\varpi}$ is the permutation of $\mathcal{R}_1, \mathcal{R}_2, \ldots, \mathcal{R}_{\varpi}$ for all $f$, where $\hbar_f = \prod_{g=1}^{f-1} \Lambda(\mathcal{R}_g)$, $\hbar'_f = \prod_{g=1}^{f-1} \Lambda(\mathcal{R}'_g)$, $f = 2, 3, \ldots, \varpi$, $\hbar_1 = \hbar'_1 = 1$. $\Lambda(\mathcal{R}_f)$ and $\Lambda(\mathcal{R}'_f)$ are the score values for $\mathcal{R}_f$ and $\mathcal{R}'_f$, respectively, for all $f$. Then, if $\mathcal{R}_f \leq \mathcal{R}'_f$ this implies that

$$PFDPA(\mathcal{R}_1, \mathcal{R}_2, \ldots, \mathcal{R}_{\varpi}) \leq PFDPA(\mathcal{R}'_1, \mathcal{R}'_2, \ldots, \mathcal{R}'_{\varpi}).$$

□

If $\mathcal{R}_f$, $f = 1, 2, \ldots, \varpi$, $\Xi = (\Xi_1, \Xi_2, \ldots, \Xi_{\varpi})^T$ such that $\Xi_f > 0$, and $\sum_{f=1}^{\varpi} \Xi_f = 1$, then based on Definition 5.4, we now define the picture fuzzy Dombi prioritized weighted averaging (PFDPWA) operator below:

**Definition 5.5.** Let $\mathcal{R}_f = (\mathcal{Y}_f, \mathcal{A}_f, \mathcal{N}_f)$, $f = 1, 2, \ldots, \varpi$, be a group of PFNs, then the accumulated value using the PFDPWA operator is also a PFN and is defined as:

$$\begin{aligned}
&PFDPWA_{\Xi}(\mathcal{R}_1, \mathcal{R}_2, \ldots, \mathcal{R}_{\varpi}) \\
&= \frac{\Xi_1 \hbar_1}{\sum_{f=1}^{\varpi} \Xi_f \hbar_f} \mathcal{R}_1 \oplus \frac{\Xi_2 \hbar_2}{\sum_{f=1}^{\varpi} \Xi_f \hbar_f} \mathcal{R}_2 \oplus \ldots \oplus \frac{\Xi_{\varpi} \hbar_{\varpi}}{\sum_{f=1}^{\varpi} \Xi_f \hbar_f} \mathcal{R}_{\varpi} \\
&= \left( 1 - \frac{1}{1+\left\{\sum_{f=1}^{\varpi} \frac{\Xi_f \hbar_f}{\sum_{f=1}^{\varpi} \Xi_f \hbar_f}\left(\frac{\mathcal{Y}_f}{1-\mathcal{Y}_f}\right)^{\varrho}\right\}^{1/\varrho}}, \frac{1}{1+\left\{\sum_{f=1}^{\varpi} \frac{\Xi_f \hbar_f}{\sum_{f=1}^{\varpi} \Xi_f \hbar_f}\left(\frac{1-\mathcal{A}_f}{\mathcal{A}_f}\right)^{\varrho}\right\}^{1/\varrho}}, \right. \\
&\left. \frac{1}{1+\left\{\sum_{f=1}^{\varpi} \frac{\Xi_f \hbar_f}{\sum_{f=1}^{\varpi} \Xi_f \hbar_f}\left(\frac{1-\mathcal{N}_f}{\mathcal{N}_f}\right)^{\varrho}\right\}^{1/\varrho}} \right),
\end{aligned} \tag{5.7}$$

where $\hbar_f = \prod_{g=1}^{f-1} \Lambda(\mathcal{R}_g)$, $f = 2, 3, \ldots, \varpi$, $\hbar_1 = 1$ and $\Lambda(\mathcal{R}_f)$ is the score value for $\mathcal{R}_f$, $f = 1, 2, \ldots, \varpi$.

**Example 5.2.** Let $\mathcal{R}_1 = (0.64, 0.10, 0.09)$, $\mathcal{R}_2 = (0.45, 0.12, 0.21)$, $\mathcal{R}_3 = (0.37, 0.08, 0.39)$, and $\mathcal{R}_4 = (0.49, 0.11, 0.14)$ be four PFNs. From Definition 5.4, then $\frac{\hbar_1}{\sum_{f=1}^4 \hbar_f} = 0.3961$, $\frac{\hbar_2}{\sum_{f=1}^4 \hbar_f} = 0.2702$, $\frac{\hbar_3}{\sum_{f=1}^4 \hbar_f} = 0.2665$, and $\frac{\hbar_4}{\sum_{f=1}^4 \hbar_f} = 0.0671$. Let $\Xi_1 = 0.25$, $\Xi_2 = 0.22$, $\Xi_3 = 0.35$, and $\Xi_4 = 0.18$ be the weight vector such that $\sum_{f=1}^4 \Xi_f = 1$, by Definition 5.5 for $\varrho = 3$:

$$PFDPA_{\Xi}(\mathcal{R}_1, \mathcal{R}_2, \ldots, p_4)$$

$$= \left( 1 - \frac{1}{1 + \left\{ \sum_{b=1}^4 \frac{\Xi_f \hbar_f}{\sum_{b=1}^4 \Xi_f \hbar_f} \left( \frac{\mathcal{Y}_f}{1 - \mathcal{Y}_f} \right)^3 \right\}^{1/3}}, \frac{1}{1 + \left\{ \sum_{b=1}^4 \frac{\Xi_f \hbar_f}{\sum_{b=1}^4 \Xi_f \hbar_f} \left( \frac{1 - \mathcal{A}_f}{\mathcal{A}_f} \right)^3 \right\}^{1/3}}, \right.$$

$$\left. \frac{1}{1 + \left\{ \sum_{b=1}^4 \frac{\Xi_f \hbar_f}{\sum_{b=1}^4 \Xi_f \hbar_f} \left( \frac{1 - N_f}{N_f} \right)^3 \right\}^{1/3}} \right)$$

$$= \left( 1 - \frac{1}{1 + \left\{ 0.3961 \left( \frac{0.64}{1-0.64} \right)^3 + 0.2702 \left( \frac{0.45}{1-0.45} \right)^3 + 0.2665 \left( \frac{0.37}{1-0.37} \right)^3 + 0.0671 \left( \frac{0.49}{1-0.49} \right)^3 \right\}^{1/3}}, \right.$$

$$\frac{1}{1 + \left\{ 0.3961 \left( \frac{1-0.10}{0.10} \right)^3 + 0.2702 \left( \frac{1-0.12}{0.12} \right)^3 + 0.2665 \left( \frac{1-0.08}{0.08} \right)^3 + 0.0671 \left( \frac{1-0.11}{0.11} \right)^3 \right\}^{1/3}},$$

$$\left. \frac{1}{1 + \left\{ 0.3961 \left( \frac{1-0.09}{0.09} \right)^3 + 0.2702 \left( \frac{1-0.21}{0.21} \right)^3 + 0.2665 \left( \frac{1-0.39}{0.39} \right)^3 + 0.0671 \left( \frac{1-0.14}{0.14} \right)^3 \right\}^{1/3}} \right)$$

$$= \left( 0.5753, 0.0960, 0.1162 \right).$$

## 5.4 Picture fuzzy Dombi prioritized geometric aggregation operators

Dombi prioritized geometric aggregation operators on PFS are defined below.

**Definition 5.6.** Let $\mathcal{R}_f = (\mathcal{Y}_f, \mathcal{A}_f, N_f)$ $(f = 1, 2, \ldots, \varpi)$ be a group of PFNs. Then, the picture fuzzy Dombi prioritized geometric (PFDPG) operator is a mapping $PFDPG : \Theta^{\varpi} \to \Theta$ such that

$$PFDPG(\mathcal{R}_1, \mathcal{R}_2, \ldots, \mathcal{R}_{\varpi}) = \bigotimes_{f=1}^{\varpi} \left( \mathcal{R}_f \right)^{\frac{\hbar_f}{\sum_{f=1}^{\varpi} \hbar_f}}$$

$$=\left(\mathcal{R}_1\right)^{\frac{\hbar_1}{\sum_{f=1}^{\varpi}\hbar_f}}\otimes\left(\mathcal{R}_2\right)^{\frac{\hbar_2}{\sum_{f=1}^{\varpi}\hbar_f}}\otimes\ldots\otimes\left(\mathcal{R}_\varpi\right)^{\frac{\hbar_\varpi}{\sum_{f=1}^{\varpi}\hbar_f}}, \tag{5.8}$$

where $\hbar_f=\prod_{g=1}^{f-1}\Lambda(\mathcal{R}_g)$, $f=2,3,\ldots,\varpi$, $\hbar_1=1$ and $\Lambda(\mathcal{R}_f)$ is the score value for $\mathcal{R}_f$, for $f=1,2,\ldots,\varpi$.

The PFDPG operator can be introduced by the following theorem.

**Theorem 5.5.** *Let $\mathcal{R}_f=(\mathcal{Y}_f,\mathcal{A}_f,\mathcal{N}_f)$ $(f=1,2,\ldots,\varpi)$ be a group of PFNs, then by the PFDPG operator to accumulate results also a PFN, and*

$$PFDPG(\mathcal{R}_1,\mathcal{R}_2,\ldots,\mathcal{R}_\varpi)$$
$$=\left(\mathcal{R}_1\right)^{\frac{\hbar_1}{\sum_{f=1}^{\varpi}\hbar_f}}\otimes\left(\mathcal{R}_2\right)^{\frac{\hbar_2}{\sum_{f=1}^{\varpi}\hbar_f}}\otimes\ldots\otimes\left(\mathcal{R}_\varpi\right)^{\frac{\hbar_\varpi}{\sum_{f=1}^{\varpi}\hbar_f}}$$
$$=\left(\frac{1}{1+\left\{\sum_{f=1}^{\varpi}\frac{\hbar_f}{\sum_{f=1}^{\varpi}\hbar_f}\left(\frac{1-\mathcal{Y}_f}{\mathcal{Y}_f}\right)^{\varrho}\right\}^{1/\varrho}},1-\frac{1}{1+\left\{\sum_{f=1}^{\varpi}\frac{\hbar_f}{\sum_{f=1}^{\varpi}\hbar_f}\left(\frac{\mathcal{A}_f}{1-\mathcal{A}_f}\right)^{\varrho}\right\}^{1/\varrho}},\right.$$
$$\left.1-\frac{1}{1+\left\{\sum_{f=1}^{\varpi}\frac{\hbar_f}{\sum_{f=1}^{\varpi}\hbar_f}\left(\frac{\mathcal{N}_f}{1-\mathcal{N}_f}\right)^{\varrho}\right\}^{1/\varrho}}\right). \tag{5.9}$$

*Proof.* This theorem can be handled by induction. (i) When $\varpi=2$, by Dombi operation on PFNs we obtain the following results:

$$PFDPG(\mathcal{R}_1,\mathcal{R}_2)=\mathcal{R}_1\bigoplus\mathcal{R}_2=(\mathcal{Y}_1,\mathcal{A}_1,\mathcal{N}_1)\bigotimes(\mathcal{Y}_2,\mathcal{A}_2,\mathcal{N}_2)$$

and the right side of Eq. (5.9) becomes

$$\left(\frac{1}{1+\left\{\frac{\hbar_1}{\sum_{f=1}^{\varpi}\hbar_f}\left(\frac{1-\mathcal{Y}_1}{\mathcal{Y}_1}\right)^{\varrho}+\frac{\hbar_2}{\sum_{f=1}^{\varpi}\hbar_f}\left(\frac{1-\mathcal{Y}_2}{\mathcal{Y}_2}\right)^{\varrho}\right\}^{1/\varrho}},\right.$$
$$1-\frac{1}{1+\left\{\frac{\hbar_1}{\sum_{f=1}^{\varpi}\hbar_f}\left(\frac{\mathcal{A}_1}{1-\mathcal{A}_1}\right)^{\varrho}+\frac{\hbar_2}{\sum_{f=1}^{\varpi}\hbar_f}\left(\frac{\mathcal{A}_2}{1-\mathcal{A}_2}\right)^{\varrho}\right\}^{1/\varrho}},$$
$$\left.1-\frac{1}{1+\left\{\frac{\hbar_1}{\sum_{f=1}^{\varpi}\hbar_f}\left(\frac{\mathcal{N}_1}{1-\mathcal{N}_1}\right)^{\varrho}+\frac{\hbar_2}{\sum_{f=1}^{\varpi}\hbar_f}\left(\frac{\mathcal{N}_2}{1-\mathcal{N}_2}\right)^{\varrho}\right\}^{1/\varrho}}\right)$$
$$=\left(\frac{1}{1+\left\{\sum_{f=1}^{2}\frac{\hbar_f}{\sum_{f=1}^{\varpi}\hbar_f}\left(\frac{1-\mathcal{Y}_f}{\mathcal{Y}_f}\right)^{\varrho}\right\}^{1/\varrho}},1-\frac{1}{1+\left\{\sum_{f=1}^{2}\frac{\hbar_f}{\sum_{f=1}^{\varpi}\hbar_f}\left(\frac{\mathcal{A}_f}{1-\mathcal{A}_f}\right)^{\varrho}\right\}^{1/\varrho}},\right.$$

$$1-\frac{1}{1+\left\{\sum_{f=1}^{2}\frac{\hbar_f}{\sum_{f=1}^{\varpi}\hbar_f}\left(\frac{N_f}{1-N_f}\right)^{\varrho}\right\}^{1/\varrho}}\Bigg).$$

Thus Eq. (5.9) holds for $\varpi \geq 2$.

(ii) Suppose that Eq. (5.9) holds for $\varpi = r$, where $r$ is an integer, then from Eq. (5.9), we have

$$PFDPA(\mathcal{R}_1,\mathcal{R}_2,\ldots,\mathcal{R}_r)$$
$$=\bigotimes_{b=1}^{r}\left(\mathcal{R}_f\right)^{\frac{\hbar_f}{\sum_{f=1}^{\varpi}\hbar_f}}$$
$$=\left(\frac{1}{1+\left\{\sum_{f=1}^{r}\frac{\hbar_f}{\sum_{f=1}^{\varpi}\hbar_f}\left(\frac{1-\mathcal{Y}_f}{\mathcal{Y}_f}\right)^{\varrho}\right\}^{1/\varrho}},1-\frac{1}{1+\left\{\sum_{f=1}^{r}\frac{\hbar_f}{\sum_{f=1}^{\varpi}\hbar_f}\left(\frac{\mathcal{A}_f}{1-\mathcal{A}_f}\right)^{\varrho}\right\}^{1/\varrho}},\right.$$
$$\left.1-\frac{1}{1+\left\{\sum_{f=1}^{r}\frac{\hbar_f}{\sum_{f=1}^{\varpi}\hbar_f}\left(\frac{N_f}{1-N_f}\right)^{\varrho}\right\}^{1/\varrho}}\right).$$

Now, for $\varpi = r+1$,

$$PFDPG(\mathcal{R}_1,\mathcal{R}_2,\ldots,\mathcal{R}_r,\mathcal{R}_{r+1})=\bigotimes_{f=1}^{r}\left(\mathcal{R}_f\right)^{\frac{\hbar_f}{\sum_{f=1}^{\varpi}\hbar_f}}\bigoplus\left(\mathcal{R}_{r+1}\right)^{\frac{\hbar_{r+1}}{\sum_{f=1}^{\varpi}\hbar_f}}$$
$$=\left(\frac{1}{1+\left\{\sum_{f=1}^{r}\frac{\hbar_f}{\sum_{f=1}^{\varpi}\hbar_f}\left(\frac{1-\mathcal{Y}_f}{\mathcal{Y}_f}\right)^{\varrho}\right\}^{1/\varrho}},1-\frac{1}{1+\left\{\sum_{f=1}^{r}\frac{\hbar_f}{\sum_{f=1}^{\varpi}\hbar_f}\left(\frac{\mathcal{A}_f}{1-\mathcal{A}_f}\right)^{\varrho}\right\}^{1/\varrho}},\right.$$
$$\left.1-\frac{1}{1+\left\{\sum_{f=1}^{r}\frac{\hbar_f}{\sum_{f=1}^{\varpi}\hbar_f}\left(\frac{N_f}{1-N_f}\right)^{\varrho}\right\}^{1/\varrho}}\right)$$
$$\bigoplus\left(\frac{1}{1+\left\{\frac{\hbar_{r+1}}{\sum_{f=1}^{\varpi}\hbar_f}\left(\frac{1-\mathcal{Y}_{r+1}}{\mathcal{Y}_{r+1}}\right)^{\varrho}\right\}^{1/\varrho}},1-\frac{1}{1+\left\{\frac{\hbar_{r+1}}{\sum_{f=1}^{\varpi}\hbar_f}\left(\frac{\mathcal{A}_{r+1}}{1-\mathcal{A}_{r+1}}\right)^{\varrho}\right\}^{1/\varrho}},\right.$$
$$\left.1-\frac{1}{1+\left\{\frac{\hbar_{r+1}}{\sum_{f=1}^{\varpi}\hbar_f}\left(\frac{N_{r+1}}{1-N_{r+1}}\right)^{\varrho}\right\}^{1/\varrho}}\right)$$
$$=\left(\frac{1}{1+\left\{\sum_{f=1}^{r+1}\frac{\hbar_f}{\sum_{f=1}^{\varpi}\hbar_f}\left(\frac{1-\mathcal{Y}_f}{\mathcal{Y}_f}\right)^{\varrho}\right\}^{1/\varrho}},1-\frac{1}{1+\left\{\sum_{f=1}^{r+1}\frac{\hbar_f}{\sum_{f=1}^{\varpi}\hbar_f}\left(\frac{\mathcal{A}_f}{1-\mathcal{A}_f}\right)^{\varrho}\right\}^{1/\varrho}},\right.$$

$$1-\frac{1}{1+\left\{\sum_{f=1}^{r+1}\frac{\hbar_f}{\sum_{f=1}^{\varpi}\hbar_f}\left(\frac{N_f}{1-N_f}\right)^{\varrho}\right\}^{1/\varrho}}\Bigg).$$

Hence, Eq. (5.9) is true for $\varpi = r+1$.

Hence, we conclude that Eq. (5.9) is true for any integer $\varpi$. □

**Example 5.3.** Let $\mathcal{R}_1 = (0.67, 0.09, 0.23)$, $\mathcal{R}_2 = (0.55, 0.15, 0.08)$, $\mathcal{R}_3 = (0.72, 0.05, 0.09)$, and $\mathcal{R}_4 = (0.40, 0.30, 0.10)$ be four PFNs. From Definition 5.6, $\frac{\hbar_1}{\sum_{f=1}^{4}\hbar_f} = 0.4760$, $\frac{\hbar_2}{\sum_{f=1}^{4}\hbar_f} = 0.3427$, $\frac{\hbar_3}{\sum_{f=1}^{4}\hbar_f} = 0.1285$, and $\frac{\hbar_4}{\sum_{f=1}^{4}\hbar_f} = 0.0527$. Then, by Theorem 5.5, for $\varrho = 4$:

$$PFDPG(\mathcal{R}_1,\mathcal{R}_2,\ldots,\mathcal{R}_{\varpi})$$

$$=\left(\frac{1}{1+\left\{\sum_{f=1}^{4}\frac{\hbar_f}{\sum_{f=1}^{4}\hbar_f}\left(\frac{1-\mathcal{Y}_f}{\mathcal{Y}_f}\right)^4\right\}^{1/4}},\ 1-\frac{1}{1+\left\{\sum_{f=1}^{4}\frac{\hbar_f}{\sum_{f=1}^{4}\hbar_f}\left(\frac{\mathcal{A}_f}{1-\mathcal{A}_f}\right)^4\right\}^{1/4}},\right.$$

$$\left.1-\frac{1}{1+\left\{\sum_{f=1}^{4}\frac{\hbar_f}{\sum_{f=1}^{4}\hbar_f}\left(\frac{N_f}{1-N_f}\right)^4\right\}^{1/4}}\right)$$

$$=\left(\frac{1}{1+\left\{0.4760\left(\frac{1-0.67}{0.67}\right)^4+0.3427\left(\frac{1-0.55}{0.55}\right)^4+0.1285\left(\frac{1-0.72}{0.72}\right)^3+0.0527\left(\frac{1-0.40}{0.40}\right)^4\right\}^{1/4}},\right.$$

$$1-\frac{1}{1+\left\{0.4760\left(\frac{0.09}{1-0.09}\right)^4+0.3427\left(\frac{0.15}{1-0.15}\right)^4+0.1285\left(\frac{0.05}{1-0.05}\right)^4+0.0527\left(\frac{0.30}{1-0.30}\right)^4\right\}^{1/4}}$$

$$\left.1-\frac{1}{1+\left\{0.4760\left(\frac{0.23}{1-0.23}\right)^4+0.3427\left(\frac{0.08}{1-0.08}\right)^4+0.1285\left(\frac{0.09}{1-0.09}\right)^4+0.0527\left(\frac{0.10}{1-0.10}\right)^4\right\}^{1/4}}\right)$$

$$=\left(0.5699, 0.1573, 0.1908\right).$$

Using the PFDWG operator we easily prove the following theorems.

**Theorem 5.6.** *(Idempotency) Let $\mathcal{R}_f = (\mathcal{Y}_f, \mathcal{A}_f, N_f)$ $(f = 1, 2, \ldots, \varpi)$ be a group of PFNs, where $\hbar_f = \prod_{g=1}^{b-1}\Lambda(\mathcal{R}_g)$, $g = 2, 3, \ldots, \varpi$, $\hbar_1 = 1$, and $\Lambda(\mathcal{R}_g)$ shows the score function for $\mathcal{R}_g$, $g = 1, 2, \ldots, \varpi$. If all $\mathcal{R}_f$ are all equal, i.e., then $\mathcal{R}_f = \mathcal{R}$ for all $f$, then*

$$PFDPG(\mathcal{R}_1,\mathcal{R}_2,\ldots,\mathcal{R}_{\varpi}) = \mathcal{R}. \tag{5.10}$$

*Proof.* Since $\mathcal{R}_f = (\mathcal{Y}_f, \mathcal{A}_f, \mathcal{N}_f) = \mathcal{R}$, $f = 1, 2, \ldots, \varpi$. Then, from Definition 5.6, $PFDPG(\mathcal{R}_1, \mathcal{R}_2, \ldots, \mathcal{R}_\varpi)$

$$
\begin{aligned}
& PFDPG(\mathcal{R}_1, \mathcal{R}_2, \ldots, \mathcal{R}_\varpi) \\
&= \bigotimes_{f=1}^{\varpi} \left(\mathcal{R}_f\right)^{\frac{\hbar_f}{\sum_{f=1}^{\varpi} \hbar_f}} \\
&= \left( \frac{1}{1 + \left\{ \sum_{f=1}^{\varpi} \frac{\hbar_f}{\sum_{f=1}^{\varpi} \hbar_f} \left( \frac{1-\mathcal{Y}_f}{\mathcal{Y}_f} \right)^{\varrho} \right\}^{1/\varrho}}, 1 - \frac{1}{1 + \left\{ \sum_{f=1}^{\varpi} \frac{\hbar_f}{\sum_{f=1}^{\varpi} \hbar_f} \left( \frac{\mathcal{A}_f}{1-\mathcal{A}_f} \right)^{\varrho} \right\}^{1/\varrho}}, \right. \\
& \left. 1 - \frac{1}{1 + \left\{ \sum_{f=1}^{\varpi} \frac{\hbar_f}{\sum_{f=1}^{\varpi} \hbar_f} \left( \frac{\mathcal{N}_f}{1-\mathcal{N}_f} \right)^{\varrho} \right\}^{1/\varrho}} \right) \\
&= \left( \frac{1}{1 + \left\{ \left( \frac{1-\mathcal{Y}}{\mathcal{Y}} \right)^{\varrho} \right\}^{1/\varrho}}, 1 - \frac{1}{1 + \left\{ \left( \frac{\mathcal{A}}{1-\mathcal{A}} \right)^{\varrho} \right\}^{1/\varrho}}, 1 - \frac{1}{1 + \left\{ \left( \frac{\mathcal{N}}{1-\mathcal{N}} \right)^{\varrho} \right\}^{1/\varrho}} \right) \\
&= \left( \frac{1}{1 + \frac{\mathcal{Y}}{1-\mathcal{Y}}}, 1 - \frac{1}{1 + \frac{\mathcal{A}}{1-\mathcal{A}}}, 1 - \frac{1}{1 + \frac{\mathcal{N}}{1-\mathcal{N}}} \right) = (\mathcal{Y}, \mathcal{A}, \mathcal{N}) = \mathcal{R}.
\end{aligned}
$$

Thus $PFDPG(\mathcal{R}_1, \mathcal{R}_2, \ldots, \mathcal{R}_\varpi) = \mathcal{R}$ holds. □

**Theorem 5.7.** *(Boundedness Property) Let $\mathcal{R}_f = (\mathcal{Y}_f, \mathcal{A}_f, \mathcal{N}_f)$ $(f = 1, 2, \ldots, \varpi)$ be a group of PFNs. Let $\mathcal{R}^- = \min(\mathcal{R}_1, \mathcal{R}_2, \ldots, \mathcal{R}_\varpi) = (\mathcal{Y}'^-, \mathcal{A}'^-, \mathcal{N}'^-)$ and $\mathcal{R}^+ = \max(\mathcal{R}_1, \mathcal{R}_2, \ldots, \mathcal{R}_\varpi) = (\mathcal{Y}'^+, \mathcal{A}'^+, \mathcal{N}'^+)$.*

*Then, $\mathcal{R}^- \leq PFDPA(\mathcal{R}_1, \mathcal{R}_2, \ldots, \mathcal{R}_\varpi) \leq \mathcal{R}^+$.*

*Proof.* The proof of this theorem is similar to that for Theorem 5.3. □

**Theorem 5.8.** *(Monotonicity Property) Let $\mathcal{R}_f$ and $\mathcal{R}'_f$, $f = 1, 2, \ldots, \varpi$ be two groups of PFNs, if $\mathcal{R}_f \leq \mathcal{R}'_f$ for all $f$, where $\hbar_f = \prod_{g=1}^{f-1} \Lambda(\mathcal{R}_g)$, $\hbar'_f = \prod_{g=1}^{f-1} \Lambda(\mathcal{R}'_g)$, $f = 2, 3, \ldots, \varpi$, $\hbar_1 = \hbar'_1 = 1$, and $\Lambda(\mathcal{R}_f)$ is the score value of $\mathcal{R}_f$, $g = 2, \ldots, \varpi$, $\Lambda(\mathcal{R}'_g)$ is the score value of $\mathcal{R}'_f$, $f = 1, 2, \ldots, \varpi$. Then,*

$$PFDPG(\mathcal{R}_1, \mathcal{R}_2, \ldots, \mathcal{R}_\varpi) \leq PFDPG(\mathcal{R}'_1, \mathcal{R}'_2, \ldots, \mathcal{R}'_\varpi).$$

*Proof.* The proof of this theorem is similar to that for Theorem 5.4. □

If $\mathcal{R}_f$, $f = 1, 2, \ldots, \varpi$, $\Xi = (\Xi_1, \Xi_2, \ldots, \Xi_\varpi)^T$ such that $\Xi_f > 0$, and $\sum_{f=1}^{\varpi} \Xi_f = 1$, then based on Definition 5.4, we define the PFDPWG operator as:

**Definition 5.7.** Let $\mathcal{R}_f = (\mathcal{Y}_f, \mathcal{A}_f, \mathcal{N}_f)$ $(f = 1, 2, \ldots, \varpi)$ be a group of PFNs, then the accumulated value of the PFDPWG operator is also a PFN and is defined as:

$$PFDPWG_{\Xi}(\mathcal{R}_1, \mathcal{R}_2, \ldots, \mathcal{R}_\varpi) = \left(\mathcal{R}_1\right)^{\frac{\Xi_1 \hbar_1}{\sum_{f=1}^{\varpi} \Xi_f \hbar_f}} \otimes \left(\mathcal{R}_2\right)^{\frac{\Xi_2 \hbar_2}{\sum_{f=1}^{\varpi} \Xi_f \hbar_f}} \otimes \ldots \otimes \left(\mathcal{R}_\varpi\right)^{\frac{\Xi_\varpi \hbar_\varpi}{\sum_{f=1}^{\varpi} \Xi_f \hbar_f}}$$

$$= \left( \frac{1}{1 + \left\{ \sum_{f=1}^{\varpi} \frac{\Xi_f \hbar_f}{\sum_{f=1}^{\varpi} \Xi_f \hbar_f} \left( \frac{1 - \mathcal{Y}_f}{\mathcal{Y}_f} \right)^{\varrho} \right\}^{1/\varrho}}, 1 - \frac{1}{1 + \left\{ \sum_{f=1}^{\varpi} \frac{\Xi_f \hbar_f}{\sum_{f=1}^{\varpi} \Xi_f \hbar_f} \left( \frac{\mathcal{A}_f}{1 - \mathcal{A}_f} \right)^{\varrho} \right\}^{1/\varrho}}, \right.$$

$$\left. 1 - \frac{1}{1 + \left\{ \sum_{f=1}^{\varpi} \frac{\Xi_f \hbar_f}{\sum_{f=1}^{\varpi} \Xi_f \hbar_f} \left( \frac{\mathcal{N}_f}{1 - \mathcal{N}_f} \right)^{\varrho} \right\}^{1/\varrho}} \right), \tag{5.11}$$

where $\hbar_f = \prod_{g=1}^{f-1} \Lambda(\mathcal{R}_g)$, $f = 2, 3, \ldots, \varpi$, $\hbar_1 = 1$, and $\Lambda(\mathcal{R}_f)$ is the score value of $\mathcal{R}_f$ for $f = 1, 2, \ldots, \varpi$.

**Example 5.4.** Let $\mathcal{R}_1 = (0.64, 0.10, 0.09)$, $\mathcal{R}_2 = (0.45, 0.12, 0.21)$, $\mathcal{R}_3 = (0.37, 0.08, 0.39)$, and $\mathcal{R}_4 = (0.49, 0.11, 0.14)$ be four PFNs. Using Definition 5.4, then $\frac{\hbar_1}{\sum_{f=1}^{4} \hbar_f} = 0.3961$, $\frac{\hbar_2}{\sum_{\theta=1}^{4} \hbar_f} = 0.2702$, $\frac{\hbar_3}{\sum_{f=1}^{4} \hbar_f} = 0.2665$, and $\frac{\hbar_4}{\sum_{f=1}^{4} \hbar_f} = 0.0671$. Let $\Xi_1 = 0.25$, $\Xi_2 = 0.22$, $\Xi_3 = 0.35$, and $\Xi_4 = 0.18$ be the weights such that $\sum_{f=1}^{4} \Xi_f = 1$. Then, using Definition 5.7 for $\varrho = 4$:

$$PFDPWG_{\mho}(\mathcal{R}_1, \mathcal{R}_2, \ldots, \mathcal{R}_\varpi)$$

$$= \left( \frac{1}{1 + \left\{ \sum_{f=1}^{4} \frac{\Xi_f \hbar_f}{\sum_{f=1}^{4} \Xi_f \hbar_f} \left( \frac{1 - \mathcal{Y}_f}{\mathcal{Y}_f} \right)^{4} \right\}^{1/4}}, 1 - \frac{1}{1 + \left\{ \sum_{f=1}^{4} \frac{\Xi_f \hbar_f}{\sum_{f=1}^{4} \Xi_f \hbar_f} \left( \frac{\mathcal{N}_f}{1 - \mathcal{N}_f} \right)^{4} \right\}^{1/4}}, \right.$$

$$\left. 1 - \frac{1}{1 + \left\{ \sum_{f=1}^{4} \frac{\Xi_f \hbar_f}{\sum_{f=1}^{4} \Xi_f \hbar_f} \left( \frac{\mathcal{N}_f}{1 - \mathcal{N}_f} \right)^{4} \right\}^{1/4}} \right)$$

$$= \left( \frac{1}{1 + \left\{ 0.3961 \left( \frac{1-0.64}{0.64} \right)^4 + 0.2702 \left( \frac{1-0.45}{0.45} \right)^4 + 0.2665 \left( \frac{1-0.37}{0.37} \right)^4 + 0.0671 \left( \frac{1-0.49}{0.49} \right)^4 \right\}^{1/4}}, \right.$$

$$1 - \frac{1}{1 + \left\{ 0.3961 \left( \frac{0.10}{1-0.10} \right)^4 + 0.2702 \left( \frac{0.12}{1-0.12} \right)^4 + 0.2665 \left( \frac{0.08}{1-0.08} \right)^4 + 0.0671 \left( \frac{0.11}{1-0.11} \right)^4 \right\}^{1/4}},$$

$$\left. 1 - \frac{1}{1 + \left\{ 0.3961 \left( \frac{0.09}{1-0.09} \right)^4 + 0.2702 \left( \frac{0.21}{1-0.21} \right)^4 + 0.2665 \left( \frac{0.39}{1-0.39} \right)^4 + 0.0671 \left( \frac{0.14}{1-0.14} \right)^4 \right\}^{1/4}} \right)$$

$$= \left(0.4444, 0.1033, 0.2970\right).$$

## 5.5 Model for MADM using picture fuzzy Dombi operator

In this section, we develop the MADM method using PF-Dombi PAOs, where the weights of attributes are real numbers and attributes information are PFNs. Let $\aleph = \{\aleph_1, \aleph_2, \ldots, \aleph_\xi\}$ be a set of alternatives, and $A = \{A_1, A_2, \ldots, A_\varpi\}$ be a set of attributes having linear ordering $A_1 \succ A_2 \succ A_3 \ldots \succ A_r$, where attribute $A_g$ has higher priority than $A_f$, if $g < f$. Let $\Xi = (\Xi_1, \Xi_2, \ldots, \Xi_\varpi)^T$ be the weight vector for $\aleph_f$, $f = 1, 2, \ldots, \varpi$ such that $\Xi_f > 0$ and $\sum_{f=1}^{\varpi} \Xi_f = 1$. Suppose that $R = \left(\mathcal{R}_{gf}\right)_{\xi \times \varpi} = (\mathcal{Y}_{gf}, \mathcal{A}_{gf}, \mathcal{N}_{gf})_{\xi \times \varpi}$ executes as a decision matrix, where $\mathcal{Y}_{gf}$ is a positive membership for $\aleph_g$ that satisfies $A_f$, $\mathcal{A}_{gf}$ is the neutral degrees of membership for $\aleph_g$ that satisfies $A_f$, and $\mathcal{N}_{gf}$ is the nonmembership for $\aleph_g$ that does not satisfy $A_f$ introduced by the decision maker, where $\mathcal{Y}_{gf} \in [0, 1]$, $\mathcal{A}_{gf} \in [0, 1]$, and $\mathcal{N}_{gf} \in [0, 1]$, where $0 \leq \mathcal{Y}_{gf} + \mathcal{A}_{gf} + \mathcal{N}_{gf} \leq 1$, $g = 1, 2, \ldots, \xi$, $f = 1, 2, \ldots, \varpi$.

Applying the PFDPWA and PFDPWG operators to the process of solving the MADM problem is described in the following algorithm.

**Algorithm.**

**Step 1.** Compute $\hbar_{gf}$, $g = 1, 2, \ldots, \xi$; $f = 1, 2, \ldots, \varpi$ as follows:

$$\hbar_{gf} = \prod_{g=1}^{f-1} \Lambda(\mathcal{R}_{gf}),\ g = 1, 2, \ldots, \xi;\ f = 1, 2, \ldots, \varpi \tag{5.12}$$

$$\hbar_{g1} = 1,\ g = 1, 2, \ldots, \xi.$$

**Step 2.** Applying the operator PFDPWA

$$\begin{aligned}
&\Upsilon_g = PFDPWA_{\Xi}(\mathcal{R}_{g1}, \mathcal{R}_{g2}, \ldots, \mathcal{R}_{g\varpi})\\
&= \bigoplus_{f=1}^{\varpi}\left(\frac{\Xi_f \hbar_{gf}}{\sum_{f=1}^{\varpi} \Xi_f \hbar_{gf}} \mathcal{R}_{gf}\right)\\
&= \frac{\Xi_1 \hbar_{g1}}{\sum_{f=1}^{\varpi} \Xi_f \hbar_{gf}} \mathcal{R}_1 \oplus \frac{\Xi_2 \hbar_{g2}}{\sum_{f=1}^{\varpi} \Xi_f \hbar_{gf}} \mathcal{R}_2 \oplus \frac{\Xi_3 \hbar_{g3}}{\sum_{f=1}^{\varpi} \Xi_f \hbar_{gf}} \mathcal{R}_3 \oplus \ldots \oplus \frac{\Xi_\varpi \hbar_{gr}}{\sum_{f=1}^{\varpi} \Xi_f \hbar_{gf}} \mathcal{R}_\varpi\\
&= \left(1 - \frac{1}{1 + \left\{\sum_{f=1}^{\varpi} \frac{\Xi_f \hbar_{gf}}{\sum_{f=1}^{\varpi} \Xi_f \hbar_{gf}} \left(\frac{\mathcal{Y}_{gf}}{1 - \mathcal{Y}_{gf}}\right)^{\varrho}\right\}^{1/\varrho}}, \frac{1}{1 + \left\{\sum_{f=1}^{\varpi} \frac{\Xi_f \hbar_{gf}}{\sum_{f=1}^{\varpi} \Xi_f \hbar_{gf}} \left(\frac{1 - \mathcal{N}_{gf}}{\mathcal{N}_{gf}}\right)^{\varrho}\right\}^{1/\varrho}},\right.\\
&\left.\frac{1}{1 + \left\{\sum_{f=1}^{\varpi} \frac{\Xi_f \hbar_{gf}}{\sum_{f=1}^{\varpi} \Xi_f \hbar_{gf}} \left(\frac{1 - \mathcal{N}_{gf}}{\mathcal{N}_{gf}}\right)^{\varrho}\right\}^{1/\varrho}}\right),
\end{aligned} \tag{5.13}$$

or

$$\Upsilon_g = PFDPWG_{\Xi}(\mathcal{R}_{g1}, \mathcal{R}_{g2}, \ldots, \mathcal{R}_{g\varpi})$$

$$=\bigotimes_{f=1}^{\varpi}\left(\mathcal{R}_{gf}\right)^{\frac{\Xi_f\hbar_{gf}}{\sum_{f=1}^{\varpi}\Xi_f\hbar_{gf}}}$$

$$=\left(\mathcal{R}_1\right)^{\frac{\Xi_1\hbar_{g1}}{\sum_{f=1}^{\varpi}\Xi_f\hbar_{gf}}}\otimes\left(\mathcal{R}_2\right)^{\frac{\Xi_2\hbar_{\phi 2}}{\sum_{f=1}^{\varpi}\Xi_f\hbar_{gf}}}\otimes\left(\mathcal{R}_3\right)^{\frac{\Xi_3\hbar_{g3}}{\sum_{f=1}^{\varpi}\Xi_f\hbar_{gf}}}\otimes\ldots\otimes\left(\mathcal{R}_{\varpi}\right)^{\frac{\Xi_{\varpi}\hbar_{g\varpi}}{\sum_{f=1}^{\varpi}\Xi_f\hbar_{gf}}}$$

$$=\left(\frac{1}{1+\left\{\sum_{f=1}^{\varpi}\frac{\Xi_f\hbar_{gf}}{\sum_{f=1}^{\varpi}\Xi_f\hbar_{gf}}\left(\frac{1-\mathcal{Y}_f}{\mathcal{Y}_f}\right)^{\varrho}\right\}^{1/\varrho}},1-\frac{1}{1+\left\{\sum_{f=1}^{\varpi}\frac{\Xi_f\hbar_{gf}}{\sum_{f=1}^{\varpi}\Xi_f\hbar_{gf}}\left(\frac{\mathcal{A}_{gf}}{1-N_{gf}}\right)^{\varrho}\right\}^{1/\varrho}},\right.$$

$$\left.1-\frac{1}{1+\left\{\sum_{f=1}^{\varpi}\frac{\Xi_f\hbar_{gf}}{\sum_{f=1}^{\varpi}\Xi_f\hbar_{gf}}\left(\frac{N_{gf}}{1-N_{gf}}\right)^{\varrho}\right\}^{1/\varrho}}\right) \tag{5.14}$$

to obtain the accumulated values of $\Upsilon_g$, $g=1,2,\ldots,\xi$ of $\aleph_g$.

**Step 3.** Compute the value of $\Phi(\Upsilon_g)$, $g=1,2,\ldots,\varpi$ based on comprehensive PFNs $\Upsilon_g$, $g=1,2,\ldots,\varpi$ in order to rank the $\aleph_g$, $g=1,2,\ldots,\varpi$. If there is no variation between the scores of $\Lambda(\Upsilon_g)$ and $\Lambda(\Upsilon_g)$, then proceed to obtain the accuracy degrees of $\Xi(\Upsilon_g)$ and $\Xi(\Upsilon_a)$ based on the overall PFNs $\Upsilon_a$ and $\Upsilon_a$, for the raking of the $\aleph_g$ depending on the degree of accuracy $\Xi(\Upsilon_g)$ and $\Xi(\Upsilon_g)$.

**Step 4.** Rank all $\aleph_g$, $g=1,2,\ldots,\xi$ in order to select the favorable one(s) in accordance with $\Phi(\Upsilon_g)$, $g=1,2,\ldots,\xi$.

**Step 5.** End.

## 5.6 Numerical example and comparative analysis

In order to promote agricultural development and enhance the production of crops, the school of crops management in an Indian agricultural university wants to select the best insecticide company's product to protects the crops. It is well known that insecticides always ruin the production of the crop, as a result, the gross domestic product (GDP) of a country will be degraded because GDP and the economy from crop production have an interrelationship. In the market, there are available different branded company's insecticides; they are made from different chemical compositions, different duration of effectiveness, and their costs are different. Some of them pollute the environment seriously (in smell, soil, etc.). In our study, we select the best insecticide company products that are less polluting, cheap, and effective among the five possible companies with the advice of decision makers. The expert team have been introduced under great attention from the farmer, a chemical technologist, the Dean of the Indian drug manufacturers association, and a human resource development officer sets up the panel of decision makers that are responsible for this project. They will judge five insecticide companies $\aleph_g$ $g=1,2,3,4,5$ under the evaluation of four attributes:

$A_1$: Economical;
$A_2$: Duration of effectiveness;
$A_3$: Ecofriendliness;
$A_4$: Composition of the insecticide and technical advancement.

The Dean of the Indian drug manufacturers association has absolute priority for decision making, the chemical technologist comes next. The prioritization relation for the criteria is given as $\aleph_1 \succ \aleph_2 \succ \aleph_3 \succ \aleph_4$, which is evaluated by the decision makers for the attributes whose weight vector $\Xi = (0.2, 0.1, 0.3, 0.4)^T$. The five companies $\aleph_g$, $g = 1, 2, 3, 4, 5$ and the used decision matrix $\mathcal{R} = (\beta_{gf})_{5\times 4}$ are given in Table 5.1. We utilize the PFDPWA and PFDPWG operators to develop the MADM method with PF information, which can be evaluated as follows:

**Table 5.1** Picture fuzzy elements.

| | $\aleph_1$ | $\aleph_2$ | $\aleph_3$ | $\aleph_4$ | $\aleph_5$ |
|---|---|---|---|---|---|
| $A_1$ | (0.56,0.34,0.10) | (0.70,0.10,0.09) | (0.88,0.09,0.03) | (0.80,0.07,0.04) | (0.85,0.06,0.03) |
| $A_2$ | (0.90,0.07,0.03) | (0.10,0.66,0.20) | (0.08,0.10,0.06) | (0.70,0.15,0.11) | (0.64,0.07,0.22) |
| $A_3$ | (0.40,0.33,0.19) | (0.06,0.81,0.12) | (0.05,0.83,0.09) | (0.03,0.88,0.05) | (0.06,0.88,0.05) |
| $A_4$ | (0.09,0.79,0.03) | (0.72,0.14,0.09) | (0.65,0.25,0.07) | (0.07,0.82,0.05) | (0.13,0.77,0.09) |

We utilize the PFDPWA and PFDPWG operators to develop the MADM method with picture fuzzy information, which can be evaluated as follows:

- **Step 1.** Utilize Eq. (5.12) to calculate the values of $\hbar_{gf}$, $g = 1, 2, \ldots, \xi$; $f = 1, 2, \ldots, \varpi$ as follows:

$$\hbar_{\phi\theta} = \begin{bmatrix} 1.000 & 0.730 & 0.935 & 0.605 \\ 1.000 & 0.805 & 0.450 & 0.470 \\ 1.000 & 0.925 & 0.510 & 0.480 \\ 1.000 & 0.880 & 0.795 & 0.490 \\ 1.000 & 0.910 & 0.710 & 0.520 \end{bmatrix}.$$

- **Step 2.** For $\varrho = 1$, using the PFDPWA operator to accumulate the overall preferences values of the objects $\aleph_g$, $g = 1, 2, \ldots, 5$

$$\Upsilon_1 = (0.5853, 0.2854, 0.0569),\ \Upsilon_2 = (0.6159, 0.1662, 0.1034),\ \Upsilon_3 = (0.7426, 0.1555, 0.0504),$$
$$\Upsilon_4 = (0.5871, 0.1827, 0.0516),\ \Upsilon_5 = (0.6530, 0.1384, 0.0522).$$

- **Step 3.** Calculate the score values of $(\Upsilon_g)$, $g = 1, 2, \ldots, 5$ of the overall PFNs $\Upsilon_g$, $g = 1, 2, \ldots, 5$ as follows:

$$\Lambda(\Upsilon_1) = 0.7642,\ \Lambda(\Upsilon_2) = 0.7563,\ \Lambda(\Upsilon_3) = 0.8461,\ \Lambda(\Upsilon_4) = 0.7678,\ \Lambda(\Upsilon_5) = 0.8004.$$

- **Step 4.** Rank all the alternatives $\aleph_g$, $g = 1, 2, \ldots, 5$ for the score functions $\Lambda(\Upsilon_g)$, $g = 1, 2, \ldots, 5$ of the overall PFNs as $\aleph_3 \succ \aleph_5 \succ \aleph_4 \succ \aleph_1 \succ \aleph_2$.

- **Step 5.** $\aleph_3$ is selected as the best company for the insecticides.

  If the PFDPWG operator is implemented instead, the problem can be solved similarly.

- **Step 1.** For $\varrho = 1$, using the PFDPWA operator to accumulate the overall preferences values $\Upsilon_g$ of the objects $\aleph_g$, $g = 1, 2, \ldots, 5$

$$\Upsilon_1 = (0.2078, 0.6663, 0.1094), \Upsilon_2 = (0.1675, 0.5653, 0.1130), \Upsilon_3 = (0.1345, 0.5688, 0.0614),$$
$$\Upsilon_4 = (0.0649, 0.7872, 0.0569), \Upsilon_5 = (0.1288, 0.7619, 0.0821).$$

- **Step 2.** Calculate the score values of $(\Upsilon_g)$, $g = 1, 2, \ldots, 5$ as follows:

$$\Lambda(\Upsilon_1) = 0.5492, \Lambda(\Upsilon_2) = 0.5273, \Lambda(\Upsilon_3) = 0.5366, \Lambda(\Upsilon_4) = 0.5040, \Lambda(\Upsilon_5) = 0.5234.$$

- **Step 3.** Rank all the alternatives $\aleph_g$, $g = 1, 2, \ldots, 5$ for the score functions $\Lambda(\Upsilon_g)$, $g = 1, 2, \ldots, 5$ of the overall PFNs as $\aleph_1 \succ \aleph_3 \succ \aleph_2 \succ \aleph_5 \succ \aleph_4$.
- **Step 4.** $\aleph_1$ is selected as the best company for the insecticides.

According to the comparison formula based on the score function, the ordering of alternatives is obtained under "means preferred to", as we depend on the aggregation operators used ordering of alternatives slightly differently, but the most desirable alternative is $\aleph_3$ for the PFDPWA operator and $\aleph_1$ for the PFDPWG operator.

We analyze the effect of the parameter $\varrho \in [1, 10]$ on the ranking of the alternatives using the BFDPWA and BFDPWG operators, the results of which are shown in Tables 5.2 and 5.3.

## 5.7 Analysis on the effect of parameter $\varrho$ on decision making results

To illustrate the effect of the operating parameter $\varrho$ on the decision-making results, we shall use different values to rank the alternatives. The ranking order of the alternatives $\aleph_g$, $g = 1, 2, 3, 4, 5$ arranged on the values of score function within the range $1 \leq \varrho \leq 10$ based on the PFDPWA and PFDPWG operators are shown in Table 5.2 and Table 5.3, respectively. In Table 5.2, it is evident that when the parameter value is changed for the PFDPWA operator, the ranking results are unstable, and the corresponding best alternative is not identical. When $1 \leq \varrho \leq 2$, the ranking order is $\aleph_3 \succ \aleph_5 \succ \aleph_4 \succ \aleph_1 \succ \aleph_2$, and $\aleph_3 \succ \aleph_5 \succ \aleph_1 \succ \aleph_4 \succ \aleph_2$. For $3 \leq \varrho \leq 5$, the order is $\aleph_3 \succ \aleph_1 \succ \aleph_5 \succ \aleph_4 \succ \aleph_2$. Although the ranking order is different in all these cases, the best alternative is $\aleph_3$. Again, for $6 \leq \varrho \leq 10$, the ranking order is stable, and the order is $\aleph_1 \succ \aleph_3 \succ \aleph_5 \succ \aleph_4 \succ \aleph_2$. Thus it is a stable ranking order, and the best company is $\aleph_1$. From Table 5.3, it is also seen that when the value of $1 \leq \varrho 2$ is changed for the PFDPWG operator, the ranking results also slightly change as the orders are $\aleph_1 \succ \aleph_3 \succ \aleph_2 \succ \aleph_5 \succ \aleph_4$ and $\aleph_3 \succ \aleph_1 \succ \aleph_5 \succ \aleph_2 \succ \aleph_4$ and the corresponding best alternative is additionally nonidentical. For, $3 \leq \varrho \leq 10$ the corresponding stable ranking is $\aleph_3 \succ \aleph_4 \succ \aleph_1 \succ \aleph_2 \succ \aleph_5$, the best one is $\aleph_3$. The PFDPWA and PFDPWG operators affect the variation of the values of parameter $\varrho$. We see that on changing the value of $\varrho$ using the PFDPWA operator, which is more responsive to the ranking orders of the alternatives, while for the various values of the operator PFDPWG and affect the ranking orders, which is less responsive to $\varrho$ in this MADM model.

**Table 5.2** The effect of the parameter $\varrho$ on the PFDPWA operator.

| $\varrho$ | $\Lambda(\Upsilon_1)$ | $\Lambda(\Upsilon_2)$ | $\Lambda(\Upsilon_3)$ | $\Lambda(\Upsilon_4)$ | $\Lambda(\Upsilon_5)$ | Ranking order |
|---|---|---|---|---|---|---|
| 1 | 0.7642 | 0.7563 | 0.8461 | 0.7678 | 0.8004 | $\aleph_3 \succ \aleph_5 \succ \aleph_4 \succ \aleph_1 \succ \aleph_2$ |
| 2 | 0.8493 | 0.7812 | 0.8818 | 0.8215 | 0.8543 | $\aleph_3 \succ \aleph_5 \succ \aleph_1 \succ \aleph_4 \succ \aleph_1$ |
| 3 | 0.8814 | 0.7903 | 0.8960 | 0.8397 | 0.8733 | $\aleph_3 \succ \aleph_1 \succ \aleph_5 \succ \aleph_4 \succ \aleph_2$ |
| 4 | 0.8973 | 0.7951 | 0.9035 | 0.8490 | 0.8828 | $\aleph_3 \succ \aleph_1 \succ \aleph_5 \succ \aleph_4 \succ \aleph_2$ |
| 5 | 0.9061 | 0.7981 | 0.9080 | 0.8548 | 0.8885 | $\aleph_3 \succ \aleph_1 \succ \aleph_5 \succ \aleph_4 \succ \aleph_2$ |
| 6 | 0.9117 | 0.8002 | 0.9110 | 0.8588 | 0.8922 | $\aleph_1 \succ \aleph_3 \succ \aleph_5 \succ \aleph_4 \succ \aleph_2$ |
| 7 | 0.9154 | 0.8017 | 0.9131 | 0.8617 | 0.8949 | $\aleph_1 \succ \aleph_3 \succ \aleph_5 \succ \aleph_4 \succ \aleph_2$ |
| 8 | 0.9181 | 0.8030 | 0.9147 | 0.8639 | 0.8969 | $\aleph_1 \succ \aleph_3 \succ \aleph_5 \succ \aleph_4 \succ \aleph_2$ |
| 9 | 0.9202 | 0.8040 | 0.9159 | 0.8656 | 0.8948 | $\aleph_1 \succ \aleph_3 \succ \aleph_5 \succ \aleph_4 \succ \aleph_2$ |
| 10 | 0.9218 | 0.8047 | 0.9168 | 0.8670 | 0.8996 | $\aleph_1 \succ \aleph_3 \succ \aleph_5 \succ \aleph_4 \succ \aleph_2$ |

**Table 5.3** The effect of the parameter $\varrho$ on the PFDPWG operator.

| $\varrho$ | $\Lambda(\Upsilon_1)$ | $\Lambda(\Upsilon_2)$ | $\Lambda(\Upsilon_3)$ | $\Lambda(\Upsilon_4)$ | $\Lambda(\Upsilon_5)$ | Ranking order |
|---|---|---|---|---|---|---|
| 1 | 0.5492 | 0.5273 | 0.5366 | 0.5040 | 0.5234 | $\aleph_1 \succ \aleph_3 \succ \aleph_2 \succ \aleph_5 \succ \aleph_4$ |
| 2 | 0.5049 | 0.4946 | 0.5115 | 0.4935 | 0.4954 | $\aleph_3 \succ \aleph_1 \succ \aleph_5 \succ \aleph_2 \succ \aleph_4$ |
| 3 | 0.4914 | 0.4814 | 0.5047 | 0.4879 | 0.4784 | $\aleph_3 \succ \aleph_1 \succ \aleph_4 \succ \aleph_2 \succ \aleph_5$ |
| 4 | 0.4816 | 0.4728 | 0.4986 | 0.4837 | 0.4668 | $\aleph_3 \succ \aleph_1 \succ \aleph_4 \succ \aleph_2 \succ \aleph_5$ |
| 5 | 0.4756 | 0.4663 | 0.4957 | 0.4803 | 0.4586 | $\aleph_3 \succ \aleph_1 \succ \aleph_4 \succ \aleph_2 \succ \aleph_5$ |
| 6 | 0.4714 | 0.4613 | 0.4936 | 0.4776 | 0.4527 | $\aleph_3 \succ \aleph_1 \succ \aleph_4 \succ \aleph_2 \succ \aleph_5$ |
| 7 | 0.4684 | 0.4573 | 0.4920 | 0.4755 | 0.4484 | $\aleph_3 \succ \aleph_1 \succ \aleph_4 \succ \aleph_2 \succ \aleph_5$ |
| 8 | 0.4661 | 0.4542 | 0.4907 | 0.4738 | 0.4451 | $\aleph_3 \succ \aleph_1 \succ \aleph_4 \succ \aleph_2 \succ \aleph_5$ |
| 9 | 0.4643 | 0.4517 | 0.4897 | 0.4723 | 0.4424 | $\aleph_3 \succ \aleph_1 \succ \aleph_4 \succ \aleph_2 \succ \aleph_5$ |
| 10 | 0.4629 | 0.4496 | 0.4888 | 0.4712 | 0.4403 | $\aleph_3 \succ \aleph_1 \succ \aleph_4 \succ \aleph_2 \succ \aleph_5$ |

## 5.8 Comparative analysis

In order to compare the feasibility of the proposed MADM method the rankings of the proposed MADM method are compared with the existing MADM methods, as presented in Table 5.4, introduced by Wei [20], where the PFWA and PFWG operators are used to aggregate the picture fuzzy input arguments. Next, we compared the existing MADM method proposed by Wei [21] as presented in Table 5.5, where Hamacher aggregation operators PFHWA and PFHWG are considered. Finally, we compared this MADM model with the existing MADM method proposed by Jana et al. [8] including Dombi aggregation PFDWA and PFDWG operators provided in Table 5.6. From Tables 5.4, 5.5, and 5.6, we can see the best alternative obtained from the existing methods [8,20,21], where alternative $\aleph_3$ is consistent

**Table 5.4** Comparison results with the PFWA (PFWG) operator.

| Methods | Ranking ordered |
|---|---|
| Wei [20] PFWA operator | $\aleph_3 \succ \aleph_2 \succ \aleph_1 \succ \aleph_5 \succ \aleph_4$ |
| Wei [20] PFWG operator | $\aleph_3 \succ \aleph_2 \succ \aleph_1 \succ \aleph_5 \succ \aleph_4$ |
| Proposed PFDPWA | $\aleph_3 \succ \aleph_5 \succ \aleph_4 \succ \aleph_1 \succ \aleph_2$ |
| Proposed PFDPWG | $\aleph_1 \succ \aleph_3 \succ \aleph_2 \succ \aleph_5 \succ \aleph_4$ |

**Table 5.5** Comparison results with the PFHWA (PFHWG) operator.

| Methods | Ranking ordered |
|---|---|
| Wei [21] PFHWA operator | $\aleph_3 \succ \aleph_2 \succ \aleph_1 \succ \aleph_5 \succ \aleph_4$ |
| Wei [21] PFHWG operator | $\aleph_3 \succ \aleph_2 \succ \aleph_1 \succ \aleph_5 \succ \aleph_4$ |
| Proposed PFDPWA | $\aleph_3 \succ \aleph_5 \succ \aleph_4 \succ \aleph_1 \succ \aleph_2$ |
| Proposed PFDPWG | $\aleph_1 \succ \aleph_3 \succ \aleph_2 \succ \aleph_5 \succ \aleph_4$ |

**Table 5.6** Comparison results with the PFDWA (PFDWG) operator.

| Methods | Ranking ordered |
|---|---|
| Jana et al. [8] PFDWA operator | $\aleph_3 \succ \aleph_1 \succ \aleph_5 \succ \aleph_2 \succ \aleph_4$ |
| Jana et al. [8] PFDWG operator | $\aleph_1 \succ \aleph_3 \succ \aleph_5 \succ \aleph_2 \succ \aleph_4$ |
| Proposed PFDPWA | $\aleph_3 \succ \aleph_5 \succ \aleph_4 \succ \aleph_1 \succ \aleph_2$ |
| Proposed PFDPWG | $\aleph_1 \succ \aleph_3 \succ \aleph_2 \succ \aleph_5 \succ \aleph_4$ |

**Table 5.7** Characteristic comparisons of different methods.

| Methods | Have made flexible measure easier | Have a prioritized relation |
|---|---|---|
| Wei [20] | ✓ | × |
| Wei [21] | ✓ | × |
| Jana et al. [8] | ✓ | × |
| Proposed PFDPWA operator | ✓ | ✓ |
| Proposed PFDPWG operator | ✓ | ✓ |

with the existing methods [8,20,21] under PFDWA, PFWA, and PFHWA operators. However, on the other hand, the best alternative obtained by the proposed MADM method using the PFDPWG operator is $\aleph_1$ that is consistent with the PFDWG operator applied in the existing method [8] and $\aleph_3$ is the best alternative of the existing methods [20,21] based on PFWG and PFHWG operators, which is not consistent with the proposed MADM method because existing methods [8,20,21] have no prioritized relationship between the attributes and the aggregation operators. The characteristic comparison of the proposed model is depicted in Table 5.7. Hence, it is clear that for the proposed model used, PFDPWA (PFDPWG) more accurately finds prioritized relationships between the attributes by using prioritized relationship aggregation operators. On the other hand, in [8,20,21], the results utilize PFDWA,

PFDWG, PFWA, PFWG, PFHWA, and PFHWG operators, which does not consider the case of higher priority of attributes than other attributes. Thus the advanced aggregation operators implement a new flexible measure for decision makers to control picture fuzzy MADM problems.

## 5.9 Conclusions

In this chapter, we have studied a multiattribute decision-making problem using PF information. We have introduced arithmetic and geometric operators to develop some PF Dombi aggregation operators from the motivation of Dombi operations such as PF Dombi prioritized averaging operator, PF Dombi prioritized geometric operator, PF Dombi prioritized weighted averaging operator, and PF Dombi prioritized weighted geometric operator. Then, we investigated some issues of these operators. Finally, we develop a multiattribute decision-making problems for the selection of a talent search in the environment of PF information. In future, the application of our proposed model can be applied to decision-making theory, risk evaluation, and other domains within ambiguous environments.

## References


[1] K.T. Atanassov, On Intuitionistic Fuzzy Sets Theory, Studies in Fuzziness and Soft Computing, vol. 283, Springer-Verlag, Berlin Heidelberg, 2012.

[2] J. Chen, J. Ye, Some single-valued neutrosophic Dombi weighted aggregation operators for multiple attribute decision-making, Symmetry 9 (82) (2017) 1–11.

[3] B.C. Cuong, Picture fuzzy sets first results. Part 1, Seminar "Neurofuzzy systems with applications", Tech. Rep., Institute of Mathematics, Hanoi, 2013.

[4] B.C. Cuong, Picture fuzzy sets first results. Part 2, Seminar "Neurofuzzy systems with applications", Tech. Rep., Institute of Mathematics, Hanoi, 2013.

[5] J. Dombi, A general class of fuzzy operators, the demorgan class of fuzzy operators and fuzziness measures induced by fuzzy operators, Fuzzy Sets Syst. 8 (1982) 149–163.

[6] Y.D. He, Z. He, P.P. Zhou, Y.J. Deng, Scaled prioritized geometric aggregation operators and their applications to decision making, Int. J. Uncertain. Fuzziness Knowl.-Based Syst. 24 (1) (2016) 13–46.

[7] X. He, Typhoon disaster assessment based on Dombi hesitant fuzzy information aggregation operators, Nat. Hazards 90 (3) (2018) 1153–1175.

[8] C. Jana, T. Senapati, M. Pal, R.R. Yager, Picture fuzzy Dombi aggregation operators: application to MADM process, Appl. Soft Comput. (2018), https://doi.org/10.1016/j.asoc.2018.10.021.

[9] C. Jana, M. Pal, Jian-qiang Wang, Bipolar fuzzy Dombi aggregation operators and its application in multiple attribute decision making process, J. Ambient Intell. Humaniz. Comput. 10 (2019) 3533–3549.

[10] X. Lu, J. Ye, Dombi aggregation operators of linguistic cubic variables for multiple attribute decision making, Information 9 (2018) 188, https://doi.org/10.3390/info9080188.

[11] P. Singh, Correlation coefficients for picture fuzzy sets, J. Intell. Fuzzy Syst. 27 (2014) 2857–2868.

[12] L. Son, DPFCM: a novel distributed picture fuzzy clustering method on picture fuzzy sets, Expert Syst. Appl. 2 (2015) 51–66.

[13] L.H. Son, Generalized picture distance measure and applications to picture fuzzy clustering, Appl. Soft Comput. 46 (2016) 284–295.

[14] P.H. Thong, L.H. Son, A new approach to multi-variables fuzzy forecasting using picture fuzzy clustering and picture fuzzy rules interpolation method, in: 6th International Conference on Knowledge and Systems Engineering, Hanoi, Vietnam, 2015, pp. 679–690.

[15] N.T. Thong, HIFCF: an effective hybrid model between picture fuzzy clustering and intuitionistic fuzzy recommender systems for medical diagnosis, Expert Syst. Appl. 42 (7) (2015) 3682–3701.

[16] L. Wang, Y. Wang, X.D. Liu, Prioritized aggregation operators and correlated aggregation operators for hesitant 2-tuple linguistic variables, Symmetry 10 (2) (2018) 39, https://doi.org/10.3390/sym10020039.

[17] G.W. Wei, Y. Wei, Some single-valued neutrosophic Dombi prioritized weighted aggregation operators in multiple attribute decision making, J. Intell. Fuzzy Syst. 33 (2) (2018) 2001–2013.

[18] G.W. Wei, Hesitant fuzzy prioritized operators and their application to multiple attribute group decision making, Knowl.-Based Syst. 31 (2012) 176–182.

[19] G.W. Wei, Picture fuzzy cross-entropy for multiple attribute decision making problems, J. Bus. Econ. Manag. 17 (4) (2016) 491–502.

[20] G. Wei, Picture fuzzy aggregation operators and their application to multiple attribute decision making, J. Intell. Fuzzy Syst. 33 (2017) 713–724.

[21] G.W. Wei, Picture fuzzy Hamacher aggregation operators and their application to multiple attribute decision making, Fundam. Inform. 157 (3) (2018) 271–320.

[22] R.R. Yager, Prioritized aggregation operators, Int. J. Approx. Reason. 48 (2008) 263–274.

[23] L.A. Zadeh, Fuzzy sets, Inf. Control 8 (1965) 338–353.

[24] X.F. Zhao, R. Lin, G.W. Wei, Fuzzy prioritized operators and their application to multiple attribute group decision making, Appl. Math. Model. 37 (2013) 4759–4770.

[25] L.Y. Zhou, R. Lin, X.F. Zhao, G.W. Wei, Uncertain linguistic prioritized aggregation operators and their application to multiple attribute group decision making, Int. J. Uncertain. Fuzziness Knowl.-Based Syst. 21 (4) (2013) 603–627.

# 6

# Picture fuzzy power Dombi operators and their utilization in decision-making problems

## 6.1 Introduction

Researchers have observed that some cases cannot be explained by intuitionistic fuzzy [1] and fuzzy [41] environments. For example, to interpret the case of voting, human beings explained their opinion as of types: yes, no, abstain, and confusing, which cannot be presented in an intuitionistic fuzzy set. To measure this situation, Cuoung [2,3] introduced an advanced extension of a fuzzy set, namely, a picture fuzzy set (PFS). After seeing its successful applications to minimize uncertainties in real-world problems, researchers have drawn attention to its implementation in different uncertain environments. The aggregation operators (AOs) are very powerful and are used in different situations. For instance, Wei [27] developed averaging and geometric weighted AO using a PFS argument and utilized them to construct the MADM process. Then, Wei [29] used crossentropy technique based decision problems in PFS. Later, Son [22] used PFS arguments to associate distance measures. In the same environment, Son et al. [23] studied inference systems-based clustering. Using the distance measurement method, Peng and Dai [20] established an algorithm-based picture fuzzy decision making. Wei utilized weighted averaging and geometric Hamacher aggregation to develop picture fuzzy decision making. Recently, Jana et al. [13] studied novel MADM problems using Dombi norms called picture fuzzy Dombi AO to aggregate picture fuzzy arguments. Khan et al. [14] developed a weighted aggregation based decision-making method using Einstein norms. For more details on the search decision-making process to aggregate picture fuzzy numbers using various AO, we refer the reader to the following references [5,9,21,24–26,29]. Dombi [4] proposed a new operational rule in 1982 called the "Dombi operation", which has a good application in the edibility of operation in parameters. Liu et al. [17] utilized Dombi operations to IFS and developed MAGDM problem using a Dombi–Bonferroni mean operator. Jana et al. [8] developed emerging technology enterprise selection challenges by aggregating bipolar fuzzy numbers based on weighted Dombi AO and using these operators. He developed a typhoon catastrophe assessment problem using the Dombi aggregation operator in a cautious fuzzy environment [7]. Lu and Ye [19] developed the linguistic cubic variable Dombi weighted arithmetic averaging (LCVDWAA) operator and the linguistic cubic variable Dombi weighted geometric averaging (LCVDWGA) operator to aggregate the information from the cubic linguistic variable (LCV). The truth is that PFNs have the ability to determine unclear data that emerges in real-world issues in a useful way. Due to the great interest in developing our

Picture Fuzzy Logic and Its Applications in Decision Making Problems. https://doi.org/10.1016/B978-0-44-322024-1.00010-8

proposed study, several fuzzy Dombi operations and PAOs were used to construct MADM approaches. As a result, aggregating SVNNs is a fantastic application of Dombi power weighted AO, which is based on the union of Dombi and power operations. To address this issue, some picture fuzzy Dombi power AO will be defined using Dombi operations, geometric AOs [33,34], and classical arithmetic [32,38,39], and by using Dombi operations [4,7,8,10–13,28] and PAO [6,15–18,30,31,35,36,42].

The rest of this chapter is structured as follows. A few hypotheses and operations on the PFNs for the following contexts are shown in Section 6.2. Section 6.3 introduces various types of picture fuzzy Dombi power AOs and highlights some of their key characteristics. Section 6.4 introduces various types of picture fuzzy geometric Dombi power geometric operators and establishes some findings. A MADM method is developed in Section 6.5 using these aggregating operators. Section 6.6 provides an application of the developed MADM method. The sensitivity and comparative analyses are given with the available methods in Section 6.6. Some concluding remarks are given in Section 6.6.2.

## 6.2 Basic definitions and terminologies

Some basic definitions and essential concepts related to PFSs and power average operators are related to developing this chapter.

Yager [37] defines a powerful operator called PA, which is defined below.

**Definition 6.1.** Let $b_1, b_2, \ldots, b_m$ be a set of $m$ objects, then the power-averaging (PA) operator is defined as:

$$PA(b_1, b_2, \ldots, b_m) = \frac{\sum_{z=1}^{m}(1+\mathcal{T}(b_z))b_z}{\sum_{z=1}^{m}(1+\mathcal{T}(b_z))}, \tag{6.1}$$

where,

$$\mathcal{T}(b_z) = \sum_{z=1\, y\neq z}^{m} Supp(b_z, b_y).$$

The support of $b_y$ and $b_z$, is denoted by $Sport(b_z, b_y)$ and satisfies the following properties: (i) $Sport(a,b) \in [0,1]$, (ii) $Sport(a,b) = Sport(b,a)$, and (iii) $Sport(a,b) \geq Sport(x,y)$, if $|a-b| < |x-y|$. The $Support(Sport)$ similarity index is measured. When two values are closer and more similar to one another, there is stronger support between them.

**Definition 6.2.** Let $b_1, b_2, \ldots, b_m$ be a set of $m$ objects, then the power geometric (PG) operator was introduced by Yager [37] defined as follows:

$$PG(b_1, b_2, \ldots b_m) = \prod_{z=1}^{m} \left(b_z\right)^{\frac{(1+\mathcal{T}(b_z))}{\sum_{z=1}^{m}(1+\mathcal{T}(b_z))}}, \tag{6.2}$$

where

$$\mathcal{T}(b_z) = \sum_{z=1\, y\neq z}^{m} Sport(b_z, b_y).$$

Here, $Sport(a,b)$ is the support of $a$ for $b$ satisfying the following properties: (i) $Supp(a,b) \in [0,1]$, (ii) $Sport(a,b) = Sport(b,a)$, and (iii) $Sport(a,b) \geq Sport(p,q)$, if $|a-b| < |p-q|$. $Sport$ measures the similarity index. An increase in similarity between two values results in a higher degree of closeness between them, and they are more supportive of each other in this case.

## 6.3 Picture fuzzy power Dombi averaging operators

By combining the Dombi and power operations on PFNs, the PFDPWA, PFDPOWA, and PFDPHWA operators are defined.

**Definition 6.3.** Let $\mathcal{R}_f = \left\langle \mathcal{Y}_f, \mathcal{A}_f, \mathcal{N}_f \right\rangle$ be a collection of PFNs, where for $f = 1, 2, \ldots, \varpi$. The picture fuzzy power Dombi averaging (PFPDA) operator is a function $PFPDA : \Theta^{\varpi} \rightarrow \Theta$ that follows as:

$$PFPDA(\mathcal{R}_1, \mathcal{R}_2, \ldots, \mathcal{R}_{\varpi}) = \frac{\bigoplus_{f=1}^{\varpi}(1+\mathcal{T}(\mathcal{R}_f))\mathcal{R}_f}{\sum_{f=1}^{\varpi}(1+\mathcal{T}(\mathcal{R}_f))}, \tag{6.3}$$

where

$$\mathcal{T}(\mathcal{R}_g) = \sum_{f=1, g\neq f}^{\varpi} Supp(\mathcal{R}_f, \mathcal{R}_g) \tag{6.4}$$

and $Sport(\mathcal{R}_f, \mathcal{R}_g)$ represents the support between $\mathcal{R}_g$ and $\mathcal{R}_f$, and follows as:

**(a)** $Sport(\mathcal{R}_g, \mathcal{R}_f) \in [0, 1]$;
**(b)** $Sport(\mathcal{R}_f, \mathcal{R}_g) = Sport(\mathcal{R}_g, \mathcal{R}_f)$;
**(c)** $Sport(\mathcal{R}_f, \mathcal{R}_g) \geq Sport(\mathcal{R}_r, \mathcal{R}_s)$, if $d(\mathcal{R}_f, \mathcal{R}_g) < d(\mathcal{R}_r, \mathcal{R}_s)$, here $d$ represents the distance measure between PFNs.

The following helpful result is obtained by applying Dombi operations on PFNs.

**Theorem 6.1.** *Suppose* $\mathcal{R}_f = \langle \mathcal{Y}_f, \mathcal{A}_f, \mathcal{N}_f \rangle$ *is a collection of PFNs for* $f = 1, 2, \ldots, \varpi$. *Then,* $PFPDA(\mathcal{R}_1, \mathcal{R}_2 \ldots, \mathcal{R}_{\varpi})$ *is a PFN.*

*Thus*

$$PFPDA(\mathcal{R}_1, \mathcal{R}_2 \ldots, \mathcal{R}_{\varpi}) = \frac{\bigoplus_{f=1}^{\varpi}(1+\mathcal{T}(\mathcal{R}_f))\mathcal{R}_f}{\sum_{f=1}^{\varpi}(1+\mathcal{T}(\mathcal{R}_f))}$$

$$=\left(1-\frac{1}{1+\left\{\sum_{f=1}^{\varpi}\frac{(1+\mathcal{T}(\mathcal{R}_f))}{\sum_{f=1}^{\varpi}(1+\mathcal{T}(\mathcal{R}_b))}\left(\frac{\mathcal{Y}_f}{1-\mathcal{Y}_f}\right)^{\varrho}\right\}^{1/\varrho}},\frac{1}{1+\left\{\sum_{f=1}^{\varpi}\frac{(1+\mathcal{T}(\mathcal{R}_f))}{\sum_{f=1}^{\varpi}(1+\mathcal{T}(\mathcal{R}_f))}\left(\frac{1-\mathcal{A}_f}{\mathcal{A}_f}\right)^{\varrho}\right\}^{1/\varrho}},\right.$$
$$\left.\frac{1}{1+\left\{\sum_{f=1}^{\varpi}\frac{(1+\mathcal{T}(\mathcal{R}_f))}{\sum_{f=1}^{\varpi}(1+\mathcal{T}(\mathcal{R}_f))}\left(\frac{1-\mathcal{N}_f}{\mathcal{N}_f}\right)^{\varrho}\right\}^{1/\varrho}}\right), \tag{6.5}$$

*where*

$$\mathcal{T}(\mathcal{R}_f)=\sum_{f=1,g\neq f}^{\varpi} Sport(\mathcal{R}_f,\mathcal{R}_g).$$

**Theorem 6.2.** *Let $\mathcal{R}_f=\left(\mathcal{Y}_f,\mathcal{A}_f,\mathcal{N}_f\right)$ be a mass of PFNs, where $f=1,2,\ldots,\varpi$. The accumulated value of PFDWA on PFNs is also a PFN, $\Xi=(\Xi_1,\Xi_2,\ldots,\Xi_{\varpi})^T$ is the weight vector of $\mathcal{R}_f$, $f=1,2,\ldots,\varpi$, where $\Xi_f\in[0,1]$ and $\sum_{f=1}^{\varpi}\Xi_f=1$. Thus $PFPDWA:\Theta^{\varpi}\rightarrow\Theta$ obeys the following:*

$$PFPDWA_{\Xi}(\mathcal{R}_1,\mathcal{R}_2\ldots,\mathcal{R}_{\varpi})=\frac{\bigoplus_{f=1}^{\varpi}(\Xi_f(1+\mathcal{T}(\mathcal{R}_f))\mathcal{R}_f)}{\sum_{f=1}^{\varpi}\Xi_f(1+T(\mathcal{R}_f))}$$
$$=\left(1-\frac{1}{1+\left\{\sum_{f=1}^{\varpi}\frac{\Xi_f(1+\mathcal{T}(\mathcal{R}_f))}{\sum_{f=1}^{\varpi}\Xi_f(1+\mathcal{T}(\mathcal{R}_f))}\left(\frac{\mathcal{Y}_f}{1-\mathcal{Y}_f}\right)^{\varrho}\right\}^{1/\varrho}},\right.$$
$$\frac{1}{1+\left\{\sum_{f=1}^{\varpi}\frac{\Xi_f(1+\mathcal{T}(\mathcal{R}_f))}{\sum_{f=1}^{\varpi}\Xi_f(1+\mathcal{T}(\mathcal{R}_f))}\left(\frac{1-\mathcal{A}_f}{\mathcal{A}_f}\right)^{\varrho}\right\}^{1/\varrho}},$$
$$\left.\frac{1}{1+\left\{\sum_{f=1}^{\varpi}\frac{\Xi_f(1+\mathcal{T}(\mathcal{R}_f))}{\sum_{f=1}^{\varpi}\Xi_f(1+\mathcal{T}(\mathcal{R}_f))}\left(\frac{1-\mathcal{N}_f}{\mathcal{N}_f}\right)^{\varrho}\right\}^{1/\varrho}}\right), \tag{6.6}$$

*where*

$$\mathcal{T}(\mathcal{R}_f)=\sum_{f=1,g\neq f}^{\varpi}\Xi_f Sup(\mathcal{R}_f,\mathcal{R}_g). \tag{6.7}$$

*Proof.* (i) Let $f=2$. By applying the Dombi operations on PFNs and by the PA, we have

$$PFPDWA(\mathcal{R}_1,\mathcal{R}_2)=\Xi_1\mathcal{R}_1\bigoplus\Xi_2\mathcal{R}_2=\Xi_1(\mathcal{Y}_1,\mathcal{A}_1,\mathcal{N}_1)\bigoplus\Xi_2(\mathcal{Y}_2,\mathcal{A}_2,\mathcal{N}_2)$$

and from the right-hand side of Eq. (6.6), we have

$$\left(1-\frac{1}{1+\left\{\frac{\Xi_1(1+\mathcal{T}(\mathcal{R}_1))}{\sum_{f=1}^{2}\Xi_f(1+\mathcal{T}(\mathcal{R}_f))}\left(\frac{\mathcal{Y}_1}{1-\mathcal{Y}_1}\right)^{\varrho}+\frac{\Xi_2(1+\mathcal{T}(\mathcal{R}_2))}{\sum_{f=1}^{2}\Xi_f(1+\mathcal{T}(\mathcal{R}_f))}\left(\frac{\mathcal{Y}_2}{1-\mathcal{Y}_2}\right)^{\varrho}\right\}^{1/\varrho}},\right.$$
$$\frac{1}{1+\left\{\frac{\Xi_1(1+\mathcal{T}(\mathcal{R}_1))}{\sum_{f=1}^{2}\Xi_f(1+\mathcal{T}(\mathcal{R}_f))}\left(\frac{1-\mathcal{A}_1}{\mathcal{A}_1}\right)^{\varrho}+\frac{\Xi_2(1+\mathcal{T}(\mathcal{R}_2))}{\sum_{f=1}^{2}\Xi_f(1+\mathcal{T}(\mathcal{R}_f))}\left(\frac{1-\mathcal{A}_2}{\mathcal{A}_2}\right)^{\varrho}\right\}^{1/\varrho}},$$
$$\left.\frac{1}{1+\left\{\frac{\Xi_1(1+\mathcal{T}(\mathcal{R}_1))}{\sum_{f=1}^{2}\Xi_f(1+\mathcal{T}(\mathcal{R}_f))}\left(\frac{1-\mathcal{N}_1}{\mathcal{N}_1}\right)^{\varrho}+\frac{\Xi_2(1+\mathcal{T}(\mathcal{R}_2))}{\sum_{f=1}^{2}\Xi_f(1+\mathcal{T}(\mathcal{R}_f))}\left(\frac{1-\mathcal{N}_2}{\mathcal{N}_2}\right)^{\varrho}\right\}^{1/\varrho}}\right)$$
$$=\left(1-\frac{1}{1+\left\{\sum_{f=1}^{2}\frac{\Xi_f(1+\mathcal{T}(\mathcal{R}_f))}{\sum_{f=1}^{2}\Xi_f(1+\mathcal{T}(\mathcal{R}_f))}\left(\frac{\mathcal{Y}_f}{1-\mathcal{Y}_f}\right)^{\varrho}\right\}^{1/\varrho}},\right.$$
$$\frac{1}{1+\left\{\sum_{f=1}^{2}\frac{\Xi_f(1+\mathcal{T}(\mathcal{R}_f))}{\sum_{f=1}^{2}\Xi_f(1+\mathcal{T}(\mathcal{R}_f))}\left(\frac{1-\mathcal{A}_f}{\mathcal{A}_f}\right)^{\varrho}\right\}^{1/\varrho}},$$
$$\left.\frac{1}{1+\left\{\sum_{f=1}^{2}\frac{\Xi_f(1+\mathcal{T}(\mathcal{R}_f))}{\sum_{f=1}^{2}\Xi_f(1+\mathcal{T}(\mathcal{R}_f))}\left(\frac{1-\mathcal{N}_f}{\mathcal{N}_f}\right)^{\varrho}\right\}^{1/\varrho}}\right). \tag{6.8}$$

Thus Eq. (6.6) is true for $\varpi=2$.
(ii) Suppose (6.6) is true for any natural number $q$, i.e., $\varpi=q$, then by Eq. (6.6),

$$PFPDWA_{\Xi}(\mathcal{R}_1,\mathcal{R}_2\ldots,\mathcal{R}_q)=\bigoplus_{f=1}^{q}\frac{\Xi_f(1+\mathcal{T}(\mathcal{R}_f))}{\sum_{f=1}^{q}\Xi_f(1+\mathcal{T}(\mathcal{R}_f))}\mathcal{R}_f$$
$$=\left(1-\frac{1}{1+\left\{\sum_{f=1}^{q}\frac{\Xi_f(1+\mathcal{T}(\mathcal{R}_f))}{\sum_{f=1}^{q}\Xi_f(1+\mathcal{T}(\mathcal{R}_f))}\left(\frac{\mathcal{Y}_f}{1-\mathcal{Y}_f}\right)^{\varrho}\right\}^{1/\varrho}},\right.$$
$$\frac{1}{1+\left\{\sum_{f=1}^{q}\frac{\Xi_f(1+\mathcal{T}(\mathcal{R}_f))}{\sum_{f=1}^{q}\Xi_f(1+\mathcal{T}(\mathcal{R}_f))}\left(\frac{1-\mathcal{A}_f}{\mathcal{A}_f}\right)^{\varrho}\right\}^{1/\varrho}},$$
$$\left.\frac{1}{1+\left\{\sum_{f=1}^{q}\frac{\Xi_z(1+\mathcal{T}(\mathcal{R}_f))}{\sum_{f=1}^{q}\Xi_f(1+\mathcal{T}(\mathcal{R}_f))}\left(\frac{1-\mathcal{N}_f}{\mathcal{N}_f}\right)^{\varrho}\right\}^{1/\varrho}}\right). \tag{6.9}$$

Now, for $\varpi=q+1$, then

$$PFPDWA_{\Xi}(\mathcal{R}_1,\mathcal{R}_2,\ldots,\mathcal{R}_q,\mathcal{R}_{q+1})$$

$$=\bigoplus_{f=1}^{q}\frac{\Xi_f(1+\mathcal{T}(\mathcal{R}_f))}{\sum_{f=1}^{q}\Xi_b(1+\mathcal{T}(\mathcal{R}_f))}\mathcal{R}_f\bigoplus\left(\frac{\Xi_{q+1}((1+\mathcal{T}(\mathcal{R}_{q+1}))\mathcal{R}_{q+1})}{\Xi_{q+1}(1+\mathcal{T}(\mathcal{R}_{q+1}))}\mathcal{R}_{q+1}\right)$$

$$=\left(1-\frac{1}{1+\left\{\sum_{f=1}^{q}\frac{\Xi_f(1+\mathcal{T}(\mathcal{R}_f))}{\sum_{f=1}^{q}\Xi_f(1+\mathcal{T}(\mathcal{R}_f))}\left(\frac{\mathcal{Y}_f}{1-\mathcal{Y}_f}\right)^{\varrho}\right\}^{1/\varrho}},\right.$$
$$\frac{1}{1+\left\{\sum_{f=1}^{q}\frac{\Xi_f(1+\mathcal{T}(\mathcal{R}_f))}{\sum_{f=1}^{q}\Xi_f(1+\mathcal{T}(\mathcal{R}_f))}\left(\frac{1-\mathcal{A}_f}{\mathcal{A}_f}\right)^{\varrho}\right\}^{1/\varrho}},$$
$$\left.\frac{1}{1+\left\{\sum_{f=1}^{q}\frac{\Xi_f(1+\mathcal{T}(\mathcal{R}_f))}{\sum_{f=1}^{q}\Xi_f(1+\mathcal{T}(\mathcal{R}_f))}\left(\frac{1-\mathcal{N}_f}{\mathcal{N}_f}\right)^{\varrho}\right\}^{1/\varrho}}\right)$$
$$\bigoplus\left(1-\frac{1}{1+\left\{\frac{\Xi_{q+1}(1+\mathcal{T}(\mathcal{R}_{q+1}))}{\Xi_{q+1}(1+\mathcal{T}(\mathcal{R}_{q+1}))}\left(\frac{\mathcal{Y}_{q+1}}{1-\mathcal{Y}_{q+1}}\right)^{\varrho}\right\}^{1/\varrho}},\frac{1}{1+\left\{\frac{\Xi_{q+1}((1+\mathcal{T}(\mathcal{R}_{q+1}))}{\Xi_{q+1}(1+\mathcal{T}(\mathcal{R}_{q+1}))}\left(\frac{1-\mathcal{A}_{q+1}}{\mathcal{A}_{q+1}}\right)^{\varrho}\right\}^{1/\varrho}},\right.$$
$$\left.\frac{1}{1+\left\{\frac{\Xi_{q+1}((1+\mathcal{T}(\mathcal{R}_{q+1}))}{\Xi_{q+1}(1+\mathcal{T}(\mathcal{R}_{q+1}))}\left(\frac{1-\mathcal{N}_{q+1}}{\mathcal{N}_{q+1}}\right)^{\varrho}\right\}^{1/\varrho}}\right)$$
$$=\left(1-\frac{1}{1+\left\{\sum_{f=1}^{q+1}\frac{\Xi_{f+1}(1+\mathcal{T}(\mathcal{R}_f))}{\sum_{f=1}^{q+1}\Xi_{f+1}(1+\mathcal{T}(\mathcal{R}_f))}\left(\frac{\mathcal{Y}_f}{1-\mathcal{Y}_f}\right)^{\varrho}\right\}^{1/\varrho}},\right.$$
$$\frac{1}{1+\left\{\sum_{f=1}^{q+1}\frac{\Xi_{f+1}(1+\mathcal{T}(\mathcal{R}_f))}{\sum_{f=1}^{q+1}\Xi_{f+1}(1+\mathcal{T}(\mathcal{R}_f))}\left(\frac{1-\mathcal{A}_f}{\mathcal{A}_f}\right)^{\varrho}\right\}^{1/\varrho}},$$
$$\left.\frac{1}{1+\left\{\sum_{f=1}^{q+1}\frac{\Xi_{f+1}(1+\mathcal{T}(\mathcal{R}_f))}{\sum_{f=1}^{q+1}\Xi_{f+1}(1+\mathcal{T}(\mathcal{R}_f))}\left(\frac{1-\mathcal{N}_f}{\mathcal{N}_f}\right)^{\varrho}\right\}^{1/\varrho}}\right). \tag{6.10}$$

From the above result, it shows that Eq. (6.6) is true for $\varpi = q+1$, where $q$ is any natural number.

Hence, Eq. (6.6) is valid for any natural number $\varpi$. □

**Theorem 6.3.** *(Idempotent property) Let $\mathcal{R}_f = (\mathcal{Y}_f, \mathcal{A}_f, \mathcal{N}_f)$ be a PFN for $f = 1, 2, \ldots, \varpi$. If all $\mathcal{R}_f$ are equal, i.e., $\mathcal{R}_f = \mathcal{R}$ for all $f$, then*

$$PFPDWA_{\Xi}(\mathcal{R}_1, \mathcal{R}_2, \ldots, \mathcal{R}_{\varpi}) = \mathcal{R}. \tag{6.11}$$

*Proof.* Since $\mathcal{R}_f = (\mathcal{Y}_f, \mathcal{A}_f, \mathcal{N}_f) = \mathcal{R}$, $f = 1, 2, \ldots, \varpi$, then we have by Eq. (6.6):

$$PFPDWA_{\Xi}(\mathcal{R}_1, \mathcal{R}_2, \ldots, \mathcal{R}_{\varpi}) = \frac{\bigoplus_{f=1}^{\varpi} \Xi_f(1+\mathcal{T}(\mathcal{R}_f))\mathcal{R}_f}{\sum_{f=1}^{\varpi} \Xi_f(1+\mathcal{T}(\mathcal{R}_f))}$$

$$= \left(1 - \frac{1}{1+\left\{\sum_{f=1}^{\varpi} \frac{\Xi_f(1+\mathcal{T}(\mathcal{R}_f))}{\sum_{f=1}^{\varpi}\Xi_f(1+\mathcal{T}(\mathcal{R}_f))}\left(\frac{\mathcal{Y}_f}{1-\mathcal{Y}_f}\right)^{\varrho}\right\}^{1/\varrho}},\right.$$
$$\frac{1}{1+\left\{\sum_{f=1}^{\varpi} \frac{\Xi_f(1+\mathcal{T}(\mathcal{R}_f))}{\sum_{f=1}^{\varpi}\Xi_f(1+\mathcal{T}(\mathcal{R}_f))}\left(\frac{1-\mathcal{A}_f}{\mathcal{A}_f}\right)^{\varrho}\right\}^{1/\varrho}},$$
$$\left.\frac{1}{1+\left\{\sum_{f=1}^{\varpi} \frac{\Xi_f(1+\mathcal{T}(\mathcal{R}_f))}{\sum_{f=1}^{\varpi}\Xi_f(1+\mathcal{T}(\mathcal{R}_f))}\left(\frac{1-\mathcal{N}_f}{\mathcal{N}_f}\right)^{\varrho}\right\}^{1/\varrho}}\right)$$
$$= \left(1 - \frac{1}{1+\left\{\left(\frac{\mathcal{Y}}{1-\mathcal{Y}}\right)^{\varrho}\right\}^{1/\varrho}}, \frac{1}{1+\left\{\left(\frac{1-\mathcal{A}}{\mathcal{A}}\right)^{\varrho}\right\}^{1/\varrho}}, \frac{1}{1+\left\{\left(\frac{1-\mathcal{N}}{\mathcal{N}}\right)^{\varrho}\right\}^{1/\varrho}}\right)$$
$$= \left(1 - \frac{1}{1+\frac{\mathcal{Y}}{1-\mathcal{Y}}}, \frac{1}{1+\frac{1-\mathcal{A}}{\mathcal{A}}}, \frac{1}{1+\frac{1-\mathcal{N}}{\mathcal{N}}}\right) = (\mathcal{Y}, \mathcal{A}, \mathcal{N}) = \mathcal{R}.$$

Hence, $PFPDWA_{\Xi}(\mathcal{R}_1, \mathcal{R}_2, \ldots, \mathcal{R}_{\varpi}) = \mathcal{R}$ holds. □

**Theorem 6.4.** *(Boundedness property) Let $\mathcal{R}_f = \langle \mathcal{Y}_f, \mathcal{A}_f, \mathcal{N}_f \rangle$ be a PFN for $f = 1, 2, \ldots, \varpi$. Let $\mathcal{R}^- = \min(\mathcal{R}_1, \mathcal{R}_2, \ldots, \mathcal{R}_{\varpi})$ and $\mathcal{R}^+ = \max(\mathcal{R}_1, \mathcal{R}_2, \ldots, \mathcal{R}_{\varpi})$.*

*Therefore $\mathcal{R}^- \leq PFPDWA_{\Xi}(\mathcal{R}_1, \mathcal{R}_2, \ldots, \mathcal{R}_{\varpi}) \leq \mathcal{R}^+$.*

*Proof.* Let $\mathcal{R}_f = (\mathcal{Y}_f, \mathcal{A}_f, \mathcal{A}_f)$, $f = 1, 2, \ldots, \varpi$, be a group of PFNs. Let $\mathcal{R}^- = \min(\mathcal{R}_1, \mathcal{R}_2, \ldots, \mathcal{R}_{\varpi}) = (\mathcal{Y}'^-, \mathcal{A}'^-, \mathcal{N}'^-)$ and $\mathcal{R}^+ = \max(\mathcal{R}_1, \mathcal{R}_2, \ldots, \mathcal{R}_{\varpi}) = (\mathcal{Y}'^+, \mathcal{A}'^+, \mathcal{N}'^+)$, where, $\mathcal{Y}'^- = \min_f\{\mathcal{Y}_f\}$, $\mathcal{A}'^- = \max_f\{\mathcal{A}_f\}$, $\mathcal{N}'^- = \max_f\{\mathcal{N}_f\}$, $\mathcal{Y}'^+ = \max_f\{\mathcal{Y}_f\}$, $\mathcal{A}'^+ = \min_f\{\mathcal{A}_f\}$ and $\mathcal{N}'^+ = \min_f\{\mathcal{N}_f\}$. We have obtained the inequalities:

$$1 - \frac{1}{1+\left\{\sum_{f=1}^{\varpi} \frac{\Xi_f(1+\mathcal{T}(\mathcal{R}_f))}{\sum_{f=1}^{\varpi}\Xi_{f+1}(1+\mathcal{T}(\mathcal{R}_f))}\left(\frac{\mathcal{Y}'^-}{1-\mathcal{Y}'^-}\right)^{\varrho}\right\}^{1/\varrho}}$$
$$\leq 1 - \frac{1}{1+\left\{\sum_{f=1}^{\varpi} \frac{\Xi_f(1+\mathcal{T}(\mathcal{R}_f))}{\sum_{f=1}^{\varpi}\Xi_f(1+\mathcal{T}(\mathcal{R}_f))}\left(\frac{\mathcal{Y}}{1-\mathcal{Y}}\right)^{\varrho}\right\}^{1/\varrho}}$$
$$\leq 1 - \frac{1}{1+\left\{\sum_{f=1}^{\varpi} \frac{\Xi_f(1+\mathcal{T}(\mathcal{R}_f))}{\sum_{f=1}^{\varpi}\Xi_f(1+\mathcal{T}(\mathcal{R}_f))}\left(\frac{\mathcal{Y}'^+}{1-\mathcal{Y}'^+}\right)^{\varrho}\right\}^{1/\varrho}},$$

$$\frac{1}{1+\left\{\sum_{f=1}^{\varpi} \frac{\Xi_f(1+\mathcal{T}(\mathcal{R}_f))}{\sum_{f=1}^{\varpi}\Xi_f(1+\mathcal{T}(\mathcal{R}_f))}\left(\frac{1-\mathcal{A}^-}{\mathcal{A}^-}\right)^{\varrho}\right\}^{1/\varrho}} \leq \frac{1}{1+\left\{\sum_{f=1}^{\varpi} \frac{\Xi_f(1+\mathcal{T}(\mathcal{R}_f))}{\sum_{f=1}^{\varpi}\Xi_f(1+\mathcal{T}(\mathcal{R}_f))}\left(\frac{1-\mathcal{A}}{\mathcal{A}}\right)^{\varrho}\right\}^{1/\varrho}}$$

$$\leq \frac{1}{1+\left\{\sum_{f=1}^{\varpi} \frac{\Xi_f(1+\mathcal{T}(\mathcal{R}_f))}{\sum_{f=1}^{\varpi}\Xi_f(1+\mathcal{T}(\mathcal{R}_f))}\left(\frac{1-\mathcal{A}^+}{\mathcal{A}^+}\right)^{\varrho}\right\}^{1/\varrho}}$$

and

$$\frac{1}{1+\left\{\sum_{f=1}^{\varpi} \frac{\Xi_f(1+\mathcal{T}(\mathcal{R}_f))}{\sum_{f=1}^{\varpi}\Xi_f(1+\mathcal{T}(\mathcal{R}_f))}\left(\frac{1-\mathcal{N}'^-}{\mathcal{N}'^-}\right)^{\varrho}\right\}^{1/\varrho}} \leq \frac{1}{1+\left\{\sum_{f=1}^{\varpi} \frac{\Xi_f(1+\mathcal{T}(\mathcal{R}_f))}{\sum_{z=1}^{\varpi}\Xi_f(1+\mathcal{T}(\mathcal{R}_f))}\left(\frac{1-\mathcal{N}}{\mathcal{N}}\right)^{\varrho}\right\}^{1/\varrho}}$$
$$\leq \frac{1}{1+\left\{\sum_{f=1}^{\varpi} \frac{\Xi_f(1+\mathcal{T}(\mathcal{R}_f))}{\sum_{f=1}^{\varpi}\Xi_f(1+\mathcal{T}(\mathcal{R}_f))}\left(\frac{1-\mathcal{N}'^+}{\mathcal{N}'^+}\right)^{\varrho}\right\}^{1/\varrho}}.$$

Therefore

$$\mathcal{R}^- \leq PFPDWA(\mathcal{R}_1, \mathcal{R}_2, \ldots, \mathcal{R}_\varpi) \leq \mathcal{R}^+.$$

□

**Theorem 6.5.** *(Monotonicity property) Let $\mathcal{R}_f = (\mathcal{Y}_f, \mathcal{A}_f, \mathcal{N}_f)$ and $\mathcal{R}'_f = (\mathcal{Y}'_f, \mathcal{A}'_f, \mathcal{N}'_f)$, $f = 1, 2, \ldots, \varpi$, be two sets of PFNs. If $\mathcal{R}_f \leq \mathcal{R}'_f$ for all $f$, then*

$$PFPDWA_\Xi(\mathcal{R}_1, \mathcal{R}_2 \ldots, \mathcal{R}_\varpi) \leq PFPDWA_\Xi(\mathcal{R}'_1, \mathcal{R}'_2, \ldots, \mathcal{R}'_\varpi).$$

**Definition 6.4.** Let $\mathcal{R}_f = (\mathcal{Y}_f, \mathcal{A}_f, \mathcal{N}_f)$ be a PFN for $f = 1, 2, \ldots, \varpi$. Then, the PFDOWA operator of dimension $\varpi$ is a mapping $PFPDOWA : \Theta^\varpi \to \Theta$ and the corresponding weight vector $\delta = (\delta_1, \delta_2, \ldots, \delta_\varpi)^T$, where $\delta_f \in [0, 1]$, and $\sum_{f=1}^{\varpi}\delta_f = 1$. Therefore

$$\begin{aligned}
&PFPDOWA_{\delta,\Xi}(\mathcal{R}_1, \mathcal{R}_2 \ldots, \mathcal{R}_\varpi) \\
&= \frac{\bigoplus_{f=1}^{\varpi}(\Xi_f(1+\mathcal{T}(\mathcal{R}_{\sigma(f)}))\mathcal{R}_{\sigma(f)}}{\sum_{f=1}^{\varpi}\delta_f(1+\mathcal{T}(\mathcal{R}_{\sigma(f)}))} \\
&= \left(1 - \frac{1}{1+\left\{\sum_{f=1}^{\varpi} \frac{\Xi_f(1+\mathcal{T}(\mathcal{R}_{\sigma(f)}))}{\sum_{f=1}^{\varpi}\sigma_f(1+\mathcal{T}(\mathcal{R})_{\sigma(f)}))}\left(\frac{\mathcal{Y}_{\sigma(f)}}{1-\mathcal{Y}_{\sigma(f)}}\right)^{\varrho}\right\}^{1/\varrho}},\right. \\
&\quad \frac{1}{1+\left\{\sum_{f=1}^{\varpi} \frac{\delta_f(1+\mathcal{T}(\dot{\mathcal{R}}_{\sigma(f)}))}{\sum_{f=1}^{\varpi}\delta_f(1+\mathcal{T}(\mathcal{R}_{\sigma(f)}))}\left(\frac{1-\mathcal{A}_{\sigma(f)}}{\mathcal{A}_{\sigma(f)}}\right)^{\varrho}\right\}^{1/\varrho}}, \\
&\quad \left.\frac{1}{1+\left\{\sum_{f=1}^{\varpi} \frac{\delta_f(1+\mathcal{T}(\mathcal{R}_{\sigma(f)}))}{\sum_{f=1}^{\varpi}\delta_f(1+\mathcal{T}(\mathcal{R}_{\sigma(f)}))}\left(\frac{1-\mathcal{N}_{\sigma(f)}}{\mathcal{N}_{\sigma(f)}}\right)^{\varrho}\right\}^{1/\varrho}}\right).
\end{aligned} \tag{6.12}$$

Here, $(\sigma(1), \sigma(2), \ldots, \sigma(\varpi))$ is a permutation of $(1, 2, \ldots, \varpi)$. $\dot{\mathcal{R}}_{\sigma(f)}$ is the $f$th largest element of PFNs $\mathcal{R}_f$, where $\dot{\mathcal{R}}_{\sigma(f)} = \varpi\,\Xi_f\mathcal{R}_f$, $f = 1, 2, \ldots, \varpi$, and $\varpi$ is called the balancing coefficient, and $\delta_f$, $f = 1, 2, \ldots, \varpi$ is the weight vector defined as:

$$\delta_f = h\left(\frac{R_f}{TV}\right) - h\left(\frac{R_{f-1}}{TV}\right),\ R_f = \sum_{g=1}^{f} V_{\sigma(g)},\ V_{\sigma(g)} = 1 + \mathcal{T}(\mathcal{R}_{\sigma(g)}) \tag{6.13}$$

and $\mathcal{T}(\mathcal{R}_{\sigma(f)})$ is the support of the $f$th largest PFNs $\mathcal{R}_{\sigma(f)}$ from the other PFNs and $h : [0, 1] \to [0, 1]$ is the monotonic function satisfying the properties: $h(1) = 1$, $h(0) = 0$, $h(g) \geq h(f)$ if $g > f$.

Therefore

$$\mathcal{T}(\mathcal{R}_{\sigma(f)}) = \sum_{g \neq f, f=1}^{\varpi} Sport(\mathcal{R}_{\sigma(f)}, \mathcal{R}_{\sigma(g)}), \tag{6.14}$$

where $\sum_{g \neq f, f=1}^{\varpi} Sport(\mathcal{R}_{\sigma(f)}, \mathcal{R}_{\sigma(f)})$ denotes the support of the $f$th largest PFN $\mathcal{R}_{\sigma(g)}$ for the $g$th largest PFN $\mathcal{R}_{\sigma(f)}$.

The following results can easily be proved from the definition of the PFPDOWA.

**Theorem 6.6.** *(Idempotency property) If $\mathcal{R}_f = (\mathcal{Y}_f, \mathcal{A}_f, \mathcal{N}_f)$ is the mass of PFNs for $f = 1, 2, \ldots, \varpi$ such that $\mathcal{R}_f = \mathcal{R}$ for all $f$, then*

$$PFPDOWA_{\delta}(\mathcal{R}_1, \mathcal{R}_2 \ldots, \mathcal{R}_{\varpi}) = \mathcal{R}.$$

**Theorem 6.7.** *(Boundedness property) Let $\mathcal{R}_f = (\mathcal{Y}_f, \mathcal{A}_f, \mathcal{N}_f)$ be a mass of PFNs for $f = 1, 2, \ldots, \varpi$ and let*

$$\mathcal{R}^- = \min_f \mathcal{R}_f, \quad \mathcal{R}^+ = \max_f \mathcal{R}_f.$$

*Then,*

$$\mathcal{R}^- \leq PFPDOWA_{\delta}(\mathcal{R}_1, \mathcal{R}_2 \ldots, \mathcal{R}_{\varpi}) \leq \mathcal{R}^+.$$

**Theorem 6.8.** *(Monotonicity property) Let $\mathcal{R}_f = \langle \mathcal{Y}_f, \mathcal{A}_f, \mathcal{N}_f \rangle$ and $\mathcal{R}'_f = \langle \mathcal{Y}'_f, \mathcal{A}'_f, \mathcal{N}'_f \rangle$, $f = 1, 2, \ldots, \varpi$ be two sets of PFNs, if $\mathcal{R}_f \leq \mathcal{R}'_f$ for all $f$, then*

$$PFPDOWA_{\delta}(\mathcal{R}_1, \mathcal{R}_2 \ldots, \mathcal{R}_{\varpi}) \leq PFPDOWA_{\delta}(\mathcal{R}'_1, \mathcal{R}'_2, \ldots, \mathcal{R}'_{\varpi}).$$

**Theorem 6.9.** *(Commutativity property) Let $\mathcal{R}_f$, $f = 1, 2, \ldots, \varpi$ and $\mathcal{R}'_f$, $f = 1, 2, \ldots, \varpi$ be two sets of PFNs. Then,*

$$PFPDOWA_{\delta}(\mathcal{R}_1, \mathcal{R}_2 \ldots, \mathcal{R}_{\varpi}) = PFPDOWA_{\delta}(\mathcal{R}'_1, \mathcal{R}'_2, \ldots, \mathcal{R}'_{\varpi}),$$

*where $\mathcal{R}'_f$, $f = 1, 2, \ldots, \varpi$ is any permutation of $\mathcal{R}_f$, $f = 1, 2, \ldots, \varpi$.*

From Definitions 6.3 and 6.4, it is seen that the PFPDWA operator took the weights of the PFN into account, while the other PFPDOWA implies the weights of the PFNs in the provided ordered locations. As a result, the weights of PFPDWA and PFPDOWA behave differently. However, just one of them is explored. To overcome these challenges, we provide the PF hybrid power Dombi averaging (PFPDHWA) operator.

**Definition 6.5.** Let $\mathcal{R}_f = (\mathcal{Y}_f, \mathcal{A}_f, \mathcal{N}_f)$ be a mass PFNs for $f = 1, 2, \ldots, \varpi$. Then, the PFPDHWA operator of dimension $\varpi$ is a mapping $PFPDHWA : \Theta^{\varpi} \to \Theta$ with corresponding weight vector $\delta = (\delta_1, \delta_2, \ldots, \delta_{\varpi})^T$ such that $\delta_f \in [0, 1]$, and $\sum_{f=1}^{\varpi} \delta_f = 1$. Therefore

$$PFPDHWA_{\delta,\Xi}(\mathcal{R}_1, \mathcal{R}_2 \ldots, \mathcal{R}_{\varpi}) = \frac{\bigoplus_{f=1}^{\varpi}(\delta_f(1+\mathcal{T}(\dot{\mathcal{R}}_{\sigma(f)}))\dot{\mathcal{R}}_{\sigma(f)}}{\sum_{f=1}^{\varpi}\delta_f(1+\mathcal{T}(\dot{\mathcal{R}}_{\sigma(f)}))}$$

$$= \left( 1 - \frac{1}{1+\left\{\sum_{f=1}^{\varpi}\frac{\delta_f(1+\mathcal{T}(\dot{\mathcal{R}}_{\sigma(f)}))}{\sum_{f=1}^{\varpi}\delta_f(1+\mathcal{T}(\dot{\mathcal{R}}_{\sigma(f)}))}\left(\frac{\dot{\mathcal{Y}}_{\sigma(f)}}{1-\dot{\mathcal{Y}}_{\sigma(f)}}\right)^{\varrho}\right\}^{1/\varrho}}, \right.$$
$$\frac{1}{1+\left\{\sum_{f=1}^{\varpi}\frac{\delta_f(1+\mathcal{T}(\dot{\mathcal{R}}_{\sigma(f)}))}{\sum_{f=1}^{\varpi}\delta_f(1+\mathcal{T}(\dot{\mathcal{R}}_{\sigma(f)}))}\left(\frac{1-\dot{\mathcal{A}}_{\sigma(f)}}{\dot{\mathcal{A}}_{\sigma(f)}}\right)^{\varrho}\right\}^{1/\varrho}},$$
$$\left.\frac{1}{1+\left\{\sum_{f=1}^{\varpi}\frac{\delta_f(1+\mathcal{T}(\dot{\mathcal{R}}_{\sigma(f)}))}{\sum_{f=1}^{\varpi}\delta_f(1+\mathcal{T}(\dot{\mathcal{R}}_{\sigma(f)}))}\left(\frac{1-\dot{\mathcal{N}}_{\sigma(f)}}{\dot{\mathcal{N}}_{\sigma(f)}}\right)^{\varrho}\right\}^{1/\varrho}}\right), \tag{6.15}$$

where $(\sigma(1), \sigma(2), \ldots, \sigma(\varpi))$ is a permutation of $\sigma(f)$, $f = 1, 2, \ldots, \varpi$, for which $\dot{\mathcal{R}}_{\sigma(f-1)} \geq \dot{\mathcal{R}}_{\sigma(f)}$ for all $f$; $\delta_f$, $f = 1, 2, \ldots, \varpi$ is the weight vector defined as:

$$\delta_f = h\left(\frac{R_f}{TV}\right) - h\left(\frac{R_{f-1}}{TV}\right),\ R_f = \sum_{g=1}^{f} V_{\sigma(g)},\ V_{\sigma(g)} = 1 + \mathcal{T}(\dot{\mathcal{R}}_{\sigma(g)}) \tag{6.16}$$

and $\mathcal{T}(\mathcal{R}_{\sigma(f)})$ as the support of the $f$th largest PFNs $\mathcal{R}_{\sigma(f)}$ from the other PFNs, $h : [0, 1] \to [0, 1]$ is the monotonic function such that $h(1) = 1$, $h(0) = 0$, $h(g) \geq h(f)$ if $g > f$.

Therefore

$$\mathcal{T}(\mathcal{R}_{\sigma(f)}) = \sum_{g \neq f, f=1}^{\varpi} Sport(\dot{\mathcal{R}}_{\sigma(f)}, \dot{\mathcal{R}}_{\sigma(g)}), \tag{6.17}$$

where $\sum_{g\neq f,f=1}^{\varpi} Sport(\dot{\mathcal{R}}_{\sigma(f)}, \dot{\mathcal{R}}_{\sigma(g)})$ represents the support of the $g$th largest PFN $\dot{\mathcal{R}}_{\sigma(g)}$ for the $f$th largest PFN $\dot{\mathcal{R}}_{\sigma(f)}$.

## 6.4 Dombi power geometric AOs with PFNs

This section defines some new picture fuzzy power Dombi geometric AOs.

**Definition 6.6.** Let $\mathcal{R}_f = \langle \mathcal{Y}_f, \mathcal{A}_f, \mathcal{N}_f \rangle$ be a mass of PFNs for $f = 1, 2, \ldots, \varpi$. Thus the picture fuzzy power Dombi geometric (PFPDG) operator is a function $\Theta^{\varpi} \to \Theta$ that is defined below:

$$PFPDG(\mathcal{R}_1, \mathcal{R}_2, \ldots, \mathcal{R}_\varpi) = \bigotimes_{f=1}^{\varpi} (\mathcal{R}_f)^{\frac{(1+\mathcal{T}(\mathcal{R}_f))}{\sum_{f=1}^{\varpi}(1+\mathcal{T}(\mathcal{R}_f))}}, \tag{6.18}$$

where

$$\mathcal{T}(\mathcal{R}_f) = \sum_{f=1, g \neq f}^{\varpi} Sport(\mathcal{R}_f, \mathcal{R}_g) \tag{6.19}$$

and $Sport(\mathcal{R}_f, \mathcal{R}_g)$ represents the support of $\mathcal{R}_g$ for $\mathcal{R}_f$, which satisfies the following properties:

**(i)** $Sport(\mathcal{R}_f, \mathcal{R}_g) \in [0, 1]$;
**(ii)** $Sport(\mathcal{R}_b, \mathcal{R}_g) = Sport(\mathcal{R}_g, \mathcal{R}_f)$;
**(iii)** $Sport(\mathcal{R}_f, \mathcal{R}_g) \geq Sport(\mathcal{R}_r, \mathcal{R}_s)$, if $d(\mathcal{R}_f, \mathcal{R}_g) < d(\mathcal{R}_r, \mathcal{R}_s)$, where $d$ denotes the distance measure between PFNs.

The following significant result can be deduced by performing Dombi operations on PFNs.

**Theorem 6.10.** *Let $\mathcal{R}_f = (\mathcal{Y}_f, \mathcal{A}_f, \mathcal{N}_f)$, $f = 1, 2, \ldots, \varpi$ be a mass of PFNs. The value of PFPDG is a PFN. That is,*

$$PFPDG(\mathcal{R}_1, \mathcal{R}_2 \ldots, \mathcal{R}_\varpi) = \bigotimes_{f=1}^{\varpi} (\mathcal{R}_f)^{\frac{(1+\mathcal{T}(\mathcal{R}_f))}{\sum_{f=1}^{\varpi}(1+\mathcal{T}(\mathcal{R}_f))}}$$

$$= \left( \frac{1}{1 + \left\{ \sum_{f=1}^{\varpi} \frac{(1+\mathcal{T}(\mathcal{R}_f))}{\sum_{f=1}^{\varpi}(1+\mathcal{T}(\mathcal{R}_f))} \left( \frac{1-\mathcal{Y}_f}{\mathcal{Y}_f} \right)^{\varrho} \right\}^{1/\varrho}}, 1 - \frac{1}{1 + \left\{ \sum_{f=1}^{\varpi} \frac{(1+\mathcal{T}(\mathcal{R}_f))}{\sum_{f=1}^{\varpi}(1+\mathcal{T}(\mathcal{R}_f))} \left( \frac{\mathcal{A}_f}{1-\mathcal{A}_f} \right)^{\varrho} \right\}^{1/\varrho}}, \right.$$
$$\left. 1 - \frac{1}{1 + \left\{ \sum_{f=1}^{\varpi} \frac{(1+\mathcal{T}(\mathcal{R}_f))}{\sum_{f=1}^{\varpi}(1+\mathcal{T}(\mathcal{R}_f))} \left( \frac{\mathcal{N}_f}{1-\mathcal{N}_f} \right)^{\varrho} \right\}^{1/\varrho}} \right), \tag{6.20}$$

*where*

$$\mathcal{T}(\mathcal{R}_f) = \sum_{f=1, g \neq f}^{\varpi} Sup(\mathcal{R}_f, \mathcal{R}_g).$$

**Theorem 6.11.** *Let $\mathcal{R}_f = (\mathcal{Y}_f, \mathcal{A}_f, \mathcal{N}_f)$ be a PFN for $f = 1, 2, \ldots, \varpi$. Then, the aggregated value using the PFPDWG operator is a PFN, where $\Xi = (\Xi_1, \Xi_2, \ldots, \Xi_\varpi)^T$ is the weight vector* *of* $\mathcal{R}_f$, $f = 1, 2, \ldots, \varpi$, *where* $\Xi_f \in [0, 1]$, $\sum_{f=1}^{\varpi} \Xi_f = 1$. *Thus* $PFPDWG : \Theta^{\varpi} \to \Theta$ *follows:*

$$PFPDWG_{\Xi}(\mathcal{R}_1, \mathcal{R}_2 \ldots, \mathcal{R}_{\varpi}) = \bigotimes_{f=1}^{\varpi} \left(\mathcal{R}_f\right)^{\frac{(\Xi_f(1+\mathcal{T}(\mathcal{R}_f))}{\sum_{f=1}^{\varpi} \Xi_f(1+\mathcal{T}(\mathcal{R}_f))}}$$

$$= \left( \frac{1}{1 + \left\{ \sum_{f=1}^{\varpi} \frac{\Xi_f(1+\mathcal{T}(\mathcal{R}_f))}{\sum_{f=1}^{\varpi} \Xi_f(1+\mathcal{T}(\mathcal{R}_f))} \left( \frac{1-\mathcal{Y}_f}{\mathcal{Y}_f} \right)^{\varrho} \right\}^{1/\varrho}}, \right.$$
$$1 - \frac{1}{1 + \left\{ \sum_{f=1}^{\varpi} \frac{\Xi_f(1+\mathcal{T}(\mathcal{R}_f))}{\sum_{f=1}^{\varpi} \Xi_f(1+\mathcal{T}(\mathcal{R}_f))} \left( \frac{\mathcal{A}_f}{1-\mathcal{A}_f} \right)^{\varrho} \right\}^{1/\varrho}},$$
$$\left. 1 - \frac{1}{1 + \left\{ \sum_{f=1}^{\varpi} \frac{\Xi_f(1+\mathcal{T}(\mathcal{R}_f))}{\sum_{f=1}^{\varpi} \Xi_f(1+\mathcal{T}(\mathcal{R}_f))} \left( \frac{N_f}{1-N_f} \right)^{\varrho} \right\}^{1/\varrho}} \right), \tag{6.21}$$

*where*

$$\mathcal{T}(\mathcal{R}_f) = \sum_{f=1, g \neq f}^{\varpi} \Xi_f Sport(\mathcal{R}_f, \mathcal{R}_g).$$

*Proof.* The proof of this theorem is similar to that for Theorem 6.1. □

**Theorem 6.12.** *(Idempotency property) Let* $\mathcal{R}_f = (\mathcal{Y}_f, \mathcal{A}_f, N_f)$ *be a PFN for* $f = 1, 2, \ldots, \varpi$, *if all* $\mathcal{R}_f$ *are equal, i.e.,* $\mathcal{R}_f = \mathcal{R}$ *for all* $f$, *then*

$$PFPDWG_{\Xi}(\mathcal{R}_1, \mathcal{R}_2, \ldots, \mathcal{R}_{\varpi}) = \mathcal{R}.$$

**Theorem 6.13.** *(Boundedness property) Let* $\mathcal{R}_f = \langle \mathcal{Y}_f, \mathcal{A}_f, N_f \rangle$ *be a PFN for* $f = 1, 2, \ldots, \varpi$ *and also Let* $\mathcal{R}^- = \min(\mathcal{R}_1, \mathcal{R}_2, \ldots, \mathcal{R}_{\varpi})$ *and* $\mathcal{R}^+ = \max(\mathcal{R}_1, \mathcal{R}_2, \ldots, \mathcal{R}_{\varpi})$.

*Therefore* $\mathcal{R}^- \leq PFPDWG_{\Xi}(\mathcal{R}_1, \mathcal{R}_2, \ldots, \mathcal{R}_{\varpi}) \leq \mathcal{R}^+$.

**Theorem 6.14.** *(Monotonicity property) Let* $\mathcal{R}_f = (\mathcal{Y}_f, \mathcal{A}_f, N_f)$ *and* $\mathcal{R}'_f = \langle \mathcal{Y}'_f, \mathcal{A}'_f, N'_f \rangle$, $f = 1, 2, \ldots, \varpi$ *be two sets of PFNs. If* $\mathcal{R}_f \leq \mathcal{R}'_f$ *for all* $f$, *then*

$$PFPDWG_{\Xi}(\mathcal{R}_1, \mathcal{R}_2 \ldots, \mathcal{R}_{\varpi}) \leq PFPDWG_{\Xi}(\mathcal{R}'_1, \mathcal{R}'_2, \ldots, \mathcal{R}'_{\varpi}).$$

**Definition 6.7.** Let $\mathcal{R}_f = \langle \mathcal{Y}_f, \mathcal{A}_f, N_f \rangle$ be a PFN for $f = 1, 2, \ldots, \varpi$. Then, the PFPDOG operator is a function $PFPDOWG : \Theta^{\varpi} \to \Theta$ with the corresponding weight vector $\delta = (\delta_1, \delta_2, \ldots, \delta_{\varpi})^T$, and $\delta_f \in [0, 1]$, $\sum_{f=1}^{\varpi} \delta_f = 1$. Therefore

$$PFPDOWG_{\delta}(\mathcal{R}_1, \mathcal{R}_2 \ldots, \mathcal{R}_{\varpi}) = \bigotimes_{f=1}^{\varpi} \left(\mathcal{R}_{\sigma(f)}\right)^{\frac{(\delta_f(1+\mathcal{T}(\mathcal{R}_{\sigma(f)}))}{\sum_{f=1}^{\varpi} \delta_f(1+\mathcal{T}(\mathcal{R}_{\sigma(f)}))}}$$

$$= \left( \frac{1}{1+\left\{\sum_{f=1}^{\varpi} \frac{\delta_f(1+\mathcal{T}(\dot{\mathcal{R}}_{\sigma(b)}))}{\sum_{f=1}^{\varpi}\delta_f(1+\mathcal{T}(\dot{\mathcal{R}}_{\sigma(f)}))}\left(\frac{1-\mathcal{Y}_{\sigma(f)}}{\mathcal{Y}_{\sigma(f)}}\right)^{\varrho}\right\}^{1/\varrho}}, \right.$$

$$1-\frac{1}{1+\left\{\sum_{f=1}^{\varpi} \frac{\delta_f(1+\mathcal{T}(\mathcal{R}_{\sigma(f)}))}{\sum_{f=1}^{\varpi}\delta_f(1+\mathcal{T}(\dot{\mathcal{R}}_{\sigma(f)}))}\left(\frac{\mathcal{A}_{\sigma(f)}}{1-\mathcal{A}_{\sigma(f)}}\right)^{\varrho}\right\}^{1/\varrho}},$$

$$\left. 1-\frac{1}{1+\left\{\sum_{f=1}^{\varpi} \frac{\delta_f(1+\mathcal{T}(\mathcal{R}_{\sigma(f)}))}{\sum_{f=1}^{\varpi}\delta_f(1+\mathcal{T}(\mathcal{R})_{\sigma(f)}))}\left(\frac{\mathcal{N}_{\sigma(f)}}{1-\mathcal{N}_{\sigma(f)}}\right)^{\varrho}\right\}^{1/\varrho}} \right). \tag{6.22}$$

Here, $(\sigma(1), \sigma(2), \ldots, \sigma(\varpi))$ is a permutation of $(1, 2, \ldots, \varpi)$, and $\mathcal{R}_{\sigma(f-1)} \geq \mathcal{R}_{\sigma(f)}$ for all $f = 1, 2, \ldots, \varpi$; $\dot{\mathcal{R}}_{\sigma(f)}$ is the $f$th largest element of PFNs $\mathcal{R}_f$, where $\mathcal{R}_{\sigma(f)} = \varpi \Xi_f \mathcal{R}_f$, $f = 1, 2, \ldots, \varpi$ and $\varpi$ is the balancing coefficient, and $\delta_f$, $f = 1, 2, \ldots, \varpi$ is a weight vector given by

$$\delta_f = h\left(\frac{R_f}{TV}\right) - h\left(\frac{R_{f-1}}{TV}\right),\ R_f = \sum_{g=1}^{f} V_{\sigma(g)} = 1 + \mathcal{T}(\mathcal{R}_{\sigma(g)}) \tag{6.23}$$

and $\mathcal{T}(\mathcal{R}_{\sigma(f)})$ as the support of the $f$th largest PFNs $\mathcal{R}_{\sigma(f)}$ from the other PFNs. In this case also, $h : [0, 1] \to [0, 1]$ is a monotonic function satisfying $h(0) = 0$, $h(1) = 1$, $h(x) \geq h(g)$ if $x > g$.

Thus

$$\mathcal{T}(\mathcal{R}_{\sigma(f)}) = \sum_{g\neq f, f=1}^{\varpi} Sport(\mathcal{R}_{\sigma(f)}, \mathcal{R}_{\sigma(g)}), \tag{6.24}$$

where $\sum_{g\neq f, f=1}^{\varpi} Sport(\mathcal{R}_{\sigma(f)}, \mathcal{R}_{\sigma(g)})$ represents the support of the $g$th largest PFN $\mathcal{R}_{\sigma(g)}$ for the $f$th largest PFN $\mathcal{R}_{\sigma(f)}$.

Using the operator PFPDOWG, the following theorems are proved.

**Theorem 6.15.** *(Idempotency property) If $\mathcal{R}_f = \langle \mathcal{Y}_f, \mathcal{A}_f, \mathcal{N}_f \rangle$ is a PFN for $f = 1, 2, \ldots, \varpi$ such that $\mathcal{R}_f = \mathcal{R}$ for all $f$, then*

$$PFPDOWG_\delta(\mathcal{R}_1, \mathcal{R}_2 \ldots, \mathcal{R}_\varpi) = \mathcal{R}.$$

**Theorem 6.16.** *(Boundedness property) Let $\mathcal{R}_f = \langle \mathcal{Y}_f, \mathcal{A}_f, \mathcal{N}_f \rangle$ be a PFN for $f = 1, 2, \ldots, \varpi$ and also let*

$$\mathcal{R}^- = \min_f \mathcal{R}_f,\quad \mathcal{R}^+ = \max_f \mathcal{R}_f.$$

*Then,*

$$\mathcal{R}^- \leq PFPDOWG_\delta(\mathcal{R}_1, \mathcal{R}_2 \ldots, \mathcal{R}_\varpi) \leq \mathcal{R}^+.$$

**Theorem 6.17.** *(Monotonicity property) Let* $\mathcal{R}_f = \langle \mathcal{Y}_f, \mathcal{A}_f, N_f \rangle$ *and* $\mathcal{R}'_f = \left\langle \mathcal{Y}'_f, \mathcal{A}'_f, N'_f \right\rangle$*,* $f = 1, 2, \ldots, \varpi$ *be two sets of PFNs, if* $\mathcal{R}_f \leq \mathcal{R}'_f$ *for all* $f$*, then*

$$PFPDOWG_\delta(\mathcal{R}_1, \mathcal{R}_2 \ldots, \mathcal{R}_\varpi) \leq PFPDOWG_\delta(\mathcal{R}'_1, \mathcal{R}'_2, \ldots, \mathcal{R}'_\varpi).$$

**Theorem 6.18.** *(Commutativity property) Let* $\mathcal{R}_f$*,* $f = 1, 2, \ldots, \varpi$ *and* $\mathcal{R}'_f$ $(f = 1, 2, \ldots, \varpi)$ *be two sets of PFNs. Then,*

$$PFPDOWG_\delta(\mathcal{R}_1, \mathcal{R}_2 \ldots, \mathcal{R}_\varpi) = PFPDOWG_\delta(\mathcal{R}'_1, \mathcal{R}'_2, \ldots, \mathcal{R}'_\varpi),$$

*where* $\mathcal{R}'_f$*,* $f = 1, 2, \ldots, \varpi$ *is any permutation of* $\mathcal{R}_f$ $(1, 2, \ldots, \varpi)$*.*

In contrast to the other PFPDOWG, which assumes the weights of the PFNs rather than the weights of the specified ordered positions of the PFNs, we can see from Definitions 6.6 and 6.7 above that the PFPDWG operator considers the weights of the PFNs. The PFPDWG and PFPDOWG weights are consequently computed differently. However, just one of them is looked into. We offer the PF hybrid power Dombi geometric (PFPDHWG) operator to circumvent these issues.

**Definition 6.8.** Let $\mathcal{R}_f = \langle \mathcal{Y}_f, \mathcal{A}_f, N_f \rangle$ be a PFN for $f = 1, 2, \ldots, \varpi$. Then, the PFPDHWG operator is of dimension $\varpi$ is a mapping $PFPDHWG : \Theta^\varpi \to \Theta$ with the associated weight vector $\delta = (\delta_1, \delta_2, \ldots, \delta_\varpi)^T$ such that $\delta_f \in [0, 1]$, $\sum_{f=1}^{\varpi} \delta_f = 1$. Therefore

$$PFPDHWG_{\delta,\Xi}(\mathcal{R}_1, \mathcal{R}_2 \ldots, \mathcal{R}_\varpi) = \frac{\bigoplus_{b=1}^{\varpi} (\delta_f (1 + \mathcal{T}(\dot{\mathcal{R}}_{\sigma(f)})) \dot{\mathcal{R}}_{\sigma(f)}}{\sum_{f=1}^{\varpi} \delta_f (1 + \mathcal{T}(\dot{\mathcal{R}}_{\sigma(f)}))}$$

$$= \left( \frac{1}{1 + \left\{ \sum_{f=1}^{\varpi} \frac{\delta_f (1+\mathcal{T}(\dot{\mathcal{R}}_{\sigma(f)}))}{\sum_{f=1}^{\varpi} \delta_f (1+\mathcal{T}(\dot{\mathcal{R}}_{\sigma(f)}))} \left( \frac{1-\dot{\mathcal{Y}}_{\sigma(f)}}{\dot{\mathcal{Y}}_{\sigma(f)}} \right)^{\varrho} \right\}^{1/\varrho}}, \right.$$
$$1 - \frac{1}{1 + \left\{ \sum_{b=1}^{\varpi} \frac{\delta_f (1+\mathcal{T}(\dot{\mathcal{R}}_{\sigma(f)}))}{\sum_{f=1}^{\varpi} \delta_f (1+\mathcal{T}(\dot{\mathcal{R}}_{\sigma(f)}))} \left( \frac{\dot{\mathcal{A}}_{\sigma(f)}}{1-\dot{\mathcal{A}}_{\sigma(f)}} \right)^{\varrho} \right\}^{1/\varrho}},$$
$$\left. 1 - \frac{1}{1 + \left\{ \sum_{f=1}^{\varpi} \frac{\delta_f (1+\mathcal{T}(\dot{\mathcal{R}}_{\sigma(f)}))}{\sum_{f=1}^{\varpi} \delta_f (1+\mathcal{T}(\dot{\mathcal{R}}_{\sigma(f)}))} \left( \frac{\dot{N}_{\sigma(f)}}{1-\dot{N}_{\sigma(f)}} \right)^{\varrho} \right\}^{1/\varrho}} \right), \tag{6.25}$$

where $(\sigma(1), \sigma(2), \ldots, \sigma(\varpi))$ is a permutation of $\sigma(f)$, $f = 1, 2, \ldots, \varpi$, such that $\dot{\mathcal{R}}_{\sigma(f-1)} \geq \dot{\mathcal{R}}_{\sigma(f)}$ for all $f$; $\delta_f$, $f = 1, 2, \ldots, \varpi$ is a weight vector for which

$$\delta_f = h\left(\frac{R_f}{TV}\right) - h\left(\frac{R_{f-1}}{TV}\right), \; R_f = \sum_{y=1}^{f} V_\sigma(g = 1 + \mathcal{T}(\dot{\mathcal{R}}_{\sigma(g)}) \tag{6.26}$$

and $\mathcal{T}(\mathcal{R}_{\sigma(f)}$ the support of the $f$th largest PFNs $\mathcal{R}_{\sigma(b)}$ from the other PFNs. Thus

$$\mathcal{T}(\mathcal{R}_{\sigma(f)} = \sum_{\neq f, f=1}^{\varpi} Sport(\dot{\mathcal{R}}_{\sigma(f)}, \dot{\mathcal{R}}_{\sigma(g)}), \tag{6.27}$$

where $\sum_{g\neq f, f=1}^{\varpi} Sup(\dot{\mathcal{R}}_{\sigma(f)}, \dot{\mathcal{R}}_{\sigma(f)})$ represents as the support of the $g$th largest PFN $\dot{\mathcal{R}}_{\sigma(g)}$ for the $f$th largest PFN $\mathcal{R}_{\sigma(f)}$, and $h : [0, 1] \to [0, 1]$ is monotonic, nondecreasing, and bounded with bounds 0 and 1.

## 6.5 MADM approach for PFNs

In this study, we have proposed the MADM model using weighted Dombi norms where attributes are known under picture fuzzy arguments. Let $\aleph = \{\aleph_1, \aleph_2, \ldots, \aleph_\xi\}$ be a finite set of alternatives, and $A = \{A_1, A_2, \ldots, A_\varpi\}$ be a set of attributes. Let the weight vector be $\Xi = (\Xi_1, \Xi_2, \ldots, \Xi_\varpi)$ for $A_f$, $f = 1, 2, \ldots, \varpi$ are known completely, where $\Xi_f \in [0, 1]$ and $\sum_{f=1}^{\varpi} \Xi_f = 1$. Let $R = (\mathcal{Y}_{gf}, \mathcal{A}_{gf}, \mathcal{N}_{gf})_{\xi\times\varpi}$ be a picture fuzzy decision matrix, where $\mathcal{Y}_{gf}$ is the degree of truth value for which the alternative $\aleph_g$ satisfies the attribute $A_f$ given by the decision makers, $\mathcal{A}_{gf}$ represents the degree for neutrality for which the alternative $\aleph_g$ does not satisfy the attribute $A_f$, and $\mathcal{N}_{gf}$ denotes the degree for negativity such that alternative $\aleph_g$ does not satisfy the attribute $A_f$ supplied by the decision maker, where $\mathcal{Y}_{gf} \subset [0, 1]$, $\mathcal{A}_{gf} \subset [0, 1]$ and $\mathcal{N}_{gf} \subset [0, 1]$ for which $0 \leq \mathcal{Y}_{gf} + \mathcal{A}_{gf} + \mathcal{N}_{gf} \leq 1$, $g = 1, 2, \ldots, \xi$ and $f = 1, 2, \ldots, \varpi$.

The following algorithm solves the MADM problem with picture fuzzy information based on the PFPDWA and PFPDWG operators.

**Algorithm.**

**Input:** All alternatives.
**Output:** Best alternative.
**Step 1.** Compute supports:

$$Supp\ (\mathcal{R}_{gf}, \mathcal{R}_{gh}) = 1 - d(\mathcal{R}_{gf}, \mathcal{R}_{gh}), \quad f,\ h = 1, 2, \ldots, \varpi, \tag{6.28}$$

which satisfies all the conditions for support. Determine $d(\mathcal{R}_{gf}, \mathcal{R}_{gh})$ with the normalized Hamming distance:

$$d(\mathcal{R}_{gf}, \mathcal{R}_{gh}) = \frac{1}{3}\Big(|\mathcal{Y}_{gf} - \mathcal{Y}_{gh}| + |\mathcal{A}_{gf} - \mathcal{A}_{gh}| + |\mathcal{N}_{gf} - \mathcal{N}_{gh}|\Big), f, h = 1, 2, \ldots, \varpi. \tag{6.29}$$

**Step 2.** Utilize the weight $\Xi_f$, $f = 1, 2, \ldots, \varpi$ of the attribute $A_f$, $f = 1, 2, \ldots, \varpi$ to compute the weighted support $\mathcal{T}(\mathcal{R}_{gf})$ of the PFN $\mathcal{R}_{gf}$ by another PFN $\mathcal{R}_{gh}$, $h = 1, 2, \ldots, \varpi; h \neq f$ such that

$$\mathcal{T}(\mathcal{R}_{gh}) = \sum_{h=1, h\neq f}^{\varpi} \Xi_f Sport(\mathcal{R}_{gf}, \mathcal{R}_{gh}) \tag{6.30}$$

and determine the weight $\Xi_f$ $(f = 1, 2, \ldots, \varpi)$ corresponding to the PFN $\mathcal{R}_{gf}$, $g = 1, 2, \ldots, \xi$; $f = 1, 2, \ldots, \varpi$:

$$\Xi_{gf} = \frac{\Xi_f(1 + \mathcal{T}(\mathcal{T}_{gf}))}{\sum_{f=1}^{\varpi} \Xi_f(1 + \mathcal{T}(\mathcal{R}_{gf}))}, g = 1, 2, \ldots, \xi, \ f = 1, 2, \ldots, \varpi, \tag{6.31}$$

where $\Xi_{gf} \geq 0$, $g = 1, 2, \ldots, \xi$, $f = 1, 2, \ldots, \varpi$, and $\sum_{f=1}^{\varpi} \Xi_{gf} = 1$, $g = 1, 2, \ldots, \xi$.

**Step 3.** We utilize matrix $R$ and using the PFPDWA operator

$$\Upsilon_g PFPDWA_{\Xi}(\mathcal{R}_1, \mathcal{R}_2 \ldots, \mathcal{R}_{\varpi}) = \frac{\bigoplus_{f=1}^{\varpi}(\Xi_f(1 + \mathcal{T}(\mathcal{R}_f))\mathcal{R}_f)}{\sum_{f=1}^{\varpi} \Xi_f(1 + \mathcal{T}(\mathcal{R}_f))}$$

$$= \left(1 - \frac{1}{1 + \left\{\sum_{f=1}^{\varpi} \frac{\Xi_f(1+\mathcal{T}(\mathcal{R}_f))}{\sum_{f=1}^{\varpi} \Xi_f(1+\mathcal{T}(\mathcal{R}_f))}\left(\frac{\mathcal{Y}_f}{1-\mathcal{Y}_f}\right)^{\varrho}\right\}^{1/\varrho}}, \right.$$
$$\frac{1}{1 + \left\{\sum_{f=1}^{\varpi} \frac{\Xi_f(1+\mathcal{T}(\mathcal{R}_f))}{\sum_{f=1}^{\varpi} \Xi_f(1+\mathcal{T}(\mathcal{R}_f))}\left(\frac{1-\mathcal{A}_f}{\mathcal{A}_f}\right)^{\varrho}\right\}^{1/\varrho}},$$
$$\left.\frac{1}{1 + \left\{\sum_{f=1}^{\varpi} \frac{\Xi_f(1+\mathcal{T}(\mathcal{R}_f))}{\sum_{f=1}^{\varpi} \Xi_f(1+\mathcal{T}(\mathcal{R}_f))}\left(\frac{1-\mathcal{N}_f}{\mathcal{N}_f}\right)^{\varrho}\right\}^{1/\varrho}}\right), \tag{6.32}$$

or

$$\Upsilon_a = PFPDWG_{\Xi}(\mathcal{R}_1, \mathcal{R}_2 \ldots, \mathcal{R}_{\varpi}) = \bigotimes_{f=1}^{\varpi} (\mathcal{R}_f)^{\frac{(1+\mathcal{T}(\mathcal{R}_f))}{\sum_{f=1}^{\varpi}(1+\mathcal{T}(\mathcal{R}_f))}}$$

$$= \left(\frac{1}{1 + \left\{\sum_{f=1}^{\varpi} \frac{\Xi_f(1+\mathcal{T}(\mathcal{R}_f))}{\sum_{f=1}^{\varpi} \Xi_f(1+\mathcal{T}(\mathcal{R}_f))}\left(\frac{1-\mathcal{Y}_f}{\mathcal{Y}_f}\right)^{\varrho}\right\}^{1/\varrho}}, \right.$$
$$1 - \frac{1}{1 + \left\{\sum_{f=1}^{\varpi} \frac{\Xi_f(1+\mathcal{T}(\mathcal{R}_f))}{\sum_{f=1}^{\varpi} \Xi_f(1+\mathcal{T}(\mathcal{R}_f))}\left(\frac{\mathcal{A}_f}{1-\mathcal{A}_f}\right)^{\varrho}\right\}^{1/\varrho}},$$
$$\left.1 - \frac{1}{1 + \left\{\sum_{f=1}^{\varpi} \frac{\Xi_f(1+\mathcal{T}(\mathcal{R}_f))}{\sum_{f=1}^{\varpi} \Xi_f(1+\mathcal{T}(\mathcal{R}_f))}\left(\frac{\mathcal{N}_f}{1-\mathcal{N}_f}\right)^{\varrho}\right\}^{1/\varrho}}\right) \tag{6.33}$$

to compute the accumulated values $\Upsilon_g$, $g = 1, 2, \ldots, \xi$ of the alternative $\aleph_g$.

**Step 4.** Computation of the score $\Phi(\Upsilon_g)$ $(g = 1, 2, \ldots, \xi)$ depending on the overall PFN $\Upsilon_g$ $(g = 1, 2, \ldots, \xi)$, which determines the ranking of all the alternatives $\aleph_g$ for selecting the desirable choice $\aleph_g$. If the values of $\Lambda(\Upsilon_g)$ and $\Lambda(\Upsilon_f)$ are the same, then calculate the degrees of accuracy $\Xi(\Upsilon_g)$ and $\Xi(\Upsilon_g)$ depending on the overall PFN of $\Upsilon_g$ and $\Upsilon_f$, and rank the alternative $\aleph_g$ based on the accuracy degrees of $\Xi(\Upsilon_g)$ and $\Xi(\Upsilon_g)$.

**Step 5.** Rank all the alternative $\aleph_g$, $g = 1, 2, \ldots, \xi$ for selecting the best alternative(s) in accordance with $\Xi(\Upsilon_g)$, $g = 1, 2, \ldots, \xi$.
**Step 6.** End.

## 6.6 Case study and comparative analysis

Information technology is growing fast and has applications in almost all areas of real life, particularly in artificial intelligence. The selection of appropriate software for a specific problem is a difficult task. The main objective of this section is to introduce a method to identify the best software systems based on their application and performances with the available alternatives of five candidates. In this section, a numerical problem is described to provide an expert opinion for software technology systems depicted in [40] under a PFN set up to study the proposed method. Suppose that a committee is formed to select five possible software systems $\aleph_g$, $g = 1, 2, \ldots, 5$, depending on four attributes. The possible attributes are shown below:

$A_1$: Performance of organization;
$A_2$: Amount of effort needed to transform from the present system;
$A_3$: Price of software and hardware requirement;
$A_4$: Reliability for outsourcing software developer.

For these attributes, the weight vector is considered as $\Xi = (0.2, 0.1, 0.3, 0.4)^T$. The alternatives $\aleph_1$, $\aleph_2$, $\aleph_3$, $\aleph_4$, and $\aleph_5$ are determined with PFNs by decision makers. The evaluation of the decision makers is shown in Table 6.1.

**Table 6.1** Results made by the decision makers.

| | $\aleph_1$ | $\aleph_2$ | $\aleph_3$ | $\aleph_4$ | $\aleph_5$ |
|---|---|---|---|---|---|
| $A_1$ | (0.6, 0.2, 0.1) | (0.5, 0.1, 0.2) | (0.09, 0.5, 0.3) | (0.6, 0.07, 0.2) | (0.5, 0.1, 0.1) |
| $A_2$ | (0.4, 0.08, 0.3) | (0.6, 0.3, 0.1) | (0.07, 0.5, 0.2) | (0.8, 0.07, 0.1) | (0.6, 0.4, 0.08) |
| $A_3$ | (0.6, 0.07, 0.2) | (0.5, 0.07, 0.2) | (0.06, 0.7, 0.1) | (0.7, 0.1, 0.2) | (0.5, 0.2, 0.03) |
| $A_4$ | (0.5, 0.06, 0.4) | (0.06, 0.5, 0.1) | (0.8, 0.03; 0.06) | (0.6, 0.06, 0.3) | (0.8, 0.05, 0.1) |

For selecting the most suitable software $\aleph_g$, $g = 1, 2, \ldots, \xi$, the PFPDWA and PFPDWG operators are used.

**Step 1.** Calculating the values of $\sigma_{gf}$, $g = 1, 2, \ldots, \xi$; $f = 1, 2, \ldots, \varpi$ using Eq. (6.31):

$$\Xi = \begin{bmatrix} 0.2011 & 0.0990 & 0.3017 & 0.3982 \\ 0.2040 & 0.0985 & 0.3022 & 0.3954 \\ 0.1940 & 0.1001 & 0.3078 & 0.3983 \\ 0.1976 & 0.1007 & 0.3004 & 0.3009 \\ 0.1995 & 0.0981 & 0.3012 & 0.4012 \end{bmatrix}.$$

**Step 2.** Using the values of $\Xi$ and the PFNs $R_{gf}$ for $g = 1, 2, \ldots, \xi$; $f = 1, 2, \ldots, \varpi$, aggregate PFN $R_{gf}$ using the operator PFPDWA (PFPDWG) to calculate the overall PFNs $\Upsilon_g$, $g = 1, 2, \ldots, 5$ for the alternative $\aleph_g$.

**Step 3.** Calculated score values are depicted in Table 6.3. These are determined using the aggregated results depicted in Table 6.2.

**Table 6.2** Aggregated values of the alternatives using the operators PFPDWA and PFPDWG.

| Alternative ($\aleph_g$) | PFPDWA | PFPDWG |
|---|---|---|
| $\aleph_1$ | (0.5492, 0.0758, 0.2064) | (0.5314, 0.0967, 0.2886) |
| $\aleph_2$ | (0.4045, 0.1338, 0.1340) | (0.1290, 0.3253, 0.1535) |
| $\aleph_3$ | (0.6212, 0.0699, 0.0921) | (0.1085, 0.5061, 0.1437) |
| $\aleph_4$ | (0.6493, 0.0808, 0.2174) | (0.6728, 0.0698, 0.2093) |
| $\aleph_5$ | (0.6926, 0.0850, 0.0579) | (0.6001, 0.1554, 0.0780) |

**Step 4.** Using the definition of the score function, the score is determined in Table 6.3. The ranking order of the alternatives is listed in Table 6.4.

**Table 6.3** Score of alternatives for the operators PFPDWA and PFPDWG.

| Alternative ($\aleph_g$) | PFPDWA | PFPDWG |
|---|---|---|
| $\aleph_1$ | 0.6714 | 0.6214 |
| $\aleph_2$ | 0.6353 | 0.4878 |
| $\aleph_3$ | 0.7646 | 0.4824 |
| $\aleph_4$ | 0.7160 | 0.7318 |
| $\aleph_5$ | 0.8174 | 0.7611 |

**Table 6.4** Ranking of alternatives.

| Aggregation operator | Ranking ordered |
|---|---|
| $PFPDWA$ | $\aleph_5 \succ \aleph_3 \succ \aleph_4 \succ \aleph_1 \succ \aleph_2$ |
| $PFPDWG$ | $\aleph_5 \succ \aleph_4 \succ \aleph_1 \succ \aleph_2 \succ \aleph_3$ |

## 6.6.1 Significance of the parameter for decision making

In this section, a new parameter $\varrho$ is introduced to analyze the different results on ranking. The variation of this parameter produces rankings of the alternatives in order to explore the impact of the performance of such a parameter for decision-making outcomes. Tables 6.5 and 6.6 present, respectively, the score of the score function as well as the corresponding ranking sequence for the alternatives $\aleph_f$ $(f = 1, 2 \ldots, 5)$ for different values of $\varrho$ between 1 and 10 for the operators PFPDWA and PFPDWG. Table 6.5 shows that the ranking results are steady for the PFPDWA operator, and the corresponding best alternative is identical. When $1 \leq \varrho \leq 10$, the ranking order is $\aleph_5 \succ \aleph_3 \succ \aleph_4 \succ \aleph_1 \succ \aleph_2$, and shows $\aleph_5$ as the best alternative. From Table 6.6, it is observed that when the value of $\varrho$ is taken between 1 and 10, for the PFPDWG operator, the ranking order is $\aleph_5 \succ \aleph_4 \succ \aleph_1 \succ \aleph_2 \succ \aleph_3$. Moreover, $\aleph_5$ is the best alternative. Therefore both operators produced $\aleph_5$ as the favorable alternative, although the ranking order for both operators is different.

**Table 6.5** Effect of the parameter for ranking order using the PFPDWA operator.

| $\varrho$ | $\Lambda(\Upsilon_1)$ | $\Lambda(\Upsilon_2)$ | $\Lambda(\Upsilon_3)$ | $\Lambda(\Upsilon_4)$ | $\Lambda(\Upsilon_5)$ | Ranking order |
|---|---|---|---|---|---|---|
| 1 | 0.6714 | 0.6353 | 0.7646 | 0.7160 | 0.8174 | $\aleph_5 \succ \aleph_3 \succ \aleph_4 \succ \aleph_1 \succ \aleph_2$ |
| 2 | 0.6913 | 0.6673 | 0.8172 | 0.7423 | 0.8414 | $\aleph_5 \succ \aleph_3 \succ \aleph_4 \succ \aleph_1 \succ \aleph_2$ |
| 3 | 0.7046 | 0.6826 | 0.8350 | 0.7613 | 0.8528 | $\aleph_5 \succ \aleph_3 \succ \aleph_4 \succ \aleph_1 \succ \aleph_2$ |
| 4 | 0.7135 | 0.6921 | 0.8438 | 0.7767 | 0.8608 | $\aleph_5 \succ \aleph_3 \succ \aleph_4 \succ \aleph_1 \succ \aleph_2$ |
| 5 | 0.7197 | 0.6991 | 0.8492 | 0.7889 | 0.8658 | $\aleph_5 \succ \aleph_3 \succ \aleph_4 \succ \aleph_1 \succ \aleph_2$ |
| 6 | 0.7244 | 0.7047 | 0.8527 | 0.7983 | 0.8691 | $\aleph_5 \succ \aleph_3 \succ \aleph_4 \succ \aleph_1 \succ \aleph_2$ |
| 7 | 0.7277 | 0.7093 | 0.8552 | 0.8055 | 0.8715 | $\aleph_5 \succ \aleph_3 \succ \aleph_4 \succ \aleph_1 \succ \aleph_2$ |
| 8 | 0.7303 | 0.7132 | 0.8571 | 0.8112 | 0.8732 | $\aleph_5 \succ \aleph_3 \succ \aleph_4 \succ \aleph_1 \succ \aleph_2$ |
| 9 | 0.7325 | 0.7166 | 0.8586 | 0.8156 | 0.8747 | $\aleph_5 \succ \aleph_3 \succ \aleph_4 \succ \aleph_1 \succ \aleph_2$ |
| 10 | 0.7342 | 0.7195 | 0.8602 | 0.8192 | 0.8757 | $\aleph_5 \succ \aleph_3 \succ \aleph_4 \succ \aleph_1 \succ \aleph_2$ |

**Table 6.6** Effect of the parameter for ranking order using the operator PFPDWG.

| $\varrho$ | $\Lambda(\Upsilon_1)$ | $\Lambda(\Upsilon_2)$ | $\Lambda(\Upsilon_3)$ | $\Lambda(\Upsilon_4)$ | $\Lambda(\Upsilon_5)$ | Ranking order |
|---|---|---|---|---|---|---|
| 1 | 0.6214 | 0.4878 | 0.4824 | 0.7318 | 0.7611 | $\aleph_5 \succ \aleph_4 \succ \aleph_1 \succ \aleph_2 \succ \aleph_3$ |
| 2 | 0.6018 | 0.4610 | 0.4537 | 0.7122 | 0.7429 | $\aleph_5 \succ \aleph_4 \succ \aleph_1 \succ \aleph_2 \succ \aleph_3$ |
| 3 | 0.5869 | 0.4551 | 0.4364 | 0.7010 | 0.7319 | $\aleph_5 \succ \aleph_4 \succ \aleph_1 \succ \aleph_2 \succ \aleph_3$ |
| 4 | 0.5754 | 0.4495 | 0.4252 | 0.7180 | 0.7251 | $\aleph_5 \succ \aleph_4 \succ \aleph_1 \succ \aleph_2 \succ \aleph_3$ |
| 5 | 0.5660 | 0.4459 | 0.4175 | 0.7206 | 0.6869 | $\aleph_5 \succ \aleph_4 \succ \aleph_1 \succ \aleph_2 \succ \aleph_3$ |
| 6 | 0.5584 | 0.4434 | 0.4120 | 0.6821 | 0.7175 | $\aleph_5 \succ \aleph_4 \succ \aleph_1 \succ \aleph_2 \succ \aleph_3$ |
| 7 | 0.5544 | 0.4415 | 0.4078 | 0.6782 | 0.7152 | $\aleph_5 \succ \aleph_4 \succ \aleph_1 \succ \aleph_2 \succ \aleph_3$ |
| 8 | 0.5467 | 0.4401 | 0.4046 | 0.6751 | 0.7134 | $\aleph_5 \succ \aleph_4 \succ \aleph_1 \succ \aleph_2 \succ \aleph_3$ |
| 9 | 0.5422 | 0.4390 | 0.4020 | 0.6726 | 0.7120 | $\aleph_5 \succ \aleph_4 \succ \aleph_1 \succ \aleph_2 \succ \aleph_3$ |
| 10 | 0.5384 | 0.4381 | 0.3999 | 0.6704 | 0.7108 | $\aleph_5 \succ \aleph_4 \succ \aleph_1 \succ \aleph_2 \succ \aleph_3$ |

Note that for the different values of the parameter $\varrho$ corresponding to ordering of the options for the PFPDWA operator cannot be modified in the proposed MADM strategy depends on the operators PFPDWA and PFPDWG. The ranking pattern of the alternatives for the operator, PFPDWG is the same in all circumstances but distinct from averaging operators, much like modification of the parameter $\varrho$. Here, we find that $\aleph_5$ is the option that is most advantageous for both operators PFPDWA (PFPDWG).

## 6.6.2 Comparative study

Table 6.7 includes the PFWA and PFWG operators proposed by Wei [27] and the PFHWA and PFHWG operators presented by Wei [28] to obtain the aggregated data in order to

**Table 6.7** Aggregate values obtained from existing operators.

| Alternative | PFWA | PFWG | PFHWA | PFHWG |
|---|---|---|---|---|
| $\aleph_1$ | (0.5449, 0.0824, 0.2378) | (0.5360, 0.0949, 0.2791) | (0.5429, 0.0556, 0.1786) | (0.5384, 0.0600, 0.2689) |
| $\aleph_2$ | (0.3772, 0.1891, 0.1423) | (0.2207, 0.2967, 0.1520) | (0.3417, 0.3004, 0.1192) | (0.2429, 0.4996, 0.1501) |
| $\aleph_3$ | (0.4962, 0.1808, 0.1082) | (0.1849, 0.4438, 0.1381) | (0.4171, 0.0288, 0.0933) | (0.2200, 0.0300, 0.1316) |
| $\aleph_4$ | (0.6249, 0.0971, 0.2477) | (0.6810, 0.0694, 0.2047) | (0.6082, 0.0811, 0.2101) | (0.6904, 0.0538, 0.1988) |
| $\aleph_5$ | (0.6613, 0.1069, 0.0681) | (0.6147, 0.1469, 0.0775) | (0.6521, 0.0469, 0.0621) | (0.6262, 0.0500, 0.0766) |

compare the new approach with the existing method. The corresponding order of the alternatives is displayed in Table 6.9 depending on the aggregating values in Table 6.1 and the scores obtained in Table 6.8. The research shows that $\aleph_5$ is the ideal piece of software. The best software system in both techniques, $\aleph_5$, is evident despite modest differences in the ranking orders of the two methodologies [27,28]. This supports the claim that our strategy is more sophisticated and successful. The main gain of the suggested operators is that they have flexibility in the operational parameter and it is easy to aggregate fuzzy information. In contrast, the current operators PFWA, PFWG, PFHWA, and PFHWG do not yet. The characteristic comparison of the existing methods is given in Table 6.10.

**Table 6.8** Scores of the PFWA, PFWG, PFHWA, and PFHWG operators.

| Alternative ($\aleph_g$) | PFWA | PFWG | PFHWA | PFHWG |
|---|---|---|---|---|
| $\aleph_1$ | 0.6536 | 0.6285 | 0.6822 | 0.6348 |
| $\aleph_2$ | 0.6175 | 0.5344 | 0.6113 | 0.5464 |
| $\aleph_3$ | 0.6940 | 0.5234 | 0.6619 | 0.5442 |
| $\aleph_4$ | 0.6886 | 0.7382 | 0.6991 | 0.7458 |
| $\aleph_5$ | 0.7966 | 0.7686 | 0.7950 | 0.7748 |

**Table 6.9** Comparison between existing works with the proposed work.

| Aggregation operator | Ranking ordered |
|---|---|
| Wei [27] PFWA | $\aleph_5 \succ \aleph_3 \succ \aleph_4 \succ \aleph_1 \succ \aleph_2$ |
| Wei [27] PFWG | $\aleph_5 \succ \aleph_4 \succ \aleph_1 \succ \aleph_2 \succ \aleph_3$ |
| Wei [28] PFHWA | $\aleph_5 \succ \aleph_4 \succ \aleph_1 \succ \aleph_3 \succ \aleph_2$ |
| Wei [28] SNWGA | $\aleph_5 \succ \aleph_4 \succ \aleph_1 \succ \aleph_2 \succ \aleph_3$ |
| *Proposed PFPDWA* | $\aleph_5 \succ \aleph_3 \succ \aleph_4 \succ \aleph_1 \succ \aleph_2$ |
| *Proposed PFPDWG* | $\aleph_5 \succ \aleph_4 \succ \aleph_1 \succ \aleph_2 \succ \aleph_3$ |

## 6.7 Conclusion

This chapter presented a method for solving MADM issues with PFN data.

**Table 6.10** Characteristic comparisons of different methods.

| Methods | Fuzzy information easier | Aggregation flexible |
|---|---|---|
| Wei [27] | Yes | No |
| Wei [28] | Yes | No |
| Proposed method | Yes | Yes |

Motivated by some Dombi operators, namely, PFPDWA, PFPDOWA, PFPDHWA, PFPDWG, PFPDHOWG, and PFPDHWG, and using PFNs, we defined arithmetic and geometric averaging operators. Various characteristics of these newly proposed operators are investigated. We then create a strategy to resolve a MADM issue. Lastly, a representative example for selecting emerging software systems is provided to demonstrate the adaptability and efficiency of the suggested approach. The presented model can be used in risk analysis, decision-support systems, and other fields where fuzzy uncertainties are involved with additional research.

## References

[1] K.T. Atanassov, Intuitionistic Fuzzy Sets Theory, Studies in Fuzziness and Soft Computing, vol. 283, Springer-Verlag, Berlin Heidelberg, 1999.

[2] B.C. Cuong, Picture fuzzy sets first results, Part 1, Seminar "Neuro-fuzzy systems with applications", Tech. Rep., Institute of Mathematics, Hanoi, 2013.

[3] B.C. Cuong, Picture fuzzy sets first results, Part 2, Seminar "Neuro-fuzzy systems with applications", Tech. Rep., Institute of Mathematics, Hanoi, 2013.

[4] J. Dombi, A general class of fuzzy operators, the demorgan class of fuzzy operators and fuzziness measures induced by fuzzy operators, Fuzzy Sets Syst. 8 (1982) 149–163.

[5] H. Garg, Some picture fuzzy aggregation operators and their applications to multi- criteria decision-making, Arab. J. Sci. Eng. 42 (12) (2017) 5275–5290.

[6] Y.D. He, H.Y. Chen, L.G. Zhou, Generalized interval-valued Atanassov intuitionistic fuzzy power operators and their application to group decision making, Int. J. Fuzzy Syst. 15 (4) (2013) 401–411.

[7] X. He, Typhoon disaster assessment based on Dombi hesitant fuzzy information aggregation operators, Nat. Hazards 90 (3) (2018) 1153–1175.

[8] C. Jana, M. Pal, J.Q. Wang, Bipolar fuzzy Dombi aggregation operators and its application in multiple attribute decision making process, J. Ambient Intell. Humaniz. Comput. 10 (2019) 3533–3549.

[9] C. Jana, M. Pal, Assessment of enterprise performance based on picture fuzzy Hamacher aggregation operators, Symmetry 11 (1) (2019) 75.

[10] C. Jana, T. Senapati, M. Pal, Pythagorean fuzzy Dombi aggregation operators and its applications in multiple attribute decision-making, Int. J. Intell. Syst. 34 (2019) 2019–2038.

[11] C. Jana, M. Pal, J.Q. Wang, Bipolar fuzzy Dombi prioritized aggregation operators in multiple attribute decision making, Soft Comput. 24 (2020) 3631–3646.

[12] C. Jana, G. Muhiuddin, M. Pal, Some Dombi aggregation of Q-rung orthopair fuzzy numbers in multiple attribute decision-making, Int. J. Intell. Syst. 34 (2019) 3220–3240.

[13] C. Jana, T. Senapati, M. Pal, R.R. Yager, Picture fuzzy Dombi aggregation operators: application to MADM process, Appl. Soft Comput. 74 (1) (2019) 99–109.

[14] S. Khan, S. Abdullha, S. Ashraf, Picture fuzzy aggregation information based on Einstein operations and their application in decision making, Math. Sci. 13 (3) (2019) 213–229.

[15] P. Liu, Y.M. Wang, Multiple attribute group decision making methods based on intuitionistic linguistic power generalized aggregation operators, Appl. Soft Comput. 17 (1) (2014) 90–104.

[16] P. Liu, X.C. Yu, 2-dimension uncertain linguistic power generalized weighted aggregation operator and its application for multiple attribute group decision making, Knowl.-Based Syst. 57 (1) (2014) 69–80.

[17] P. Liu, Multiple attribute group decision making method based on interval-valued intuitionistic fuzzy power Heronian aggregation operators, Comput. Ind. Eng. 108 (2017) 199–212.
[18] P. Liu, X. Liu, Multiattribute group decision making methods based on linguistic intuitionistic fuzzy power Bonferroni mean operators, Complexity 2017 (2017) 3571459.
[19] X. Lu, J. Ye, Dombi aggregation operators of linguistic cubic variables for multiple attribute decision making, Information 9 (2018) 188, https://doi.org/10.3390/info9080188.
[20] X. Peng, J. Dai, Algorithm for picture fuzzy multiple attribute decision making based on new distance measure, Int. J. Uncertain. Quantificat. 7 (2017) 177–187.
[21] P. Singh, Correlation coefficients for picture fuzzy sets, J. Intell. Fuzzy Syst. 27 (2014) 2857–2868.
[22] L.H. Son, Measuring analogousness in picture fuzzy sets: from picture distance measures to picture association measures, Fuzzy Optim. Decis. Mak. 16 (3) (2017) 1–20.
[23] L.H. Son, P. Viet, P. Hai, Picture inference system: a new fuzzy inference system on picture fuzzy set, Appl. Intell. 46 (3) (2017) 652–669.
[24] L.H. Son, DPFCM: a novel distributed picture fuzzy clustering method on picture fuzzy sets, Expert Syst. Appl. 2 (2015) 51–66.
[25] P.H. Thong, L.H. Son, Picture fuzzy clustering for complex data, Eng. Appl. Artif. Intell. 56 (2016) 121–130.
[26] P.H. Thong, L.H. Son, A novel automatic picture fuzzy clustering method based on particle swarm optimization and picture composite cardinality, Knowl.-Based Syst. 109 (2016) 48–60.
[27] G.W. Wei, Picture fuzzy aggregation operators and their application to multiple attribute decision making, J. Intell. Fuzzy Syst. 33 (2017) 713–724.
[28] G.W. Wei, Picture fuzzy Hamacher aggregation operators and their application to multiple attribute decision making, Fundam. Inform. 157 (3) (2018) 271–320.
[29] G.W. Wei, Picture fuzzy cross-entropy for multiple attribute decision making problems, J. Bus. Econ. Manag. 17 (4) (2016) 491–502.
[30] G.W. Wei, Z. Zhang, Some single-valued neutrosophic Bonferroni power aggregation operators in multiple attribute decision making, J. Ambient Intell. Humaniz. Comput. 10 (2019) 863–882.
[31] G.W. Wei, M. Lu, Pythagorean fuzzy power aggregation operators in multiple attribute decision making, Int. J. Intell. Syst. 33 (1) (2018) 169–186.
[32] Z.S. Xu, Intuitionistic fuzzy aggregation operators, IEEE Trans. Fuzzy Syst. 15 (6) (2007) 1179–1187.
[33] Z.S. Xu, R.R. Yager, Some geometric aggregation operators based on intuitionistic fuzzy sets, Int. J. Gen. Syst. 35 (4) (2006) 417–433.
[34] Z.S. Xu, Q.L. Da, An overview of operators for aggregating information, Int. J. Intell. Syst. 18 (9) (2003) 953–969.
[35] Z.S. Xu, Approaches to multiple attribute group decision making based on intuitionistic fuzzy power aggregation operators, Knowl.-Based Syst. 24 (2011) 749–760.
[36] Y. Xu, J.M. Merigo, H. Wang, Linguistic power aggregation operators and their application to multiple attribute group decision making, Appl. Math. Model. 36 (2012) 5427–5444.
[37] R.R. Yager, The power average operator, IEEE Trans. Syst. Man Cybern., Part A 31 (2001) 724–731.
[38] R.R. Yager, On ordered weighted averaging aggregation operators in multicriteria decision making, IEEE Trans. Syst. Man Cybern. 18 (1) (1988) 183–190.
[39] R.R. Yager, J. Kacprzyk, The Ordered Weighted Averaging Operators: Theory and Applications, M.A., Kluwer, Boston, 1997.
[40] J. Ye, Prioritized aggregation operators of trapezoidal intuitionistic fuzzy sets and their application to multicriteria decision making, Neural Comput. Appl. 25 (6) (2014) 1447–1454.
[41] L.A. Zadeh, Fuzzy sets, Inf. Control 8 (1965) 338–353.
[42] L. Zhou, H. Chen, J. Liu, Generalized power aggregation operators and their applications in group decision making, Comput. Ind. Eng. 62 (4) (2012) 989–999.

# 7 m-Polar picture fuzzy Dombi operators and their applications in multicriteria decision-making processes

## 7.1 Introduction

Many researchers have focused on the topic of information aggregation, which led to the development of some outstanding works in the contexts of interval-valued intuitionistic fuzzy (IVIFS) and intuitionistic fuzzy (IFS) sets (see [4,18–21] is a generalization of fuzzy sets [36]). A set of fundamental values can be aggregated using some conventional works, including [32–35]. Although IFS and IVIFS can address the ambiguity of practical issues, they cannot indicate that counter properties match an object's properties. Zhang [37,38] introduced the bipolar fuzzy set (BFS), a concept of fuzzy sets (FSs) whose membership degrees belong to $[-1, 1]$, in this regard. A positive membership to $[0, 1]$ and a negative membership to $[-1, 0]$ were two functions that the BFS was linked to. Then, several MADM issues based on aggregation operators in BF environments were established. For example, Jana et al. [11,15] explored bipolar fuzzy Hamacher operator-based MADM problems, while Wei et al. [27] studied MADM problems based on Dombi norms in BF structures. Gao et al. [9] suggested a dual hesitant bipolar Hamacher aggregation-based MADM model and Xu and Wei [31] built a dual hesitant bipolar fuzzy decision-making system based on arithmetic and geometric operators. The MCDM approach based on a bipolar soft aggregation operator was proposed by Jana et al. in [16]. Researchers discovered that multipolar information was needed to analyze projects like a diesel power plant and a petrol pump at the same time. The m-polar fuzzy set (mPoFS) has been created in this way. Chen et al. [5] initially presented the idea of mFS as a generalization of BFS. Khameneh and Kilicman invented the m-polar fuzzy soft weighted aggregation operators [10]. Using various ideas, Akram and others [2,3] have developed numerous m-polar fuzzy set applications. Hamacher operators were recently used by Waseem et al. [1] to aggregate mPoFNs data and create a MADM problem. However, Coung's [6,7] initial proposal that the feeling of refusal property only is taken into account in picture fuzzy sets (PFSs) is not true today. Following a study of PFS applications in numerous fields, Wei [29] presented geometric and average aggregation operators for use in a picture fuzzy environment. Wei [28] studied a MADM technique based on an entropy measure in the same setting. Additionally, he created a MADM problem and used Hamacher aggregation operations in PFS ideas [30]. Please see the following references for further details on PFS and their use [16,17,22–26]. Dombi [8]

Picture Fuzzy Logic and Its Applications in Decision Making Problems. https://doi.org/10.1016/B978-0-44-322024-1.00011-X 

put forth the idea of the Dombi operator, which has certain drawbacks to direct aggregate linguistic data but an excellent advantage to operate parameter flexibility for aggregation of diverse fuzzy data. The development of MADM models based on Dombi operators has since received attention. In their study of Dombi–Bonferroni, mean operators, Liu et al. [18] applied Dombi operators to create MADM models to aggregate bipolar fuzzy numbers [11], picture fuzzy numbers [12], Pythagorean fuzzy numbers [13], and q-rung orthopair fuzzy numbers [14]. No MADM issues exist that aggregate mPoPF data using Dombi norms. As a result, the concern is that mPFS have assessed a different ability to model ambiguous information that happens in real-world challenges. Traditional problems [32,33,35] and MADM problems [18,8,11,13,15] employing Dombi norms in various fuzzy structures give us sufficient impetus to create the current study. This research aims to aggregate mPFNs using Dombi norms in the context of mPFS and create a MADM technique employing these operators.

## 7.2 Preliminaries

In this chapter, an $m$-polar PFS (mPoPFS) is introduced, where each element of $m$ components has three components of PFS. In FS, every element is associated with a number between 0 and 1, called the membership value. In IFS, each element is associated with two numbers known as membership and nonmembership values; their sum is less than or equal to 1. This concept is extended, and a new variety of FS called an $m$-polar fuzzy set is defined. In this new set, each element is associated with $m$ number of elements. The membership value of an element of an $m$-polar FS is a vector of dimension $m$. The formal definition is given below.

**Definition 7.1.** Let $\mathcal{R}$ be an $m$-polar fuzzy set (mPoFS) over the universe $U$ with a mapping $\mathcal{R}: U \to [0,1]^m$, where $\mathcal{R}(u) = (\mathcal{R}_1(u), \mathcal{R}_2(u), \ldots, \mathcal{R}_m(u))$. This mPoFS can also be written as $\mathcal{R}(u) = (\mathcal{P}_1 * \mathcal{R}(u), \mathcal{P}_2 * \mathcal{R}(u), \ldots, \mathcal{P}_m * \mathcal{R}(u))$, where $\mathcal{P}$ is a vector containing $m$ components and $\mathcal{P}_l * \mathcal{R}: [0,1]^m \to [0,1]$ represents the $l$th projection mapping. The vector $\mathcal{P}$ is used to extract a fixed element from the mPoFS.

Actually, $(\mathcal{P}_l * \mathcal{R}(u)$ represents the $l$th component of the mPoFS $\mathcal{R}(u)$. If all the elements of a $m$-polar set are equal to 1, it is called an $m$-polar number. A PFS is called a picture fuzzy number (PFN) if there is at least one element of the form $(1,0,0)$. Thus if in an mPoPFSs there is at least one element of the form $((1,0,0),(1,0,0),\ldots,(1,0,0))$ then such an mPoPFS is called an $m$-polar PFN (mPoPFN).

Let $\mathcal{R} = (\mathcal{R}_1, \mathcal{R}_2, \ldots, \mathcal{R}_m)$ be an $m$-polar picture fuzzy number (mPoPFN) and $\mathcal{P}_l * \mathcal{R} \in [0,1]$ for $l = 1, 2, \ldots, m$.

Here, some basic operations on $mPoPFN$ are defined below:

**Definition 7.2.** Let

$$\mathcal{R}_1 = ((\mathcal{Y}_{11}, \mathcal{A}_{11}, \mathcal{N}_{11}), (\mathcal{Y}_{12}, \mathcal{A}_{12}, \mathcal{N}_{12}), \ldots, (\mathcal{Y}_{1m}, \mathcal{A}_{1m}, \mathcal{N}_{1m})) \text{ and}$$

$$\mathcal{R}_2 = ((\mathcal{Y}_{21}, \mathcal{A}_{21}, \mathcal{N}_{21}), (\mathcal{Y}_{22}, \mathcal{A}_{22}, \mathcal{N}_{22}), \ldots, (\mathcal{Y}_{2m}, \mathcal{A}_{2m}, \mathcal{N}_{2m}))$$

be two $mPoPFNs$, and $\tau > 0$. Then,

**(1)** $\mathcal{R}_1 \oplus \mathcal{R}_2 = \Big(\big(\mathcal{Y}_{11} + \mathcal{Y}_{21} - \mathcal{Y}_{11}\mathcal{Y}_{21}, \mathcal{A}_{11}\mathcal{A}_{21}, \mathcal{N}_{11}\mathcal{N}_{21}\big), \big(\mathcal{Y}_{12} + \mathcal{Y}_{22} - \mathcal{Y}_{12}\mathcal{Y}_{22}, \mathcal{A}_{12}\mathcal{A}_{22}, \mathcal{N}_{12}\mathcal{N}_{22}\big); \ldots, \big(\mathcal{Y}_{1m} + \mathcal{Y}_{2m} - \mathcal{Y}_{1m}\mathcal{Y}_{2m}, \mathcal{A}_{1m}\mathcal{A}_{2m}, \mathcal{N}_{1m}\mathcal{N}_{2m}\big)\Big)$;

**(2)** $\mathcal{R}_1 \otimes \mathcal{R}_2 = \Big(\big(\mathcal{Y}_{11}\mathcal{Y}_{21}, \mathcal{A}_{11} + \mathcal{A}_{21} - \mathcal{A}_{11}\mathcal{A}_{21}, \mathcal{N}_{11} + \mathcal{N}_{21} - \mathcal{N}_{11}\mathcal{N}_{21}\big), \big(\mathcal{Y}_{12}\mathcal{Y}_{22}, \mathcal{A}_{12} + \mathcal{A}_{22} - \mathcal{A}_{12}\mathcal{A}_{22}, \mathcal{N}_{12} + \mathcal{N}_{22} - \mathcal{N}_{12}\mathcal{N}_{22}\big), \ldots, \big(\mathcal{Y}_{1m}\mathcal{Y}_{2m}, \mathcal{A}_{1m} + \mathcal{A}_{2m} - \mathcal{A}_{1m}\mathcal{A}_{2m}, \mathcal{N}_{1m} + \mathcal{N}_{2m} - \mathcal{N}_{1m}\mathcal{N}_{2m}\big)\Big)$;

**(3)** $\tau\mathcal{R}_1 = \Big(\big(1 - (1 - \mathcal{Y}_{11})^\tau, (\mathcal{A}_{11})^\tau, (\mathcal{N}_{11})^\tau\big), \big(1 - (1 - \mathcal{Y}_{12})^\tau, (\mathcal{A}_{12})^\tau, (\mathcal{N}_{12})^\tau\big), \ldots, \big(1 - (1 - \mathcal{Y}_{1m})^\tau, (\mathcal{A}_{1m})^\tau, (\mathcal{N}_{1m})^\tau\big)\Big)$;

**(4)** $(\mathcal{R}_1)^\tau = \Big(\big((\mathcal{Y}_{11})^\tau, 1 - (1 - \mathcal{A}_{11})^\tau, 1 - (1 - \mathcal{N}_{11})^\tau\big), \big((\mathcal{Y}_{12})^\tau, 1 - (1 - \mathcal{A}_{12})^\tau, 1 - (1 - \mathcal{N}_{12})^\tau\big), \ldots, \big((\mathcal{Y}_{1m})^\tau, 1 - (1 - \mathcal{A}_{1m})^\tau, 1 - (1 - \mathcal{N}_{1m})^\tau\big)\Big)$;

**(5)** $(\mathcal{R}_1)^c = \big((\mathcal{N}_{11}, \mathcal{A}_{11}, \mathcal{Y}_{11}), (\mathcal{N}_{12}, \mathcal{A}_{12}, \mathcal{Y}_{12}), \ldots, (\mathcal{N}_{1m}, \mathcal{A}_{1m}, \mathcal{Y}_{1m}))\big)$;

**(6)** $\mathcal{R}_1 \subseteq \mathcal{R}_2$ if and only if $\mathcal{Y}_{11} \leq \mathcal{Y}_{21}, \mathcal{Y}_{12} \leq \mathcal{Y}_{22}, \ldots, \mathcal{Y}_{1m} \leq \mathcal{Y}_{2m}, \mathcal{A}_{11} \leq \mathcal{A}_{21}, \mathcal{A}_{12} \leq \mathcal{A}_{22}, \ldots, \mathcal{A}_{1m} \leq \mathcal{A}_{2m}, \mathcal{N}_{11} \geq \mathcal{N}_{21}, \mathcal{N}_{12} \geq \mathcal{N}_{22}, \ldots, \mathcal{N}_{1m} \geq \mathcal{N}_{2m}$;

**(7)** $\mathcal{R}_1 \bigcup \mathcal{R}_2 = \Big(\big(\max(\mathcal{Y}_{11}, \mathcal{Y}_{21}), \min(\mathcal{A}_{11}, \mathcal{A}_{21}), \min(\mathcal{N}_{11}, \mathcal{N}_{21})\big), \big(\max(\mathcal{Y}_{12}, \mathcal{Y}_{22}), \min(\mathcal{A}_{12}, \mathcal{A}_{22}), \min(\mathcal{N}_{12}, \mathcal{N}_{22})\big), \ldots, \big(\max(\mathcal{Y}_{1m}, \mathcal{Y}_{2m}), \min(\mathcal{A}_{1m}, \mathcal{A}_{2m}), \min(\mathcal{N}_{1m}, \mathcal{N}_{2m})\big)\Big)$;

**(8)** $\mathcal{R}_1 \bigcap \mathcal{R}_2 = \Big(\big(\min(\mathcal{Y}_{11}, \mathcal{Y}_{21}), \max(\mathcal{A}_{11}, \mathcal{A}_{21}), \max(\mathcal{N}_{11}, \mathcal{N}_{21})\big), \big(\min(\mathcal{Y}_{12}, \mathcal{Y}_{22}), \max(\mathcal{A}_{12}, \mathcal{A}_{22}), \max(\mathcal{N}_{12}, \mathcal{N}_{22})\big), \ldots, \big(\min(\mathcal{Y}_{1m}, \mathcal{Y}_{2m}), \max(\mathcal{A}_{1m}, \mathcal{A}_{2m}), \max(\mathcal{N}_{1m}, \mathcal{N}_{2m})\big)\Big)$.

**Definition 7.3.** The score function $\Lambda$ of an mPoPFN

$$\mathcal{R}_1 = ((\mathcal{Y}_{11}, \mathcal{A}_{11}, \mathcal{N}_{11}), (\mathcal{Y}_{12}, \mathcal{A}_{12}, \mathcal{N}_{12}), \ldots, (\mathcal{Y}_{1m}, \mathcal{A}_{1m}, \mathcal{N}_{1m}))$$

is calculated as:

$$\Lambda(\mathcal{R}_1) = \frac{1}{m}\left(1 + \sum_{l=1}^{m} \mathcal{Y}_{1l} - \sum_{l=1}^{m} \mathcal{A}_{1l} - \sum_{l=1}^{m} \mathcal{N}_{1l})\right) \in [-1, 1]. \tag{7.1}$$

**Definition 7.4.** The accuracy function $\Phi$ of an mPoPFN

$$\mathcal{R}_1 = ((\mathcal{Y}_{11}, \mathcal{A}_{11}, \mathcal{N}_{11}), (\mathcal{Y}_{12}, \mathcal{A}_{12}, \mathcal{N}_{12}), \ldots, (\mathcal{Y}_{1m}, \mathcal{A}_{1m}, \mathcal{N}_{1m}))$$

is computed by the rule:

$$\Phi(\mathcal{R}_1) = \frac{1}{m}\left(\sum_{l=1}^{m} (-1)^l (\mathcal{Y}_{1l} - 1)\right) \in [-1, 1]. \tag{7.2}$$

Using score and accuracy functions, we introduced a prioritized relation between two mPoPFNs.

**Definition 7.5.** Let

$$\mathcal{R}_1 = ((\mathcal{Y}_{11}, \mathcal{A}_{11}, \mathcal{N}_{11}), (\mathcal{Y}_{12}, \mathcal{A}_{12}, \mathcal{N}_{12}), \ldots, (\mathcal{Y}_{1m}, \mathcal{A}_{1m}, \mathcal{N}_{1m})) \text{ and}$$
$$\mathcal{R}_2 = ((\mathcal{Y}_{21}, \mathcal{A}_{21}, \mathcal{N}_{21}), (\mathcal{Y}_{22}, \mathcal{A}_{22}, \mathcal{N}_{22}), \ldots, (\mathcal{Y}_{2m}, \mathcal{A}_{2m}, \mathcal{N}_{2m}))$$

be two mPoPFNs. Then,

**(i)** If $\Lambda(\mathcal{R}_1) < \Lambda(\mathcal{R}_2)$, indicates $\mathcal{R}_1 \prec \mathcal{R}_2$;
**(ii)** If $\Lambda(\mathcal{R}_1) > \Lambda(\mathcal{R}_2)$, indicates $\mathcal{R}_1 \succ \mathcal{R}_2$;
**(iii)** If $\Lambda(\mathcal{R}_1) = \Lambda(\mathcal{R}_2)$, then
  **(1)** If $\Phi(\mathcal{R}_1) < \Phi(\mathcal{R}_2)$, indicates $\mathcal{R}_1 \prec \mathcal{R}_2$;
  **(2)** If $\Phi(\mathcal{R}_1) > \Phi(\mathcal{R}_2)$, indicates $\mathcal{R}_1 \succ \mathcal{R}_2$;
  **(3)** If $\Phi(\mathcal{R}_1) = \Phi(\mathcal{R}_2)$, indicates $\mathcal{R}_1 \sim \mathcal{R}_2$.

## 7.3 Dombi operations on mPFNs

Dombi recommended the operations Dombi product and Dombi sum that are specific types of triangular norms and conorms given in the following definition.

**Definition 7.6.** [8] Assume $f$ and $g$ are any two real numbers. Then, Dombi norms and Dombi conorms are defined within the following expressions:

$$Dom(f, g) = \frac{1}{1 + \{(\frac{1-f}{f})^{\varrho} + (\frac{1-g}{g})^{\varrho}\}^{1/\varrho}} \tag{7.3}$$

$$Dom^*(f, g) = 1 - \frac{1}{1 + \{(\frac{f}{1-f})^{\varrho} + (\frac{g}{1-g})^{\varrho}\}^{1/\varrho}}, \tag{7.4}$$

where, $\varrho \geq 1$ and $(f, g) \in [0, 1] \times [0, 1]$.

Because of the Dombi norm, furthermore, the Dombi conorm, we explain Dombi operations concerning mPoPFNs.

**Definition 7.7.** Let

$$\mathcal{R}_1 = ((\mathcal{Y}_{11}, \mathcal{A}_{11}, \mathcal{N}_{11}), (\mathcal{Y}_{12}, \mathcal{A}_{12}, \mathcal{N}_{12}), \ldots, (\mathcal{Y}_{1m}, \mathcal{A}_{1m}, \mathcal{N}_{1m})) \text{ and}$$
$$\mathcal{R}_2 = ((\mathcal{Y}_{21}, \mathcal{A}_{21}, \mathcal{N}_{21}), (\mathcal{Y}_{22}, \mathcal{A}_{22}, \mathcal{N}_{22}), \ldots, (\mathcal{Y}_{2m}, \mathcal{A}_{2m}, \mathcal{N}_{2m}))$$

be two mPoPFNs, and $\tau > 0$. Now, we introduced Dombi operations on mPoPFNs:

**(1)** $\mathcal{R}_1 \oplus_D \mathcal{R}_2$

$$= \Bigg( \Bigg( 1 - \frac{1}{1+\left\{\left(\frac{\mathcal{Y}_{11}}{1-\mathcal{Y}_{11}}\right)^{\varrho} + \left(\frac{\mathcal{Y}_{21}}{1-\mathcal{Y}_{21}}\right)^{\varrho}\right\}^{1/\varrho}}, \frac{1}{1+\left\{\left(\frac{1-\mathcal{A}_{11}}{\mathcal{A}_{11}}\right)^{\varrho} + \left(\frac{1-\mathcal{A}_{21}}{\mathcal{A}_{21}}\right)^{\varrho}\right\}^{1/\varrho}}, \frac{1}{1+\left\{\left(\frac{1-\mathcal{N}_{11}}{\mathcal{N}_{11}}\right)^{\varrho} + \left(\frac{1-\mathcal{N}_{21}}{\mathcal{N}_{21}}\right)^{\varrho}\right\}^{1/\varrho}} \Bigg),$$

$$\Bigg( 1 - \frac{1}{1+\left\{\left(\frac{\mathcal{Y}_{12}}{1-\mathcal{Y}_{12}}\right)^{\varrho} + \left(\frac{\mathcal{Y}_{22}}{1-\mathcal{Y}_{22}}\right)^{\varrho}\right\}^{1/\varrho}}, \frac{1}{1+\left\{\left(\frac{1-\mathcal{A}_{12}}{\mathcal{A}_{12}}\right)^{\varrho} + \left(\frac{1-\mathcal{A}_{22}}{\mathcal{A}_{22}}\right)^{\varrho}\right\}^{1/\varrho}}, \frac{1}{1+\left\{\left(\frac{1-\mathcal{N}_{12}}{\mathcal{N}_{12}}\right)^{\varrho} + \left(\frac{1-\mathcal{N}_{22}}{\mathcal{N}_{22}}\right)^{\varrho}\right\}^{1/\varrho}} \Bigg),$$

$$\ldots, \Bigg( 1 - \frac{1}{1+\left\{\left(\frac{\mathcal{Y}_{1m}}{1-\mathcal{Y}_{1m}}\right)^{\varrho} + \left(\frac{\mathcal{Y}_{2m}}{1-\mathcal{Y}_{2m}}\right)^{\varrho}\right\}^{1/\varrho}}, \frac{1}{1+\left\{\left(\frac{1-\mathcal{A}_{1m}}{\mathcal{A}_{1m}}\right)^{\varrho} + \left(\frac{1-\mathcal{A}_{2m}}{\mathcal{A}_{2m}}\right)^{\varrho}\right\}^{1/\varrho}}, \frac{1}{1+\left\{\left(\frac{1-\mathcal{N}_{1m}}{\mathcal{N}_{1m}}\right)^{\varrho} + \left(\frac{1-\mathcal{N}_{2m}}{\mathcal{N}_{2m}}\right)^{\varrho}\right\}^{1/\varrho}} \Bigg) \Bigg)$$

**(2)** $\mathcal{R}_1 \otimes_D \mathcal{R}_2$

$$= \Bigg( \Bigg( \frac{1}{1+\left\{\left(\frac{1-\mathcal{Y}_{11}}{\mathcal{Y}_{11}}\right)^{\varrho} + \left(\frac{1-\mathcal{Y}_{21}}{\mathcal{Y}_{21}}\right)^{\varrho}\right\}^{1/\varrho}}, 1 - \frac{1}{1+\left\{\left(\frac{\mathcal{A}_{11}}{1-\mathcal{A}_{11}}\right)^{\varrho} + \left(\frac{\mathcal{A}_{21}}{1-\mathcal{A}_{21}}\right)^{\varrho}\right\}^{1/\varrho}}, 1 - \frac{1}{1+\left\{\left(\frac{\mathcal{N}_{11}}{1-\mathcal{N}_{11}}\right)^{\varrho} + \left(\frac{\mathcal{N}_{21}}{1-\mathcal{N}_{21}}\right)^{\varrho}\right\}^{1/\varrho}} \Bigg),$$

$$\Bigg( \frac{1}{1+\left\{\left(\frac{1-\mathcal{Y}_{12}}{\mathcal{Y}_{12}}\right)^{\varrho} + \left(\frac{1-\mathcal{Y}_{22}}{\mathcal{Y}_{22}}\right)^{\varrho}\right\}^{1/\varrho}}, 1 - \frac{1}{1+\left\{\left(\frac{\mathcal{A}_{12}}{1-\mathcal{A}_{12}}\right)^{\varrho} + \left(\frac{\mathcal{A}_{22}}{1-\mathcal{A}_{22}}\right)^{\varrho}\right\}^{1/\varrho}}, 1 - \frac{1}{1+\left\{\left(\frac{\mathcal{N}_{12}}{1-\mathcal{N}_{12}}\right)^{\varrho} + \left(\frac{\mathcal{N}_{22}}{1-\mathcal{N}_{22}}\right)^{\varrho}\right\}^{1/\varrho}} \Bigg),$$

$$\ldots, \Bigg( \frac{1}{1+\left\{\left(\frac{1-\mathcal{Y}_{1m}}{\mathcal{Y}_{1m}}\right)^{\varrho} + \left(\frac{1-\mathcal{Y}_{2m}}{\mathcal{Y}_{2m}}\right)^{\varrho}\right\}^{1/\varrho}}, 1 - \frac{1}{1+\left\{\left(\frac{\mathcal{A}_{1m}}{1-\mathcal{A}_{1m}}\right)^{\varrho} + \left(\frac{\mathcal{A}_{2m}}{1-\mathcal{A}_{2m}}\right)^{\varrho}\right\}^{1/\varrho}}, 1 - \frac{1}{1+\left\{\left(\frac{\mathcal{N}_{1m}}{1-\mathcal{N}_{1m}}\right)^{\varrho} + \left(\frac{\mathcal{N}_{2m}}{1-\mathcal{N}_{2m}}\right)^{\varrho}\right\}^{1/\varrho}} \Bigg) \Bigg)$$

**(3)** $$\tau.\mathcal{R}_1=\left(\left(1-\frac{1}{1+\left\{\tau\left(\frac{\mathcal{Y}_{11}}{1-\mathcal{Y}_{11}}\right)^{\varrho}\right\}^{1/\varrho}},\frac{1}{1+\left\{\tau\left(\frac{1-\mathcal{A}_{11}}{\mathcal{A}_{11}}\right)^{\varrho}\right\}^{1/\varrho}},\frac{1}{1+\left\{\tau\left(\frac{1-\mathcal{N}_{11}}{\mathcal{N}_{11}}\right)^{\varrho}\right\}^{1/\varrho}}\right),\right.$$
$$\left(1-\frac{1}{1+\left\{\tau\left(\frac{\mathcal{Y}_{12}}{1-\mathcal{Y}_{12}}\right)^{\varrho}\right\}^{1/\varrho}},\frac{1}{1+\left\{\tau\left(\frac{1-\mathcal{A}_{12}}{\mathcal{A}_{12}}\right)^{\varrho}\right\}^{1/\varrho}},\frac{1}{1+\left\{\tau\left(\frac{1-\mathcal{N}_{12}}{\mathcal{N}_{12}}\right)^{\varrho}\right\}^{1/\varrho}}\right),$$
$$\left.\ldots,\left(1-\frac{1}{1+\left\{\tau\left(\frac{\mathcal{Y}_{1m}}{1-\mathcal{Y}_{1m}}\right)^{\varrho}\right\}^{1/\varrho}},\frac{1}{1+\left\{\tau\left(\frac{1-\mathcal{A}_{1m}}{\mathcal{A}_{1m}}\right)^{\varrho}\right\}^{1/\varrho}},\frac{1}{1+\left\{\tau\left(\frac{1-\mathcal{N}_{1m}}{\mathcal{N}_{1m}}\right)^{\varrho}\right\}^{1/\varrho}}\right)\right)$$

**(4)** $$(\mathcal{R}_1)^{\tau}=\left(\left(\frac{1}{1+\left\{\tau\left(\frac{1-\mathcal{Y}_{11}}{\mathcal{Y}_{11}}\right)^{\varrho}\right\}^{1/\varrho}},1-\frac{1}{1+\left\{\tau\left(\frac{\mathcal{A}_{11}}{1-\mathcal{A}_{11}}\right)^{\varrho}\right\}^{1/\varrho}},1-\frac{1}{1+\left\{\tau\left(\frac{\mathcal{N}_{11}}{1-\mathcal{N}_{11}}\right)^{\varrho}\right\}^{1/\varrho}}\right),\right.$$
$$\left(\frac{1}{1+\left\{\tau\left(\frac{1-\mathcal{Y}_{12}}{\mathcal{Y}_{12}}\right)^{\varrho}\right\}^{1/\varrho}},1-\frac{1}{1+\left\{\tau\left(\frac{\mathcal{A}_{12}}{1-\mathcal{A}_{12}}\right)^{\varrho}\right\}^{1/\varrho}},1-\frac{1}{1+\left\{\tau\left(\frac{\mathcal{N}_{12}}{1-\mathcal{N}_{12}}\right)^{\varrho}\right\}^{1/\varrho}}\right),$$
$$\left.\ldots,\left(\frac{1}{1+\left\{\tau\left(\frac{1-\mathcal{Y}_{1m}}{\mathcal{Y}_{1m}}\right)^{\varrho}\right\}^{1/\varrho}},1-\frac{1}{1+\left\{\tau\left(\frac{\mathcal{A}_{1m}}{1-\mathcal{A}_{1m}}\right)^{\varrho}\right\}^{1/\varrho}},1-\frac{1}{1+\left\{\tau\left(\frac{\mathcal{N}_{1m}}{1-\mathcal{N}_{1m}}\right)^{\varrho}\right\}^{1/\varrho}}\right)\right).$$

## 7.4 mPoPFN Dombi arithmetic operators

**Definition 7.8.** Let $\mathcal{R}_f=((\mathcal{Y}_{f1},\mathcal{A}_{f1},\mathcal{N}_{f1}),(\mathcal{Y}_{f2},\mathcal{A}_{f2},\mathcal{N}_{f2}),\ldots,(\mathcal{Y}_{fm},\mathcal{A}_{fm},\mathcal{N}_{fm}))$ be an mPoPFN. Then, an $mPoPF$ Dombi weighted averaging (mPoPFDWA) operator is a mapping $mPoPFDWA:\Theta^{\varpi}\to\Theta$ introduced below:

$$mPoPFDWA_{\delta}(\mathcal{R}_1,\mathcal{R}_2,\ldots,\mathcal{R}_{\varpi})=\bigoplus_{f=1}^{\varpi}\left(\Xi_f\mathcal{R}_f\right),\tag{7.5}$$

where $\Xi=(\Xi_1,\Xi_2,\ldots,\Xi_{\varpi})^T$ is a set of weight vectors on $\mathcal{R}_f$ such that $\Xi_f>0$ and $\sum_{f=1}^{\varpi}\Xi_f=1$.

**Theorem 7.1.** *Let $\mathcal{R}_f=((\mathcal{Y}_{f1},\mathcal{A}_{f1},\mathcal{N}_{f1}),(\mathcal{Y}_{f2},\mathcal{A}_{f2},\mathcal{N}_{f2}),\ldots,(\mathcal{Y}_{fm},\mathcal{A}_{fm},\mathcal{N}_{fm}))$, $f=1,2,\ldots,\varpi$. Then, the group of all mPoPFNs under an mPoPFDWA operator is also an mPoPFN number, which is evaluated as:*

$$mPoPFDWA_{\Xi}(\mathcal{R}_1,\mathcal{R}_2,\ldots,\mathcal{R}_{\varpi})=\bigoplus_{f=1}^{\varpi}\left(\Xi_f\mathcal{R}_f\right)$$
$$=\left(\left(1-\frac{1}{1+\left\{\sum_{f=1}^{\varpi}\Xi_f\left(\frac{\mathcal{Y}_{11}}{1-\mathcal{Y}_{11}}\right)^{\varrho}\right\}^{1/\varrho}},\frac{1}{1+\left\{\sum_{f=1}^{\varpi}\Xi_f\left(\frac{1-\mathcal{A}_{11}}{\mathcal{A}_{11}}\right)^{\varrho}\right\}^{1/\varrho}},\right.\right.$$

$$\left.\frac{1}{1+\left\{\sum_{f=1}^{\varpi}\Xi_f\left(\frac{1-\mathcal{N}_{11}}{\mathcal{N}_{11}}\right)^{\varrho}\right\}^{1/\varrho}}\right),$$

$$\ldots\left(1-\frac{1}{1+\left\{\sum_{f=1}^{\varpi}\Xi_f\left(\frac{\mathcal{Y}_{1m}}{1-\mathcal{Y}_{1m}}\right)^{\varrho}\right\}^{1/\varrho}},\frac{1}{1+\left\{\sum_{f=1}^{\varpi}\Xi_f\left(\frac{1-\mathcal{A}_{1m}}{\mathcal{A}_{1m}}\right)^{\varrho}\right\}^{1/\varrho}},\right.$$

$$\left.\left.\frac{1}{1+\left\{\sum_{f=1}^{\varpi}\Xi_f\left(\frac{1-\mathcal{N}_{1m}}{\mathcal{N}_{1m}}\right)^{\varrho}\right\}^{1/\varrho}}\right)\right), \tag{7.6}$$

*where* $\Xi=(\Xi_1,\Xi_2,\ldots,\Xi_\varsigma)$ *is the set of weight vectors of* $\mathcal{R}_f$, $f=1,2,\ldots,\varpi$, *such that* $\Xi_f>0$, *and* $\sum_{f=1}^{\varpi}\Xi_f=1$.

*Proof.* This theorem can be proved by the mathematical induction technique.

If for $f=2$ and $\Xi=2$, then the left side of the theorem becomes

$$mPoPFDWA_{\Xi}(\mathcal{R}_1,\mathcal{R}_2,\ldots,\mathcal{R}_\varpi)=\bigoplus_{f=1}^{\varpi}\left(\Xi_f\mathcal{R}_f\right)=\Xi_1\mathcal{R}_1\oplus\Xi_2\mathcal{R}_2$$

and the right is

$$\Xi_1\mathcal{R}_1\oplus\Xi_2\mathcal{R}_2$$

$$=\left(\left(1-\frac{1}{1+\left\{\Xi_1\left(\frac{\mathcal{Y}_{11}}{1-\mathcal{Y}_{11}}\right)^{\varrho}+\Xi_2\left(\frac{\mathcal{Y}_{21}}{1-\mathcal{Y}_{21}}\right)^{\varrho}\right\}^{1/\varrho}},\frac{1}{1+\left\{\Xi_1\left(\frac{1-\mathcal{A}_{11}}{\mathcal{A}_{11}}\right)^{\varrho}+\Xi_2\left(\frac{1-\mathcal{A}_{21}}{\mathcal{A}_{21}}\right)^{\varrho}\right\}^{1/\varrho}},\right.\right.$$

$$\left.\frac{1}{1+\left\{\Xi_1\left(\frac{1-\mathcal{N}_{11}}{\mathcal{N}_{11}}\right)^{\varrho}+\Xi_2\left(\frac{1-\mathcal{N}_{21}}{\mathcal{N}_{21}}\right)^{\varrho}\right\}^{1/\varrho}}\right),$$

$$\ldots\left(1-\frac{1}{1+\left\{\Xi_1\left(\frac{\mathcal{Y}_{1m}}{1-\mathcal{Y}_{1m}}\right)^{\varrho}+\Xi_2\left(\frac{\mathcal{Y}_{2m}}{1-\mathcal{Y}_{2m}}\right)^{\varrho}\right\}^{1/\varrho}},\frac{1}{1+\left\{\Xi_1\left(\frac{1-\mathcal{A}_{1m}}{\mathcal{A}_{1m}}\right)^{\varrho}+\Xi_2\left(\frac{1-\mathcal{A}_{2m}}{\mathcal{A}_{2m}}\right)^{\varrho}\right\}^{1/\varrho}},\right.$$

$$\left.\left.\frac{1}{1+\left\{\Xi_1\left(\frac{1-\mathcal{N}_{1m}}{\mathcal{N}_{1m}}\right)^{\varrho}+\Xi_2\left(\frac{1-\mathcal{N}_{2m}}{\mathcal{N}_{2m}}\right)^{\varrho}\right\}^{1/\varrho}}\right)\right)$$

$$=\left(\left(1-\frac{1}{1+\left\{\sum_{f=1}^{2}\Xi_f\left(\frac{\mathcal{Y}_{f1}}{1-\mathcal{Y}_{f1}}\right)^{\varrho}\right\}^{1/\varrho}},\frac{1}{1+\left\{\sum_{f=1}^{2}\Xi_f\left(\frac{1-\mathcal{A}_{f1}}{\mathcal{A}_{f1}}\right)^{\varrho}\right\}^{1/\varrho}},\right.\right.$$

$$\left.\frac{1}{1+\left\{\sum_{f=1}^{2}\Xi_f\left(\frac{1-\mathcal{N}_{f1}}{\mathcal{N}_{f1}}\right)^{\varrho}\right\}^{1/\varrho}}\right)$$

$$\ldots\left(1-\frac{1}{1+\left\{\sum_{f=1}^{2}\Xi_f\left(\frac{\mathcal{Y}_{fm}}{1-\mathcal{Y}_{fm}}\right)^{\varrho}\right\}^{1/\varrho}},\frac{1}{1+\left\{\sum_{f=1}^{2}\Xi_f\left(\frac{1-\mathcal{A}_{fm}}{\mathcal{A}_{fm}}\right)^{\varrho}\right\}^{1/\varrho}},\frac{1}{1+\left\{\sum_{f=1}^{2}\Xi_f\left(\frac{1-N_{fm}}{N_{fm}}\right)^{\varrho}\right\}^{1/\varrho}}\right)\bigg). \tag{7.7}$$

Thus the above result is true for $f=2$.

Let us assume that Theorem 7.1 holds for $f \geq r$, where $r \in \mathbb{N}$ (the set of natural numbers), which gives:

$$\begin{aligned}
&mPoPFDWA_{\Xi}(\mathcal{R}_1,\mathcal{R}_2,\ldots,\mathcal{R}_r)\\
&=\bigoplus_{b=1}^{r}\left(\Xi_f\mathcal{R}_f\right)\\
&=\Bigg(\Bigg(1-\frac{1}{1+\left\{\sum_{f=1}^{r}\Xi_f\left(\frac{\mathcal{Y}_{f1}}{1-\mathcal{Y}_{f1}}\right)^{\varrho}\right\}^{1/\varrho}},\frac{1}{1+\left\{\sum_{f=1}^{r}\Xi_f\left(\frac{1-\mathcal{A}_{f1}}{\mathcal{A}_{f1}}\right)^{\varrho}\right\}^{1/\varrho}},\\
&\qquad\frac{1}{1+\left\{\sum_{f=1}^{r}\Xi_f\left(\frac{1-N_{f1}}{N_{f1}}\right)^{\varrho}\right\}^{1/\varrho}}\Bigg),\\
&\ldots\Bigg(1-\frac{1}{1+\left\{\sum_{f=1}^{r}\Xi_f\left(\frac{\mathcal{Y}_{fm}}{1-\mathcal{Y}_{fm}}\right)^{\varrho}\right\}^{1/\varrho}},\frac{1}{1+\left\{\sum_{f=1}^{r}\Xi_f\left(\frac{1-\mathcal{A}_{fm}}{\mathcal{A}_{fm}}\right)^{\varrho}\right\}^{1/\varrho}},\\
&\qquad\frac{1}{1+\left\{\sum_{f=1}^{r}\Xi_f\left(\frac{1-N_{fm*}}{N_{fm}}\right)^{\varrho}\right\}^{1/\varrho}}\Bigg)\Bigg).
\end{aligned}$$

Now, for $f=r+1$, then

$$\begin{aligned}
&mPFDWA_{\Xi}(\mathcal{R}_1,\mathcal{R}_2,\mathcal{R}_r\ldots,\mathcal{R}_{r+1})\\
&=\bigoplus_{f=1}^{r}(\Xi_f\mathcal{R}_f)\oplus\Xi_{r+1}\mathcal{R}_{r+1}\\
&=\Bigg(\Bigg(1-\frac{1}{1+\left\{\sum_{f=1}^{r}\Xi_f\left(\frac{\mathcal{Y}_{f1}}{1-\mathcal{Y}_{f1}}\right)^{\varrho}\right\}^{1/\varrho}},\frac{1}{1+\left\{\sum_{f=1}^{r}\Xi_f\left(\frac{1-\mathcal{A}_{f1}}{\mathcal{A}_{f1}}\right)^{\varrho}\right\}^{1/\varrho}},
\end{aligned}$$

$$\left.\frac{1}{1+\left\{\sum_{f=1}^{r}\Xi_f\left(\frac{1-\mathcal{N}_{f1}}{\mathcal{N}_{f1}}\right)^{\varrho}\right\}^{1/\varrho}}\right),$$

$$\ldots,\left(1-\frac{1}{1+\left\{\sum_{f=1}^{r}\Xi_f\left(\frac{\mathcal{Y}_{fm}}{1-\mathcal{Y}_{fm}}\right)^{\varrho}\right\}^{1/\varrho}},\frac{1}{1+\left\{\sum_{f=1}^{r}\Xi_f\left(\frac{1-\mathcal{A}_{fm}}{\mathcal{A}_{fm}}\right)^{\varrho}\right\}^{1/\varrho}},\right.$$

$$\left.\left.\frac{1}{1+\left\{\sum_{f=1}^{r}\Xi_f\left(\frac{1-\mathcal{N}_{fm}}{\mathcal{N}_{fm}}\right)^{\varrho}\right\}^{1/\varrho}}\right)\right)$$

$$\bigoplus\left(\left(1-\frac{1}{1+\left\{\Xi_{r+1}\left(\frac{\mathcal{Y}_{(r+1)1}}{1-\mathcal{Y}_{(r+1)1}}\right)^{\varrho}\right\}^{1/\varrho}},\frac{1}{1+\left\{\Xi_{r+1}\left(\frac{1-\mathcal{A}_{(r+1)1}}{\mathcal{A}_{(r+1)1}}\right)^{\varrho}\right\}^{1/\varrho}},\right.\right.$$

$$\left.\frac{1}{1+\left\{\Xi_{r+1}\left(\frac{1-\mathcal{N}_{(r+1)1}}{\mathcal{N}_{(r+1)1}}\right)^{\varrho}\right\}^{1/\varrho}}\right),$$

$$\ldots,\left(1-\frac{1}{1+\left\{\Xi_{r+1}\left(\frac{\mathcal{Y}_{(r+1)m}}{1-\mathcal{Y}_{(r+1)m}}\right)^{\varrho}\right\}^{1/\varrho}},\frac{1}{1+\left\{\Xi_{r+1}\left(\frac{1-\mathcal{A}_{(r+1)m}}{\mathcal{A}_{(r+1)m}}\right)^{\varrho}\right\}^{1/\varrho}},\right.$$

$$\left.\left.\frac{1}{1+\left\{\Xi_{r+1}\left(\frac{1-\mathcal{N}_{(r+1)m}}{\mathcal{N}_{(r+1)m}}\right)^{\varrho}\right\}^{1/\varrho}}\right)\right)$$

$$=\left\{\left(1-\frac{1}{1+\left\{\sum_{f=1}^{r+1}\Xi_f\left(\frac{\mathcal{Y}_{f1}}{1-\mathcal{Y}_{f1}}\right)^{\varrho}\right\}^{1/\varrho}},\frac{1}{1+\left\{\sum_{f=1}^{r+1}\Xi_f\left(\frac{1-\mathcal{A}_{f1}}{\mathcal{A}_{f1}}\right)^{\varrho}\right\}^{1/\varrho}},\right.\right.$$

$$\left.\frac{1}{1+\left\{\sum_{f=1}^{r+1}\Xi_f\left(\frac{1-\mathcal{N}_{f1}}{\mathcal{N}_{f1}}\right)^{\varrho}\right\}^{1/\varrho}}\right),$$

$$\ldots,\left(1-\frac{1}{1+\left\{\sum_{f=1}^{r+1}\Xi_f\left(\frac{\mathcal{Y}_{fm}}{1-\mathcal{Y}_{fm}}\right)^{\varrho}\right\}^{1/\varrho}},\frac{1}{1+\left\{\sum_{f=1}^{r+1}\Xi_f\left(\frac{1-\mathcal{A}_{fm}}{\mathcal{A}_{fm}}\right)^{\varrho}\right\}^{1/\varrho}},\right.$$

$$\left.\left.\frac{1}{1+\left\{\sum_{f=1}^{r+1}\Xi_f\left(\frac{1-\mathcal{N}_{fm}}{\mathcal{N}_{fm}}\right)^{\varrho}\right\}^{1/\varrho}}\right)\right\}.$$

Hence, Theorem 7.1 is true for all natural numbers. □

The mPoPFDWA operators provided the following properties.

**Theorem 7.2.** *(Idempotency property) Let $\mathcal{R}_f = ((\mathcal{Y}_{f1}, \mathcal{A}_{f1}, \mathcal{N}_{f1}), (\mathcal{Y}_{f2}, \mathcal{A}_{f2}, \mathcal{N}_{f2}), \ldots, (\mathcal{Y}_{fm}, \mathcal{A}_{fm}, \mathcal{N}_{fm}))$ $f = 1, 2, \ldots, \varpi$ be a set of mPoPFNs. If all $mPoPFNs$ are equal, i.e., $\mathcal{R}_f = \mathcal{R}$ for all $f = 1, 2, \ldots, \varpi$, then*

$$mPoPFDWA_{\Xi}(\mathcal{R}_1, \mathcal{R}_2, \ldots, \mathcal{R}_{\varpi}) = \mathcal{R}. \tag{7.8}$$

**Theorem 7.3.** *(Boundedness property) Let $\mathcal{R}_f = ((\mathcal{Y}_{f1}, \mathcal{A}_{f1}, \mathcal{N}_{f1}), (\mathcal{Y}_{f2}, \mathcal{A}_{f2}, \mathcal{N}_{f2}), \ldots, (\mathcal{Y}_{fm}, \mathcal{A}_{fm}, \mathcal{N}_{fm}))$, $f = 1, 2, \ldots, \varpi$ be a mPoPFNs. If $\mathcal{R}^- = \min\{\mathcal{R}_1, \mathcal{R}_2, \ldots, \mathcal{R}_{\varpi}\}$ and $\mathcal{R}^+ = \max\{\mathcal{R}_1, \mathcal{R}_2, \ldots, \mathcal{R}_{\varpi}\}$, then*

$$\mathcal{R}^- \leq mPoPFDWA_{\Xi}(\mathcal{R}_1, \mathcal{R}_2, \ldots, \mathcal{R}_{\varpi}) \leq \mathcal{R}^+. \tag{7.9}$$

**Theorem 7.4.** *(Monotonicity property) Let $\mathcal{R}_f = ((\mathcal{Y}_{f1}, \mathcal{A}_{f1}, \mathcal{N}_{f1}), (\mathcal{Y}_{f2}, \mathcal{A}_{f2}, \mathcal{N}_{f2}), \ldots, (\mathcal{Y}_{fm}, \mathcal{A}_{fm}, \mathcal{N}_{fm}))$ and $\mathcal{R}'_f = ((\mathcal{Y}'_{f1}, \mathcal{A}'_{f1}, \mathcal{N}'_{f1}), (\mathcal{Y}'_{f2}, \mathcal{A}'_{f2}, \mathcal{N}'_{f2}), \ldots, (\mathcal{Y}'_{fm}, \mathcal{A}'_{fm}, \mathcal{N}'_{fm}))$, $f = 1, 2, \ldots, \varpi$ be two mPoPFNs such that $\mathcal{R}_f \leq \mathcal{R}'_f$ for all $f$, then*

$$mPoPFDWA_{\Xi}(\mathcal{R}_1, \mathcal{R}_2, \ldots, \mathcal{R}_{\varpi}) \leq mPoPFDWA_{\Xi}(\mathcal{R}'_1, \mathcal{R}'_2, \ldots, \mathcal{R}'_{\varpi}). \tag{7.10}$$

Next, proceed with the mPoPFDOWA operator.

**Definition 7.9.** Let $\mathcal{R}_f = ((\mathcal{Y}_{f1}, \mathcal{A}_{f1}, \mathcal{N}_{f1}), (\mathcal{Y}_{f2}, \mathcal{A}_{f2}, \mathcal{N}_{f2}), \ldots, (\mathcal{Y}_{fm}, \mathcal{A}_{fm}, \mathcal{N}_{fm}))$, $f = 1, 2, \ldots, \varpi$ be a set of mPoPFNs. Then, the Dombi ordered weighted averaging of mPoPFN is an mPoPFDOWA operator introducing a mapping $mPoPFDOWA : \Theta^{\varpi} \to \Theta$ defined below:

$$mPoPFDOWA_{\Xi}(\mathcal{R}_1, \mathcal{R}_2, \ldots, \mathcal{R}_{\varpi}) = \bigoplus_{f=1}^{\varpi} \left(\delta_f \mathcal{R}_{\sigma(f)}\right), \tag{7.11}$$

where the weight vector $\delta = (\delta_1, \delta_2, \ldots, \delta_{\varpi})^T$ of $\mathcal{R}_f$ such that $\delta_f > 0$ and $\sum_{f=1}^{\varpi} \delta_f = 1$. Also, $(\sigma(1), \sigma(2), \ldots, \sigma(\varpi))$ is a permutation of $(1, 2, \ldots, \varpi)$ for which $\mathcal{R}_{\sigma(f-1)} \geq \mathcal{R}_{\sigma(f)}$ for all $f = 1, 2, \ldots, \varpi$.

**Theorem 7.5.** *Let $\mathcal{R}_f = ((\mathcal{Y}_{f1}, \mathcal{A}_{f1}, \mathcal{N}_{f1}), (\mathcal{Y}_{f2}, \mathcal{A}_{f2}, \mathcal{N}_{f2}), \ldots, (\mathcal{Y}_{fm}, \mathcal{A}_{fm}, \mathcal{N}_{fm}))$, $f = 1, 2, \ldots, \varpi$ be mPoPFNs. Then, the collected value of all mPoPFNs under an mPoPFDOWA operator is also an mPoPFN, which follows as:*

$$\begin{aligned}
&mPoPFDOWA_{\delta}(\mathcal{R}_1, \mathcal{R}_2, \ldots, \mathcal{R}_{\varpi}) \\
&= \bigoplus_{f=1}^{\zeta} \left(\delta_f \mathcal{R}_{\sigma(f)}\right) \\
&= \Bigg(\Bigg(1 - \frac{1}{1 + \left\{\sum_{f=1}^{\varpi} \delta_f \left(\frac{\mathcal{Y}_{\sigma(f1)}}{1 - \mathcal{Y}_{\sigma(f1)}}\right)^{\varrho}\right\}^{1/\varrho}}, \frac{1}{1 + \left\{\sum_{f=1}^{\varpi} \delta_f \left(\frac{1 - \mathcal{A}_{\sigma(f1)}}{\mathcal{A}_{\sigma(f1)}}\right)^{\varrho}\right\}^{1/\varrho}},
\end{aligned}$$

$$\left.\frac{1}{1+\left\{\sum_{f=1}^{\varpi}\delta_f\left(\frac{1-\mathcal{N}_{\sigma(f1)}}{\mathcal{N}_{\sigma(f1)}}\right)^{\varrho}\right\}^{1/\varrho}}\right),$$

$$\ldots,\left(1-\frac{1}{1+\left\{\sum_{f=1}^{\varpi}\delta_f\left(\frac{\mathcal{Y}_{\sigma(fm)}}{1-\mathcal{Y}_{\sigma(f)}}\right)^{\varrho}\right\}^{1/\varrho}},\frac{1}{1+\left\{\sum_{f=1}^{\varpi}\delta_f\left(\frac{1-\mathcal{A}_{\sigma(fm)}}{\mathcal{A}_{\sigma(fm)}}\right)^{\varrho}\right\}^{1/\varrho}},\right.$$

$$\left.\left.\frac{1}{1+\left\{\sum_{f=1}^{\varpi}\delta_f\left(\frac{1-\mathcal{N}_{\sigma(fm)}}{\mathcal{N}_{\sigma(fm)}}\right)^{\varrho}\right\}^{1/\varrho}}\right)\right),$$

*where $\delta=(\delta_1,\delta_2,\ldots,\delta_\varpi)$ is the weight vector of $\mathcal{R}_f$, $f=1,2,\ldots,\varpi$ such that $\delta_f>0$, and $\sum_{f=1}^{\varpi}\delta_f=1$. Also, $\sigma(1),\sigma(2),\ldots,\sigma(\varpi)$ is a permutation for $(1,2,\ldots,\varpi)$ for which $\mathcal{R}_{\sigma(f-1)}\geq\mathcal{R}_{\sigma(f)}$ for all $f=1,2,\ldots,\varpi$.*

The following properties of mPoPFDOWA can be proved easily.

**Theorem 7.6.** *(Idempotency property) Let $\mathcal{R}_f=((\mathcal{Y}_{f1},\mathcal{A}_{f1},\mathcal{N}_{f1}),(\mathcal{Y}_{f2},\mathcal{A}_{f2},\mathcal{N}_{f2}),\ldots,(\mathcal{Y}_{fm},\mathcal{A}_{fm},\mathcal{N}_{fm}))$, $f=1,2,\ldots,\varpi$ be a mPoPFN. If $\mathcal{R}_f=\mathcal{R}$ for all $f=1,2,\ldots,\varpi$, then*

$$mPoPFDWA_\delta(\mathcal{R}_1,\mathcal{R}_2,\ldots,\mathcal{R}_\varpi)=\mathcal{R}. \tag{7.12}$$

**Theorem 7.7.** *(Boundedness property) Let $\mathcal{R}_f=((\mathcal{Y}_{f1},\mathcal{A}_{f1},\mathcal{N}_{f1}),(\mathcal{Y}_{f2},\mathcal{A}_{f2},\mathcal{N}_{f2}),\ldots,(\mathcal{Y}_{fm},\mathcal{A}_{fm},\mathcal{N}_{fm}))$ be a mPoPFN. If $\mathcal{R}^-=\min\{\mathcal{R}_1,\mathcal{R}_2,\ldots,\mathcal{R}_\varpi\}$ and $\mathcal{R}^+=\max\{\mathcal{R}_1,\mathcal{R}_2,\ldots,\mathcal{R}_\varpi\}$, then*

$$\mathcal{R}^-\leq mPoPFDWA_\delta(\mathcal{R}_1,\mathcal{R}_2,\ldots,\mathcal{R}_\varpi)\leq\mathcal{R}^+. \tag{7.13}$$

**Theorem 7.8.** *(Monotonicity property) Let $\mathcal{R}_f=((\mathcal{Y}_{f1},\mathcal{A}_{f1},\mathcal{N}_{f1}),(\mathcal{Y}_{f2},\mathcal{A}_{f2},\mathcal{N}_{f2}),\ldots,(\mathcal{Y}_{fm},\mathcal{A}_{fm},\mathcal{N}_{fm}))$ and $\mathcal{R}'_f=((\mathcal{Y}'_{f1},\mathcal{A}'_{f1},\mathcal{N}'_{f1}),(\mathcal{Y}'_{f2},\mathcal{A}'_{f2},\mathcal{N}'_{f2}),\ldots,(\mathcal{Y}'_{fm},\mathcal{A}'_{fm},\mathcal{N}'_{fm}))$, $f=1,2,\ldots,\varpi$ be two sets of mPoPFNs such that $\mathcal{R}_f\leq\mathcal{R}'_f$ for all $f$, then*

$$mPoPFDWA_\delta(\mathcal{R}_1,\mathcal{R}_2,\ldots,\mathcal{R}_\varpi)\leq mPoPFDWA_\delta(\mathcal{R}'_1,\mathcal{R}'_2,\ldots,\mathcal{R}'_\varpi). \tag{7.14}$$

**Theorem 7.9.** *(Commutative property) Let $\mathcal{R}_f=((\mathcal{Y}_{f1},\mathcal{A}_{f1},\mathcal{N}_{f1}),(\mathcal{Y}_{f2},\mathcal{A}_{f2},\mathcal{N}_{f2}),\ldots,(\mathcal{Y}_{fm},\mathcal{A}_{fm},\mathcal{N}_{fm}))$ and $\mathcal{R}'_f=((\mathcal{Y}'_{f1},\mathcal{A}'_{f1},\mathcal{N}'_{f1}),(\mathcal{Y}'_{f2},\mathcal{A}'_{f2},\mathcal{N}'_{f2}),\ldots,(\mathcal{Y}'_{fm},\mathcal{A}'_{fm},\mathcal{N}'_{fm}))$, $f=1,2,\ldots,\varpi$ be two sets of mPoPFNs such that*

$$mPoPFDWA_\delta(\mathcal{R}_1,\mathcal{R}_2,\ldots,\mathcal{R}_\varpi)=mPoPFDWA_\delta(\mathcal{R}'_1,\mathcal{R}'_2,\ldots,\mathcal{R}'_\varpi), \tag{7.15}$$

*where $\mathcal{R}'_f$ is arbitrary permutation of $\mathcal{R}_f$ for all $f=1,2,\ldots,\varpi$.*

In Definitions 7.8 and 7.9, the mPoPFDWA operator used the weights of the mPoPFN; however, this time, the weight of the mPoPFDOWA operator indicates the ordered position of the mPoPFN rather than the weights of the mPoPFN themselves. As a result, we define another operator, the mPoPF Dombi hybrid averaging operator (mPoPFDHWA), which makes qualitative use of both the mPoPFDWA and mPoPFDOWA operators.

**Definition 7.10.** Let $\mathcal{R}_f = ((\mathcal{Y}_{f1}, \mathcal{A}_{f1}, \mathcal{N}_{f1}), (\mathcal{Y}_{f2}, \mathcal{A}_{f2}, \mathcal{N}_{f2}), \ldots, (\mathcal{Y}_{fm}, \mathcal{A}_{fm}, \mathcal{N}_{fm}))$, $f = 1, 2, \ldots, \varpi$ be a set of mPoPFNs. Then, the Dombi hybrid weighted averaging (mPoPFD-HWA) operator of mPoPFNs is a function $mPoPFDHWA : \Theta^{\varpi} \to \Theta$ that is defined below:

$$mPoPFDHWA_{\Xi,\delta}(\mathcal{R}_1, \mathcal{R}_2, \ldots, \mathcal{R}_{\varpi}) = \bigoplus_{f=1}^{\varpi} \left( \delta_f \dot{\mathcal{R}}_{\sigma(fg)} \right). \tag{7.16}$$

Also, $(\sigma(1), \sigma(2), \ldots, \sigma(\varpi))$ is a permutation for $(1, 2, \ldots, \varpi)$ for which $\mathcal{R}_{\sigma(f-1)} \geq \mathcal{R}_{\sigma(f)}$ for all $f = 1, 2, \ldots, \varpi$ for mPoPFNs $\mathcal{R}_f$ and $\delta = (\delta_1, \delta_2, \ldots, \delta_{\varpi})^T$ is the associated weighted vector of the mPoPFNs $(\mathcal{R}_1, \mathcal{R}_2, \ldots, \mathcal{R}_{\varpi})$ such that $\delta_f > 0$ and $\sum_{f=1}^{\varpi} \delta_f = 1$. $\dot{\mathcal{R}}_f$ is the largest mPoPFNs, where $\dot{\mathcal{R}}_f = (\Xi)\mathcal{R}_f$, $f = 1, 2, \ldots, \varpi$ for which $\Xi = (\Xi_1, \Xi_2, \ldots, \Xi_{\varpi})^T$ is the weight vector such that $\Xi_f > 0$ and $\sum_{f=1}^{\varpi} \Xi_f = 1$.

**Theorem 7.10.** *Let $\mathcal{R}_f = ((\mathcal{Y}_{f1}, \mathcal{A}_{f1}, \mathcal{N}_{f1}), (\mathcal{Y}_{f2}, \mathcal{A}_{f2}, \mathcal{N}_{f2}), \ldots, (\mathcal{Y}_{fm}, \mathcal{A}_{fm}, \mathcal{N}_{fm}))$, $f = 1, 2, \ldots, \varpi$ be a set of mPoPFNs. Then, the collected values of mPoPFNs $\mathcal{R}_f$ using the mPoPFDHWA operator are also an mPoPFN. Further:*

$$\begin{aligned}
&mPoPFDHWA_{\Xi,\delta}(\mathcal{R}_1, \mathcal{R}_2, \ldots, \mathcal{R}_{\varpi}) \\
&= \bigoplus_{f=1}^{\zeta} \left( \delta_f \dot{\mathcal{R}}_{\sigma(fg)} \right) \\
&= \Bigg( \Bigg( 1 - \frac{1}{1 + \left\{ \sum_{f=1}^{\varpi} \delta_f \left( \frac{\dot{\mathcal{Y}}_{\sigma(f1)}}{1 - \dot{\mathcal{Y}}_{\sigma(f1)}} \right)^{\varrho} \right\}^{1/\varrho}}, \frac{1}{1 + \left\{ \sum_{f=1}^{\varpi} \delta_f \left( \frac{1 - \dot{\mathcal{A}}_{\sigma(f1)}}{\dot{\mathcal{A}}_{\sigma(f1)}} \right)^{\varrho} \right\}^{1/\varrho}}, \\
&\qquad \frac{1}{1 + \left\{ \sum_{f=1}^{\varpi} \delta_f \left( \frac{1 - \dot{\mathcal{N}}_{\sigma(f1)}}{\dot{\mathcal{N}}_{\sigma(f)}} \right)^{\varrho} \right\}^{1/\varrho}} \Bigg), \\
&\qquad \ldots, \Bigg( 1 - \frac{1}{1 + \left\{ \sum_{f=1}^{\varpi} \delta_f \left( \frac{\dot{\mathcal{Y}}_{\sigma(fm)}}{1 - \dot{\mathcal{Y}}_{\sigma(fm)}} \right)^{\varrho} \right\}^{1/\varrho}}, \frac{1}{1 + \left\{ \sum_{f=1}^{\varpi} \delta_f \left( \frac{1 - \dot{\mathcal{A}}_{\sigma(fm)}}{\dot{\mathcal{A}}_{\sigma(fm)}} \right)^{\varrho} \right\}^{1/\varrho}}, \\
&\qquad \frac{1}{1 + \left\{ \sum_{f=1}^{\varpi} \delta_f \left( \frac{1 - \dot{\mathcal{N}}_{\sigma(f1)}}{\dot{\mathcal{N}}_{\sigma(fm)}} \right)^{\varrho} \right\}^{1/\varrho}} \Bigg) \Bigg).
\end{aligned}$$

## 7.5 mPoPFN Dombi geometric operators

Now, we discuss Dombi geometric operators.

**Definition 7.11.** Let $(\mathcal{R}_1, \mathcal{R}_2, \ldots, \mathcal{R}_\varpi)$ be a set of mPoPFNs. Then, an $mPoPF$ Dombi weighted geometric (mPoPFDWG) operator is a mapping $mPoPFDWG : \Theta^\varpi \to \Theta$ given by:

$$mPoPFDWG_\delta(\mathcal{R}_1, \mathcal{R}_2, \ldots, \mathcal{R}_\varpi) = \bigotimes_{f=1}^{\varpi} \left(\mathcal{R}_f\right)^{\Xi_f}, \tag{7.17}$$

where $\Xi = (\Xi_1, \Xi_2, \ldots, \Xi_\varsigma)^T$ is a set of weight vectors on $\mathcal{R}_f$ such that $\Xi_f > 0$ and $\sum_{f=1}^{\varpi} \Xi_f = 1$.

**Theorem 7.11.** *Let $(\mathcal{R}_1, \mathcal{R}_2, \ldots, \mathcal{R}_\varpi)$ be a set of mPoPFNs. Then, the group of all mPoPFNs under the mPoPFDWG operator is also the mPoPFN number, which is evaluated as:*

$$mPoPFDWG_\Xi(\mathcal{R}_1, \mathcal{R}_2, \ldots, \mathcal{R}_\varpi) = \bigotimes_{f=1}^{\varpi} \left(\mathcal{R}_f\right)^{\Xi_f}$$

$$= \Bigg(\Bigg(\frac{1}{1+\left\{\sum_{f=1}^{\varpi} \Xi_f \left(\frac{1-\mathcal{Y}_{11}}{\mathcal{Y}_{11}}\right)^\varrho\right\}^{1/\varrho}}, 1 - \frac{1}{1+\left\{\sum_{f=1}^{\varpi} \Xi_f \left(\frac{\mathcal{A}_{11}}{1-\mathcal{A}_{11}}\right)^\varrho\right\}^{1/\varrho}},$$

$$1 - \frac{1}{1+\left\{\sum_{f=1}^{\varpi} \Xi_f \left(\frac{\mathcal{N}_{11}}{1-\mathcal{N}_{11}}\right)^\varrho\right\}^{1/\varrho}}\Bigg),$$

$$\ldots, \Bigg(\frac{1}{1+\left\{\sum_{f=1}^{\varpi} \Xi_f \left(\frac{1-\mathcal{Y}_{1m}}{\mathcal{Y}_{1m}}\right)^\varrho\right\}^{1/\varrho}}, 1 - \frac{1}{1+\left\{\sum_{f=1}^{\varpi} \Xi_f \left(\frac{\mathcal{A}_{1m}}{1-\mathcal{A}_{1m}}\right)^\varrho\right\}^{1/\varrho}},$$

$$1 - \frac{1}{1+\left\{\sum_{f=1}^{\varpi} \Xi_f \left(\frac{\mathcal{N}_{1m}}{1-\mathcal{N}_{1m}}\right)^\varrho\right\}^{1/\varrho}}\Bigg)\Bigg),$$

*where $\Xi = (\Xi_1, \Xi_2, \ldots, \Xi_\varpi)$ is the set of weight vectors of $\mathcal{R}_f$, $f = 1, 2, \ldots, \varpi$ such that $\Xi_f >$ 0, and $\sum_{f=1}^{\varpi} \Xi_f = 1$.*

*Proof.* For $f = 2$ and $\Xi = 2$, the left side of the theorem becomes $mPoPFDWG_\Xi(\mathcal{R}_1, \mathcal{R}_2, \ldots, \mathcal{R}_\varpi) = \left(\mathcal{R}_1\right)^{\Xi_1} \otimes \left(\mathcal{R}_2\right)^{\Xi_2}$.

The right-hand side of the theorem is

$$\left(\mathcal{R}_1\right)^{\Xi_1} \otimes \left(\mathcal{R}_2\right)^{\Xi_2}$$

$$= \Bigg(\Bigg(\frac{1}{1+\left\{\Xi_1\left(\frac{1-\mathcal{Y}_{11}}{\mathcal{Y}_{11}}\right)^\varrho + \Xi_2\left(\frac{\mathcal{Y}_{21}}{1-\mathcal{Y}_{21}}\right)^\varrho\right\}^{1/\varrho}}, 1 - \frac{1}{1+\left\{\Xi_1\left(\frac{\mathcal{A}_{11}}{1-\mathcal{A}_{11}}\right)^\varrho + \Xi_2\left(\frac{\mathcal{A}_{21}}{1-\mathcal{A}_{21}}\right)^\varrho\right\}^{1/\varrho}},$$

$$1-\frac{1}{1+\left\{\Xi_1\left(\frac{\mathcal{N}_{11}}{1-\mathcal{N}_{11}}\right)^{\varrho}+\Xi_2\left(\frac{\mathcal{N}_{21}}{1-\mathcal{N}_{21}}\right)^{\varrho}\right\}^{1/\varrho}}\Bigg),$$

$$\ldots,\left(\frac{1}{1+\left\{\Xi_1\left(\frac{1-\mathcal{Y}_{1m}}{\mathcal{Y}_{1m}}\right)^{\varrho}+\Xi_2\left(\frac{1-\mathcal{Y}_{2m}}{\mathcal{Y}_{2m}}\right)^{\varrho}\right\}^{1/\varrho}},1-\frac{1}{1+\left\{\Xi_1\left(\frac{1-\mathcal{A}_{1m}}{\mathcal{A}_{1m}}\right)^{\varrho}+\Xi_2\left(\frac{\mathcal{A}_{2m}}{1-\mathcal{A}_{2m}}\right)^{\varrho}\right\}^{1/\varrho}},\right.$$

$$\left.1-\frac{1}{1+\left\{\Xi_1\left(\frac{\mathcal{N}_{1m}}{1-\mathcal{N}_{1m}}\right)^{\varrho}+\Xi_2\left(\frac{\mathcal{N}_{2m}}{1-\mathcal{N}_{2m}}\right)^{\varrho}\right\}^{1/\varrho}}\right)\Bigg)$$

$$=\Bigg(\Bigg(\frac{1}{1+\left\{\sum_{f=1}^{2}\Xi_f\left(\frac{1-\mathcal{Y}_{f1}}{\mathcal{Y}_{f1}}\right)^{\varrho}\right\}^{1/\varrho}},1-\frac{1}{1+\left\{\sum_{f=1}^{2}\Xi_f\left(\frac{\mathcal{A}_{f1}}{1-\mathcal{A}_{f1}}\right)^{\varrho}\right\}^{1/\varrho}},$$

$$1-\frac{1}{1+\left\{\sum_{f=1}^{2}\Xi_f\left(\frac{\mathcal{N}_{f1}}{1-\mathcal{N}_{f1}}\right)^{\varrho}\right\}^{1/\varrho}},\Bigg)$$

$$\ldots,\left(\frac{1}{1+\left\{\sum_{f=1}^{2}\Xi_f\left(\frac{1-\mathcal{Y}_{fm}}{\mathcal{Y}_{fm}}\right)^{\varrho}\right\}^{1/\varrho}},1-\frac{1}{1+\left\{\sum_{f=1}^{2}\Xi_f\left(\frac{\mathcal{A}_{fm}}{1-\mathcal{A}_{fm}}\right)^{\varrho}\right\}^{1/\varrho}},\right.$$

$$\left.1-\frac{1}{1+\left\{\sum_{f=1}^{2}\Xi_f\left(\frac{\mathcal{N}_{fm}}{1-\mathcal{N}_{fm}}\right)^{\varrho}\right\}^{1/\varrho}}\right)\Bigg).$$

Thus the above result is valid for $f=2$.
Let us assume that the theorem holds for $f \geq r$, where $r \in \mathbb{N}$:

$$mPoPFDWG_{\Xi}(\mathcal{R}_1,\mathcal{R}_2,\ldots,\mathcal{R}_r)=\bigotimes_{f=1}^{r}\left(\mathcal{R}_f\right)^{\Xi_f}$$

$$=\Bigg(\Bigg(\frac{1}{1+\left\{\sum_{f=1}^{r}\Xi_f\left(\frac{1-\mathcal{Y}_{f1}}{\mathcal{Y}_{f1}}\right)^{\varrho}\right\}^{1/\varrho}},1-\frac{1}{1+\left\{\sum_{f=1}^{r}\Xi_f\left(\frac{\mathcal{A}_{f1}}{1-\mathcal{A}_{f1}}\right)^{\varrho}\right\}^{1/\varrho}},$$

$$1-\frac{1}{1+\left\{\sum_{f=1}^{r}\Xi_f\left(\frac{\mathcal{N}_{f1}}{1-\mathcal{N}_{f1}}\right)^{\varrho}\right\}^{1/\varrho}},\Bigg)$$

$$\ldots,\left(\frac{1}{1+\left\{\sum_{f=1}^{r}\Xi_f\left(\frac{1-\mathcal{Y}_{fm}}{\mathcal{Y}_{fm}}\right)^{\varrho}\right\}^{1/\varrho}},1-\frac{1}{1+\left\{\sum_{f=1}^{r}\Xi_f\left(\frac{\mathcal{A}_{fm}}{1-\mathcal{A}_{fm}}\right)^{\varrho}\right\}^{1/\varrho}},\right.$$

$$1-\frac{1}{1+\left\{\sum_{f=1}^{r}\Xi_f\left(\frac{\mathcal{N}_{fm*}}{1-\mathcal{N}_{fm}}\right)^{\varrho}\right\}^{1/\varrho}}\Bigg)\Bigg).$$

Now, for $f=r+1$,

$$mPoPFDWG_{\Xi}(\mathcal{R}_1,\mathcal{R}_2,\mathcal{R}_r\ldots,\mathcal{R}_{r+1})=\bigotimes_{f=1}^{r}\left(\mathcal{R}_f\right)^{\Xi_f}\otimes\left(\mathcal{R}_{r+1}\right)^{\Xi_{r+1}}$$

$$=\Bigg(\Bigg(\frac{1}{1+\left\{\sum_{f=1}^{r}\Xi_f\left(\frac{1-\mathcal{Y}_{f1}}{\mathcal{Y}_{f1}}\right)^{\varrho}\right\}^{1/\varrho}},1-\frac{1}{1+\left\{\sum_{f=1}^{r}\Xi_f\left(\frac{\mathcal{A}_{f1}}{1-\mathcal{A}_{f1}}\right)^{\varrho}\right\}^{1/\varrho}},$$

$$1-\frac{1}{1+\left\{\sum_{f=1}^{r}\Xi_f\left(\frac{\mathcal{N}_{f1}}{1-\mathcal{N}_{f1}}\right)^{\varrho}\right\}^{1/\varrho}}\Bigg),$$

$$\ldots,\Bigg(\frac{1}{1+\left\{\sum_{f=1}^{r}\Xi_f\left(\frac{1-\mathcal{Y}_{fm}}{\mathcal{Y}_{fm}}\right)^{\varrho}\right\}^{1/\varrho}},1-\frac{1}{1+\left\{\sum_{f=1}^{r}\Xi_f\left(\frac{\mathcal{A}_{fm}}{1-\mathcal{A}_{fm}}\right)^{\varrho}\right\}^{1/\varrho}},$$

$$1-\frac{1}{1+\left\{\sum_{f=1}^{r}\Xi_f\left(\frac{\mathcal{N}_{fm}}{1-\mathcal{N}_{fm}}\right)^{\varrho}\right\}^{1/\varrho}}\Bigg)\Bigg)$$

$$\bigoplus\Bigg(\Bigg(\frac{1}{1+\left\{\Xi_{r+1}\left(\frac{1-\mathcal{Y}_{(r+1)1}}{\mathcal{Y}_{(r+1)1}}\right)^{\varrho}\right\}^{1/\varrho}},1-\frac{1}{1+\left\{\Xi_{r+1}\left(\frac{\mathcal{A}_{(r+1)1}}{1-\mathcal{A}_{(r+1)1}}\right)^{\varrho}\right\}^{1/\varrho}},$$

$$1-\frac{1}{1+\left\{\Xi_{r+1}\left(\frac{\mathcal{N}_{(r+1)1}}{1-\mathcal{N}_{(r+1)1}}\right)^{\varrho}\right\}^{1/\varrho}},\Bigg)$$

$$\ldots,\Bigg(\frac{1}{1+\left\{\Xi_{r+1}\left(\frac{1-\mathcal{Y}_{(r+1)m}}{\mathcal{Y}_{(r+1)m}}\right)^{\varrho}\right\}^{1/\varrho}},1-\frac{1}{1+\left\{\Xi_{r+1}\left(\frac{\mathcal{A}_{(r+1)m}}{1-\mathcal{A}_{(r+1)m}}\right)^{\varrho}\right\}^{1/\varrho}},$$

$$1-\frac{1}{1+\left\{\Xi_{r+1}\left(\frac{\mathcal{N}_{(r+1)m}}{1-\mathcal{N}_{(r+1)m}}\right)^{\varrho}\right\}^{1/\varrho}}\Bigg)\Bigg)$$

$$=\Bigg(\Bigg(\frac{1}{1+\left\{\sum_{f=1}^{r+1}\Xi_f\left(\frac{1-\mathcal{Y}_{f1}}{\mathcal{Y}_{f1}}\right)^{\varrho}\right\}^{1/\varrho}},1-\frac{1}{1+\left\{\sum_{f=1}^{r+1}\Xi_f\left(\frac{\mathcal{A}_{f1}}{1-\mathcal{A}_{f1}}\right)^{\varrho}\right\}^{1/\varrho}},$$

$$1-\frac{1}{1+\left\{\sum_{f=1}^{r+1}\Xi_f\left(\frac{\mathcal{N}_{f1}}{1-\mathcal{N}_{f1}}\right)^{\varrho}\right\}^{1/\varrho}}\Bigg),$$

$$\ldots,\left(\frac{1}{1+\left\{\sum_{f=1}^{r+1}\Xi_f\left(\frac{1-\mathcal{Y}_{fm}}{\mathcal{Y}_{fm}}\right)^{\varrho}\right\}^{1/\varrho}},1-\frac{1}{1+\left\{\sum_{f=1}^{r+1}\Xi_f\left(\frac{\mathcal{A}_{fm}}{1-\mathcal{A}_{fm}}\right)^{\varrho}\right\}^{1/\varrho}},\right.$$

$$\left.\left.1-\frac{1}{1+\left\{\sum_{f=1}^{r+1}\Xi_f\left(\frac{\mathcal{N}_{fm}}{1-\mathcal{N}_{fm}}\right)^{\varrho}\right\}^{1/\varrho}}\right)\right).$$

Hence, Theorem 7.11 is true for all natural numbers. □

The mPoPFDWG operators obey the following properties.

**Theorem 7.12.** *(Idempotency property) Let* $\mathcal{R}_f = ((\mathcal{Y}_{f1},\mathcal{A}_{f1},\mathcal{N}_{f1}),(\mathcal{Y}_{f2},\mathcal{A}_{f2},\mathcal{N}_{f2}),\ldots,(\mathcal{Y}_{fm},\mathcal{A}_{fm},\mathcal{N}_{fm}))$, $f=1,2,\ldots,\varpi$ *be a set of mPoPFNs. If all* $mPoPFNs$ *are equal, i.e.,* $\mathcal{R}_f=\mathcal{R}$ *for all* $f=1,2,\ldots,\varpi$*, then:*

$$mPoPFDWG_{\Xi}(\mathcal{R}_1,\mathcal{R}_2,\ldots,\mathcal{R}_{\varpi})=\mathcal{R}. \tag{7.18}$$

**Theorem 7.13.** *(Boundedness property) Let* $\mathcal{R}_f = ((\mathcal{Y}_{f1},\mathcal{A}_{f1},\mathcal{N}_{f1}),(\mathcal{Y}_{f2},\mathcal{A}_{f2},\mathcal{N}_{f2}),\ldots,(\mathcal{Y}_{fm},\mathcal{A}_{fm},\mathcal{N}_{fm}))$, $f=1,2,\ldots,\varpi$ *be a set of mPoPFNs. Let* $\mathcal{R}^-=\min\{\mathcal{R}_1,\mathcal{R}_2,\ldots,\mathcal{R}_{\varpi}\}$ *and* $\mathcal{R}^+=\max\{\mathcal{R}_1,\mathcal{R}_2,\ldots,\mathcal{R}_{\varpi}\}$.

*Then,*

$$\mathcal{R}^-\le mPoPFDWG_{\Xi}(\mathcal{R}_1,\mathcal{R}_2,\ldots,\mathcal{R}_{\varpi})\le\mathcal{R}^+. \tag{7.19}$$

**Theorem 7.14.** *(Monotonicity property) Let* $\mathcal{R}_f = ((\mathcal{Y}_{f1},\mathcal{A}_{f1},\mathcal{N}_{f1}),(\mathcal{Y}_{f2},\mathcal{A}_{f2},\mathcal{N}_{f2}),\ldots,(\mathcal{Y}_{fm},\mathcal{A}_{fm},\mathcal{N}_{fm}))$ *and* $\mathcal{R}'_f=((\mathcal{Y}'_{f1},\mathcal{A}'_{f1},\mathcal{N}'_{f1}),(\mathcal{Y}'_{f2},\mathcal{A}'_{f2},\mathcal{N}'_{f2}),\ldots,(\mathcal{Y}'_{fm},\mathcal{A}'_{fm},\mathcal{N}'_{fm}))$, $f=1,2,\ldots,\varpi$ *be two sets of mPoPFNs such that* $\mathcal{R}_f\le\mathcal{R}'_f$ *for all* $f$*, then*

$$mPoPFDWG_{\Xi}(\mathcal{R}_1,\mathcal{R}_2,\ldots,\mathcal{R}_{\varpi})\le mPoPFDWG_{\Xi}(\mathcal{R}'_1,\mathcal{R}'_2,\ldots,\mathcal{R}'_{\varpi}). \tag{7.20}$$

Next, we proceed with the properties of the mPoPFDOWG operator.

**Definition 7.12.** Let $\mathcal{R}_f = ((\mathcal{Y}_{f1},\mathcal{A}_{f1},\mathcal{N}_{f1}),(\mathcal{Y}_{f2},\mathcal{A}_{f2},\mathcal{N}_{f2}),\ldots,(\mathcal{Y}_{fm},\mathcal{A}_{fm},\mathcal{N}_{fm}))$, $f=1,2,\ldots,\varpi$ be a set of mPoPFNs. Then, the Dombi ordered weighted geometric (mPoPF-DOWG) operator of mPoPFNs introduces a mapping $mPoPFDOWG:\Theta^{\varpi}\to\Theta$ defined as:

$$mPoPFDOWG_{\delta}(\mathcal{R}_1,\mathcal{R}_2,\ldots,\mathcal{R}_{\varpi})=\bigotimes_{f=1}^{\varpi}\left(\mathcal{R}_{\sigma(f)}\right)^{\delta_f}, \tag{7.21}$$

where $\delta = (\delta_1, \delta_2, \ldots, \delta_\varpi)^T$ is a set of weight vectors on $\mathcal{R}_f$ such that $\delta_f > 0$ and $\sum_{f=1}^{\varpi} \delta_f = 1$. Also, $(\sigma(1), \sigma(2), \ldots, \sigma(\varpi))$ is the permutation of $(1, 2, \ldots, \varpi)$ for which $\mathcal{R}_{\sigma(f-1)} \geq \mathcal{R}_{\sigma(f)}$ for all $f = 1, 2, \ldots, \varpi$.

**Theorem 7.15.** *Let $\mathcal{R}_f = ((\mathcal{Y}_{f1}, \mathcal{A}_{f1}, \mathcal{N}_{f1}), (\mathcal{Y}_{f2}, \mathcal{A}_{f2}, \mathcal{N}_{f2}), \ldots, (\mathcal{Y}_{fm}, \mathcal{A}_{fm}, \mathcal{N}_{fm}))$, $f = 1, 2, \ldots, \varpi$ be a set of mPoPFNs. Then, the collected value of all mPoPFNs under the mPoPF-DOWG operator is an mPoPFN, which is given by:*

$$
\begin{aligned}
& mPoPFDOWG_\delta(\mathcal{R}_1, \mathcal{R}_2, \ldots, \mathcal{R}_\varpi) \\
&= \bigotimes_{f=1}^{\varpi} \left(\mathcal{R}_{\sigma(f)}\right)^{\delta_f} \\
&= \left( \left( \frac{1}{1+\left\{\sum_{f=1}^{\varpi} \delta_f \left(\frac{1-\mathcal{Y}_{\sigma(f1)}}{\mathcal{Y}_{\sigma(f1)}}\right)^{\varrho}\right\}^{1/\varrho}}, 1-\frac{1}{1+\left\{\sum_{f=1}^{\varpi} \delta_f \left(\frac{\mathcal{A}_{\sigma(f1)}}{1-\mathcal{A}_{\sigma(f1)}}\right)^{\varrho}\right\}^{1/\varrho}}, \right.\right. \\
&\quad \left. 1-\frac{1}{1+\left\{\sum_{f=1}^{\varpi} \delta_f \left(\frac{\mathcal{N}_{\sigma(f1)}}{1-\mathcal{N}_{\sigma(f1)}}\right)^{\varrho}\right\}^{1/\varrho}} \right), \\
&\quad \ldots, \left( \frac{1}{1+\left\{\sum_{f=1}^{\varpi} \delta_f \left(\frac{1-\mathcal{Y}_{\sigma(fm)}}{\mathcal{Y}_{\sigma(f)}}\right)^{\varrho}\right\}^{1/\varrho}}, 1-\frac{1}{1+\left\{\sum_{f=1}^{\varpi} \delta_f \left(\frac{\mathcal{A}_{\sigma(fm)}}{1-\mathcal{A}_{\sigma(fm)}}\right)^{\varrho}\right\}^{1/\varrho}}, \right. \\
&\quad \left.\left. 1-\frac{1}{1+\left\{\sum_{f=1}^{\varpi} \delta_f \left(\frac{\mathcal{N}_{\sigma(fm)}}{1-\mathcal{N}_{\sigma(fm)}}\right)^{\varrho}\right\}^{1/\varrho}} \right)\right), \qquad (7.22)
\end{aligned}
$$

*where $\delta = (\delta_1, \delta_2, \ldots, \delta_\varpi)$ is the weight vector of $\mathcal{R}_f$, $f = 1, 2, \ldots, \varpi$ such that $\delta_f > 0$, and $\sum_{f=1}^{\varpi} \delta_f = 1$. Also, $(\sigma(1), \sigma(2), \ldots, \sigma(\zeta))$ is the permutation for $(1, 2, \ldots, \varpi)$ for which $\mathcal{R}_{\sigma(f-1)} \geq \mathcal{R}_{\sigma(f)}$ for all $f = 1, 2, \ldots, \varpi$.*

The mPoPFDOWG operator displayed the following properties.

**Theorem 7.16.** *(Idempotency property) Let $(\mathcal{R}_1, \mathcal{R}_2, \ldots, \mathcal{R}_\varpi)$ be a set of mPoPFNs. If $\mathcal{R}_f = \mathcal{R}$ for all $f = 1, 2, \ldots, \varpi$, then*

$$mPoPFDWG_\delta(\mathcal{R}_1, \mathcal{R}_2, \ldots, \mathcal{R}_\varpi) = \mathcal{R}. \qquad (7.23)$$

**Theorem 7.17.** *(Boundedness property) Let $(\mathcal{R}_1, \mathcal{R}_2, \ldots, \mathcal{R}_\varpi)$ be a set of mPoPFNs and $\mathcal{R}^- = \min\{\mathcal{R}_1, \mathcal{R}_2, \ldots, \mathcal{R}_\varpi\}$ and $\mathcal{R}^+ = \max\{\mathcal{R}_1, \mathcal{R}_2, \ldots, \mathcal{R}_\varpi\}$, then*

$$\mathcal{R}^- \leq mPoPFDWG_\delta(\mathcal{R}_1, \mathcal{R}_2, \ldots, \mathcal{R}_\varpi) \leq \mathcal{R}^+. \qquad (7.24)$$

**Theorem 7.18.** *(Monotonicity property) Let* $(\mathcal{R}_1, \mathcal{R}_2, \ldots, \mathcal{R}_\varpi)$ *and* $(\mathcal{R}'_1, \mathcal{R}'_2, \ldots, \mathcal{R}'_\varpi)$ *be two sets of mPoPFNs. If for all* $f = 1, 2, \ldots, \varpi$, $\mathcal{R}_f \leq \mathcal{R}'_f$, *then*

$$mPoPFDWG_\delta(\mathcal{R}_1, \mathcal{R}_2, \ldots, \mathcal{R}_\varpi) \leq mPoPFDWG_\delta(\mathcal{R}'_1, \mathcal{R}'_2, \ldots, \mathcal{R}'_\varpi). \tag{7.25}$$

**Theorem 7.19.** *(Commutative property) Let* $(\mathcal{R}_1, \mathcal{R}_2, \ldots, \mathcal{R}_\varpi)$ *and* $(\mathcal{R}'_1, \mathcal{R}'_2, \ldots, \mathcal{R}'_\varpi)$ *be two sets of mPoPFNs. Then,*

$$mPoPFDWG_\delta(\mathcal{R}_1, \mathcal{R}_2, \ldots, \mathcal{R}_\varpi) = mPoPFDWG_\delta(\mathcal{R}'_1, \mathcal{R}'_2, \ldots, \mathcal{R}'_\varpi), \tag{7.26}$$

*where* $\mathcal{R}'_f$ *is an arbitrary permutation of* $\mathcal{R}_f$ *for all* $(1, 2, \ldots, \varpi)$.

In Definitions 7.11 and 7.12, the mPoPFDWG operator used the weights of the mPoPFN; once more, the weight of the mPoPFDOWG operator suggests the ordered position of the mPoPFN rather than the weights of the mPoPFN themselves. As a result, we define another operator, the mPoPFN Dombi hybrid averaging operator (mPoPFDHWG), which makes qualitative use of both the mPoPFDWG and mPoPFDOWG operators.

**Definition 7.13.** Let $(\mathcal{R}_1, \mathcal{R}_2, \ldots, \mathcal{R}_\varpi)$ be a set of mPoPFNs. Then, the Dombi hybrid weighted geometric (mPoPFDHWG) operator is a function $mPoPFDHWG : \Theta^\varpi \to \Theta$ that is defined as:

$$mPoPFDHWG_{\Xi,\delta}(\mathcal{R}_1, \mathcal{R}_2, \ldots, \mathcal{R}_\varpi) = \bigotimes_{f=1}^{\varpi} \left(\dot{\mathcal{R}}_{\sigma(f)}\right)^{\delta_f}. \tag{7.27}$$

$(\sigma(1), \sigma(2), \ldots, \sigma(\varpi))$ is a permutation on $1, 2, \ldots, \varpi$ for which $\mathcal{R}_{\sigma(f-1)} \geq \mathcal{R}_{\sigma(f)}$ for all $f = 1, 2, \ldots, \varpi$. $\delta = (\delta_1, \delta_2, \ldots, \delta_\varpi)^T$ is the weight vector for $(\mathcal{R}_1, \mathcal{R}_2, \ldots, \mathcal{R}_\varpi)$ such that $\delta_f > 0$ and $\sum_{f=1}^{\varpi} \delta_f = 1$. $\dot{\mathcal{R}}_f$ is the largest mPoPFN, where, $\dot{\mathcal{R}}_f = (\varpi\, \Xi)\mathcal{R}_\varpi$, $f = 1, 2, \ldots, \varpi$ for which the weight vector $\Xi = (\Xi_1, \Xi_2, \ldots, \Xi_\varsigma)^T$ such that $\Xi_f > 0$ and $\sum_{f=1}^{\varpi} \Xi_f = 1$.

**Theorem 7.20.** *Let* $\mathcal{R}_f = ((\mathcal{Y}_{f1}, \mathcal{A}_{f1}, \mathcal{N}_{f1}), (\mathcal{Y}_{f2}, \mathcal{A}_{f2}, \mathcal{N}_{f2}), \ldots, (\mathcal{Y}_{fm}, \mathcal{A}_{fm}, \mathcal{N}_{fm}))$, $f = 1, 2, \ldots, \varpi$ *be a set of mPoPFNs. Then, the collected values of mPoPFNs using the mPoPFDHWG operator are also an mPoPFN. Further:*

$$\begin{aligned}
&mPoPFDHWG_{\Xi,\delta}(\mathcal{R}_1, \mathcal{R}_2, \ldots, \mathcal{R}_\varpi) \\
&= \bigotimes_{f=1}^{\varpi} \left(\dot{\mathcal{R}}_{\sigma(fg)}\right)^{\delta_f} \\
&= \left( \left( \frac{1}{1 + \left\{ \sum_{f=1}^{\varpi} \delta_f \left( \frac{1 - \dot{\mathcal{Y}}_{\sigma(f1)}}{\dot{\mathcal{Y}}_{\sigma(f1)}} \right)^{\varrho} \right\}^{1/\varrho}}, 1 - \frac{1}{1 + \left\{ \sum_{f=1}^{\varpi} \delta_f \left( \frac{\dot{\mathcal{A}}_{\sigma(f1)}}{1 - \dot{\mathcal{A}}_{\sigma(f1)}} \right)^{\varrho} \right\}^{1/\varrho}}, \right.\right. \\
&\qquad \left. 1 - \frac{1}{1 + \left\{ \sum_{f=1}^{\varpi} \delta_f \left( \frac{\dot{\mathcal{N}}_{\sigma(f1)}}{1 - \dot{\mathcal{N}}_{\sigma(f)}} \right)^{\varrho} \right\}^{1/\varrho}} \right),
\end{aligned}$$

$$\ldots,\left(\frac{1}{1+\left\{\sum_{f=1}^{\varpi}\delta_f\left(\frac{1-\dot{\mathcal{Y}}_{\sigma(fm)}}{\dot{\mathcal{Y}}_{\sigma(fm)}}\right)^{\varrho}\right\}^{1/\varrho}},1-\frac{1}{1+\left\{\sum_{f=1}^{\varpi}\delta_f\left(\frac{\dot{\mathcal{A}}_{\sigma(fm)}}{1-\dot{\mathcal{A}}_{\sigma(fm)}}\right)^{\varrho}\right\}^{1/\varrho}},\right.$$

$$\left.\left.1-\frac{1}{1+\left\{\sum_{f=1}^{\varpi}\delta_f\left(\frac{\dot{\mathcal{N}}_{\sigma(f)}}{1-\dot{\mathcal{N}}_{\sigma(fm)}}\right)^{\varrho}\right\}^{1/\varrho}}\right)\right).$$

## 7.6 Model for MADM using mPoPF data

Let $\aleph=\{\aleph_1,\aleph_2,\ldots,\aleph_{\varpi}\}$ be the alternatives set and $A=\{A_1,A_2,\ldots,A_{\xi}\}$ be a set of attributes. Let $\Xi=(\Xi_1,\Xi_2,\ldots,\Xi_{\varpi})$ be the weighting vector of the attributes $A_f$, $f=1,2,\ldots,\varpi$ is assigned by DMs such that $\Xi_f>0$ and $\sum_{f=1}^{\varpi}\Xi_f=1$. Suppose that $\widetilde{D}=((\mathcal{Y}_{gf1},\mathcal{A}_{gf1},\mathcal{N}_{gf1}),(\mathcal{Y}_{gf2},\mathcal{A}_{gf2},\mathcal{N}_{gf2}),\ldots,(\mathcal{Y}_{gfm},\mathcal{A}_{gfm},\mathcal{N}_{gfm}))_{(\xi\times\varpi)}$ is an mPoPFN decision matrix, where $\mathcal{Y}_r$ is the supported membership degree, $\mathcal{A}_r$ is the neutral membership degree, and $\mathcal{N}_r$ is the nonmembership degree for alternatives $\aleph_f$ that fulfils the attribute $A_{\varpi}$ and $\mathcal{R}_r * \aleph$.

In the following algorithm, we propose to solve a MADM problem using $mPFS$ data using mPFDWA and mPFDWG operators.

**Step 1.** We employ the decision matrix $D$, and the operator mPoPFDWA:

$$mPoPFDWA_{\Xi}(\mathcal{R}_{g1},\mathcal{R}_{g2},\ldots,\mathcal{R}_{gf})=\bigoplus_{f=1}^{\varpi}\left(\Xi_f\mathcal{R}_f\right)$$

$$=\left(\left(1-\frac{1}{1+\left\{\sum_{f=1}^{\varpi}\Xi_f\left(\frac{\mathcal{Y}_{11}}{1-\mathcal{Y}_{11}}\right)^{\varrho}\right\}^{1/\varrho}},\frac{1}{1+\left\{\sum_{f=1}^{\varpi}\Xi_f\left(\frac{1-\mathcal{A}_{11}}{\mathcal{A}_{11}}\right)^{\varrho}\right\}^{1/\varrho}},\frac{1}{1+\left\{\sum_{f=1}^{\varpi}\Xi_f\left(\frac{1-\mathcal{N}_{11}}{\mathcal{N}_{11}}\right)^{\varrho}\right\}^{1/\varrho}}\right),\right.$$

$$\left.\ldots,\left(1-\frac{1}{1+\left\{\sum_{f=1}^{\varpi}\Xi_f\left(\frac{\mathcal{Y}_{1m}}{1-\mathcal{Y}_{1m}}\right)^{\varrho}\right\}^{1/\varrho}},\frac{1}{1+\left\{\sum_{f=1}^{\varpi}\Xi_f\left(\frac{1-\mathcal{A}_{1m}}{\mathcal{A}_{1m}}\right)^{\varrho}\right\}^{1/\varrho}},\frac{1}{1+\left\{\sum_{f=1}^{\varpi}\Xi_f\left(\frac{1-\mathcal{N}_{1m}}{\mathcal{N}_{1m}}\right)^{\varrho}\right\}^{1/\varrho}}\right)\right),$$

or

$$mPoPFDWG_{\Xi}(\mathcal{R}_{g1},\mathcal{R}_{g2},\ldots,\mathcal{R}_{gf})=\bigotimes_{f=1}^{\varpi}\left(\mathcal{R}_f\right)^{\Xi_f}$$

$$= \left( \left( \frac{1}{1+\left\{\sum_{f=1}^{\varpi} \Xi_f \left(\frac{1-\mathcal{Y}_{11}}{\mathcal{Y}_{11}}\right)^{\varrho}\right\}^{1/\varrho}}, 1 - \frac{1}{1+\left\{\sum_{f=1}^{\varpi} \Xi_f \left(\frac{\mathcal{A}_{11}}{1-\mathcal{A}_{11}}\right)^{\varrho}\right\}^{1/\varrho}}, \right.\right.$$

$$\left. 1 - \frac{1}{1+\left\{\sum_{f=1}^{\varpi} \Xi_f \left(\frac{\mathcal{N}_{11}}{1-\mathcal{N}_{11}}\right)^{\varrho}\right\}^{1/\varrho}} \right),$$

$$\ldots, \left( \frac{1}{1+\left\{\sum_{f=1}^{\varpi} \Xi_f \left(\frac{1-\mathcal{Y}_{1m}}{\mathcal{Y}_{1m}}\right)^{\varrho}\right\}^{1/\varrho}}, 1 - \frac{1}{1+\left\{\sum_{f=1}^{\varpi} \Xi_f \left(\frac{\mathcal{A}_{1m}}{1-\mathcal{A}_{1m}}\right)^{\varrho}\right\}^{1/\varrho}}, \right.$$

$$\left.\left. 1 - \frac{1}{1+\left\{\sum_{f=1}^{\varpi} \Xi_f \left(\frac{\mathcal{N}_{1m}}{1-\mathcal{N}_{1m}}\right)^{\varrho}\right\}^{1/\varrho}} \right)\right)$$

to compute the accumulated values of $\mathcal{R}_g, g = 1, 2, \ldots, \xi$ of the alternatives $\aleph_g$.

Let $\Upsilon_g = mPoPFDWA_{\Xi}(\mathcal{R}_{g1}, \mathcal{R}_{g2}, \ldots, \mathcal{R}_{gf})$ and $\Upsilon_g = mPoPFDWG_{\Xi}(\mathcal{R}_{g1}, \mathcal{R}_{g2}, \ldots, \mathcal{R}_{gf})$.

**Step 2.** Compute the score function $\Lambda(\Upsilon_g)$, $g = 1, 2, \ldots, \xi$ based on the overall mPoPF information $\Upsilon_g$, $g = 1, 2, \ldots, \xi$ in order to rank all the alternative $\aleph_g$, $g = 1, 2, \ldots, \xi$ to select the best choice $\aleph_g$. If there is no variation of $\Lambda(\Upsilon_g)$ and $\Lambda(\Upsilon_f)$, then we compute the value of $\Phi(\Upsilon_g)$ and $\Phi(\Upsilon_f)$ based on the overall mPoPF information of $\Upsilon_g$ and $\Upsilon_f$, and rank the alternatives $\aleph_g$ depending on the accuracy degrees of $\Phi(\Upsilon_g)$ and $\Phi(\Upsilon_f)$.

**Step 3.** Rank all the alternatives $\aleph_g$, $g = 1, 2, \ldots, \xi$ in order to choose the leading one(s) in a manner conforming to $\Lambda(\Upsilon_g, g = 1, 2, \ldots, \xi$.

**Step 4.** Stop.

## 7.7 Numerical example

We utilize the proposed approach to solve a MADM problem in this part.

### 7.7.1 Selection of suitable location for a thermal power station

Thermal power stations are used to generate electricity by changing chemical energy into mechanical energy. A company selects a suitable location for that purpose. Assume that the company has five possible locations $\aleph_1$, $\aleph_2$, $\aleph_3$, $\aleph_4$, and $\aleph_5$ treated as alternatives for possible thermal power stations. With an expert team of engineers, the company selects the best alternative. The expert team of engineers select the best location under the following main criteria:

$A_1$: Infrastructures;
$A_2$: Environmental conditions;
$A_3$: Social impact;
$A_4$: Governmental policies.

Then, each of the criteria is subdivided into three subcriteria to form a 3-polar fuzzy set. The Infrastructures depend on the availability of coal, water, and transportation facilities. The Environmental conditions depend on ambient temperature, humidity, and air velocities. The Social impact depends on education facilities, hospital facilities, and healthcare facilities. The Governmental policies subdivide into licensing policies, institutional finance, and Government subsidies. They have no dominant power over each other; decision makers (DM) are required to rank the five possible locations $\aleph_g$, $g = 1, 2, \ldots, 4, 5$ under the mentioning attributes whose weights $\Xi = (0.4, 0.3, 0.1, 0.2)^T$ addressed by DM, the decision matrix $D = (\Upsilon_{gf})_{5\times 4}$, which is provided in Table 7.1, where $\Upsilon_{gf}$ are in the form of 3PoFPNs.

In order to chose suitable location $\aleph_g$, $g = 1, 2, \ldots, \xi$, we apply the mPoPFDWA and mPoPFDWG operators to model MADM, which can be computed as follows:

- **Step 1.** Let $\varrho = 1$, using the mPoPFDWA operator overall accumulated values $\Upsilon_g$ of $\aleph_g$, $g = 1, 2, \ldots, 5$, are:

$$\Upsilon_1 = \Big((0.4815, 0.1481, 0.1200), (0.4070, 0.1429, 0.2273), (0.4581, 0.1714, 0.1333)\Big),$$
$$\Upsilon_2 = \Big((0.5839, 0.1022, 0.08433), (0.2877, 0.2500, 0.2105), (0.3135, 0.1818, 0.2105)\Big),$$
$$\Upsilon_3 = \Big((0.5200, 0.1657, 0.1072), (0.4917, 0.1475, 0.1622), (0.3300, 0.3000, 0.1244)\Big),$$
$$\Upsilon_4 = \Big((0.3636, 0.1096, 0.0703), (0.3406, 0.2667, 0.0622), (0.3703, 0.2000, 0.2439)\Big),$$
$$\Upsilon_5 = \Big((0.3708, 0.0833, 0.1840), (0.6591, 0.0929, 0.0759), (0.3000, 0.1538, 0.1176)\Big).$$

- **Step 2.** Score values of $\Lambda(\Upsilon_g)$, $g = 1, 2, \ldots, 5$ are as follows:

$$\Lambda(\Upsilon_1) = 0.4679,\ \Lambda(\Upsilon_2) = 0.3819,\ \Lambda(\Upsilon_3) = 0.4449,$$
$$\Lambda(\Upsilon_4) = 0.4268,\ \Lambda(\Upsilon_5) = 0.5408.$$

- **Step 3.** Ranking results of $\aleph_g$, $g = 1, 2, \ldots, 5$ in accordance with the score values of $\Lambda(\Upsilon_g)$, $g = 1, 2, \ldots, 5$ of the overall mPoPFNs as $\aleph_5 \succ \aleph_1 \succ \aleph_3 \succ \aleph_4 \succ \aleph_2$.
- **Step 4.** $\aleph_5$ is suggested as the most favorable location.

If the mPoPFDWG operator is used instead in a similar manner, the problem is solved as follows:

- **Step 1.** Let $\varrho = 1$, using the mPoPFDWG operator to compute the overall accumulated values of $\aleph_g$, $g = 1, 2, \ldots, 5$

$$\Upsilon_1 = \Big((0.4268, 0.1708, 0.1523), (0.2486, 0.1628, 0.3054), (0.3429, 0.2186, 0.1805)\Big),$$
$$\Upsilon_2 = \Big((0.2809, 0.1244, 0.2655), (0.1402, 0.2941, 0.2653), (0.1799, 0.1910, 0.2704)\Big),$$
$$\Upsilon_3 = \Big((0.5000, 0.3285, 0.1416), (0.2449, 0.1883, 0.1962), (0.1585, 0.4337, 0.2092)\Big),$$

**Table 7.1** 3-polar picture fuzzy decision matrix.

| | $\aleph_1$ | $\aleph_2$ | $\aleph_3$ |
|---|---|---|---|
| $A_1$ | ((0.5, 0.2, 0.1), (.4, 0.1, 0.2), (0.6, 0.3, 0.1)) | ((0.7, 0.1, 0.06), (0.3, 0.4, 0.2), (0.4, 0.2, 0.2)) | ((0.5, 0.4, 0.07), (0.6, 0.2, 0.2), (0.3, 0.3, 0.2)) |
| $A_2$ | ((0.3, 0.2, 0.2), (0.5, 0.2, 0.3), (0.3, 0.2, 0.3)) | ((0.5, 0.07, 0.08), (0.07, 0.2, 0.4), (.2, 0.2, 0.3)) | ((0.6, 0.07, 0.2), (0.4, 0.1, 0.1), (0.2, 0.6, 0.07)) |
| $A_3$ | ((0.5, 0.08, 0.3), (0.5, 0.2, 0.1), (0.4, 0.1, 0.2)) | ((0.6, 0.2, 0.09), (0.09, 0.2, 0.1), (0.5, 0.1, 0.08)) | ((0.5, 0.4, 0.09), (0.6, 0.09, 0.2), (0.7, 0.1, 0.08)) |
| $A_4$ | ((0.6, 0.1, 0.08), (0.09, 0.2, 0.5), (0.2, 0.1, 0.1)) | ((0.09, 0.2, 0.6), (0.5, 0.2, 0.2), (0.07, 0.2, 0.4)) | ((0.4, 0.4, 0.2), (0.08, 0.3, 0.3), (0.06, 0.4, 0.4)) |
| | $\aleph_4$ | $\aleph_5$ | |
| | ((0.6, 0.07, 0.08), (0.3, 0.3, 0.03), (0.5, 0.2, 0.2)) | ((0.4, 0.06, 0.3), (0.6, 0.05, 0.04), (0.2, 0.1, 0.1)) | |
| | ((0.4, 0.05, 0.06), (0.08, 0.4, 0.4), (0.09, 0.3, 0.5)) | ((0.07, 0.2, 0.5), (0.5, 0.5, 0.3), (0.3, 0.2, 0.1)) | |
| | ((0.3, 0.4, 0.1), (0.7, 0.1, 0.1), (0.2, 0.1, 0.2)) | ((0.5, 0.2, 0.1), (0.7, 0.6, 0.6), (0.4, 0.3, 0.2)) | |
| | ((0.2, 0.05, 0.05), (0.3, 0.3, 0.2), (0.4, 0.2, 0.2)) | ((0.5, 0.06, 0.08), (0.8, 0.1, 0.1), (0.4, 0.3, 0.2)) | |

$$\Upsilon_4 = \Big((0.3636, 0.1096, 0.0703), (0.1697, 0.3189, 0.2148), (0.1548, 0.2246, 0.3220)\Big),$$
$$\Upsilon_5 = \Big((0.1699, 0.1215, 0.3333), (0.6026, 0.3303, 0.2410), (0.2667, 0.1987, 0.1325)\Big).$$

- **Step 2.** Compute the score values $\Lambda(\Upsilon_a)$ of $\Upsilon_g$, $g = 1, 2, \ldots, 5$ as:

$$\Lambda(\Upsilon_1) = 0.2766,\ \Lambda(\Upsilon_2) = 0.0643,\ \Lambda(\Upsilon_3) = 0.1353,$$
$$\Lambda(\Upsilon_4) = 0.1559,\ \Lambda(\Upsilon_5) = 0.2273.$$

- **Step 3.** Ranking all $\aleph_g$, $g = 1, 2, \ldots, 5$ in the value of the score functions $\Lambda(\Upsilon_g)$, $g = 1, 2, \ldots, 5$ of the overall mPFNs as $\aleph_1 \succ \aleph_5 \succ \aleph_4 \succ \aleph_3 \succ \aleph_2$.
- **Step 4.** Return $\aleph_1$ is selected as the suitable location.

From the above computation, it shows from the ranking order of the alternatives that $\aleph_2$ is the most desirable location for the mPoPFDWA operator, and $\aleph_1$ is the suitable location for the mPoPFDWG operator.

We diagnose the sensitivity of the parameter $\varrho \in [1, 10]$ on the ranking order of the alternatives $\aleph$ using the mPoPFDWA and mPoPFDWG operators, which are given in Tables 7.2 and 7.3.

### 7.7.2 Analysis on the effect of parameter $\varrho$ on decision-making results

Here, the operational behaviors of the working parameter $\varrho$ on MADM based on the mPoPFDWA and mPoPFDWG operators are addressed in Table 7.2 and Table 7.3. Table 7.2 shows that when $\varrho$ is changed for the mPoPFDWA operator, the corresponding favorable alternative does not remain the same. Thus when $\varrho = 1$, then $\aleph_5 \succ \aleph_1 \succ \aleph_3 \succ \aleph_4 \succ \aleph_2$. When $2 \leq \varrho \leq 3$, then $\aleph_5 \succ \aleph_3 \succ \aleph_1 \succ \aleph_4 \succ \aleph_2$, and it shows that $\aleph_5$ is the best alternative. When, $3 \leq \varrho \leq 10$, it shows the ranking $\aleph_3 \succ \aleph_5 \succ \aleph_4 \succ \aleph_1 \succ \aleph_2$, in this case, the best one is $\aleph_3$. In Table 7.3, when $\varrho = 1, 2$ is changed for the mPoPFDWG operator, the ranking order is $\aleph_1 \succ \aleph_5 \succ \aleph_4 \succ \aleph_3 \succ \aleph_2$. Also, for $3 \leq \varrho \leq 10$ then the ranking orders is as $\aleph_1 \succ \aleph_5 \succ \aleph_3 \succ \aleph_4 \succ \aleph_2$. Thus $\aleph_1$ is again the best alternative for a suitable location.

The proposed MADM problems based on the mPoPFDWA and mPoPFDWG operators, show that for different values of the working parameter $\varrho$ can change the corresponding ranking orders of the alternatives for the mPoPFDWA operator, which is more responsive to $\varrho$, while various values of the parameter $\varrho$ could not change the ranking forms corresponding to the mPoPFDWG operator, which is less effective by $\varrho$ for the so-called MADM model.

## 7.8 Conclusion

This chapter used fuzzy m-polar picture information to study MADM issues. We introduce the Dombi weighted, order weighted, hybrid weighted, weighted geometric, order weighted geometric, and hybrid weighted geometric operators. In order to create new aggregation operators employing Dombi norms, such as mPoPFNs, we presented weighted

**Table 7.2** Effect of parameters on ranking orders using the mPoPFDWA operator.

| $\varrho$ | $\Lambda(\Upsilon_1)$, $\Lambda(\Upsilon_2)$, $\Lambda(\Upsilon_3)$, $\Lambda(\Upsilon_4)$, $\Lambda(\Upsilon_5)$ | Ranking order |
|---|---|---|
| 1 | 0.4679, 0.3819, 0.4449, 0.4268, 0.5408 | $\aleph_5 \succ \aleph_1 \succ \aleph_3 \succ \aleph_4 \succ \aleph_2$ |
| 2 | 0.5273, 0.4577, 0.5560, 0.5207, 0.6050 | $\aleph_5 \succ \aleph_3 \succ \aleph_1 \succ \aleph_4 \succ \aleph_2$ |
| 3 | 0.5675, 0.6244, 0.5078, 0.5836, 0.6413 | $\aleph_5 \succ \aleph_3 \succ \aleph_1 \sim \aleph_4 \succ \aleph_2$ |
| 4 | 0.5950, 0.5432, 0.6651, 0.6235, 0.6650 | $\aleph_3 \succ \aleph_5 \succ \aleph_4 \succ \aleph_1 \succ \aleph_2$ |
| 5 | 0.6145, 0.5688, 0.6909, 0.6495, 0.6713 | $\aleph_3 \succ \aleph_5 \succ \aleph_4 \succ \aleph_1 \succ \aleph_2$ |
| 6 | 0.6287, 0.5877, 0.7086, 0.6672, 0.6918 | $\aleph_3 \succ \aleph_5 \succ \aleph_4 \succ \aleph_1 \succ \aleph_2$ |
| 7 | 0.6389, 0.6022, 0.7214, 0.6801, 0.7005 | $\aleph_3 \succ \aleph_5 \succ \aleph_4 \succ \aleph_1 \succ \aleph_2$ |
| 8 | 0.6393, 0.6134, 0.7310, 0.6899, 0.7060 | $\aleph_3 \succ \aleph_5 \succ \aleph_4 \succ \aleph_1 \succ \aleph_2$ |
| 9 | 0.6548, 0.6224, 0.7385, 0.6974, 0.7123 | $\aleph_3 \succ \aleph_5 \succ \aleph_4 \succ \aleph_1 \succ \aleph_2$ |
| 10 | 0.6604, 0.6297, 0.7446, 0.7025, 0.7167 | $\aleph_3 \succ \aleph_5 \succ \aleph_4 \succ \aleph_1 \succ \aleph_2$ |

**Table 7.3** The effect of parameters on ranking order using the mPoPFDWG operator.

| $\varrho$ | $\Lambda(\Upsilon_1)$, $\Lambda(\Upsilon_2)$, $\Lambda(\Upsilon_3)$, $\Lambda(\Upsilon_4)$, $\Lambda(\Upsilon_5)$ | Ranking order |
|---|---|---|
| 1 | 0.2766, 0.0634, 0.1353, 0.1559, 0.2273 | $\aleph_1 \succ \aleph_5 \succ \aleph_4 \succ \aleph_3 \succ \aleph_2$ |
| 2 | 0.1899, −0.0704, 0.0349, 0.0446, 0.1028 | $\aleph_1 \succ \aleph_5 \succ \aleph_4 \succ \aleph_3 \succ \aleph_2$ |
| 3 | 0.1315, −0.1378, −0.0238, −0.0263, 0.0278 | $\aleph_1 \succ \aleph_5 \succ \aleph_3 \succ \aleph_4 \succ \aleph_2$ |
| 4 | 0.0919, −0.1793, −0.0613, −0.0706, −0.0192 | $\aleph_1 \succ \aleph_5 \succ \aleph_3 \succ \aleph_4 \succ \aleph_2$ |
| 5 | 0.0642, −0.2058, −0.0875, −0.1003, −0.0515 | $\aleph_1 \succ \aleph_5 \succ \aleph_3 \succ \aleph_4 \succ \aleph_2$ |
| 6 | 0.0439, −0.2243, −0.1075, −0.1215, −0.0749 | $\aleph_1 \succ \aleph_5 \succ \aleph_3 \succ \aleph_4 \succ \aleph_2$ |
| 7 | 0.0287, −0.2380, −0.1212, −0.1373, −0.0925 | $\aleph_1 \succ \aleph_5 \succ \aleph_3 \succ \aleph_4 \succ \aleph_2$ |
| 8 | 0.0169, −0.2483, −0.1326, −0.1495, −0.1062 | $\aleph_1 \succ \aleph_5 \succ \aleph_3 \succ \aleph_4 \succ \aleph_2$ |
| 9 | 0.0076, −0.2526, −0.1419, −0.1593, −0.1172 | $\aleph_1 \succ \aleph_5 \succ \aleph_3 \succ \aleph_4 \succ \aleph_2$ |
| 10 | −0.0001, −0.2633, −0.1493, −0.1673, −0.1261 | $\aleph_1 \succ \aleph_5 \succ \aleph_3 \succ \aleph_4 \succ \aleph_2$ |

averaging and weighted geometric aggregation operators with mPoPFNs. Based on these aggregation operators, a multicriteria decision-making issue was developed. We believe the presented model can be used to create economic models, business and management issues, intelligent diagnosis, three-way decision making, and other uncertain contexts.

# References

[1] N. Waseem, M. Akram, J. Carlose, R. Alcantud, Multi-attribute decision-making based on m-polar fuzzy Hamacher aggregation operators, Symmetry 11 (2019) 1498.
[2] M. Akram, N. Waseem, Novel approach in decision-making with m-polar fuzzy ELECTRE-I, Int. J. Fuzzy Syst. 21 (2019) 1117–1129.
[3] M. Akram, Springer m-polar fuzzy graphs, in: Studies in Fuzziness and Soft Computing, Springer, Cham, Switzerland, 2019, p. 371.
[4] K.T. Atanassov, On Intuitionistic Fuzzy Sets Theory, Studies in Fuzziness and Soft Computing, vol. 283, Springer-Verlag, Berlin Heidelberg, 1999.
[5] J. Chen, S. Li, S. Ma, X. Wang, m-polar fuzzy sets: an extension of bipolar fuzzy sets, Sci. World J. 2014 (2014) 416530.

[6] B.C. Cuong, Picture fuzzy sets -first results. Part 1, Seminar "Neuro-fuzzy systems with applications", Tech. Rep., Institute of Mathematics, Hanoi, 2013.
[7] B.C. Cuong, Picture fuzzy sets -first results. Part 2, Seminar "Neuro-fuzzy systems with applications", Tech. Rep., Institute of Mathematics, Hanoi, 2013.
[8] J. Dombi, A general class of fuzzy operators, the demorgan class of fuzzy operators and fuzziness measures induced by fuzzy operators, Fuzzy Sets Syst. 8 (1982) 149–163.
[9] H. Gao, G.W. Wei, Y.H. Huang, Dual hesitant bipolar fuzzy Hamacher prioritized aggregation operators in multiple attribute decision making, IEEE Access 6 (1) (2018) 11508–11522.
[10] A.Z. Khameneh, A. Kilicman, m-polar fuzzy soft weighted aggregation operators and their applications in group decision-making, Symmetry 10 (2018) 636.
[11] C. Jana, M. Pal, J.Q. Wang, Bipolar fuzzy Dombi aggregation operators and its application in multiple attribute decision making process, J. Ambient Intell. Humaniz. Comput. 10 (2019) 3533–3549.
[12] C. Jana, T. Senapati, M. Pal, R.R. Yager, Picture fuzzy Dombi aggregation operators: application to MADM process, Appl. Soft Comput. 74 (1) (2019) 99–109.
[13] C. Jana, T. Senapati, M. Pal, Pythagorean fuzzy Dombi aggregation operators and its applications in multiple attribute decision-making, Int. J. Intell. Syst. 34 (9) (2019) 2019–2038.
[14] C. Jana, G. Muhiuddin, M. Pal, Some Dombi aggregation of Q-rung orthopair fuzzy numbers in multiple-attribute decision making, Int. J. Intell. Syst. 34 (12) (2019) 3220–3240.
[15] C. Jana, M. Pal, J.Q. Wang, Bipolar fuzzy Dombi prioritized aggregation operators in multiple attribute decision making, Soft Comput. 24 (2020) 3631–3646.
[16] C. Jana, M. Pal, J.Q. Wang, A robust aggregation operator for multi-criteria decision-making method with bipolar fuzzy soft environment, Iran. J. Fuzzy Syst. 16 (6) (2019) 1–16.
[17] C. Jana, M. Pal, Assessment of enterprise performance based on picture fuzzy Hamacher aggregation operators, Symmetry 11 (1) (2019) 75.
[18] P. Liu, J.L. Liu, S.M. Chen, Some intuitionistic fuzzy Dombi Bonferroni mean operators and their application to multi-attribute group decision making, J. Oper. Res. Soc. 69 (1) (2018) 1–24.
[19] P. Liu, Z. Liu, X. Zhang, Some intuitionistic uncertain linguistic Heronian mean operators and their application to group decision making, Appl. Math. Comput. 230 (2014) 570–586.
[20] P. Liu, Multiple attribute group decision making method based on interval-valued intuitionistic fuzzy power Heronian aggregation operators, Comput. Ind. Eng. 108 (2017) 199–212.
[21] D. Rani, H. Garg, Complex intuitionistic fuzzy power aggregation operators and their applications in multicriteria decision-making, Expert Syst. (2018), https://doi.org/10.1111/exsy.12325.
[22] L.H. Son, DPFCM: a novel distributed picture fuzzy clustering method on picture fuzzy sets, Expert Syst. Appl. 2 (2015) 51–66.
[23] L.H. Son, Generalized picture distance measure and applications to picture fuzzy clustering, Appl. Soft Comput. 46 (2016) 284–295.
[24] L.H. Son, Measuring analogousness in picture fuzzy sets: from picture distance measures to picture association measures, Fuzzy Optim. Decis. Mak. 16 (3) (2017) 1–20.
[25] P.H. Thong, L.H. Son, A new approach to multi-variables fuzzy forecasting using picture fuzzy clustering and picture fuzzy rules interpolation method, in: 6th International Conference on Knowledge and Systems Engineering, Hanoi, Vietnam, 2015, pp. 679–690.
[26] P.H. Thong, L.H. Son, H. Fujita, Interpolative picture fuzzy rules: a novel forecast method for weather nowcasting, in: Fuzzy Systems (FUZZ-IEEE), 2016 IEEE International Conference on IEEE, Vancouver, Canada, July 24-29, 2016, pp. 86–93.
[27] G.W. Wei, F.E. Alsaadi, H. Tasawar, A. Alsaedi, Bipolar fuzzy Hamacher aggregation operators in multiple attribute decision making, Int. J. Fuzzy Syst. 20 (1) (2018) 1–12.
[28] G.W. Wei, Picture fuzzy cross-entropy for multiple attribute decision making problems, J. Bus. Econ. Manag. 17 (4) (2016) 491–502.
[29] G.W. Wei, Picture fuzzy aggregation operators and their application to multiple attribute decision making, J. Intell. Fuzzy Syst. 33 (2017) 713–724.
[30] G.W. Wei, Picture fuzzy Hamacher aggregation operators and their application to multiple attribute decision making, Fundam. Inform. 157 (3) (2018) 271–320.
[31] X.R. Xu, G.W. Wei, Dual hesitant bipolar fuzzy aggregation operators in multiple attribute decision making, Int. J. Knowl. Based Intell. Eng. Syst. 21 (2017) 155–164.
[32] Z.S. Xu, Intuitionistic fuzzy aggregation operators, IEEE Trans. Fuzzy Syst. 15 (6) (2007) 1179–1187.

[33] Z.S. Xu, R.R. Yager, Some geometric aggregation operators based on intuitionistic fuzzy sets, Int. J. Gen. Syst. 35 (4) (2006) 417–433.
[34] R.R. Yager, On ordered weighted averaging aggregation operators in multicriteria decision making, IEEE Trans. Syst. Man Cybern. 18 (1) (1998) 183–190.
[35] R.R. Yager, J. Kacprzyk, The Ordered Weighted Averaging Operators: Theory and Applications, M.A., Kluwer, Boston, 1997.
[36] L.A. Zadeh, Fuzzy sets, Inf. Control 8 (1965) 338–353.
[37] W.R. Zhang, Bipolar fuzzy sets and relations: a computational frame work for cognitive modelling and multiagent decision analysis, in: Proceedings of IEEE Conference, 1994, pp. 305–309.
[38] W.R. Zhang, Bipolar fuzzy sets, in: Proceedings of FUZZYIEEE, 1998, pp. 835–840.

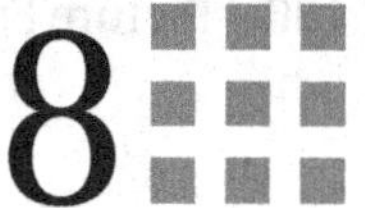

# Picture fuzzy MABAC approach and its application in multi-attribute group decision-making processes

## 8.1 Introduction

To circumvent some of the issues that real-world situations present we use fuzzy sets (FSs) [46] and intuitionistic fuzzy sets (IFSs) [1,2]. Cuong presented the picture fuzzy set (PFS) as a new idea for computational intelligence problems to handle this circumstance [4,5]. Due to its significance, several researchers have added to the idea of PFS information in actual decision making. Wei [39] defined weighted AOs, in this manner, to aggregate PFS data. Wei [38] analyzed the entropy of PFS information and applied this idea to create a MADM model. Later, Wei employed the Hamacher AOs and built picture fuzzy Hamacher AOs using this concept [40]. Additionally, Peng and Dai [15] created an approach for creating distance-based MADM issues in a PFS. Singh [22] studied the coefficient of correlation measure using PFS arguments. New inference systems for PFS were proposed by Son et al. [24,26]. Son [23] used distribution-based picture fuzzy clustering when developing decisions. In their research on the clustering approach to analyze complicated dates and weather forecasting in PFS environments, Thong and Son cited the following papers: [28–32]. In order to create customer requirements, Ping et al. [16] used a novel quality function deployment (QFD) environment. Projections-based MADM issues in a PFS are discussed by Wei et al. in [35]. A dice similarity measure approach was employed in PFSs by Wei and Gao to create MADM models [36]. Wei proposed new similarity measures in a picture fuzzy environment in [37]. In the same setting, weighted Dombi AOs were proposed by Jana et al. [7] and utilized to build a MADM technique. Readers are referred to the following sources for more details on the decision-making process connected to the PFS [17,18,24–26,33,43]. There are several research methodologies to simulate MADM issues from which we correctly identified a good option. In the earlier, well-known studies, the researchers looked at a range of MADM approaches, including the TODIM technique [9], the TOPOSIS approach [3,6], the CODAS approach [20], the VIKOR approach [21], the ELECTRE approach [8], the PROMETHEE approach [45], and others. The multiattributive border approximation area comparison (MABAC), first proposed by Pamučar and Ćirović [11], is a novel MADM approach that can take opposing qualities into account while making decisions. The intangibility of decision makers (DMs) and the haziness of the decision-theoretic environment is taken into account in a model of MABAC structures in order to obtain more tangible and workable aggregate information.

Picture Fuzzy Logic and Its Applications in Decision Making Problems. https://doi.org/10.1016/B978-0-44-322024-1.00012-1

We have observed some creative MADM approaches in that direction. According to this perspective, Wu and Liao [41] studied a novel outranking approach that quantified dominance score gains and losses. (GLDS). The “group utility” and the individual regret score were both measured simultaneously by the GLDS approach. When researching the issue of choosing a logistics center for transportation and distribution resources, Pamučar and Ćirović [11] developed the MABAC technique. Later, Xue et al. [42] suggested a unique MABAC technique using IVIFNs for material-selection purposes. Later, Peng and Yang examined the expanded MABAC approach in a Pythagorean setting [14]. Following that, Pamuvcar et al. refined the original MABAC technique [12,13]. Once more, Pamuvcar et al. studied the hybrid IR-AHP-MABAC model. A novel MCDM method combining the AHP and MABAC methodology was developed by Roy et al. [19]. The extended MABAC technique with HFLNs was proposed by Sun et al. [27] and used to determine patient priority levels. Yu et al. [44] developed a likelihood-based MABAC technique using intuitionistic trapezoidal linguistics (ITLNs). Mishra et al. [10] introduced the MCDM model, which is based on the MABAC method and IVIFS. Wang et al. [34] investigated the MABAC technique in a fuzzy Q-rung orthopair setting. Zhang et al. [47] explored the MABAC strategy in a pictorial 2-tuple linguistic context. This study aims to close the knowledge gap about the MABAC technique for MAGDM with PFNs data. This study set up a MAGDM process, and an MABAC model based on a traditional MABAC model with PFS information was built. First, we apply the MABAC technique to PFNs in this model. The original MABAC strategy is then used for MAGDM using PFNs. In order to compare the proposed PFS MABAC model with the existing operators such as the picture fuzzy weighted average (PFWA) operator, the weighted geometric (PFWG) operator, the picture fuzzy Dombi weighted average (PFDWA) operator, the weighted geometric (PFDWG) operator, the picture fuzzy Hamacher weighted average (PFHWA) operator, and the weighted geometric (PFHWG) operator, we conclude by introducing a numerical example to discuss the novel approach with PFNs

The remainder of the chapter is organized as follows: In Section 8.2 we reexamine some earlier literature. In Section 8.3 we discuss the original MABAC strategy and development procedure. In Section 8.4 we update the MABAC model suggestion for MAGDM using PFNs. In Section 8.5 we provide a sample estimate and choose a renewable energy power generation project to explore the suggested approach. Section 8.6 demonstrates some analysis of this strategy using specific preexisting models. Section 8.7 concludes the chapter with some future working directions also mentioned.

## 8.2 Some results of picture fuzzy sets

In this section, we express some essential concepts related to PFSs over the universe of discourse $X$.

**Definition 8.1.** [7] Let $\mathcal{R}_f = (\mathcal{Y}_f, \mathcal{A}_f, \mathcal{N}_f)$ $(f = 1, 2, \ldots \varpi)$ be a group of PFNs. A picture fuzzy weighted averaging (PFWA) operator of dimension $\varpi$ is a function $\Theta^{\varpi} \to \Theta$ that correlated with weight vector $\Xi = (\Xi_1, \Xi_2, \ldots, \Xi_\varpi)^T$ for which $\Xi > 0$ and $\sum_{f=1}^{\varpi} \Xi_f = 1$, as:

$$PFWA_\Xi(\mathcal{R}_1, \mathcal{R}_2, \ldots, \mathcal{R}_\varpi) = \bigoplus_{f=1}^{\varpi} (\Xi_f \mathcal{R}_f)$$

$$=\left(1 - \prod_{f=1}^{\varpi} (1 - \mathcal{Y}_f)^{\Xi_f}, \prod_{f=1}^{\varpi} \mathcal{A}_f{}^{\Xi_f}, \prod_{f=1}^{\varpi} \mathcal{N}_f{}^{\Xi_f}\right). \tag{8.1}$$

**Definition 8.2.** [7] Let $\mathcal{R}_f = (\mathcal{Y}_f, \mathcal{A}_f, \mathcal{N}_f)$ $(f = 1, 2, \ldots \varpi)$ be a group of PFNs. A picture fuzzy weighted geometric (PFWG) operator of dimension $\varpi$ is a function $\Theta^\varpi \to \Theta$ that correlated with weight vector $\Xi = (\Xi_1, \Xi_2, \ldots, \Xi_\varpi)^T$ for which $\Xi > 0$ and $\prod_{f=1}^{\varpi} \Xi_f = 1$, as:

$$PFWG_\Xi(\mathcal{R}_1, \mathcal{R}_2, \ldots, \mathcal{R}_\varpi) = \bigoplus_{f=1}^{\varpi} (\Xi_f \mathcal{R}_f)$$

$$=\left(1 - \prod_{f=1}^{\varpi} \mathcal{Y}_f{}^{\Xi_f}, 1 - \prod_{f=1}^{\varpi} (1 - \mathcal{A}_f)^{\Xi_f}, 1 - \prod_{f=1}^{\varpi} (1 - \mathcal{N}_f)^{\Xi_f}\right). \tag{8.2}$$

**Definition 8.3.** Let $\mathcal{R}_1 = (\mathcal{Y}_1, \mathcal{A}_1, \mathcal{N}_1)$ and $\mathcal{R}_2 = (\mathcal{Y}_2, \mathcal{A}_2, \mathcal{N}_2)$ be any two PFNs, then the picture fuzzy normalized Hamming distance (PFNHD) is defined as:

$$d_{PFNHD}(\mathcal{R}_1, \mathcal{R}_2) = \frac{1}{3}(|\mathcal{Y}_1 - \mathcal{Y}_2| + |\mathcal{A}_1 - \mathcal{A}_2| + |\mathcal{N}_1 - \mathcal{N}_2|). \tag{8.3}$$

## 8.3 Conventional MABAC model

Let there be a set of $\xi$ alternatives $\{\aleph_1, \aleph_2, \ldots, \aleph_\xi\}$, and $\varpi$ attributes $\{A_1, A_2, \ldots, A_\varpi\}$ with correlated sets of weight vectors $\{\Xi_1, \Xi_2, \ldots, \Xi_\varpi\}$, $(f = 1, 2, \ldots, \varpi)$ and $\delta$ experts $\{\gamma_1, \gamma_2, \ldots, \gamma_\delta\}$ with weighting vector $\{\theta_1, \theta_2, \ldots, \theta_\delta\}$, then the form of a conventional MABAC decision-making model follows the expression:

**Step 1:** The matrix evaluate $P = [\aleph_{gf}^\delta]_{\xi \times \varpi}$, where $g = 1, 2, \ldots, \xi$, $f = 1, 2, \ldots, \varpi$ as:

$$P_{g \times f} = \begin{array}{c} \\ \aleph_1 \\ \aleph_2 \\ \vdots \\ \aleph_\varpi \end{array} \begin{array}{c} \begin{array}{cccc} A_1 & A_2 & \cdots & A_\xi \end{array} \\ \begin{bmatrix} \aleph_{11}^\delta & \aleph_{12}^\delta & \cdots & \aleph_{1\varpi}^\delta \\ \aleph_{21}^\delta & \aleph_{22}^\delta & \cdots & \aleph_{2\varpi}^\delta \\ \vdots & \vdots & \ddots & \vdots \\ \aleph_{\xi 1}^\delta & \aleph_{\xi 2}^\delta & \cdots & \aleph_{\xi\varpi}^\delta \end{bmatrix} \end{array}$$

where, $\aleph_{gf}^\delta$ $(g = 1, 2, \ldots, \xi; f = 1, 2, \ldots, \varpi)$ represents the assessment formula of alternative $\aleph_g$ based on the attributes $A_f$ by the experts $\gamma_\delta$.

**Step 2:** Based on some aggregation operators, determine the aggregate overall $\aleph_{gf}^\delta$ to $\aleph_{gf}$.

**Step 3:** Normalize the fuse resultant matrix $M = [\aleph_{gf}]_{\xi \times \varpi}, g = 1, 2, \ldots, \xi; f = 1, 2, \ldots, \varpi$ based on the nature of each attribute by the following formula:

The formula for benefit attributes:

$$P_{gf} = \aleph_{gf}, g = 1, 2, \ldots, \xi, f = 1, 2, \ldots, \varpi. \tag{8.4}$$

The formula for cost attributes:

$$P_{gf} = 1 - \aleph_{gf}, g = 1, 2, \ldots, \xi, f = 1, 2, \ldots, \varpi. \tag{8.5}$$

**Step 4:** For the normalized matrix $P_{gf}$, $g = 1, 2, \ldots, \xi$, $f = 1, 2, \ldots, \varpi$ and attribute's weight $\Xi_f$, $(f = 1, 2, \ldots, \varpi)$, we normalize the weighted matrix $\Xi P_{gf}$ $(g = 1, 2, \ldots, \xi, f = 1, 2, \ldots, \varpi)$ by the following formula:

$$\Xi P_{gf} = \Xi_f P_{gf}, \quad g = 1, 2, \ldots, \xi, f = 1, 2, \ldots, \varpi. \tag{8.6}$$

**Step 5:** Evaluate the values of the border approximation areas (BAA) and the BAA matrix $T = [t_f]_{1 \times \varpi}$ can be computed as:

$$t_f = \left( \prod_{g=1}^{\xi} \Xi P_{gf} \right)^{1/\xi}, g = 1, 2, \ldots, \xi, f = 1, 2, \ldots, \varpi. \tag{8.7}$$

**Step 6:** Compute the distance $D = [d_{gf}]_{\xi \times \varpi}$ between each alternative and BAA by the following equation:

$$d_{gf} = \begin{cases} d\left(\Xi P_{gf}, t_f\right), & \text{if } \Xi M_{gf} > t_f \\ 0, & \text{if } \Xi P_{gf} = t_f \\ -d\left(\Xi P_{gf}, t_f\right), & \text{if } \Xi M_{gf} < t_f \end{cases}, \tag{8.8}$$

where $d\left(\Xi P_{gf}, t_f\right)$ is the mean distance from $\Xi P_{gf}$ to $t_f$.

Based on the values of $d_{gf}$, we can find the following:

- $d_{gf} > 0$, which implies that alternatives belong to the upper approximation region $A^+(UAA)$;
- $d_{gf} = 0$, which implies that alternatives belong to the border approximation region $A^+(BAA)$;
- $d_{gf} < 0$, which implies that alternatives belong to the lower approximation region $A^-(LAA)$.

It is obvious that the best alternatives belong to $A^+(UAA)$, and the worst alternatives belong to $A^-(LAA)$.

**Step 7:** Add the values of each alternative's $d_{gf}$ by the following equation:

$$S_g = \sum_{f=1}^{\varpi} d_{gf}. \tag{8.9}$$

## 8.4 MABAC model with PFNs

Let there be a set of $\xi$ alternatives $\{\aleph_1, \aleph_2, \ldots, \aleph_\xi\}$, and $\varpi$ attributes $\{A_1, A_2, \ldots, A_\varpi\}$ with a correlated set of weight vectors $\{\Xi_1, \Xi_2, \ldots, \Xi_\varpi\}$, $(f = 1, 2, \ldots, \varpi)$ and $\delta$ experts $\{\gamma_1, \gamma_2, \ldots, \gamma_\delta\}$ with weighting vector $\{\theta_1, \theta_2, \ldots, \theta_\delta\}$, then the PFS decision matrix $P = [\aleph^\delta_{gf}]_{\xi\times\varpi} = \left(\mathcal{Y}^\delta_{gf}, \mathcal{A}^\delta_{gf}, N^\delta_{gf}\right)_{\xi\times\varpi}$, $g = 1, 2, \ldots, \xi$, $f = 1, 2, \ldots, \varpi$, $\mathcal{Y}^\delta_{ab} \in [0, 1]$ represents the positive membership degree, $\mathcal{A}^\delta_{gf} \in [0, 1]$ denotes the neutral degree of membership, and $\mathcal{Y}^\delta_{gf} \in [0, 1]$ denotes the nonmembership degree, then the PFS-MABAC decision-making model follows the expression:

**Step 1.** Evaluate the PFM matrix $P = [\aleph^\delta_{gf}]_{\xi\times\varpi} = \left(\mathcal{Y}^\delta_{gf}, \mathcal{A}^\delta_{gf}, N^\delta_{gf}\right)_{\xi\times\varpi}$, $g = 1, 2, \ldots, \xi$, $f = 1, 2, \ldots, \varpi$ given as

$$P = [\aleph^\delta_{gf}]_{\xi\times\varpi} = \begin{array}{c} \\ \aleph_1 \\ \aleph_2 \\ \vdots \\ \aleph_\varpi \end{array} \begin{array}{c} \begin{array}{cccc} A_1 & A_2 & \cdots & A_\xi \end{array} \\ \left[\begin{array}{cccc} \left(\mathcal{Y}^\delta_{11}, \mathcal{A}^\delta_{11}, N^\delta_{11}\right) & \left(\mathcal{Y}^\delta_{12}, \mathcal{A}^\delta_{12}, N^\delta_{12}\right) & \cdots & \left(\mathcal{Y}^\delta_{1\varpi}, \mathcal{A}^\delta_{1\varpi}, N^\delta_{1\varpi}\right) \\ \left(\mathcal{Y}^\delta_{21}, \mathcal{A}^\delta_{21}, N^\delta_{21}\right) & \left(\mathcal{Y}^\delta_{22}, \mathcal{A}^\delta_{22}, N^\delta_{22}\right) & \cdots & \left(\mathcal{Y}^\delta_{2\varpi}, \mathcal{A}^\delta_{2\varpi}, N^\delta_{2\varpi}\right) \\ \vdots & \vdots & \ddots & \vdots \\ \left(\mathcal{Y}^\delta_{\xi1}, \mathcal{A}^\delta_{\xi1}, N^\delta_{\xi1}\right) & \left(\mathcal{Y}^\delta_{\xi2}, \mathcal{A}^\delta_{\xi2}, N^\delta_{\xi2}\right) & \cdots & \left(\mathcal{Y}^\delta_{\xi\varpi}, \mathcal{A}^\delta_{\xi\varpi}, N^\delta_{\xi\varpi}\right) \end{array}\right] \end{array}, \tag{8.10}$$

where, $\aleph^\delta_{gf} = \left(\mathcal{Y}^\delta_{gf}, \mathcal{A}^\delta_{gf}, N^\delta_{gf}\right)$, $g = 1, 2, \ldots, \xi$; $f = 1, 2, \ldots, \varpi$, denotes the formula for picture fuzzy information of $\aleph_g$ based on $A_f$ by the experts $\gamma_\delta$.

**Step 2.** For the PFN, AOs PFWA, or PFWG, aggregate overall $\aleph^\delta_{gf}$ to $\aleph_{gf}$, then the fused PFNs matrix is shown as:

$$P = [\aleph_{gf}]_{\xi\times\varpi} = \begin{array}{c} \\ \aleph_1 \\ \aleph_2 \\ \vdots \\ \aleph_\varpi \end{array} \begin{array}{c} \begin{array}{cccc} A_1 & A_2 & \cdots & A_\xi \end{array} \\ \left[\begin{array}{cccc} \left(\mathcal{Y}_{11}, \mathcal{A}_{11}, N_{11}\right) & \left(\mathcal{Y}_{12}, \mathcal{A}_{12}, N_{12}\right) & \cdots & \left(\mathcal{Y}_{1\varpi}, \mathcal{A}_{1\varpi}, N_{1\varpi}\right) \\ \left(\mathcal{Y}_{21}, \mathcal{A}_{21}, N_{21}\right) & \left(\mathcal{Y}_{22}, \mathcal{A}_{22}, N_{22}\right) & \cdots & \left(\mathcal{Y}_{2\varpi}, \mathcal{A}_{2\varpi}, N_{2\varpi}\right) \\ \vdots & \vdots & \ddots & \vdots \\ \left(\mathcal{Y}_{\xi1}, \mathcal{A}_{\xi1}, N_{\xi1}\right) & \left(\mathcal{Y}_{\xi2}, \mathcal{A}_{\xi2}, N_{\xi2}\right) & \cdots & \left(\mathcal{Y}_{\xi\varpi}, \mathcal{A}_{\xi\varpi}, N_{\xi\varpi}\right) \end{array}\right] \end{array}, \tag{8.11}$$

where, $\aleph_{gf} = \left(\mathcal{Y}_{gf}, \mathcal{A}_{gf}, N_{gf}\right)$ depicts a formula for PFN information of $\aleph_g$ based on $A_f$ by the experts $\gamma_\delta$.

**Step 3.** Normalize matrix $P = [\aleph_{gf}]_{\xi\times\varpi}$, $g = 1, 2, \ldots, \xi$; $f = 1, 2, \ldots, \varpi$ based on the nature of each attribute by the following formulas:

For benefit attributes:

$$P_{gf} = \aleph_{gf} = \left(\mathcal{Y}_{gf}, \mathcal{A}_{gf}, N_{gf}\right), g = 1, 2, \ldots, \xi, f = 1, 2, \ldots, \varpi. \tag{8.12}$$

For cost attributes:

$$\mathcal{Q}_{gf} = \left(\aleph_{gf}\right)^{c} = \left(\mathcal{N}_{gf}, \mathcal{A}_{gf}, \mathcal{Y}_{gf}\right), g = 1, 2, \ldots, \xi, f = 1, 2, \ldots, \varpi. \tag{8.13}$$

**Step 4.** For the normalized matrix $P_{gf} = \left(\mathcal{Y}_{gf}, \mathcal{A}_{gf}, \mathcal{N}_{gf}\right)$, $(g = 1, 2, \ldots, \xi, f = 1, 2, \ldots, \varpi)$, and the attribute's weight $\Xi_f, f = 1, 2, \ldots, \varpi$, we compute the normalized PFWDM $\Xi P_{gf} = \left(\mathcal{Y}'_{gf}, \mathcal{A}'_{gf}, \mathcal{N}'_{gf}\right)$, $g = 1, 2, \ldots, \xi, f = 1, 2, \ldots, \varpi$, by the following rule:

$$\begin{aligned} \Xi P_{gf} &= \Xi_f \oplus P_{gf}, \quad g = 1, 2, \ldots, \xi, f = 1, 2, \ldots, \varpi \\ &= \left(1 - \prod_{f=1}^{\varpi} (1 - \mathcal{Y}_{gf})^{\Xi_f}, \prod_{f=1}^{\varpi} \mathcal{A}_{gf}{}^{\Xi_f}, \prod_{f=1}^{\varpi} \mathcal{N}_{gf}{}^{\Xi_f}\right). \end{aligned} \tag{8.14}$$

**Step 5:** Evaluate BAA. The BAA matrix $T = [t_f]_{1\times\varpi}$ is computed as

$$\begin{aligned} t_f &= \left(\prod_{g=1}^{\xi} \Xi P_{gf}\right)^{1/\xi}, g = 1, 2, \ldots, \xi, f = 1, 2, \ldots, \varpi \\ &= \left\{\left(\prod_{g=1}^{\xi} \mathcal{Y}_{gf}\right)^{1/\xi}, 1 - \prod_{g=1}^{\xi}\left(1 - \mathcal{A}_{gf}\right)^{1/\xi}, 1 - \prod_{g=1}^{\xi}\left(1 - \mathcal{N}_{gf}\right)^{1/\xi}\right\}. \end{aligned} \tag{8.15}$$

**Step 6.** Compute the distance $D = [d_{gf}]_{\xi\times\varpi}$ between each alternative and BAA by the following equation:

$$d_{gf} = \begin{cases} d\left(\Xi P_{gf}, t_b\right), & \text{if } \Xi P_{gf} > t_f \\ 0, & \text{if } \Xi P_{gf} = t_f \\ -d\left(\Xi P_{gf}, t_f\right), & \text{if } \Xi P_{gf} < t_f \end{cases}, \tag{8.16}$$

where $d\left(\Xi P_{gf}, t_f\right)$ is the mean distance from $\Xi P_{gf}$ to $t_f$, and can be calculated by Definition 8.3.

**Step 7.** Add the values of each alternative's $d_{gf}$ by the following equation:

$$S_g = \sum_{f=1}^{\varpi} d_{gf}. \tag{8.17}$$

## 8.5 Case study

Our civilization is currently in danger due to various environmental problems and the use of fossil fuels. Therefore renewed energy power generation project (REPGP) development is currently in full swing. To choose the right projects, we need more thorough evaluation techniques. This will allow us to identify their strengths and weaknesses and present some fresh ideas for achieving the goal. For REPGP, a single method-oriented strategy has been directly advised in previous studies. A multiobjective, comprehensive evaluation technique with straightforward guiding principles has not yet been created, nevertheless. The method for choosing the best renewable energy power generation projects utilizing the MABAC methodology and PFNs are provided in this section. Here, we introduce three experts and their weight vector $\delta = (0.39, 0.28, 0.33)^T$ to select the five possible REPGP $\aleph_g$ $(g = 1, 2, 3, 4, 5)$ that are evaluated under four attributes (1) $A_1$ value at risk, (2) $A_2$ is the static duration for investment return, (3) $A_3$ is the return on investment, and (4) $A_4$ is the rate of internal return. The five possible REPGP $\aleph_g$ $(g = 1, 2, 3, 4, 5)$ are evaluated using four attribute weights $\Xi = (0.16, 0.36, 0.25, 0.23)^T$ proposed by three experts.

**Step 1.** For the PFD matrix $P = [P^{\delta}_{gf}]_{\xi\times\varpi} = (\mathcal{Y}^{\delta}_{gf}, \mathcal{A}^{\delta}_{gf}, \mathcal{N}^{\delta}_{gf})$ $g = 1, 2, \ldots, \xi,$ $f = 1, 2, \ldots, \varpi$.

$$P_1 = \begin{bmatrix} (0.5, 0.3, 0.1) & (0.2, 0.3, 0.1) & (0.4, 0.2, 0.1) & (0.1, 0.7, 0.04) \\ (0.6, 0.2, 0.1) & (0.6, 0.1, 0.1) & (0.06, 0.5, 0.1) & (0.4, 0.3, 0.1) \\ (0.5, 0.1, 0.2) & (0.2, 0.2, 0.3) & (0.5, 0.1, 0.09) & (0.5, 0.1, 0.2) \\ (0.1, 0.2, 0.4) & (0.09, 0.2, 0.4) & (0.4, 0.2, 0.05) & (0.05, 0.3, 0.4) \\ (0.7, 0.1, 0.1) & (0.6, 0.09, 0.2) & (0.2, 0.07, 0.4) & (0.4, 0.6, 0.1) \end{bmatrix},$$

$$P_2 = \begin{bmatrix} (0.4, 0.1, 0.2) & (0.6, 0.1, 0.1) & (0.4, 0.2, 0.1) & (0.5, 0.06, 0.1) \\ (0.5, 0.1, 0.2) & (0.5, 0.3, 0.2) & (0.08, 0.5, 0.1) & (0.3, 0.2, 0.1) \\ (0.4, 0.1, 0.2) & (0.4, 0.3, 0.2) & (0.5, 0.1, 0.08) & (0.5, 0.1, 0.2) \\ (0.3, 0.2, 0.4) & (0.7, 0.02, 0.1) & (0.4, 0.2, 0.09) & (0.07, 0.3, 0.2) \\ (0.6, 0.1, 0.05) & (0.5, 0.1, 0.2) & (0.6, 0.08, 0.1) & (0.4, 0.1, 0.2) \end{bmatrix},$$

$$P_3 = \begin{bmatrix} (0.4, 0.2, 0.1) & (0.6, 0.2, 0.1) & (0.4, 0.2, 0.2) & (0.7, 0.2, 0.08) \\ (0.5, 0.2, 0.1) & (0.5, 0.05, 0.1) & (0.09, 0.4, 0.1) & (0.4, 0.1, 0.2) \\ (0.4, 0.1, 0.2) & (0.3, 0.09, 0.3) & (0.5, 0.2, 0.1) & (0.5, 0.2, 0.07) \\ (0.3, 0.3, 0.2) & (0.4, 0.1, 0.4) & (0.3, 0.2, 0.1) & (0.6, 0.3, 0.09) \\ (0.6, 0.1, 0.1) & (0.4, 0.08, 0.2) & (0.4, 0.09, 0.3) & (0.2, 0.3, 0.1) \end{bmatrix}.$$

**Step 2.** According to the PFWA operator and using the expert's weight vector, we obtain $P^{\delta}_{gf}$ to $P_{gf}$. Therefore we compute the matrix as follows:

$$M = \begin{bmatrix} (0.4412, 0.1929, 0.1214) & (0.4758, 0.1929, 0.1000) & (0.4000, 0.2000, 0.1257) & (0.4687, 0.2327, 0.0650) \\ (0.5417, 0.1647, 0.1214) & (0.5417, 0.1082, 0.1214) & (0.0756, 0.4645, 0.1000) & (0.3735, 0.1864, 0.1257) \\ (0.4412, 0.1000, 0.2000) & (0.2937, 0.1721, 0.2678) & (0.5000, 0.1257, 0.0902) & (0.5000, 0.1257, 0.1414) \\ (0.2279, 0.2286, 0.3182) & (0.4187, 0.0835, 0.2713) & (0.3687, 0.2000, 0.0741) & (0.2901, 0.3000, 0.2014) \\ (0.6425, 0.1000, 0.0824) & (0.5133, 0.0892, 0.2000) & (0.4008, 0.0790, 0.2467) & (0.3402, 0.2890, 0.1214) \end{bmatrix}.$$

**Step 3:** Normalize the resultant matrix $M' = [P_{gf}]$ $g = 1, 2, \ldots, \xi,$ $f = 1, 2, \ldots, \varpi$ based on the character of the attributes using Eqs. (8.12) and (8.13), here $A_2$ is the cost attribute. As $A_2$ is the cost attribute, $\aleph_{12}$, $\aleph_{22}$, $\aleph_{32}$, $\aleph_{42}$, and $\aleph_{52}$ are to be normalized by the formula

$P_{12} = (\aleph_{12})^c = (0.4758, 0.1929, 0.1000)^c = (0.1000, 0.1929, 0.4758)$. Therefore the normalized matrix $P_{gf}$ is given as:

$$M' = \begin{bmatrix} (0.4412,0.1929,0.1214) & (0.1000,0.1929,0.4758) & (0.4000,0.2000,0.1257) & (0.4687,0.2327,0.0650) \\ (0.5417,0.1647,0.1214) & (0.1214,0.1082,0.5417) & (0.0756,0.4645,0.1000) & (0.3735,0.1864,0.1257) \\ (0.4412,0.1000,0.2000) & (0.2678,0.1721,0.2937) & (0.5000,0.1257,0.0902) & (0.5000,0.1257,0.1414) \\ (0.2279,0.2286,0.3182) & (0.2713,0.0835,0.4187) & (0.3687,0.2000,0.0741) & (0.2901,0.3000,0.2014) \\ (0.6425,0.1000,0.0824) & (0.2000,0.0892,0.5133) & (0.4008,0.0790,0.2467) & (0.3402,0.2890,0.1214) \end{bmatrix}.$$

**Step 4.** For the normalized matrix $P_{gf}$, $g = 1, 2, \ldots, \xi$, $f = 1, 2, \ldots, \varpi$, the weighted normalized matrix $\Xi P_{gf} = \Xi_f \oplus P_{gf} = \left(1 - \prod_{f=1}^{\varpi} (1 - \mathcal{Y}_{gf})^{\Xi_f}, \prod_{f=1}^{\varpi} \mathcal{A}_{gf}{}^{\Xi_f}, \prod_{f=1}^{\varpi} N_{gf}{}^{\Xi_f}\right)$, now is given as follows:

$$\Xi P_{\xi\varpi} = \begin{bmatrix} (0.0889,0.7685,0.7136) & (0.0372,0.5529,0.7654) & (0.1199,0.6687,0.5954) & (0.1354,0.7151,0.5333) \\ (0.1174,0.7493,0.7136) & (0.0455,0.4490,0.8019) & (0.0195,0.8256,0.5623) & (0.1019,0.6795,0.6207) \\ (0.0889,0.6918,0.7730) & (0.1061,0.5307,0.6433) & (0.1591,0.5954,0.5480) & (0.1474,0.6206,0.6377) \\ (0.0405,0.7897,0.8325) & (0.1077,0.4091,0.7309) & (0.1086,0.6687,0.5217) & (0.0758,0.7581,0.6917) \\ (0.1517,0.6918,0.6707) & (0.0772,0.4189,0.7866) & (0.1202,0.5302,0.7048) & (0.0912,0.7516,0.6157) \end{bmatrix}.$$

**Step 5:** Evaluate the values of the BAA and the BAA matrix using Eq. (8.15), which follow as $t_1 = (0.0894, 0.7413, 0.7475)$, $t_2 = (0.0684, 0.4755, 0.7513)$ $t_3 = (0.0865, 0.6748, 0.5920)$, $t_4 = (0.1071, 0.7091, 0.6232)$.

**Step 6:** Evaluate the distance $d$ by using Eq. (8.3) between the alternatives and BAA. See Table 8.1.

**Table 8.1** Distance between alternatives and BAA.

| Alternatives | $A_1$ | $A_2$ | $A_3$ | $A_4$ |
|---|---|---|---|---|
| $\aleph_1$ | 0.0036 | −0.0302 | 0.0154 | 0.0278 |
| $\aleph_2$ | 0.0011 | −0.0004 | −0.0271 | −0.0187 |
| $\aleph_3$ | −0.0123 | 0.0076 | 0.0254 | 0.0169 |
| $\aleph_4$ | −0.0423 | 0.0238 | 0.0272 | −0.0431 |
| $\aleph_5$ | 0.0320 | −0.0067 | −0.0010 | −0.0096 |

**Step 7:** Compute the sums of the distances $S_\rho$ for each alternative using Eq. (8.17) as follows:

$$S_1 = \sum_{f=1}^{\varpi} d_{1b} = d_{11} + d_{12} + d_{13} + d_{14}$$
$$= 0.0036 - 0.0302 + 0.0154 + 0.0278 = 0.0166, \text{ for alternative } \aleph_1.$$

$$S_2 = \sum_{f=1}^{\varpi} d_{2b} = d_{21} + d_{22} + d_{23} + d_{24}$$
$$= 0.0011 - 0.0004 - 0.0271 - 0.0187 = -0.0451, \text{ for alternative } \aleph_2.$$

$$S_3 = \sum_{f=1}^{\varpi} d_{3b} = d_{31} + d_{32} + d_{33} + d_{34}$$
$$= -0.0123 + 0.0076 + 0.0254 + 0.0169 = 0.0376, \text{ for alternative } \aleph_3.$$

$$S_4 = \sum_{f=1}^{\varpi} d_{4b} = d_{41} + d_{42} + d_{43} + d_{44}$$
$$= -0.0423 + 0.0238 + 0.0272 - 0.0431 = -0.0344, \text{ for alternative } \aleph_4.$$

$$S_5 = \sum_{\tau=1}^{\mathcal{A}} d_{5\tau} = d_{51} + d_{52} + d_{53} + d_{54}$$
$$= 0.0320 - 0.0067 - 0.0010 - 0.0096 = 0.0147, \text{ for alternative } \aleph_5.$$

From the comprehensive evaluation results of $S_\rho$ for detecting the best choice, obtain the order list as $\aleph_3 \succ \aleph_1 \succ \aleph_5 \succ \aleph_4 \succ \aleph_2$, and $\aleph_3$ is the favorable solution.

## 8.6 Compare PFNs MABAC approach with some PFNs operators

In this section, we compare the proposed PFNs MABAC technique with other currently existing picture fuzzy operators, such as the picture fuzzy weighted aggregation operator (PFWA) put forward by Wei [39], other picture fuzzy Dombi weighted operators proposed by Jana et al. [7], and the picture fuzzy Hamacher weighted operator (PFHWA) introduced by Wei et al. [40]. As a result, the aggregated values of the alternative under the attributes using the operators PFWA, PFDWA, and PFHWA are given in Table 8.2. Table 8.3 displays the ranking comparison of the alternatives with existing operators and the suggested MABAC model.

**Table 8.2** Some existing picture fuzzy aggregation operators.

| Alternative | PFWA | PFDWA | PFHWA |
|---|---|---|---|
| $B_1$ | (0.3325, 0.2032, 0.1734) | (0.3489, 0.2027, 0.1315) | (0.3162, 0.1776, 0.1368) |
| $B_2$ | (0.2582, 0.1888, 0.1997) | (0.2839, 0.1647, 0.1584) | (0.2368, 0.1493, 0.1531) |
| $B_3$ | (0.4161, 0.1357, 0.1738) | (0.4246, 0.1331, 0.1557) | (0.4100, 0.1077, 0.1398) |
| $B_4$ | (0.2947, 0.1638, 0.2196) | (0.2964, 0.1423, 0.1701) | (0.2932, 0.2143, 0.1663) |
| $B_5$ | (0.3741, 0.1155, 0.2289) | (0.3988, 0.1042, 0.1801) | (0.3600, 0.2086, 0.1710) |

## 8.7 Conclusions

In this chapter, we examine several fundamental concepts of PFNs and the traditional MABAC paradigm. For MAGDM, we suggest an MABAC technique with PFNs. In this study, we consider the definition of PFNs as well as their operating laws, scoring function, and

**Table 8.3** Ranking order of alternatives for some BFNs operators.

| Methods | Ranking order |
|---|---|
| [39] PFWA | $\aleph_3 \succ \aleph_1 \succ \aleph_5 \succ \aleph_4 \succ \aleph_2$ |
| [7] PFDWA | $\aleph_3 \succ \aleph_5 \succ \aleph_1 \succ \aleph_4 \succ \aleph_2$ |
| [40] PFHWA | $\aleph_3 \succ \aleph_5 \succ \aleph_1 \succ \aleph_4 \succ \aleph_2$ |
| Proposed PFNs MABAC model | $\aleph_3 \succ \aleph_1 \succ \aleph_5 \succ \aleph_4 \succ \aleph_2$ |

accuracy function. Next, two weighted aggregation procedures for PFNs are defined. The original MABAC technique is then combined with PFNs to create a PFNs MABAC model for MAGDM. Additionally, we use the suggested approach to assess a particular renewable energy power generation plant as an example. Finally, we evaluate the proposed model and demonstrate its effectiveness by comparing it to various current PFNs operators. We can use the suggested approach in various uncertain and fuzzy contexts [10,39,42].

## References

[1] K. Atanassov, Intuitionistic Fuzzy Sets: Theory and Applications, Studies in Fuzziness and Soft Computing, vol. 35, Physica-Verlag, Heidelberg, 1999.
[2] K.T. Atanassov, G. Gargov, Interval-valued intuitionistic fuzzy sets, Fuzzy Sets Syst. 31 (1989) 343–349.
[3] P. Biswas, S. Pramanik, B.C. Giri, Neutrosophic TOPSIS with group decision making, in: C. Kahraman, İ. Otay (Eds.), Fuzzy Multicriteria Decision-Making Using Neutrosophic Sets, in: Studies in Fuzziness and Soft Computing, vol. 369, Springer, Cham, 2019.
[4] B.C. Cuong, Picture fuzzy sets - first results. Part 1, Seminar "Neuro-fuzzy systems with applications", Tech. Rep., Institute of Mathematics, Hanoi, 2013.
[5] B.C. Cuong, Picture fuzzy sets - first results. Part 2, Seminar "Neuro-fuzzy systems with applications", Tech. Rep., Institute of Mathematics, Hanoi, 2013.
[6] T. Gebrehiwet, H.B. Luo, Risk level evaluation on construction project lifecycle using fuzzy comprehensive evaluation and TOPSIS, Symmetry 11 (1) (2018) 12.
[7] C. Jana, T. Senapati, M. Pal, R.R. Yager, Picture fuzzy Dombi aggregation operators: application to MADM process, Appl. Soft Comput. 74 (1) (2019) 99–109.
[8] P. Liu, H. Li, P. Wang, J. Liu, ELECTRE method and its application in multiple attribute decision making based on INS, J. Shandong Univ. Financ. Econ. 28 (2016) 80–87.
[9] R. Lourenzuttia, R.A. Krohlingb, A study of TODIM in a intuitionistic fuzzy and random environment, Expert Syst. Appl. 40 (16) (2013) 6459–6468.
[10] A.R. Mishra, A. Chandel, D.J.G.C. Motwani, Extended MABAC method based on divergence measures for multicriteria assessment of programming language with interval-valued intuitionistic fuzzy sets, Granul. Comput. 5 (2020) 97–117.
[11] D. Pamučar, G. Ćirović, The selection of transport and handling resources in logistics centers using MultiAttributive border approximation area comparison (MABAC), Expert Syst. Appl. 42 (2015) 3016–3028.
[12] D. Pamučar, I. Petrovic, G. Ćirović, Modification of the best- worst and MABAC methods: a novel approach based on interval-valued fuzzy-rough numbers, Expert Syst. Appl. 91 (2018) 89–106.
[13] D. Pamučar, Z. Stević, E.K. Zavadskas, Integration of interval rough AHP and interval rough MABAC methods for evaluating university web pages, Appl. Soft Comput. 67 (2018) 141–163.
[14] X.D. Peng, Y. Yang, Pythagorean fuzzy Choquet integral based MABAC method for multiple attribute group decision making, Int. J. Intell. Syst. 31 (2016) 989–1020.
[15] X. Peng, J. Dai, Algorithm for picture fuzzy multiple attribute decision making based on new distance measure, Int. J. Uncertain. Quantificat. 7 (2017) 177–187.

[16] Y.J. Ping, R. Liu, W. Lin, H.C. Liu, A new integrated approach for engineering characteristic prioritisation in quality function deployment, Adv. Eng. Inform. (ISSN 1474-0346) 45 (2020).
[17] P.H. Phong, D.T. Hieu, R.T.H. Ngan, P.T. Them, Some compositions of picture fuzzy relations, in: Proceedings of the 7th National Conference on Fundamental and Applied Information Technology Research, FAIR'7, Thai Nguyen, 2014, pp. 19–20.
[18] P.T.M. Phuong, P.H. Thong, L.H. Son, Theoretical analysis of picture fuzzy clustering, J. Comput. Sci. Cybern. 34 (1) (2018) 17–31.
[19] J. Roy, K. Chatterjee, A. Bandhopadhyay, S. Kar, Evaluation and selection of medical tourism sites: a rough AHP based MABAC approach, https://doi.org/10.1111/exsy.12232.
[20] J. Roy, S. Das, S. Kar, D. Pamucar, An extension of the CODAS approach using interval-valued intuitionistic fuzzy set for sustainable material selection in construction projects with incomplete weight information, Symmetry 11 (3) (2019) 393.
[21] K.W. Shen, J.Q. Wang, Z-VIKOR method based on a new comprehensive weighted distance measure of Z-number and its application, IEEE Trans. Fuzzy Syst. 26 (6) (2018) 3232–3245.
[22] P. Singh, Correlation coefficients for picture fuzzy sets, J. Intell. Fuzzy Syst. 27 (2014) 2857–2868.
[23] L.H. Son, DPFCM: a novel distributed picture fuzzy clustering method on picture fuzzy sets, Expert Syst. Appl. 2 (2015) 51–66.
[24] L.H. Son, Generalised picture distance measure and applications to picture fuzzy clustering, Appl. Soft Comput. 46 (2016) 284–295.
[25] L.H. Son, Measuring analogousness in picture fuzzy sets: from picture distance measures to picture association measures, Fuzzy Optim. Decis. Mak. 16 (3) (2017) 1–20.
[26] L.H. Son, P. Viet, P. Hai, Picture inference system: a new fuzzy inference system on picture fuzzy set, Appl. Intell. 46 (3) (2017) 652–669.
[27] R. Sun, J. Hu, J. Zhou, X. Chen, A hesitant fuzzy linguistic projection-based MABAC method for patients's prioritisation, Int. J. Fuzzy Syst. 20 (2017) 1–17.
[28] P.H. Thong, L.H. Son, Picture fuzzy clustering for complex data, Eng. Appl. Artif. Intell. 56 (2016) 121–130.
[29] P.H. Thong, L.H. Son, A novel automatic picture fuzzy clustering method based on particle swarm optimisation and picture composite cardinality, Knowl.-Based Syst. 109 (2016) 48–60.
[30] P.H. Thong, L.H. Son, A new approach to multi-variables fuzzy forecasting using picture fuzzy clustering and picture fuzzy rules interpolation method, in: 6th International Conference on Knowledge and Systems Engineering, Hanoi, Vietnam, 2015, pp. 679–690.
[31] P.H. Thong, L.H. Son, H. Fujita, Interpolative picture fuzzy rules: a novel forecast method for weather nowcasting, in: Fuzzy Systems (FUZZ-IEEE), 2016 IEEE International Conference on IEEE, Vancouver, Canada, July 24-29, 2016, pp. 86–93.
[32] N.T. Thong, L.H. Son, HIFCF: an effective hybrid model between picture fuzzy clustering and intuitionistic fuzzy recommender systems for medical diagnosis, Expert Syst. Appl. 42 (7) (2015) 3682–3701.
[33] P.V. Viet, H.T.M. Chau, L.H. Son, P.V. Hai, Some extensions of membership graphs for picture inference systems, in: Knowledge and Systems Engineering (KSE), 2015 Seventh International Conference on, Ho Chi Minh City, Vietnam, IEEE, October 8-10, 2015, pp. 192–197.
[34] J. Wang, G.W. Wei, C. Wei, Y. Wei, MABAC method for multiple attribute group decision making under q-rung orthopair fuzzy environment, Def. Technol. 16 (2020) 208–216.
[35] G.W. Wei, F.E. Alsaadi, T. Hayat, A. Alsaedi, Projection models for multiple attribute decision making with picture fuzzy information, Int. J. Mach. Learn. Cybern. 9 (4) (2018) 713–719.
[36] G.W. Wei, H. Gao, The generalised Dice similarity measures for picture fuzzy sets and their applications, Informatica 29 (1) (2018) 1–18.
[37] G.W. Wei, Some similarity measures for picture fuzzy sets and their applications, Iran. J. Fuzzy Syst. 15 (1) (2018) 77–89.
[38] G.W. Wei, Picture fuzzy cross-entropy for multiple attribute decision making problems, J. Bus. Econ. Manag. 17 (4) (2016) 491–502.
[39] G.W. Wei, Picture fuzzy aggregation operators and their application to multiple attribute decision making, J. Intell. Fuzzy Syst. 33 (2017) 713–724.
[40] G.W. Wei, Picture fuzzy Hamacher aggregation operators and their application to multiple attribute decision making, Fundam. Inform. 157 (3) (2018) 271–320.

[41] X.L. Wu, H.C. Liao, A consensus-based probabilistic linguistic gained and lost dominance score method, Eur. J. Oper. Res. 272 (2019) 1017–1027.
[42] Y.X. Xue, J.X. You, X.D. Lai, H.C. Liu, An interval-valued intuitionistic fuzzy MABAC approach for material selection with incomplete weight information, Appl. Soft Comput. 38 (2016) 703–713.
[43] Y. Yang, C. Liang, S. Ji, T. Liu, Adjustable soft discernibility matrix based on picture fuzzy soft sets and its applications in decision making, J. Intell. Fuzzy Syst. 29 (4) (2015) 1711–1722.
[44] S.M. Yu, J. Wang, J.Q. Wang, An interval type-2 fuzzy likelihood-based MABAC approach and its application in selecting hotels on a tourism website, Int. J. Fuzzy Syst. 19 (2017) 47–61.
[45] P. Ziemba, NEAT F-PROMETHEE-a new fuzzy multiple criteria decision making method based on the adjustment of mapping trapezoidal fuzzy numbers, Expert Syst. Appl. 110 (2018) 363–380.
[46] L.A. Zadeh, Fuzzy sets, Inf. Control 8 (1965) 338–353.
[47] S. Zhang, G.W. Wei, Fuad E. Alsaadi, T. Hayat, C. Wei, Z. Zhang, MABAC method for multiple attribute group decision making under picture 2-tuple linguistic environment, Soft Comput. 24 (2020) 5819–5829.

# 9

# Linear programming problem in a picture fuzzy environment

## Abbreviations

| | |
|---|---|
| **IFS** | Intuitionistic fuzzy set |
| **IFN** | Intuitionistic fuzzy number |
| **IF** | Intuitionistic fuzzy |
| **LPP** | Linear programming problem |
| **MOLPP** | Multiobjective LPP |
| **PF** | Picture fuzzy |
| **PFS** | Picture fuzzy set |
| **PFN** | Picture fuzzy number |
| **TrPFN** | Trapezoidal PFN |

## 9.1 Introduction

In 1965, Zadeh [15] proposed the idea of fuzzy set theory by generalizing the concept of the crisp set theory with the involvement of incomplete and uncertain information. Later, the said topic became known as fuzzy set theory. Then, the idea to solve any decision-making problem was proposed by Bellman and Zadeh [4] in a fuzzy environment. Conversely, several objective functions were handled in fuzzy linear programming by Zimmerman [16]. A method was considered by Lotfi et al. [10] to solve a complete fuzzy linear programming (LPP) with the use of a fuzzy approximate solution and a lexicography method. Kaur and Kumar [9] have presented Mehar's method to solve such types of programming problems that are fully fuzzy with the use of LR fuzzy parameters. For the sense of extension or exploration of the fuzzy set, Atanassov [3] has proposed the idea of the intuitionistic fuzzy set (IFS), which has two types of characteristic functions, i.e., membership and nonmembership. Then, several types of work have been done on IFS. Chakraborty et al. [5] proposed an intuitionistic fuzzy (IF) optimization technique. The intuitionistic fuzzy linear programming problem (IFLPP) is defined by Nagoorgani and Ponnalago [12] and Dubey and Mehera [7]. In both these works, triangular intuitionistic fuzzy numbers (TIFN) are used for solving LPP. Nachammai and Thangaraj [11] also discussed IFLPP using similarity measures. Parvathi and Malathi [13] developed a decisive set method on IFLPP. Other than linear programming in the IF environment, Chakraborty et al. [5] described an IF nonlinear programming problem. Ghosh and Roy [8] proposed and solved the IF goal geometric programming problem. Also, Pythagorean fuzzy linear programming problems were introduced by Akram et al. [2] with equality constraints. Any decision maker (DM) also wishes to use the neutral grades of any fuzzy parameter for any critical challenge of real-life problems, along with membership and nonmembership. To overcome such a difficulty, the

Picture Fuzzy Logic and Its Applications in Decision Making Problems. https://doi.org/10.1016/B978-0-44-322024-1.00013-3

novel concept of a picture fuzzy set (PFS) was first introduced by Cuong [6]. Each element of the universal set of any PFS has individual membership, nonmembership, and neutral grades. This unique concept of PFS can handle incomplete and uncertain information in the best process when any decision maker suggests the grades as yes to some extent, no to some extent, and abstain to some context. Nowadays, many researchers are interested in working with picture fuzzy sets globally.

According to the literature, all existing methods have the variables or parameters as either fuzzy numbers or intuitionistic fuzzy numbers (IFNs). They do not have the facility to choose a neutral membership function along with both the membership and nonmembership functions. Such types of limitations motivate us to work on the development of a new methodology for handling such uncertain information. Lastly, we have to note that Akram et al. [1] removed maximum drawbacks in the existing definitions, and by fixing all these, they defined a more general type of picture fuzzy numbers (PFNs), which are known as LR flat PFNs. Also, they have proposed a solution procedure for the full-picture fuzzy linear programming problems (FPFLPPs) with the use of LR flat PFNs as parameters and variables. The dealing technique is better than the fuzzy set and IFS in an uncertain environment.

The remainder of the chapter is organized as follows: In Section 9.2 we reexamine some earlier literature. In Section 9.3 we discuss the picture fuzzy linear programming problem and its development. In Section 9.4 we update the picture fuzzy optimization model. In Section 9.5 we give some examples of PFLPP that are solved to explore the suggested approach. Section 9.6 concludes the chapter with some future working directions also mentioned.

## 9.2 Preliminaries

In this section, some basic definitions are discussed.

**Definition 9.1.** [6] On a universe of discourse $X$, a PFS $Z$ is an object of the form $Z = \{(x, \mathcal{Y}_Z(x), \mathcal{A}_Z(x), \mathcal{N}_Z(x) | x \in X)\}$, where $\mathcal{Y}_Z(x), \mathcal{A}_Z(x), \mathcal{N}_Z(x)$ are membership, neutral, and nonmembership functions, respectively, of element $x \in X$ such that $\mathcal{Y}_Z(x), \mathcal{A}_Z(x), \mathcal{N}_Z(x) \in [0, 1]$ and $0 \leq \mathcal{Y}_Z(x) + \mathcal{A}_Z(x) + \mathcal{N}_Z(x) \leq 1$, for every $x \in X$. Also, the term $\{1 - \mathcal{Y}_Z(x) - \mathcal{A}_Z(x) - \mathcal{N}_Z(x)\}$ is said to be the refusal degree of $x$ to the set $Z$.

**Definition 9.2.** [14] A triplet $C = (([p_1, q, n, l_1], a), ([p_2, q, n, l_2], b), ([p_3, q, n, l_3], c))$ is a picture fuzzy number (PFN) of trapezoidal type if

$$\mathcal{Y}_C(z) = \begin{cases} \dfrac{a(z - p_1)}{q - p_1} & \text{if } \; p_1 \leq z < q \\ a & \text{if } \; q \leq z < n \\ \dfrac{a(l_1 - z)}{l_1 - n} & \text{if } \; n \leq z \leq l_1 \\ 0 & \text{if } \; z < p_1 \text{ or, } z > l_1, \end{cases}$$

$$\mathcal{A}_C(z) = \begin{cases} \dfrac{q - z + b(z - p_2)}{q - p_2} & \text{if } \; p_2 \le z < q \\ b & \text{if } \; q \le z < n \\ \dfrac{z - n + b(l_2 - z)}{l_2 - n} & \text{if } \; n \le z \le l_2 \\ 0 & \text{if } \; z < p_2 \text{ or, } z > l_2, \end{cases}$$

$$\mathcal{N}_C(z) = \begin{cases} \dfrac{q - z + c(z - p_3)}{q - p_3} & \text{if } \; p_3 \le z < q \\ c & \text{if } \; q \le z < n \\ \dfrac{z - n + c(l_3 - z)}{l_3 - n} & \text{if } \; n \le z \le l_3 \\ 0 & \text{if } \; z < p_3 \text{ or, } z > l_3 \end{cases}$$

are its membership, neutral, and nonmembership functions, respectively, such that $a, b, c \in [0, 1]$, and $0 \le a + b + c \le 1$.

However, if we consider the following number,

$$D = (([-1, 2, 4, 6], 0.5), ([-1, 2, 4, 6], 0.1), ([-1, 2, 4, 6], 0.2))$$

then, with the help of Definition 9.2, the membership, neutral, and nonmembership degrees are calculated, and the sum of these degrees is 1.4, i.e., greater than 1. This is a major drawback of the existing definition surely. Hence, the definition is modified, and the following definition is made.

**Definition 9.3.** [1] A PFS of the form, $P = [(l_3, l_2, l_1, x, y, r_1, r_2, r_3); a, b, c]$ is called a simple trapezoidal picture fuzzy number (TrPFN), whose membership ($\mathcal{Y}_P$), neutral membership ($\mathcal{A}_P$), and nonmembership ($\mathcal{N}_P$) functions are, respectively, defined as:

$$\mathcal{Y}_P(z) = \begin{cases} \dfrac{(z - l_1)a}{x - l_1} & \text{if } \; l_1 \le z \le x \\ a & \text{if } \; x \le z \le y \\ \dfrac{a(r_1 - z)}{r_1 - y} & \text{if } \; y \le z \le r_1 \\ 0 & otherwise \end{cases},$$

$$\mathcal{A}_P(z) = \begin{cases} \dfrac{(z - l_2)b}{x - l_2} & \text{if } \; l_2 \le z \le x \\ b & \text{if } \; x \le z \le y \\ \dfrac{(r_2 - z)b}{r_2 - y} & \text{if } \; y \le z \le r_2 \\ 0 & otherwise \end{cases},$$

$$\mathcal{N}_P(z) = \begin{cases} \dfrac{x - z + c(z - l_3)}{x - l_3} & \text{if } l_3 \leq z \leq x \\ c & \text{if } x \leq z \leq y \\ \dfrac{z - y + c(r_3 - z)}{r_3 - y} & \text{if } y \leq z \leq r_3 \\ 1 & \text{if } otherwise \end{cases},$$

where $l_3 \leq l_2 \leq l_1 \leq x \leq y \leq r_1 \leq r_2 \leq r_3$. The maximum degree of membership ($\mathcal{Y}_P$), maximum degree of neutral ($\mathcal{A}_P$), and minimum degree of nonmembership ($\mathcal{N}_P$), are denoted by the values $a$, $b$, and $c$, respectively, such that $a, b, c \in [0, 1]$, and $0 \leq a + b + c \leq 1$.

Note that the three quantities $a, b, c$ play a very crucial role in representing a PFN. These three quantities are not independent; they are interconnected, only one quantity is independent, and their sum is less than or equal to 1.

A TrPFN is depicted in Fig. 9.1.

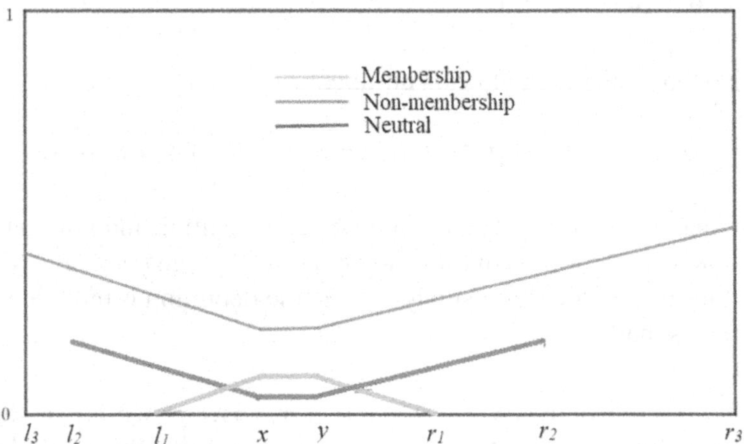


**FIGURE 9.1** Trapezoidal picture fuzzy number ($l_3 < l_2 < l_1 < x < y < r_1 < r_2 < r_3$).

In Fig. 9.1, the range of membership value is less than the neutral value, and the neutral value is less than the nonmembership value. That is, the intervals are nested with others. In this type of situation, a fewer number of DMs accepts a proposition, and most of them reject the same. However, this may not always happen. The three intervals may start and end from the same points, shown in Fig. 9.2. Here, only two possibilities of the intervals for membership, neutral, and nonmembership functions, are depicted. There are several possibilities among such intervals.

Figs. 9.1 and 9.2 represent the TrPFN, and all the lines are straight. However, it may happen that the lines may be any arbitrary curve or a specific curve. Any arbitrary curve with specific left and right sides is known as an LR PFS, as shown in Fig. 9.3.

If $x = y$, the TrPFN becomes a triangular picture fuzzy number (TPFN) by the above definition. A triangular-shaped PFN is depicted in Fig. 9.4.

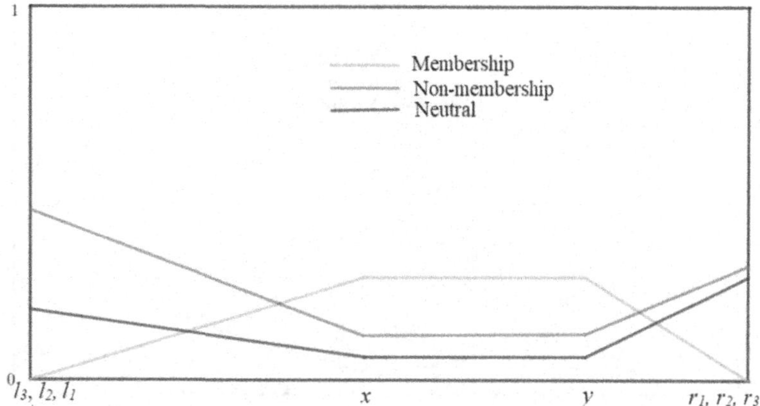


**FIGURE 9.2** Trapezoidal picture fuzzy number ($l_3 = l_2 = l_1 < x < y < r_1 = r_2 = r_3$).

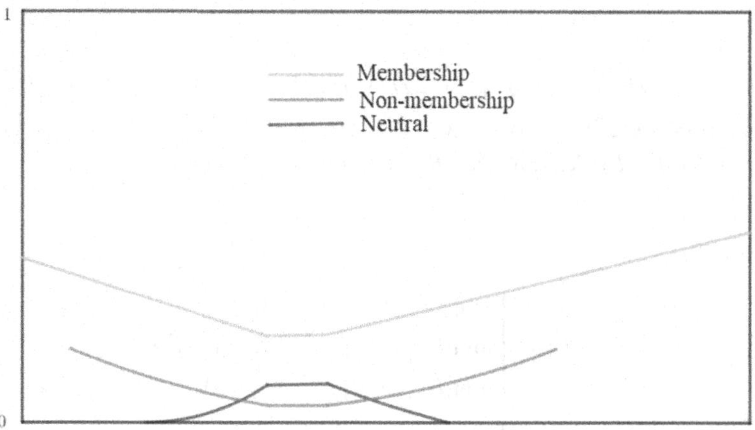


**FIGURE 9.3** LR picture fuzzy number.

The lines in Fig. 9.4 are curves, so it is not a TPFN but a triangular-shaped/typed PFN.

**Definition 9.4.** [1] A TrPFN $P = [(l_3, l_2, l_1, x, y, r_1, r_2, r_3); a, b, c]$ is nonnegative (respectively, nonpositive), denoted by $P \geq 0$ (respectively, $P \leq 0$), if $l_3 \geq 0$ (respectively, $r_3 \leq 0$) and $P$ is unrestricted if $l_3$ is a real number.

**Definition 9.5.** [1] Let $P_1 = [(l_3, l_2, l_1, x, y, r_1, r_2, r_3); a, b, c]$, and $P_2 = [(l'_3, l'_2, l'_1, x', y', r'_1, r'_2, r'_3); a', b', c']$ be two TrPFNs and $\chi$ be a real number, then:

**(i)** $P_1 \oplus P_2 = [(l_3 + l'_3, l_2 + l'_2, l_1 + l'_1, x + x', y + y', r_1 + r'_1, r_2 + r'_2, r_3 + r'_3); a + a' - aa', bb', cc']$;
**(ii)** $-P_1 = [(-r_3, -r_2, -r_1, -y, -x, -l_1, -l_2, -l_3); a, b, c]$;
**(iii)** $P_1 \ominus P_2 = P_1 \oplus (-P_2)[(l_3 - r'_3, l_2 - r'_2, l_1 - r'_1, x - y', y - x', r_1 - l'_1, r_2 - l'_2, r_3 - l'_3); a + a' - aa', bb', cc']$;

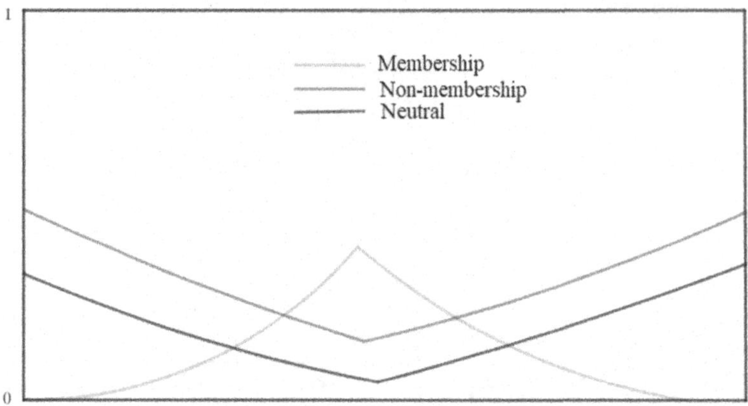


**FIGURE 9.4** Triangular-shaped picture fuzzy number.

**(iv)** $\chi P_1 = [(\chi l_3, \chi l_2, \chi l_1, \chi x, \chi y, \chi r_1, \chi r_2, \chi r_3); a^{\chi}, b^{\chi}, 1-(1-c)^{\chi}]$, for $\chi \geq 0$
and $\chi P_1 = [(\chi r_3, \chi r_2, \chi r_1, \chi y, \chi x, \chi l_1, \chi l_2, \chi l_3); a^{-\chi}, b^{-\chi}, 1-(1-c)^{-\chi}]$, for $\chi < 0$;

**(v)** $P_1 \otimes P_2 = [(L_3, L_2, L_1, X, Y, R_1, R_2, R_3); aa', bb', c+c'-cc']$,
where,

$$L_1 = \begin{cases} \min\{l_1 l_1', r_1 l_1'\}, & l_1 \geq 0 \\ \min\{l_1 r_1', r_1 l_1'\}, & l_1 < 0, r_1 \geq 0 \\ \min\{l_1 r_1', r_1 r_1'\}, & r_1 < 0, \end{cases}$$

$$L_2 = \begin{cases} \min\{l_2 l_2', r_2 l_2'\}, & l_2 \geq 0 \\ \min\{l_2 r_2', r_2 l_2'\}, & l_2 < 0, r_2 \geq 0 \\ \min\{l_2 r_2', r_2 r_2'\}, & r_2 < 0, \end{cases}$$

$$L_3 = \begin{cases} \min\{l_3 l_3', r_3 l_3'\}, & l_3 \geq 0 \\ \min\{l_3 r_3', r_3 l_3'\}, & l_3 < 0, r_3 \geq 0 \\ \min\{l_3 r_3', r_3 r_3'\}, & r_3 < 0, \end{cases}$$

$$X = \begin{cases} \min\{xx', yx'\}, & x \geq 0 \\ \min\{xy', yx'\}, & x < 0, y \geq 0 \\ \min\{xy', yy'\}, & y < 0, \end{cases}$$

$$Y = \begin{cases} \max\{xy', yy'\}, & x \geq 0 \\ \max\{xx', yy'\}, & x < 0, y \geq 0 \\ \max\{xx', yx'\}, & y < 0, \end{cases}$$

$$R_1 = \begin{cases} \max\{l_1 r'_1, r_1 r'_1\}, & l_1 \geq 0 \\ \max\{l_1 l'_1, r_1 r'_1\}, & l_1 < 0, r_1 \geq 0 \\ \max\{l_1 l'_1, r_1 l'_1\}, & r_1 < 0, \end{cases}$$

$$R_2 = \begin{cases} \max\{l_2 r'_2, r_2 r'_2\}, & l_2 \geq 0 \\ \max\{l_2 l'_2, r_2 r'_2\}, & l_2 < 0, r_2 \geq 0 \\ \max\{l_2 l'_2, r_2 l'_2\}, & r_2 < 0, \end{cases}$$

$$R_3 = \begin{cases} \max\{l_3 r'_3, r_3 r'_3\}, & l_3 \geq 0 \\ \max\{l_3 l'_3, r_3 r'_3\}, & l_3 < 0, r_3 \geq 0 \\ \max\{l_3 l'_3, r_3 l'_3\}, & r_3 < 0. \end{cases}$$

**Definition 9.6.** [1] A PFN $Z = [(x, y; m_1, n_1; m_2, n_2; m_3, n_3); a, b, c]_{LR}$ is defined as a simple LR flat PFN, if its membership ($\mathcal{Y}_Z$), neutral ($\mathcal{A}_Z$), and nonmembership ($\mathcal{N}_Z$) functions are given by:

$$\mathcal{Y}_Z(z) = \begin{cases} L_1\left(\left(\frac{x-z}{m_1}\right)a\right) & \text{if } \ z \leq x, m_1 > 0 \\ a & \text{if } \ x \leq z \leq y \\ R_1\left(\left(\frac{z-y}{n_1}\right)a\right) & \text{if } \ z \leq y,\ n_1 > 0, \end{cases}$$

$$\mathcal{A}_Z(z) = \begin{cases} L_2\left(\left(\frac{x-z}{m_2}\right)b\right) & \text{if } \ z \leq x,\ m_2 > 0 \\ b & \text{if } \ x \leq z \leq y \\ R_2\left(\left(\frac{z-y}{n_2}\right)b\right) & \text{if } \ z \geq y,\ n_2 > 0, \end{cases}$$

$$\mathcal{N}_Z(z) = \begin{cases} L_3\left(\left(\frac{x-z}{m_3}\right)c\right) & \text{if } \ z \leq x,\ m_3 > 0 \\ c & \text{if } \ x \leq z \leq y \\ R_3\left(\left(\frac{z-y}{n_3}\right)c\right) & \text{if } \ z \geq y,\ n_3 > 0, \end{cases}$$

where, $m_1 \leq m_2 \leq m_3, n_1 \leq n_2 \leq n_3$ and $0 \leq \mathcal{Y}_Z(z) + \mathcal{A}_Z(z) + \mathcal{N}_Z(z) \leq 1$.

$L_1$, $R_1$ are monotone, decreasing, and continuous functions from $[0, \infty)$ to $[0, a]$, $L_2$, $R_2$ are decreasing, monotone, and continuous functions from $[0, \infty)$ to $[0, b]$ and $L_3$, $R_3$ are monotone, increasing, and continuous functions from $[0, \infty)$ to $[c, 1]$ such that

- **(i)** $L_1(0) = R_1(0) = a$;
- **(ii)** $\lim_{\beta\to\infty} R_1(\beta) = \lim_{\beta\to\infty} L_1(\beta) = 0$;
- **(iii)** $L_2(0) = R_2(0) = b$;
- **(iv)** $\lim_{\beta\to\infty} R_2(\beta) = \lim_{\beta\to\infty} L_2(\beta) = 0$;
- **(v)** $L_3(0) = R_3(0) = c$;
- **(vi)** $\lim_{\beta\to\infty} R_3(\beta) = \lim_{\beta\to\infty} L_3(\beta) = 1$.

**Note:** The above-described definition will be called a general LR flat if $L_1$, $R_1$ are monotone, decreasing, and continuous functions from $[0, \infty)$ to $[0, a]$, $L_3$, $R_3$ are monotone, increasing, and continuous functions from $[0, \infty)$ to $[c, 1]$. Whereas, $L_2$ is a continuous, monotone, increasing/decreasing function from $[0, \infty)$ to $[b, \min\{1-(a+c), 1-\mathcal{N}_Z(x-m_2)\}]/[\min\{1-(a+c), 1-\mathcal{N}_Z(x-m_2), b\}]$, $R_2$ is a continuous, monotone, increasing/decreasing function from $[0, \infty)$ to $[b, \min\{1-(a+c), 1-\mathcal{N}_Z(y+n_2)\}]/[min\{1-(a+c), 1-\mathcal{N}_Z(y+n_2), b\}]$.

**Remark 9.1.** If we take the following,

$$L_1(\beta) = R_1(\beta) = \begin{cases} a-\beta, & 0 \leq \beta \leq a \\ 0, & \text{otherwise,} \end{cases}$$

$$L_2(\beta) = R_2(\beta) = \begin{cases} b-\beta, & 0 \leq \beta \leq b \\ 0, & \text{otherwise,} \end{cases}$$

$$L_3(\beta) = R_3(\beta) = \begin{cases} \dfrac{\beta - c\beta + c^2}{c}, & 0 \leq \beta \leq c \\ 1, & \text{otherwise,} \end{cases}$$

then by Definition 9.6 this represents a TrPFN.

**Definition 9.7.** [1] An LR flat PFN, $Z = [(x, y; m_1, n_1; m_2, n_2; m_3, n_3); a, b, c]_{LR}$ is said to be nonnegative (respectively, nonpositive), denoted as, $Z \geq 0$ (respectively, $Z \leq 0$) if $x - m_3 \geq 0$ (respectively, $y + n_3 \leq 0$) and $Z$ is unrestricted if $x$ or, $y$ belong to real numbers.

**Definition 9.8.** [1] An LR flat PFN, $Z = [(x, y; m_1, n_1; m_2, n_2; m_3, n_3); a, b, c]_{LR}$ is said to be positive if $x - m_3 > 0$ and negative if $y + n_3 < 0$.

**Definition 9.9.** [1] An LR flat PFN, $Z = [(x, y; m_1, n_1; m_2, n_2; m_3, n_3); a, b, c]_{LR}$ is said to be zero if and only if, $x = 0$, $y = 0$, $m_1 = 0$, $n_1 = 0$, $m_2 = 0$, $n_2 = 0$, $m_3 = 0$, $n_3 = 0$.

**Definition 9.10.** [1] Two LR flat PFNs, $Z_1 = [(x, y; m_1, n_1; m_2, n_2; m_3, n_3); a, b, c]_{LR}$ and $Z_2 = [(x', y'; m'_1, n'_1; m'_2, n'_2; m'_3, n'_3); a', b', c']_{LR}$ are said to be equal if, $x = x'$, $y = y'$, $m_1 = m'_1$, $n_1 = n'_1$, $m_2 = m'_2$, $n_2 = n'_2$, $m_3 = m'_3$, $n_3 = n'_3$, $a = a'$, $b = b'$, $c = c'$.

## 9.3 Some results

In this section, some interesting results with their compelling proofs are illustrated.

**Theorem 9.1.** *Let* $Z_1 = [(x, y; m_1, n_1; m_2, n_2; m_3, n_3); a, b, c]_{LR}$ *and* $Z_2 = [(x', y'; m'_1, n'_1; m'_2, n'_2; m'_3, n'_3); a', b', c']_{LR}$ *be two LR flat PFNs, then* $Z_1 \oplus Z_2 = [(x + x', y + y'; m_1 + m'_1, n_1 + n'_1; m_2 + m'_2, n_2 + n'_2; m_3 + m'_3, n_3 + n'_3); a + a' - aa', bb', cc']_{LR}$.

*Proof.* Let $Z_1 = [(x, y; m_1, n_1; m_2, n_2; m_3, n_3); a, b, c]_{LR}$ and $Z_2 = [(x', y'; m_1', n_1'; m_2', n_2'; m_3', n_3'); a', b', c']_{LR}$ be two LR flat PFNs, then their $\chi$-cut, $\phi$-cut, $\psi$-cut, $\chi'$-cut, $\phi'$-cut, $\psi'$-cut are given as:

$$Z_{1\chi} = \left[x - \frac{m_1}{a} L_1^{-1}(\chi), y + \frac{n_1}{a} R_1^{-1}(\chi)\right],$$
$$Z_1^{\phi} = \left[x - \frac{m_2}{b} L_2^{-1}(\phi), y + \frac{n_2}{b} R_2^{-1}(\phi)\right],$$
$${}^{\psi}Z_1 = \left[x - \frac{m_3}{c} L_3^{-1}(\psi), y + \frac{n_3}{c} R_3^{-1}(\psi)\right],$$
$$Z_{2\chi'} = \left[x' - \frac{m_1'}{a'} L_1^{-1}(\chi'), y' + \frac{n_1'}{a'} R_1^{-1}(\chi')\right],$$
$$Z_2^{\phi'} = \left[x' - \frac{m_2'}{b'} L_2^{-1}(\phi'), y' + \frac{n_2'}{b'} R_2^{-1}(\phi')\right],$$
$${}^{\psi'}Z_2 = \left[x' - \frac{m_3'}{c'} L_3^{-1}(\psi'), y' + \frac{n_3'}{c'} R_3^{-1}(\psi')\right].$$

Thus

$$Z_{1\chi} + Z_{2\chi'} = \left[x - \frac{m_1}{a} L_1^{-1}(\chi) + x' - \frac{m_1'}{a'} L_1^{-1}(\chi'), y + \frac{n_1}{a} R_1^{-1}(\chi) + y' + \frac{n_1'}{a'} R_1^{-1}(\chi')\right]. \quad (9.1)$$

By taking $\chi = a$, and $\chi' = a'$ in Eq. (9.1), we obtain

$$Z_{1\chi=a} + Z_{2\chi'=a'} = [x + x', y + y']. \quad (9.2)$$

By taking $\chi = 0$, and $\chi' = 0$ in Eq. (9.1), we have

$$Z_{1\chi=0} + Z_{2\chi'=0} = [x + x' - m_1 - m_1', y + y' + n_1 + n_1']. \quad (9.3)$$

Now,

$$Z_1^{\phi} + Z_2^{\phi'} = \left[x - \frac{m_2}{b} L_2^{-1}(\phi) + x' - \frac{m_2'}{b'} L_2^{-1}(\phi'), y + \frac{n_2}{b} R_2^{-1}(\phi) + y' - \frac{n_2'}{b'} R_2^{-1}(\phi')\right]. \quad (9.4)$$

By taking $\phi = b$, and $\phi' = b'$ in Eq. (9.4), we obtain

$$Z_1^{\phi=b} + Z_2^{\phi'=b'} = [x + x', y + y']. \quad (9.5)$$

By taking $\phi = 0$, and $\phi' = 0$ in Eq. (9.4), we have

$$Z_1^{\phi=0} + Z_2^{\phi'=0} = [x + x' - m_2 - m_2', y + y' + n_2 + n_2']. \quad (9.6)$$

Now,

$${}^{\psi}Z_1 +{}^{\psi'} Z_2 = \left[x - \frac{m_3}{c} L_3^{-1}(\psi) + x' - \frac{m_3'}{c'} L_3^{-1}(\psi'), y + \frac{n_3}{c} R_3^{-1}(\psi) + y' + \frac{n_3'}{c'} R_3^{-1}(\psi')\right]. \quad (9.7)$$

By placing $\psi = c$, and $\psi' = c'$ in Eq. (9.7), we obtain

$$^{\psi=c}Z_1 +^{\psi'=c'} Z_2 = [x + x', y + y']. \tag{9.8}$$

By choosing $\psi = 1$, and $\psi' = 1$ in Eq. (9.7), we obtain

$$^{\psi=1}Z_1 +^{\psi'=1} Z_2 = [x + x' - m_3 - m_3', y + y' + n_3 + n_3']. \tag{9.9}$$

We have combined Eqs. (9.2)–(9.9), and the result is deduced. □

**Theorem 9.2.** *Let* $Z_1 = [(x, y; m_1, n_1; m_2, n_2; m_3, n_3); a, b, c]_{LR}$, *and* $Z_2 = [(x', y'; m_1', n_1'; m_2', n_2'; m_3', n_3'); a', b', c']_{LR}$ *be two LR flat PFNs, then* $Z_1 \ominus Z_2 = [(x - y', y - x'; m_1 + n_1', n_1 + m_1'; m_2 + n_2', n_2 + m_2'; m_3 + n_3', n_3 + m_3'); a + a' - aa', bb', cc']_{LR}$.

*Proof.* The result can be proved easily by following a similar process and steps as those for Theorem 9.1. □

**Theorem 9.3.** *Let* $Z_1 = [(x, y; m_1, n_1; m_2, n_2; m_3, n_3); a, b, c]_{LR}$ *be an LR flat PFN in which* $x - m_3 < 0$, $x - m_2 \geq 0$, *and* $Z_2 = [(x', y'; m_1', n_1'; m_2', n_2'; m_3', n_3'); a', b', c']_{LR}$ *be an unrestricted LR flat PFN, then*

$$Z_1 \otimes Z_2 = [(X, Y; M_1, N_1; M_2, N_2; M_3, N_3); aa', bb', c + c' - cc'] - LR,$$

*where,*

$$\begin{aligned}
&X = \min\{xx', yx'\},\ Y = \max\{xy', yy'\},\\
&M_1 = \min\{xx', yx'\} - \min\{xx' - xm_1' - x'm_1 + m_1m_1', yx' - ym_1' + x'n_1 - n_1m_1'\},\\
&N_1 = \max\{xy' + xn_1' - n'm_1 - m_1n_1', yy' + yn_1' + y'n_1 + n_1n_1'\} - \max\{xy', yy'\},\\
&M_2 = \min\{xx', yx'\} - \min\{xx' - xm_2' - x'm_2 + m_2m_2', yx' - ym_2' + x'n_2 - n_2m_2'\},\\
&N_2 = \max\{xy' + xn_2' - y'm_2 - m_2n_2', yy' + yn_2' + y'n_2 + n_2n_2'\} - \max\{xy', yy'\},\\
&M_3 = \min\{xx', yx'\} - \min\{xy' + xn_3' - y'm_3 - m_3n_3', yx' - ym_3' + x'n_3 - m_3'n_3\},\\
&N_3 = \max\{xx' - xm_3' - x'm_3 + m_3m_3', yy' + yn_3' + y'n_3 + n_3n_3'\} - \max\{xy', yy'\}.
\end{aligned}$$

*Proof.* Let $Z_1 = [(x, y; m_1, n_1; m_2, n_2; m_3, n_3); a, b, c]_{LR}$ and $Z_2 = [(x', y'; m_1', n_1'; m_2', n_2'; m_3', n_3'); a', b', c']_{LR}$ be two LR flat PFNs, then their $\chi$-cut, $\phi$-cut, $\psi$-cut, $\chi'$-cut, $\phi'$-cut, $\psi'$-cut are given as:

$$\begin{aligned}
Z_{1\chi} &= \left[x - \frac{m_1}{a}L_1^{-1}(\chi), y + \frac{n_1}{a}R_1^{-1}(\chi)\right],\\
Z_1^{\phi} &= \left[x - \frac{m_2}{b}L_2^{-1}(\phi), y + \frac{n_2}{b}R_2^{-1}(\phi)\right],\\
{}^{\psi}Z_1 &= \left[x - \frac{m_3}{c}L_3^{-1}(\psi), y + \frac{n_3}{c}R_3^{-1}(\psi)\right],\\
Z_{2\chi'} &= \left[x' - \frac{m_1'}{a'}L_1^{-1}(\chi'), y' + \frac{n_1'}{a'}R_1^{-1}(\chi')\right],
\end{aligned}$$

$$Z_2^{\phi'} = \left[x' - \frac{m_2'}{b'}L_2^{-1}(\phi'), y' + \frac{n_2'}{b'}R_2^{-1}(\phi')\right],$$

$$^{\psi'}Z_2 = \left[x' - \frac{m_3'}{c'}L_3^{-1}(\psi'), y' + \frac{n_3'}{c'}R_3^{-1}(\psi')\right].$$

Thus

$$\begin{aligned} &Z_{1\chi} Z_{2\chi'} \\ &= \Bigg[\min\left\{\left(x - \frac{m_1}{a}L_1^{-1}(\chi)\right)\left(x' - \frac{m_1'}{a'}L_1^{-1}(\chi')\right), \left(y + \frac{n_1}{a}R_1^{-1}(\chi)\right)\left(x' - \frac{m_1'}{a'}L_1^{-1}(\chi')\right)\right\}, \\ &\max\left\{\left(x - \frac{m_1}{a}L_1^{-1}(\chi)\right)\left(y' + \frac{n_1'}{a'}R_1^{-1}(\chi')\right), \left(y + \frac{n_1}{a}R_1^{-1}(\chi)\right)\left(y' + \frac{n_1'}{a'}R_1^{-1}(\chi')\right)\right\}\Bigg]. \end{aligned} \tag{9.10}$$

By taking $\chi = a$, and $\chi' = a'$ in Eq. (9.10), we obtain

$$Z_{1\chi=a} Z_{2\chi'=a'} = [\min\{xx', yx'\}, \max\{xy', yy'\}]. \tag{9.11}$$

By taking $\chi = 0$, and $\chi' = 0$ in Eq. (9.10), we obtain

$$\begin{aligned} Z_{1\chi=0} Z_{2\chi'=0} = [&\min\{xx' - xm_1' - x'm_1 + m_1m_1', yx' - ym_1' + x'n_1 - n_1m_1'\}, \\ &\max\{xy' + xn_1' - y'm_1 - m_1n_1', yy' + yn_1' + y'n_1 + n_1n_1'\}]. \end{aligned} \tag{9.12}$$

Now,

$$\begin{aligned} Z_1^{\phi} Z_2^{\phi'} = \Bigg[&\min\left\{\left(x - \frac{m_2}{b}L_2^{-1}(\phi)\right)\left(x' - \frac{m_2'}{b'}L_2^{-1}(\phi')\right), \left(y + \frac{n_2}{b}R_2^{-1}(\phi)\right)\left(x' - \frac{m_2'}{b'}L_2^{-1}(\phi')\right)\right\}, \\ &\max\left\{\left(x - \frac{m_2}{b}L_2^{-1}(\phi)\right)\left(y' + \frac{n_2'}{b'}R_2^{-1}(\phi')\right), \left(y + \frac{n_2}{b}R_2^{-1}(\phi)\right)\left(y' + \frac{n_2'}{b'}R_2^{-1}(\phi')\right)\right\}\Bigg]. \end{aligned} \tag{9.13}$$

By putting $\phi = b$, and $\phi' = b'$ in Eq. (9.13), we obtain

$$Z_1^{\phi=b} Z_2^{\phi'=b'} = [\min\{xx', yx'\}, \max\{xy', yy'\}]. \tag{9.14}$$

By considering $\phi = 0$, and $\phi' = 0$ in Eq. (9.13), we obtain

$$\begin{aligned} Z_1^{\phi=0} Z_2^{\phi'=0} = [&\min\{xx' - xm_2' - x'm_2 + m_2m_2', yx' - ym_2' + x'n_2 - n_2m_2'\}, \\ &\max\{xy' + xn_2' - y'm_2 - m_2n_2', yy' + yn_2' + y'n_2 + n_1n_1'\}]. \end{aligned} \tag{9.15}$$

Now, these two cases may arise:

$$x - \frac{m_3}{c}L_3^{-1}(\psi) < 0 \text{ for } \psi > L_3(\frac{xn_1}{m_3}), \text{ and}$$
$$x - \frac{m_3}{c}L_3^{-1}(\psi) \geq 0 \text{ for } \psi \leq L_3(\frac{xn_1}{m_3}).$$

For the case $\left(x - \frac{m_3}{c} L_3^{-1}(\psi) < 0 \text{ for } \psi > L_3\left(\frac{xn_1}{m_3}\right)\right)$:

$$\begin{aligned} &{}^{\psi}Z_1^{\psi'} Z_2 \\ &= \left[\min\left\{\left(x - \frac{m_3}{c} L_3^{-1}(\psi)\right)\left(x' - \frac{m_3'}{c'} L_3^{-1}(\psi')\right), \left(y + \frac{n_3}{c} R_3^{-1}(\psi)\right)\left(x' - \frac{m_3'}{c'} L_3^{-1}(\psi')\right)\right\},\right. \\ &\left.\max\left\{\left(x - \frac{m_3}{c} L_3^{-1}(\psi)\right)\left(y' + \frac{n_3'}{c'} R_3^{-1}(\psi')\right), \left(y + \frac{n_3}{c} R_3^{-1}(\psi)\right)\left(y' + \frac{n_3'}{c'} R_3^{-1}(\psi')\right)\right\}\right]. \end{aligned} \tag{9.16}$$

By putting $\psi = c$, and $\psi' = c'$ in Eq. (9.16), we obtain

$$^{\psi=c}Z_1^{\psi'=c'} Z_2 = [\min\{xx', yx'\}, \max\{xy', yy'\}]. \tag{9.17}$$

For the case $(x - \frac{m_3}{c} L_3^{-1}(\psi) \geq 0 \text{ for } \psi \leq L_3(\frac{xn_1}{m_3}))$:

$$\begin{aligned} &{}^{\psi}Z_1^{\psi'} Z_2 \\ &= \left[\min\left\{\left(x - \frac{m_3}{c} L_3^{-1}(\psi)\right)\left(y' + \frac{n_3'}{c'} R_3^{-1}(\psi')\right), \left(y + \frac{n_3}{c} R_3^{-1}(\psi)\right)\left(x' - \frac{m_3'}{c'} L_3^{-1}(\psi')\right)\right\},\right. \\ &\left.\max\left\{\left(x - \frac{m_3}{c} L_3^{-1}(\psi)\right)\left(x' - \frac{m_3'}{c'} L_3^{-1}(\psi')\right), \left(y + \frac{n_3}{c} R_3^{-1}(\psi)\right)\left(y' + \frac{n_3'}{c'} R_3^{-1}(\psi')\right)\right\}\right]. \end{aligned} \tag{9.18}$$

By taking $\psi = 1$, and $\psi' = 1$ in Eq. (9.18), we obtain

$$\begin{aligned} ^{\psi=1}Z_1^{\psi'=1} Z_2 = [&\min\{xy' + xn_3' - y'm_3 - m_3n_3', yx' - ym_3' + x'n_3 - m_3'n_3\}, \\ &\max\{xx' - xm_3' - x'm_3 + m_3m_3', yy' + yn_3' + y'n_3 + n_3n_3'\}]. \end{aligned} \tag{9.19}$$

By combining Eqs. (9.11)–(9.19), the required results are proved. □

The following results can be easily deduced by following the same procedure as used in Theorem 9.3.

**Theorem 9.4.** *Let* $Z_1 = [(x, y; m_1, n_1; m_2, n_2; m_3, n_3); a, b, c]_{LR}$ *and* $Z_2 = [(x', y'; m_1', n_1'; m_2', n_2'; m_3', n_3'); a', b', c']_{LR}$ *be two nonnegative LR flat PFNs, then*

$$Z_1 \otimes Z_2 = [(X, Y; M_1, N_1; M_2, N_2; M_3, N_3); aa', bb', c + c' - cc']_{LR},$$

*where,*

$$\begin{aligned} &X = xx', \ Y = yy', \\ &M_1 = xm_1' + m_1x' - m_1m_1', \ N_1 = yn_1' + n_1y' + n_1n_1', \\ &M_2 = xm_2' + m_2x' - m_2m_2', \ N_2 = yn_2' + n_2y' + n_2n_2', \\ &M_3 = xm_3' + m_3x' - m_3m_3', \ N_3 = yn_3' + n_3y' + n_3n_3'. \end{aligned}$$

**Theorem 9.5.** *Let* $Z_1 = [(x, y; m_1, n_1; m_2, n_2; m_3, n_3); a, b, c]_{LR}$ *be a nonnegative LR flat PFN, and* $Z_2 = [(x', y'; m'_1, n'_1; m'_2, n'_2; m'_3, n'_3); a', b', c']_{LR}$ *be a nonpositive LR flat PFN, then*

$$Z_1 \otimes Z_2 = [(X, Y; M_1, N_1; M_2, N_2; M_3, N_3); aa', bb', c + c' - cc']_{LR},$$

*where,*

$$\begin{aligned}
&X = yx',\ Y = xy',\\
&M_1 = ym'_1 - n_1x' + n_1m'_1,\ N_1 = xn'_1 - m_1y' - m_1n'_1,\\
&M_2 = ym'_2 - n_2x' + n_2m'_2,\ N_2 = xn'_2 - m_2y' - m_2n'_2,\\
&M_3 = ym'_3 - n_3x' + n_3m'_3,\ N_3 = xn'_3 - m_3y' - m_3n'_3.
\end{aligned}$$

**Theorem 9.6.** *Let* $Z = [(x, y; m_1, n_1; m_2, n_2; m_3, n_3); a, b, c]_{LR}$ *be an LR flat PFN and* $d$ *be any real number, then*

$$dZ = \begin{cases} [(dx, dy; dm_1, dn_1; dm_2, dn_2; dm_3, dn_3); a^d, b^d, 1 - (1 - c)^d]_{LR} & \text{for } d \geq 0 \\ [(dy, dx; dn_1, dm_1; dn_2, dm_2; dn_3, dm_3); a^{-d}, b^{-d}, 1 - (1 - c)^{-d}]_{LR} & \text{for } d < 0. \end{cases}$$

**Definition 9.11.** [1] Let $Z = [(x, y; m_1, n_1; m_2, n_2; m_3, n_3); a, b, c]_{LR}$ be an LR flat PFN, then the ranking of $Z$, denoted by $\Re(Z)$, is defined as follows:

$$\Re(Z) = \frac{1}{2}\Bigg[\int_0^a (x - \frac{m_1}{a}L_1^{-1}(z))dz + \int_0^a (y + \frac{n_1}{a}R_1^{-1}(z))dz + \int_0^b (x - \frac{m_2}{b}L_2^{-1}(z))dz$$
$$+ \int_0^b (y + \frac{n_2}{b}R_2^{-1}(z))dz + \int_c^1 (x - \frac{m_3}{c}L_3^{-1}(z))dz + \int_c^1 (y + \frac{n_3}{c}R_3^{-1}(z))dz\Bigg].$$

**Definition 9.12.** [1] Let $Z_1$ and $Z_2$ be two LR flat PFNs, then

**(i)** $Z_1 \prec Z_2$ if $\Re(Z_1) < \Re(Z_2)$;
**(ii)** $Z_1 \succ Z_2$ if $\Re(Z_1) > \Re(Z_2)$;
**(iii)** $Z_1 \approx Z_2$ if $\Re(Z_1) = \Re(Z_2)$.

**Remark 9.2.** For TrPFN ($L_1(z) = R_1(z) = \max\{0, a - z\}$, $L_2(z) = R_2(z) = \max\{0, b - z\}$, $L_3(z) = R_3(z) = \min\left\{1, \dfrac{z - cz + c^2}{c}\right\}$), the ranking function reduces to:

$$\frac{1}{4}\{a(l_1 + x + y + r_1) + b(l_2 + x + y + r_2) + (1 - c)(l_3 + x + y + r_3)\},$$

where, $m_1 = x - l_1, n_1 = r_1 - y, m_2 = x - l_2, n_2 = r_2 - y, m_3 = x - l_3, n_3 = r_3 - y$.

## 9.4 Methodology for solving FPFLPP with LR flat PFNs

Let us consider FPFLPP with LR flat PFNs:

$$\text{Max/Min} \sum_{j=1}^{m} D_j \otimes Z_j \tag{9.20}$$

subject to

$$\sum_{j=1}^{m} B_{ij} \otimes Z_j \prec, \approx, \succ A_i \forall i = 1, 2, ....n;$$

where, $B_{ij}$, $Z_j$, $A_i$, and $D_j$ are all LR flat PFNs.

Now, separating the considered constraints into three conditions:

$$\sum_{j=1}^{p} B_{kj} \otimes Z_j \prec A_k, \forall k \in X_1,$$
$$\sum_{j=1}^{p} B_{uj} \otimes Z_j \approx A_u, \forall u \in X_2,$$
$$\sum_{j=1}^{p} B_{vj} \otimes Z_j \succ A_v, \forall v \in X_3,$$

the above-considered FPFLPP (9.20) can be rewritten as:

$$\text{Max/Min} \sum_{j=1}^{p} D_j \otimes Z_j \tag{9.21}$$

subject to

$$\sum_{j=1}^{p} B_{kj} \otimes Z_j \prec A_k, \forall k \in X_1,$$
$$\sum_{j=1}^{p} B_{uj} \otimes Z_j \approx A_u, \forall u \in X_2,$$
$$\sum_{j=1}^{p} B_{vj} \otimes Z_j \succ A_v, \forall v \in X_3,$$

where $Z_j$ are LR flat PFNs, and

$$X_1 = \{i; 1 \leq i \leq n, \sum_{j=1}^{p} B_{ij} \otimes Z_j \prec A_i\},$$

$$X_2 = \{i; 1 \leq i \leq n, \sum_{j=1}^{p} B_{ij} \otimes Z_j \approx A_i\},$$

$$X_3 = \{i; 1 \leq i \leq n, \sum_{j=1}^{p} B_{ij} \otimes Z_j \succ A_i\}.$$

Now, we assume that,

$$\begin{aligned} B_{ij} &= [(x_{ij}^n, x_{ij}^m; m_{ij}^1, n_{ij}^1; m_{ij}^2, n_{ij}^2; m_{ij}^3, n_{ij}^3); a_{ij}, b_{ij}, c_{ij}]_{LR}, \\ Z_j &= [(z_j^n, z_j^m; y_j^1, h_j^1; y_j^2, h_j^2; y_j^3, h_j^3); \xi_j, \phi_j, \theta_j]_{LR}, \\ A_i &= [(q_i^n, q_i^m; g_i^1, f_i^1; g_i^2, f_i^2; g_i^3, f_i^3); \sigma_i, \delta_i, \lambda_i]_{LR} \end{aligned}$$

and

$$D_j = [(d_j^n, d_j^m; N_j^1, e_j^1; N_j^2, e_j^2; N_j^3, e_j^3); \varepsilon_j, \kappa_j, \omega_j]_{LR}.$$

Then, FPFLPP (9.21) can be rewritten as:

$$\text{Max/Min} \sum_{j=1}^{p} [(d_j^n, d_j^m; N_j^1, e_j^1; N_j^2, e_j^2; N_j^3, e_j^3); \varepsilon_j, \kappa_j, \omega_j]_{LR} \tag{9.22}$$

$$\otimes [(z_j^n, z_j^m; y_j^1, h_j^1; y_j^2, h_j^2; y_j^3, h_j^3); \xi_j, \phi_j, \theta_j]_{LR} \tag{9.23}$$

subject to

$$\begin{aligned} &\sum_{j=1}^{p} [(x_{lj}^n, x_{lj}^m; m_{lj}^1, n_{lj}^1; m_{lj}^2, n_{lj}^2; m_{lj}^3, n_{lj}^3); a_{lj}, b_{lj}, c_{lj}]_{LR} \\ &\otimes [(z_l^n, z_l^m; y_l^1, h_l^1; y_l^2, h_l^2; y_l^3, h_l^3); \xi_l, \phi_l, \theta_l]_{LR} \\ &\preceq [(q_l^n, q_l^m; g_l^1, f_l^1; g_l^2, f_l^2; g_l^2, f_l^2); \sigma_l, \delta_l, \lambda_l]_{LR} \forall l \in X_1, \\ &\sum_{j=1}^{p} [(x_{sj}^n, x_{sj}^m; m_{sj}^1, n_{sj}^1; m_{sj}^2, n_{sj}^2; m_{sj}^3, n_{sj}^3); a_{sj}, b_{sj}, c_{sj}]_{LR} \\ &\otimes [(z_s^n, z_s^m; y_s^1, h_s^1; y_s^2, h_s^2; y_s^3, h_s^3); \xi_s, \phi_s, \theta_s]_{LR} \\ &\approx [(q_s^n, q_s^m; g_s^1, f_s^1; g_s^2, f_s^2; g_s^3, f_s^3); \sigma_s, \delta_s, \lambda_s]_{LR} \forall s \in X_2, \\ &\sum_{j=1}^{p} [(x_{rj}^n, x_{rj}^m; m_{rj}^1, n_{rj}^1; m_{rj}^2, n_{rj}^2; m_{rj}^3, n_{rj}^3); a_{rj}, b_{rj}, c_{rj}]_{LR} \\ &\otimes [(z_r^n, z_r^m; y_r^1, h_r^1; y_r^2, h_r^2; y_r^3, h_r^3); \xi_r, \phi_r, \theta_r]_{LR} \\ &\succeq [(q_r^n, q_r^m; g_r^1, f_r^1; g_r^2, f_r^2; g_r^3, f_r^3); \sigma_r, \delta_r, \lambda_r]_{LR} \forall r \in X_3. \end{aligned}$$

Then, the total programming problem reduces to the following form:

$$\text{Max/Min} \sum_{j=1}^{p} [(z_j^{n'}, z_j^{m'}; y_j^{1'}, h_j^{1'}; y_j^{2'}, h_j^{2'}; y_j^{3'}, h_j^{3'}); \xi'_j, \phi'_j, \theta'_j]_{LR} \tag{9.24}$$

subject to

$$\begin{aligned}
&\sum_{j=1}^{p} [(x_{lj}^{n'}, x_{lj}^{m'}; m_{lj}^{1'}, n_{lj}^{1'}; m_{lj}^{2'}, n_{lj}^{2'}; m_{lj}^{3'}, n_{lj}^{3'}); a'_{lj}, b'_{lj}, c'_{lj}]_{LR} \\
&\preceq [(q_l^n, q_l^m; g_l^1, f_l^1; g_l^3, f_l^3; g_l^3, f_l^3); \sigma_l, \delta_l, \lambda_l]_{LR} \forall l \in X_1, \\
&\sum_{j=1}^{p} [(x_{sj}^{n'}, x_{sj}^{m'}; m_{sj}^{1'}, n_{sj}^{1'}; m_{sj}^{2'}, n_{sj}^{2'}; m_{sj}^{3'}, n_{sj}^{3'}); a'_{sj}, b'_{sj}, c'_{sj}]_{LR} \\
&\approx [(q_s^n, q_s^m; g_s^1, f_s^1; g_s^2, f_s^2; g_s^3, f_s^3); \sigma_s, \delta_s, \lambda_s]_{LR} \forall s \in X_2, \\
&\sum_{j=1}^{p} [(x_{rj}^{n'}, x_{rj}^{m'}; m_{rj}^{1'}, n_{rj}^{1'}; m_{rj}^{2'}, n_{rj}^{2'}; m_{rj}^{3'}, n_{rj}^{3'}); a'_{rj}, b'_{rj}, c'_{rj}]_{LR} \\
&\succeq [(q_r^n, q_r^m; g_r^1, f_r^1; g_r^2, f_r^2; g_r^3, f_r^3); \sigma_r, \delta_r, \lambda_r]_{LR} \forall r \in X_3,
\end{aligned}$$

where,

$$\begin{aligned}
&[(z_j^{n'}, z_j^{m'}; y_j^{1'}, h_j^{1'}; y_j^{2'}, h_j^{2'}; y_j^{3'}, h_j^{3'}); \xi'_j, \phi'_j, \theta'_j]_{LR} \\
&= [(d_j^n, d_j^m; N_j^1, e_j^1; N_j^2, e_j^2; N_j^3, e_j^3); \varepsilon_j, \kappa_j, \omega_j]_{LR} \\
&\otimes [(z_j^n, z_j^m; y_j^1, h_j^1; y_j^2, h_j^2; y_j^3, h_j^3); \xi_j, \phi_j, \theta_j]_{LR}, \\
&[(x_{lj}^{n'}, x_{lj}^{m'}; m_{lj}^{1'}, n_{lj}^{1'}; m_{lj}^{2'}, n_{lj}^{2'}; m_{lj}^{3'}, n_{lj}^{3'}); a'_{lj}, b'_{lj}, c'_{lj}]_{LR} \\
&= [(x_{lj}^n, x_{lj}^m; m_{lj}^1, n_{lj}^1; m_{lj}^2, n_{lj}^2; m_{lj}^3, n_{lj}^3); a_{lj}, b_{lj}, c_{lj}]_{LR} \\
&\otimes [(z_l^n, z_l^m; y_l^1, h_l^1; y_l^2, h_l^2; y_l^3, h_l^3); \xi_l, \phi_l, \theta_l]_{LR}, \\
&[(x_{sj}^{n'}, x_{sj}^{m'}; m_{sj}^{1'}, n_{sj}^{1'}; m_{sj}^{2'}, n_{sj}^{2'}; m_{sj}^{3'}, n_{sj}^{3'}); a'_{sj}, b'_{sj}, c'_{sj}]_{LR} \\
&= [(x_{sj}^n, x_{sj}^m; m_{sj}^1, n_{sj}^1; m_{sj}^2, n_{sj}^2; m_{sj}^3, n_{sj}^3); a_{sj}, b_{sj}, c_{sj}]_{LR} \\
&\otimes [(z_s^n, z_s^m; y_s^1, h_s^1; y_s^2, h_s^2; y_s^3, h_s^3); \xi_s, \phi_s, \theta_s]_{LR}, \\
&[(x_{rj}^{n'}, x_{rj}^{m'}; m_{rj}^{1'}, n_{rj}^{1'}; m_{rj}^{2'}, n_{rj}^{2'}; m_{rj}^{3'}, n_{rj}^{3'}); a'_{rj}, b'_{rj}, c'_{rj}]_{LR} \\
&= [(x_{rj}^n, x_{rj}^m; m_{rj}^1, n_{rj}^1; m_{rj}^2, n_{rj}^2; m_{rj}^3, n_{rj}^3); a_{rj}, b_{rj}, c_{rj}]_{LR} \\
&\otimes [(z_r^n, z_r^m; y_r^1, h_r^1; y_r^2, h_r^2; y_r^3, h_r^3); \xi_r, \phi_r, \theta_r]_{LR}.
\end{aligned}$$

Now, we have applied the ranking method of LR flat PFNs as discussed above for all objective functions and subject to conditions to obtain the following version of FPFLPP:

$$\text{Max/Min } \Re\{\sum_{j=1}^{p} [(z_j^{n'}, z_j^{m'}; y_j^{1'}, h_j^{1'}; y_j^{2'}, h_j^{2'}; y_j^{3'}, h_j^{3'}); \xi'_j, \phi'_j, \theta'_j]_{LR}\} \tag{9.25}$$

subject to

$$\Re\{\sum_{j=1}^{p}[(x_{lj}^{n'}, x_{lj}^{m'}; m_{lj}^{1'}, n_{lj}^{1'}; m_{lj}^{2'}, n_{lj}^{2'}; m_{lj}^{3'}, n_{lj}^{3'}); a'_{lj}, b'_{lj}, c'_{lj}]_{LR}\} \preceq$$
$$\Re\{[(q_l^n, q_l^m; g_l^1, f_l^1; g_l^2, f_l^2; g_l^3, f_l^3); \sigma_l, \delta_l, \lambda_l]_{LR}\} \forall l \in X_1,$$
$$\Re\{\sum_{j=1}^{p}[(x_{rj}^{n'}, x_{rj}^{m'}; m_{rj}^{1'}, n_{rj}^{1'}; m_{rj}^{2'}, n_{rj}^{2'}; m_{rj}^{3'}, n_{rj}^{3'}); a'_{rj}, b'_{rj}, c'_{rj}]_{LR}\} \succeq$$
$$\Re\{[(q_r^n, q_r^m; g_r^1, f_r^1; g_r^2, f_r^2; g_r^3, f_r^3); \sigma_r, \delta_r, \lambda_r]_{LR}\} \forall r \in X_3,$$
$$\sum_{j=1}^{p} x_{rj}^{n'} = q_r^n \forall r \in X_2, \sum_{j=1}^{p} x_{rj}^{m'} = q_r^n \forall r \in X_2,$$
$$\sum_{j=1}^{p} m_{rj}^{1'} = g_r^1 \forall r \in X_2, \sum_{j=1}^{p} n_{rj}^{1'} = f_r^1 \forall r \in X_2,$$
$$\sum_{j=1}^{p} m_{rj}^{2'} = g_r^2 \forall r \in X_2, \sum_{j=1}^{p} n_{rj}^{2'} = f_r^2 \forall r \in X_2,$$
$$\sum_{j=1}^{p} m_{rj}^{3'} = g_r^3 \forall r \in X_2, \sum_{j=1}^{p} n_{rj}^{3'} = f_r^3 \forall r \in X_2,$$
$$\sum_{j=1}^{p} a'_{rj} = \sigma_r \forall r \in X_2, \sum_{j=1}^{p} b'_{rj} = \delta_r \forall r \in X_2, \sum_{j=1}^{p} c'_{rj} = \lambda_r \forall r \in X_2,$$
$$z_j^n \le z_j^m, y_j^1 \ge 0, y_j^1 \le y_j^2 \le y_j^3; h_j^1 \ge 0, h_j^1 \le h_j^2 \le h_j^3, \xi_j \ge 0, \phi_j \ge 0, \theta_j \ge 0;$$
$$0 \le \xi_j + \phi_j + \theta_j \le 1 \forall j = 1, 2, ..., k.$$

Now, we solve the crisp version of the linear/nonlinear deduced programming problem (9.25) by any software to find all optimal solutions as:

$$\{z_j^{n*}, z_j^{m*}, y_j^{1*}, y_j^{2*}, y_j^{3*}, h_j^{1*}, h_j^{2*}, h_j^{3*}, \xi*_j, \phi*_j, \theta*_j\}.$$

Therefore we have calculated the LR flat picture fuzzy optimal solution $Z*_j$ of the considered FPFLPP by putting in all the values component-wise, and finally, we have obtained the optimized objective function as a picture fuzzy optimal solution, with the help of the following expression:

$$\sum_{j=1}^{p} D_j \otimes Z_j.$$

The deduced optimal solution of the considered FPFLPP with LR flat PFNs as parameters as well as variables obtained by our proposed methodology is also an LR flat PFN in the picture fuzzy environment.

## 9.5 Numerical example of FPFLPP

A restaurant needs around $[(450, 495; 18, 19; 21, 23; 24, 26); 0.99, 0.025, 0.025]_{LR}$ kg of nonveg items per month, and the total nonveg items consist of three different items; fish, chicken, and mutton. The cost of fish is $\$[(250, 280; 3, 4; 4, 5; 5, 6); 0.7, 0.5, 0.4]_{LR}$ per kg, the cost of chicken is $\$[(180, 200; 2, 2; 5, 5; 7, 7); 0.6, 0.4, 0.3]_{LR}$ per kg, and the cost of mutton is $\$[(650, 700; 3, 4; 4, 5; 5, 6); 0.5, 0.3, 0.3]_{LR}$ per kg. Generally, the restaurant requires at least $[(40, 60; 3, 3; 5, 5; 7, 7); 0.5, 0.4, 0.2]_{LR}$ kg of fish, not more than $[(170, 210; 2, 2; 4, 4; 6, 6); 0.5, 0.2, 0.2]_{LR}$ kg of chicken, and at least $[(110, 140; 4, 4; 5, 5; 8, 8); 0.5, 0.4, 0.3]_{LR}$ kg of mutton. How many kg of each item should be purchased to minimize the total cost?

In this problem, we consider linear $L$ and $R$ functions, i.e., if

$$Z = [(x, y; m, n; m, n; m, n); a, b, c]_{LR},$$

then

$$L_1(\beta) = R_1(\beta) = \max\{0, a - \beta\},\ L_2(\beta) = R_2(\beta) = \max\{0, b - \beta\},$$
$$L_3(\beta) = R_3(\beta) = \max\left\{1, \frac{\beta - c\beta + c^2}{c}\right\}.$$

Here, all the amounts of nonveg items in kg are LR flat PFNs and not crisp ones. They explain the fuzzy quantities. For example, $[(450, 495; 18, 19; 21, 23; 24, 26); 0.99, 0.025, 0.025]_{LR}$ kg of nonveg items vary from 432 kg to 521 kg having certainty, confusion, and negativity given by the membership, neutral, and nonmembership functions. Similarly, all the quantities are fuzzy, and their memberships represent how much fuzziness they contain.

Let $Z_1$, $Z_2$, and $Z_3$ be the amount in kg of fish, chicken, and mutton, respectively. This problem can be formulated as FPFLPP:

$$\text{Min } [[(250, 280; 3, 4; 4, 5; 5, 6); 0.7, 0.5, 0.4]_{LR}] \otimes Z_1$$
$$\oplus[[(180, 200; 2, 2; 5, 5; 7, 7); 0.6, 0.4, 0.3]_{LR}] \otimes Z_2$$
$$\oplus[[(650, 700; 3, 4; 4, 5; 5, 6); 0.5, 0.3, 0.3]_{LR}] \otimes Z_3$$

subject to

$$Z_1 \succeq [(40, 60; 3, 3; 5, 5; 7, 7); 0.5, 0.4, 0.2]_{LR},$$
$$Z_2 \preceq [(170, 210; 2, 2; 4, 4; 6, 6); 0.5, 0.2, 0.2]_{LR},$$
$$Z_3 \succeq [(110, 140; 4, 4; 5, 5; 8, 8); 0.5, 0.4, 0.3]_{LR},$$
$$Z_1 \oplus Z_2 \oplus Z_3 \approx [(450, 495; 18, 19; 21, 23; 24, 26); 0.99, 0.025, 0.025]_{LR},$$

where, $Z_1$, $Z_2$ and $Z_3$ are nonnegative LR flat PFNs.

Let

$$Z_1 = [(x^1, y^1; m_1^1, n_1^1; m_2^1, n_2^1; m_3^1, n_3^1); a^1, b^1, c^1]_{LR},$$

$$Z_2 = [(x^2, y^2; m_1^2, n_1^2; m_2^2, n_2^2; m_3^2, n_3^2); a^2, b^2, c^2]_{LR},$$
$$Z_3 = [(x^3, y^3; m_1^3, n_1^3; m_2^3, n_2^3; m_3^3, n_3^3); a^3, b^3, c^3]_{LR}.$$

Then, we have the reduced form:

$$\begin{aligned}
&\text{Min } [[(250, 280; 3, 4; 4, 5; 5, 6); 0.7, 0.5, 0.4]_{LR}]\\
&\otimes[(x^1, y^1; m_1^1, n_1^1; m_2^1, n_2^1; m_3^1, n_3^1); a^1, b^1, c^1]_{LR} \oplus [[(180, 200; 2, 2; 5, 5; 7, 7); 0.6, 0.4, 0.3]_{LR}]\\
&\otimes[(x^2, y^2; m_1^2, n_1^2; m_2^2, n_2^2; m_3^2, n_3^2); a^2, b^2, c^2]_{LR} \oplus [[(650, 700; 3, 4; 4, 5; 5, 6); 0.5, 0.3, 0.3]_{LR}]\\
&\otimes[(x^3, y^3; m_1^3, n_1^3; m_2^3, n_2^3; m_3^3, n_3^3); a^3, b^3, c^3]_{LR}
\end{aligned}$$

subject to

$$\begin{aligned}
&[(x^1, y^1; m_1^1, n_1^1; m_2^1, n_2^1; m_3^1, n_3^1); a^1, b^1, c^1]_{LR} \succeq [(40, 60; 3, 3; 5, 5; 7, 7); 0.5, 0.4, 0.2]_{LR},\\
&[(x^2, y^2; m_1^2, n_1^2; m_2^2, n_2^2; m_3^2, n_3^2); a^2, b^2, c^2]_{LR} \preceq [(170, 210; 2, 2; 4, 4; 6, 6); 0.5, 0.2, 0.2]_{LR},\\
&[(x^3, y^3; m_1^3, n_1^3; m_2^3, n_2^3; m_3^3, n_3^3); a^3, b^3, c^3]_{LR} \succeq [(110, 140; 4, 4; 5, 5; 8, 8); 0.5, 0.4, 0.3]_{LR},\\
&Z_1 \oplus Z_2 \oplus Z_3 \approx [(450, 495; 18, 19; 21, 23; 24, 26); 0.99, 0.025, 0.025]_{LR},
\end{aligned}$$

where,

$$[(x^1, y^1; m_1^1, n_1^1; m_2^1, n_2^1; m_3^1, n_3^1); a^1, b^1, c^1]_{LR},\ [(x^2, y^2; m_1^2, n_1^2; m_2^2, n_2^2; m_3^2, n_3^2); a^2, b^2, c^2]_{LR},$$
$$[(x^3, y^3; m_1^3, n_1^3; m_2^3, n_2^3; m_3^3, n_3^3); a^3, b^3, c^3]_{LR}$$

are nonnegative LR flat PFNs.

Now, simplifying some calculations, we have:

$$\begin{aligned}
&\text{Min } [(250x^1, 280y^1; 3x^1 + 247m_1^1, 4y^1 + 284n_1^1; 4x^1 + 246m_2^1, 5y^1 + 285n_2^1;\\
&5x^1 + 245m_3^1, 6y^1 + 286n_3^1); 0.7a^1, 0.5b^1, 0.4 + c^1 - 0.4c^1]_{LR}\\
&\oplus[(180x^2, 200y^2; 2x^2 + 178m_1^2, 2y^2 + 202n_1^2; 5x^2 + 175m_2^2, 5y^2 + 205n_2^2;\\
&7x^2 + 173m_3^2, 7y^2 + 207n_3^2); 0.6a^2, 0.4b^2, 0.3 + c^2 - 0.3c^2]_{LR}\\
&\oplus[(650x^3, 700y^3; 3x^3 + 647m_1^3, 4y^3 + 704n_1^3; 4x^3 + 646m_2^3, 5y^3 + 705n_2^3;\\
&5x^3 + 645m_3^3, 6y^3 + 706n_3^3); 0.5a^3, 0.3b^3, 0.3 + c^3 - 0.3c^3]_{LR}
\end{aligned}$$

subject to

$$\begin{aligned}
&[(x^1, y^1; m_1^1, n_1^1; m_2^1, n_2^1; m_3^1, n_3^1); a^1, b^1, c^1]_{LR} \succeq [(40, 60; 3, 3; 5, 5; 7, 7); 0.5, 0.4, 0.2]_{LR},\\
&[(x^2, y^2; m_1^2, n_1^2; m_2^2, n_2^2; m_3^2, n_3^2); a^2, b^2, c^2]_{LR} \preceq [(170, 210; 2, 2; 4, 4; 6, 6); 0.5, 0.2, 0.2]_{LR},\\
&[(x^3, y^3; m_1^3, n_1^3; m_2^3, n_2^3; m_3^3, n_3^3); a^3, b^3, c^3]_{LR} \succeq [(110, 140; 4, 4; 5, 5; 8, 8); 0.5, 0.4, 0.3]_{LR},\\
&Z_1 \oplus Z_2 \oplus Z_3 \approx [(450, 495; 18, 19; 21, 23; 24, 26); 0.99, 0.025, 0.025]_{LR},
\end{aligned}$$

where,

$$[(x^1, y^1; m_1^1, n_1^1; m_2^1, n_2^1; m_3^1, n_3^1); a^1, b^1, c^1]_{LR},\ [(x^2, y^2; m_1^2, n_1^2; m_2^2, n_2^2; m_3^2, n_3^2); a^2, b^2, c^2]_{LR},$$
$$[(x^3, y^3; m_1^3, n_1^3; m_2^3, n_2^3; m_3^3, n_3^3); a^3, b^3, c^3]_{LR}$$

are nonnegative LR flat PFNs.

In this part, we have performed the $\oplus$ operation for specific types of LR flat PFNs, and after calculation, we obtain:

$$\begin{aligned}
&\text{Min } [(250x^1+180x^2+650x^3, 280y^1+200y^2+700y^3;\\
&3x^1+247m_1^1+2x^2+178m_1^2+3x^3+647m_1^3, 4y^1+284n_1^1+2y^2+202n_1^2+4y^3+704n_1^3;\\
&4x^1+246m_2^1+5x^2+175m_2^2+4x^3+646m_2^3, 5y^1+285n_2^1+5y^2+205n_2^2+5y^3+705n_2^3;\\
&5x^1+245m_3^1+7x^2+173m_3^2+5x^3+645m_3^3, 6y^1+286n_3^1+7y^2+207n_3^2+6y^3+706n_3^3);\\
&0.7a^1+0.6a^2+0.5a^3-0.42a^1a^2-0.3a^2a^3-0.35a^1a^3+0.21a^1a^2a^3, 0.6b^1b^2b^3,\\
&0.036+0.054c^1+0.084c^2+0.084c^3+0.12c^1c^2+0.196c^2c^3+0.126c^1c^3+0.294c^1c^2c^3]_{LR}
\end{aligned}$$

subject to

$$\begin{aligned}
&[(x^1, y^1; m_1^1, n_1^1; m_2^1, n_2^1; m_3^1, n_3^1); a^1, b^1, c^1]_{LR} \succeq [(40, 60; 3, 3; 5, 5; 7, 7); 0.5, 0.4, 0.2]_{LR},\\
&[(x^2, y^2; m_1^2, n_1^2; m_2^2, n_2^2; m_3^2, n_3^2); a^2, b^2, c^2]_{LR} \preceq [(170, 210; 2, 2; 4, 4; 6, 6); 0.5, 0.2, 0.2]_{LR},\\
&[(x^3, y^3; m_1^3, n_1^3; m_2^3, n_2^3; m_3^3, n_3^3); a^3, b^3, c^3]_{LR} \succeq [(110, 140; 4, 4; 5, 5; 8, 8); 0.5, 0.4, 0.3]_{LR},\\
&Z_1 \oplus Z_2 \oplus Z_3 \approx [(450, 495; 18, 19; 21, 23; 24, 26); 0.99, 0.025, 0.025]_{LR},
\end{aligned}$$

where,

$$[(x^1, y^1; m_1^1, n_1^1; m_2^1, n_2^1; m_3^1, n_3^1); a^1, b^1, c^1]_{LR},$$
$$[(x^2, y^2; m_1^2, n_1^2; m_2^2, n_2^2; m_3^2, n_3^2); a^2, b^2, c^2]_{LR},$$
$$[(x^3, y^3; m_1^3, n_1^3; m_2^3, n_2^3; m_3^3, n_3^3); a^3, b^3, c^3]_{LR}$$

are nonnegative LR flat PFNs.

Using the ranking method for defuzzification as in Definition 9.11, we obtain:

$$\begin{aligned}
&\text{Min } \mathfrak{R}\{[(250x^1+180x^2+650x^3, 280y^1+200y^2+700y^3;\\
&3x^1+247m_1^1+2x^2+178m_1^2+3x^3+647m_1^3, 4y^1+284n_1^1+2y^2+202n_1^2+4y^3+704n_1^3;\\
&4x^1+246m_2^1+5x^2+175m_2^2+4x^3+646m_2^3, 5y^1+285n_2^1+5y^2+205n_2^2+5y^3+705n_2^3;\\
&5x^1+245m_3^1+7x^2+173m_3^2+5x^3+645m_3^3, 6y^1+286n_3^1+7y^2+207n_3^2+6y^3+706n_3^3);\\
&0.7a^1+0.6a^2+0.5a^3-0.42a^1a^2-0.3a^2a^3-0.35a^1a^3+0.21a^1a^2a^3, 0.6b^1b^2b^3,\\
&0.036+0.054c^1+0.084c^2+0.084c^3+0.12c^1c^2+0.196c^2c^3+0.126c^1c^3+0.294c^1c^2c^3]_{LR}\}
\end{aligned}$$

subject to

$$\mathfrak{R}[(x^1, y^1; m_1^1, n_1^1; m_2^1, n_2^1; m_3^1, n_3^1); a^1, b^1, c^1]_{LR} \succeq \mathfrak{R}[(40, 60; 3, 3; 5, 5; 7, 7); 0.5, 0.4, 0.2]_{LR},$$
$$\mathfrak{R}[(x^2, y^2; m_1^2, n_1^2; m_2^2, n_2^2; m_3^2, n_3^2); a^2, b^2, c^2]_{LR} \preceq \mathfrak{R}[(170, 210; 2, 2; 4, 4; 6, 6); 0.5, 0.2, 0.2]_{LR},$$
$$\mathfrak{R}[(x^3, y^3; m_1^3, n_1^3; m_2^3, n_2^3; m_3^3, n_3^3); a^3, b^3, c^3]_{LR} \succeq \mathfrak{R}[(110, 140; 4, 4; 5, 5; 8, 8); 0.5, 0.4, 0.3]_{LR},$$
$$x^1 + x^2 + x^3 = 450, y^1 + y^2 + y^3 = 495, m_1^1 + m_1^2 + m_1^3 = 18, n_1^1 + n_1^2 + n_1^3 = 19,$$
$$m_2^1 + m_2^2 + m_2^3 = 21, n_2^1 + n_2^2 + n_2^3 = 23, m_3^1 + m_3^2 + m_3^3 = 24, n_3^1 + n_3^2 + n_3^3 = 26,$$
$$a^1 + a^2 + a^3 - a^1a^2 - a^1a^3 - a^2a^3 + a^1a^2a^3 = 0.99, b^1b^2b^3 = 0.025, c^1c^2c^3 = 0.025,$$
$$x^1 \le y^1, m_1^1 \le m_2^1 \le m_3^1, n_1^1 \ge 0, n_1^1 \le n_2^1 \le n_3^1, a^1 \ge 0, b^1 \ge 0, c^1 \ge 0, a^1 + b^1 + c^1 \ge 0;$$
$$x^2 \le y^2, m_1^2 \le m_2^2 \le m_3^2, n_1^2 \ge 0, n_1^2 \le n_2^2 \le n_3^2, a^2 \ge 0, b^2 \ge 0, c^2 \ge 0, a^2 + b^2 + c^2 \ge 0;$$
$$x^3 \le y^3, m_1^3 \le m_2^3 \le m_3^3, n_1^3 \ge 0, n_1^3 \le n_2^3 \le n_3^3, a^3 \ge 0, b^3 \ge 0, c^3 \ge 0, a^3 + b^3 + c^3 \ge 0.$$

In the next step, we have simplified the formulae and obtain the following version of the programming problem:

$$\text{Min } \frac{1}{2}[(0.7a^1 + 0.6a^2 + 0.5a^3 - .42a^1a^2 - 0.3a^2a^3 - 0.35a^1a^3 + 0.21a^1a^2a^3)$$
$$\{250x^1 + 180x^2 + 650x^3 + 280y^1 + 200y^2 + 700y^3$$
$$+\frac{1}{2}(4y^1 + 284n_1^1 + 2y^2 + 202n_1^2 + 4y^3 + 704n_1^3 - 3x^1 - 247m_1^1 - 2x^2 - 178m_1^2 - 3x^3 - 647m_1^3)\}$$
$$+0.6b^1b^2b^3\{250x^1 + 180x^2 + 650x^3 + 280y^1 + 200y^2 + 700y^3$$
$$+\frac{1}{2}(5y^1 + 285n_2^1 + 5y^2 + 205n_2^2 + 5y^3 + 705n_2^3 - 4x^1 - 246m_2^1 - 5x^2 - 175m_2^2 - 4x^3 - 646m_2^3)\}$$
$$+\{1 - (0.036 + 0.054c^1 + 0.084c^2 + 0.084c^3 + 0.12c^1c^2 + 0.19c^2c^3 + 0.126c^1c^3 + 0.294c^1c^2c^3)\}$$
$$\{250x^1 + 180x^2 + 650x^3 + 280y^1 + 200y^2 + 700y^3$$
$$+\frac{1}{2}(6y^1 + 286n_3^1 + 7y^2 + 207n_3^2 + 6y^3 + 706n_3^3 - 5x^1 - 245m_3^1 - 7x^2 - 173m_3^2 - 5x^3 - 645m_3^3)\}]$$

subject to

$$\frac{1}{2}[a^1(x^1 + y^1 - \frac{m_1^1}{2} + \frac{n_1^1}{2}) + b^1(x^1 + y^1 - \frac{m_2^1}{2} + \frac{n_2^1}{2}) + (1 - c^1)(x^1 + y^1 - \frac{m_3^1}{2} + \frac{n_3^1}{2})] \ge 95.8,$$
$$\frac{1}{2}[a^2(x^2 + y^2 - \frac{m_1^2}{2} + \frac{n_1^2}{2}) + b^2(x^2 + y^2 - \frac{m_2^2}{2} + \frac{n_2^2}{2}) + (1 - c^2)(x^2 + y^2 - \frac{m_3^2}{2} + \frac{n_3^2}{2})] \le 272.5,$$
$$\frac{1}{2}[a^3(x^3 + y^3 - \frac{m_1^3}{2} + \frac{n_1^3}{2}) + b^3(x^3 + y^3 - \frac{m_2^3}{2} + \frac{n_2^3}{2}) + (1 - c^3)(x^3 + y^3 - \frac{m_3^3}{2} + \frac{n_3^3}{2})] \ge 196,$$
$$x^1 + x^2 + x^3 = 450, y^1 + y^2 + y^3 = 495, m_1^1 + m_1^2 + m_1^3 = 18, n_1^1 + n_1^2 + n_1^3 = 19,$$
$$m_2^1 + m_2^2 + m_2^3 = 21, n_2^1 + n_2^2 + n_2^3 = 23, m_3^1 + m_3^2 + m_3^3 = 24, n_3^1 + n_3^2 + n_3^3 = 26,$$
$$a^1 + a^2 + a^3 - a^1a^2 - a^1a^3 - a^2a^3 + a^1a^2a^3 = 0.99, b^1b^2b^3 = 0.025, c^1c^2c^3 = 0.025,$$

$$x^1 \leq y^1, m_1^1 \leq m_2^1 \leq m_3^1, n_1^1 \geq 0, n_1^1 \leq n_2^1 \leq n_3^1, a^1 \geq 0, b^1 \geq 0, c^1 \geq 0, a^1 + b^1 + c^1 \geq 0;$$
$$x^2 \leq y^2, m_1^2 \leq m_2^2 \leq m_3^2, n_1^2 \geq 0, n_1^2 \leq n_2^2 \leq n_3^2, a^2 \geq 0, b^2 \geq 0, c^2 \geq 0, a^2 + b^2 + c^2 \geq 0;$$
$$x^3 \leq y^3, m_1^3 \leq m_2^3 \leq m_3^3, n_1^3 \geq 0, n_1^3 \leq n_2^3 \leq n_3^3, a^3 \geq 0, b^3 \geq 0, c^3 \geq 0, a^3 + b^3 + c^3 \geq 0.$$

In the next step, we used the mathematical software 'Mathematica' to find the optimal solution to this programming problem in a crisp version, and the values are:

$x^1 = 407, y^1 = 445, m_1^1 = 0.907, n_1^1 = 1.606, m_2^1 = 2.492, n_2^1 = 3.082, m_3^1 = 17.181, n_3^1 = 17.921,$
$a^1 = 0.203, b^1 = 0.038, c^1 = 0.235;$
$x^2 = 36, y^2 = 40, m_1^2 = 0.774, n_1^2 = 0.936, m_2^2 = 1.819, n_2^2 = 1.858, m_3^2 = 3.369, n_3^2 = 3.204,$
$a^2 = 0.243 \times 10^{-2}, b^2 = 0.327, c^2 = 0.475;$
$x^3 = 7, y^3 = 10, m_1^3 = 1.903, n_1^3 = 2.409, m_2^3 = 1.679, n_2^3 = 3.481, m_3^3 = 3.448, n_3^3 = 4.875,$
$a^3 = 0.853, b^3 = 0.459, c^3 = 0.234.$

Putting all these values in

$$Z_1 = [(x^1, y^1; m_1^1, n_1^1; m_2^1, n_2^1; m_3^1, n_3^1); a^1, b^1, c^1]_{LR},$$
$$Z_2 = [(x^2, y^2; m_1^2, n_1^2; m_2^2, n_2^2; m_3^2, n_3^2); a^2, b^2, c^2]_{LR},$$
$$Z_3 = [(x^3, y^3; m_1^3, n_1^3; m_2^3, n_2^3; m_3^3, n_3^3); a^3, b^3, c^3]_{LR};$$

the actual solution in the form of LR flat picture fuzzy type is:

$$Z_1 = [(407, 445; 0.907, 1.606; 2.492, 3.082; 17.181, 17.921); 0.203, 0.038, 0.235]_{LR},$$
$$Z_2 = [(36, 40; 0.774, 0.936; 1.819, 1.858; 3.369, 3.204); 0.243 \times 10^{-2}, 0.327, 0.475]_{LR},$$
$$Z_3 = [(7, 10; 1.903, 2.409; 1.679, 3.481; 3.448, 4.875); 0.853, 0.459, 0.234]_{LR}.$$

Substituting the values of $Z_1$, $Z_2$, and $Z_3$, obtained in the previous step, into the objective function component-wise, then the picture fuzzy optimal value is:

$$[(450, 495; 3.584, 4.951; 5.99, 8.421; 23.998, 26); 0.42 \times 10^{-4}, 0.0057, 0.692]_{LR}.$$

Thus the owner of the restaurant should purchase or order

$[(407, 445; 0.907, 1.606; 2.492, 3.082; 17.181, 17.921); 0.203, 0.038, 0.235]_{LR}$ kg of fish items,
$[(36, 40; 0.774, 0.936; 1.819, 1.858; 3.369, 3.204); 0.243 \times 10^{-2}, 0.327, 0.475]_{LR}$ kg of chicken,

and

$[(7, 10; 1.903, 2.409; 1.679, 3.481; 3.448, 4.875); 0.853, 0.459, 0.234]_{LR}$ kg of mutton

at a minimum cost of $\$[(450, 495; 3.584, 4.951; 5.99, 8.421; 23.998, 26); 0.42 \times 10^{-4}, 0.0057, 0.692]_{LR}$.

Here, we have mentioned that $[(407, 445; 0.907, 1.606; 2.492, 3.082; 17.181, 17.921); 0.203, 0.038, 0.235]_{LR}$ kg of fish items means that the owner of the restaurant should buy the fish items ranging from 389.819 kg to 462.921 kg with the risk given by membership, neutral, and nonmembership functions, respectively. Similarly, all the quantities are fuzzy and show how large the quantity range should be to minimize the risk of obtaining an optimal picture fuzzy solution. Also, the neutral and nonmembership functions represent how much we are confused and do not favor the deduced minimum cost, respectively.

## 9.6 Conclusions

This is the first time that LPP is considered in the context of a picture fuzzy set, which was just introduced. The PFS has many opportunities to collect inconsistent and inaccurate information, which significantly contributes to decision-making issues. Therefore we introduce PFLPP in this chapter in a picture fuzzy environment, which entails maximizing membership and minimizing neutral and nonmembership functions in the picture fuzzy decision set. With the help of membership, neutral, and nonmembership degrees, the deterministic form of PFLPP is obtained. Additionally, several numerical examples are provided to demonstrate the applicability of the suggested SFLPP solution approach. The PFLPP also solves two distinct real-world issues: purchasing strategy and production planning.

By presenting the LPP duality theory in a picture fuzzy environment, numerous research projects in the picture fuzzy domain can be studied in the future. Additionally, the suggested approach can be expanded to include fractional programming issues, nonlinear programming issues, etc. Researchers have several opportunities due to the application of PFLPP to practical issues like supply chain, inventory control, portfolio management, and transportation issues.

## References

[1] M. Akram, T. Ullah, T. Allahviranloo, A new method to solve linear programming problems in the environment of picture fuzzy sets, Iranian Journal of Fuzzy Systems 19 (6) (2022) 29–49.

[2] M. Akram, I. Ulah, S.A. Edalatpanah, T. Allahviranloo, Fully Pythagorean fuzzy linear programming problems with equality constraints, Computational & Applied Mathematics 40 (2021) 120.

[3] K.T. Atanassov, Intuitionistic fuzzy sets, Fuzzy Sets and Systems 20 (1986) 87–96.

[4] R.E. Bellman, L.A. Zadeh, Decision making in a fuzzy environment, Management Science 17 (1970) 141–164.

[5] S. Chakraborty, M. Pal, P.K. Nayak, Intuitionistic fuzzy optimization technique for the solution of an EOQ model, Notes on Intuitionistic Fuzzy Sets 17 (2) (2011) 52–64.

[6] B.C. Cuong, Picture fuzzy sets-first results, in: Seminar Neuro-Fuzzy Systems with Applications, Institute of Mathematics, Vietnam Academy of Science and Technology, Hanoi-Vietnam, 2013, Preprint 03/2013.

[7] D. Dubey, A. Mehra, Linear programming with triangular intuitionistic fuzzy number, Advances in Intelligent Systems Research 1 (1) (2011) 563–569.

[8] P. Ghosh, T.K. Roy, Intuitionistic fuzzy goal geometric programming problem, Notes on Intuitionistic Fuzzy Sets 20 (1) (2014) 63–78.

[9] J. Kaur, A. Kumar, Mehar's method for solving fully fuzzy linear programming problems with LR fuzzy parameters, Applied Mathematical Modeling 37 (2013) 7142–7153.
[10] F.H. Lotfi, T. Allahviranloo, M.A. Jondabeh, L.A. Zadeh, Solving a full fuzzy linear programming using lexicography method and fuzzy approximate solution, Applied Mathematical Modeling 33 (7) (2009) 3151–3156.
[11] A.L. Nachammai, P. Thangaraj, Solving intuitionistic fuzzy linear programming problem by using similarity measures, European Journal of Scientific Research 72 (2) (2012) 204–210.
[12] A. Nagoorgani, K. Ponnalagu, A new approach on solving intuitionistic fuzzy linear programming problem, Applied Mathematical Sciences 6 (2012) 3467–3474.
[13] R. Parvathi, C. Malathi, Intuitionistic fuzzy linear programming problems, World Applied Sciences Journal 17 (12) (2012) 1802–1807.
[14] M. Qiyas, S. Abdullah, S. Ashraf, S. Khan, A. Khan, Triangular picture fuzzy linguistic induced ordered weighted aggregation operators and its application on decision-making problems, Mathematical Foundations of Computing 2 (3) (2019) 183–201.
[15] L.A. Zadeh, Fuzzy sets, Information and Control 8 (1965) 338–353.
[16] H.J. Zimmerman, Fuzzy programming and linear programming with several objective functions, Fuzzy Sets and Systems 1 (1) (1978) 45–55.

# 10 Multiobjective linear programming problem in a picture fuzzy environment

## Abbreviations

| | |
|---|---|
| **FS** | Fuzzy set |
| **IFS** | Intuitionistic fuzzy set |
| **IFN** | Intuitionistic fuzzy number |
| **TIFN** | Triangular intuitionistic fuzzy number |
| **IF** | Intuitionistic fuzzy |
| **LPP** | Linear programming problem |
| **MOLPP** | Multiobjective LPP |
| **IFMOLPP** | IF multiobjective LPP |
| **IFPA** | IF programming approach |
| **PF** | Picture fuzzy |
| **PFS** | Picture fuzzy set |
| **PFN** | Picture fuzzy number |
| **PFMOLPP** | PF multiobjective LPP |

## 10.1 Introduction

The linear programming problem (LPP) is a very classical mathematical tool to solve many real-life problems. That is, many real-life problems are modeled as an LPP, and by solving such an LPP, the original problem is solved. Normally, in a conventional LPP, only one objective function and multiple constraints occur. However, in some cases, it is seen that multiple objective functions are present, and they have to be optimized. Some of them are minimization type, and others are of maximization type, or all of them are similar optimization type. Several methods are available to solve such problems.

We observed that many issues of the real world are not certain, i.e., uncertain. These may be random or nonrandom uncertain. The well-known mathematical tool known as fuzzy mathematics is used to handle nonrandom uncertainty. In an uncertain domain, the coefficients of objective functions, coefficients of constraints, and/or right-hand vector of an LPP are uncertain. The decision variable may or may not be uncertain. In 1978, Zimmermann [40] introduced the fuzzy programming approach (FPA) for multiobjective programming problems (MOPP) in which membership functions represent the marginal evaluation of each objective. In this approach, the satisfaction level of decision makers is achieved by maximizing the membership function of each objective. The FPA is extended for solving fuzzy interval programming, fuzzy goal programming, fuzzy stochastic programming, etc. In 1980, Dubois and Prade [16] first introduced the concept of fuzzy set theory to the multiobjective LPPs. They stipulated that "When the complexity of the system increases, our ability to formulate a precise and yet meaningful statements on this system decreases up to a threshold beyond which precision and significance become mutually exclusive characteristics", which is also instructive in this regard. After this work, the fuzzy programming methods were developed by Zimmermann [41] and Rommelfanger [28]. In these methods, the uncertain and imprecise nature of the decision maker is considered, which is very applicable and promising. These fundamental works played a vital role in further investigation of FMPPs [21,26,42]. In [37], a multiobjective programming problem with fuzzy objective functions was considered, which is an ill-defined problem. This paper used a new ranking of fuzzy numbers and was properly tailored to consider the fuzziness surrounding the problem.

Nowadays, many efficient FPAs are developed with several variations. Jiménez et al. [24,25] proposed a fuzzy ranking procedure for solving fuzzy mathematical programming problems (FMPPs) in which the coefficients of the objective function, coefficients of the constraints, and right-hand vector are fuzzy numbers. Although there are several methods with different assumptions for FMPPs by employing fuzzy ranking methods, generally, the FMPPs are converted into conventional mathematical programming problems, and then these crisp programming problems are solved by conventional techniques.

Due to the growth of online marketing, increasing population, increasing pollution, type of products, etc., and as per social needs and criteria, the modern problems are large, complex, and contain multiobjective functions, and some of them are conflicting in nature (maximizing profit, minimizing cost, minimizing pollution during production and transport, etc.). The most difficult case is that such a problem does not always have a completely optimal solution that optimizes all the objective functions together. In this situation, the Pareto-optimal solution or noninferior solution is defined. A solution is called **Pareto-optimal** if any improvement of one objective function can be achieved only at the expense of at least one of the other objective functions.

Tong [38] and Gasimov and Yenilmez [18] solved single-objective FMPPs. Tong [38] discussed the fuzzy LPP (FLPP) along with fuzzy constraints. This problem was converted to the defuzzified problem and then solved by the fuzzy decisive set method introduced by Sakawa and Yano [29]. In [18], Gasimov and Yenilmez considered FLPP with one objective function along with less than type constraints only and fuzzy parameters. The type of problem is solved by the fuzzy decisive set method and modified subgradient method. In 2006, Ganesan and Veeramani [19] studied FLPP with trapezoidal fuzzy numbers. Chanas [10] proposed an FPP as MOLPP and it was solved by a parametric approach.

In a fuzzy setup, only the membership value of an object is considered, and hence some other information is not considered, and for this reason, fuzzy logic is not sufficient to solve all types of uncertain mathematics problems. In 1983, Atanassov [5] extended the concept of a fuzzy set to an intuitionistic fuzzy (IF) set (IFS) by incorporating the nonmembership value.

Using the concept of IFS, Angelov [3] proposed the IF programming approach (IFPA) for solving a MOPP, where the marginal evaluation of each objective function was determined by the membership as well as nonmembership functions. In IFPA, the satisfaction level of the decision maker is achieved by maximizing the membership and minimizing the nonmembership functions of each objective together. Later, this approach was extended, and IF stochastic programming, IF goal programming, etc., were developed and applied to the appropriate decision-making problems. Many approaches were developed to find the best possible solution of multiobjective linear programming problems (MOLPPs). In the last few years, the generalized concept of a fuzzy set was used for solving the fuzzy MOLPPs (FMOLPPs). Several researchers proposed IFPA for solving intuitionistic fuzzy MOLPPs (IF-MOLPPs) and applied it to solve some real-life problems. Unfortunately, one is superior to the other.

Rani et al. [27] proposed a method for solving MOPP under an IF setup. They also compared their method with other existing approaches. Singh and Yadav [31–33] developed the mathematical background for IFLPP and applied it to transportation and manufacturing problems. Singh et al. [35] modeled MOPP for IF setup and proposed a framework for solving IFMLOPP from optimistic and pessimistic points of view. Singh et al. [36] designed a framework for solving an IF multiobjective mixed-integer programming problem and it was applied to the supply-chain problem. Apart from the above references, some more works on fuzzy and IF theory have been used successfully in [4,7–9,14,15,22,23,39]. In [34], Singh and Yadav considered IFMOLPP with mixed constraints and the coefficients of objectives as well as constraint functions and the right-hand sides of constraints are taken as TIFNs. Then, using the accuracy function IFMOLPP was converted into equivalent crisp MOLPP. Recently, Ahmadini et al. [1] studied a preference scheme for multiobjective goal programming problems over different types of membership functions. In [2], Ahmadini et al. investigated the MOLPP in a neutrosophic environment.

To the best of our knowledge, no work has been done on picture fuzzy MOLPP (PF-MOLPP).

## 10.2 Preliminaries

We begin this section with some basic definitions and terminologies related to IFMOLPP. Then, we discuss necessary terms related to PFMOLPP.

The definition of IFS is given below as a ready reference for the reader.

**Definition 10.1.** Let $X$ be a universal set. Then, an IFS $\tilde{F}$ in $X$ is an expression

$$\tilde{F} = \{< y, \mathcal{Y}_{\tilde{F}}(y), \mathcal{N}_{\tilde{F}}(y) >: y \in X\},$$

where $\mathcal{Y}_{\tilde{F}}(y), \mathcal{N}_{\tilde{F}}(y) : X \to [0,1]$ are two functions satisfying the condition $0 \leq \mathcal{Y}_{\tilde{F}}(y) + \mathcal{N}_{\tilde{F}}(y) \leq 1$ for all $y \in X$.

$\mathcal{Y}_{\tilde{F}}(y)$ and $\mathcal{N}_{\tilde{F}}(y)$ represent the degree membership and nonmembership of the element $y \in X$. The expression $\pi(y) = 1 - \mathcal{Y}_{\tilde{F}}(y) - \mathcal{N}_{\tilde{F}}(y)$ represents the degree of hesitation for the element $y \in \tilde{F}$.

Note that this is the extension of the fuzzy set by incorporating the degree of nonmembership values.

The IF number (IFN) is defined as a fuzzy number. As per the definition, an IFS is an IF number if for at least one $y \in X$ such that $\mathcal{Y}(y) + \mathcal{N}(y) = 1$.

The triangular IFN (TIFN) is an IFS with membership and nonmembership functions, i.e., $\mathcal{Y}_{\tilde{F}}(y)$ and $\mathcal{N}_{\tilde{F}}(y)$, which satisfy the following conditions:

**(i)** There is a real number $m$ such that $\mathcal{Y}_{\tilde{F}}(m) = 1$ and $\mathcal{N}_{\tilde{F}}(m) = 0$, ($m$ is generally known as the mean value);

**(ii)** $\mathcal{Y}_{\tilde{F}}(y)$ and $\mathcal{N}_{\tilde{F}}(y)$ are piecewise-continuous functions from a set of reals to [0, 1] and $0 \leq \mathcal{Y}_{\tilde{F}}(y) + \mathcal{N}_{\tilde{F}}(y) \leq 1$ for all $y$ and the values of $\mathcal{Y}_{\tilde{F}}(y)$ and $\mathcal{N}_{\tilde{F}}(y)$ are given by:

$$\mathcal{Y}_{\tilde{F}}(y) = \begin{cases} f_1(y), & m - a \leq y < m \\ 1, & y = m \\ g_1(y), & m < y \leq m + b \\ 0, & \text{otherwise} \end{cases}$$

$$\mathcal{N}_{\tilde{F}}(y) = \begin{cases} f_2(y), & m - a' \leq y < m \\ 0, & y = m \\ g_2(y), & m < y \leq m + b' \\ 1, & \text{otherwise} \end{cases},$$

where $m$ is the mean value and the possibility of attainment of $y$ to $m$ is very high, i.e., close to 1, and $a \geq 0$, $b \geq 0$ are the left and right spreads of the membership function, and similarly, $a' \geq 0$ and $b' \geq 0$ are the left and right spreads of the nonmembership function. The function $f_1$ ($g_1$) are piecewise-continuous, strictly increasing (strictly decreasing) functions in $[m - a, m)$ and $(m, m + b]$, respectively; similarly $f_2$ ($g_2$) are piecewise-continuous, strictly decreasing (strictly increasing) functions in $[m - a', m)$ and $(m, m + b']$, respectively.

To fulfil the condition $\mathcal{Y}_{\tilde{F}}(y) + \mathcal{N}_{\tilde{F}}(y) \leq 1$, $a' \geq a$ and $b' \geq b$ must be satisfied.

For different values of $\mathcal{Y}_{\tilde{F}}(y)$ and $\mathcal{N}_{\tilde{F}}(y)$, different types of IFN need to be constructed.

For example, the triangular intuitionistic fuzzy number (TIFN) is an IFN if its membership and nonmembership functions are defined as:

$$\mathcal{Y}_{\tilde{F}}(y) = \begin{cases} \frac{y - f_1}{f_2 - f_1}, & f_1 \leq y < f_2 \\ 1, & y = f_2 \\ \frac{f_2 - y}{f_3 - f_2}, & f_2 < y \leq f_3 \\ 0, & \text{otherwise} \end{cases}$$

$$\mathcal{N}_{\tilde{F}}(y) = \begin{cases} \frac{f_2 - y}{f_2 - f_1'}, & f_1' \leq y < f_2 \\ 0, & y = f_2 \\ \frac{y - f_2}{f_3' - f_2}, & f_2 < y \leq f_3' \\ 1, & \text{otherwise} \end{cases},$$

where $f_1' \leq f_1 \leq f_2 \leq f_3 \leq f_3'$. These are the membership and nonmembership values of the IFTFN $\tilde{F}$ and it is denoted by $(f_1, f_2, f_3; f_1', f_2, f_3')$ or it can be written in mean-spread form as $(m, 0, q, p', q')$, where $f_1 = m - p$, $f_2 = m$, $f_3 = m + q$, $f_1' = m - p'$, $f_3' = m + q'$.

The TIFN can be defuzzified in many different ways, one of them is the use of an accuracy function. The accuracy function and associated score functions are defined below for any TIFN.

**Definition 10.2.** Let $\tilde{F} = (f_1, f_2, f_3; f_1', f_2, f_3')$ be a TIFN. The score functions for membership and nonmembership functions are denoted by $C(\mathcal{Y}_{\tilde{F}})$ and $C(\mathcal{N}_{\tilde{F}})$, respectively. These are defined by:

$$C(\mathcal{Y}_{\tilde{F}}) = \frac{f_1 + 2f_2 + f_3}{4} \text{ and } C(\mathcal{N}_{\tilde{F}}) = \frac{f_1' + 2f_2 + f_3'}{4}.$$

The accuracy function of the TIFN $\tilde{F}$ is denoted by $\mathcal{R}$ and is defined by:

$$\mathcal{R}(\tilde{F}) = \frac{C(\mathcal{Y}_{\tilde{F}}) + C(\mathcal{N}_{\tilde{F}})}{2}.$$

Few works on MOLPP on IF environments have been published on fuzzy and IF environments [11,20,23]. To the best of our knowledge, no work has been done for the same problem in a picture fuzzy environment. In picture fuzzy logic, one new parameter called the neutral function is associated with a PFN. A PFS is defined below.

**Definition 10.3.** [12,13] A PFS $P$ over the fixed set $X$ is written as:

$$P = \{(\mathcal{Y}_P(t), \mathcal{A}_P(t), \mathcal{N}_P(t)) : t \in X\},$$

$\mathcal{Y}_P(t) : X \to [0,1]$, $\mathcal{A}_P(t) : X \to [0,1]$ and $\mathcal{N}_P(t) : X \to [0,1]$ are, respectively, presented membership degree, neutral membership degree, and a nonmembership degree in a fuzzy set $P$, where $0 \leq \mathcal{Y}_P(t) + \mathcal{A}_P(t) + \mathcal{N}_P(t) \leq 1$ for $t \in X$. Also, the term $\pi_P(t) = 1 - (\mathcal{Y}_P(t) + \mathcal{A}_P(t) + \mathcal{N}_P(t))$ is called the refusal membership degree for $t$.

By definition, a PFS carries more information than FS and IFS. Note that all three degrees/functions are not independent, as their sum is less than or equal to 1.

## 10.3 Multiobjective linear programming problem

A single-objective LPP is a very well-known mathematical problem. However, in many real-life situations, multiple objectives come with some linear constraints. These types of problems are written as follows:

$$\begin{aligned} &\text{Minimize } Z(y) = [Z_1(y), Z_2(y), \ldots, Z_\kappa(y)] \\ &\text{subject to } A_i y \,(\leq \text{ or } = \geq)\, b_i, i = 1, 2, \ldots, m. \\ &y \geq 0 \end{aligned} \tag{10.1}$$

In more detail, this problem is written as:

$$\begin{aligned} &\text{Minimize } Z(y) = [Z_1(y), Z_2(y), \ldots, Z_\kappa(y)] \\ &\text{subject to } \sum_{j=1}^{n} a_{ij} y_j \geq b_i, i = 1, 2, \ldots, m_1 \\ &\sum_{j=1}^{n} a_{ij} y_j \leq b_i, i = m_1 + 1, m_1 + 2, \ldots, m_2 \\ &\sum_{j=1}^{n} a_{ij} y_j = b_i, i = m_2 + 1, m_2 + 2, \ldots, m, \end{aligned} \tag{10.2}$$

where the $s$th objective function $Z_s(y)$ can be written as

$$Z_s(y) = \sum_{j=1}^{n} c_{sj} y_j, \ s = 1, 2, \ldots, \kappa$$

and $b_i$, $i = 1, 2, \ldots, m$ are the right-hand side numbers and $y_j$, $j = 1, 2, \ldots, n$, is a set of decision variables.

Here, we assume that the first $m_1$ number of constraints are $\geq$ type, the next $m_2 - m_1$ constraints are $\leq$ type variables, and the remaining $m - m_2$ constraints are $=$ type.

In Eq. (10.2), it is assumed that all the objective functions are of minimization type, but practically some of them may be of maximization type. It is obvious that any minimization (maximization) problem can be converted by multiplying the objective function by $-1$.

There is difficulty in determining a feasible solution for a MOLPP. Now, the solution is defined below:

Suppose $X$ is the solution space of the problem (10.1). A feasible solution $y' \in X$ is called a **complete optimal solution** if for all $y \in X$:

$$Z_s(y) \leq Z_s(y') \text{ for all } s = 1, 2, \ldots, \kappa.$$

A complete solution may not only minimize all the objective functions together because some objective functions may conflict with each other. For this reason, a new type of solution is defined known as a Pareto-optimal solution or efficient solution, defined as:

**Definition 10.4.** [30] (Pareto-optimal solution) A feasible solution $y^* \in X$ is said to be a Pareto-optimal one if there is no $y \in X$ such that $Z_s(y) \geq Z_s(y^*)$ for all $s = 1, 2, \ldots, \kappa$ and for at least one $s$, $Z_s(y) > Z_s(y*)$.

The set of all Pareto-optimal solutions of Eq. (10.2) is denoted by $X^P$.

**Definition 10.5.** [30] (Weakly Pareto-optimal solution) A feasible solution $y^* \in X$ is said to be a weakly Pareto-optimal solution if there is no $y \in X$ such that $Z_s(y) > Z_s(y^*)$ for all $s = 1, 2, \ldots, \kappa$.

The set of all weakly Pareto-optimal solutions is denoted by $X^{\omega p}$.

Optimizing all objective functions simultaneously is very difficult or sometimes impossible. Also, there may be some preferences to optimize certain objective functions. For example, $Z_1$ and $Z_2$ are two objective functions, where $Z_1$ represents the profit and $Z_2$ represents the cost of a problem. Our apparent aim is to maximize $Z_1$ and minimize $Z_2$. Note that these are conflicting with each other. If it is possible for a given set of constraints, $Z_1$ is maximum, and $Z_2$ is minimum, then it is the best solution. If it is impossible, then our first target is maximizing $Z_1$, and the second is minimizing $Z_2$. That is, the preference of $Z_1$ is more than $Z_2$. Hence, motivated by this analogy, several methods were developed for finding MOLPPs. Among these methods, the weighted sum method is simple.

In the weighted sum method, each objective function is multiplied by a suitable weight and adding them to produce a single objective function. That is:

$$\min_{y\in X} \sum_{s=1}^{\kappa} \omega_s Z_\kappa(y). \tag{10.3}$$

Theorem 10.1 shows the difference between the Pareto-optimal solution of Eq. (10.2) and the optimal solution of Eq. (10.3).

**Theorem 10.1.** *[17] Let $\check{y} \in X$ be an optimal solution of Eq.* (10.3). *Therefore*

**(i)** *if $\omega = (\omega_1, \omega_2, \ldots, \omega_\kappa) > 0$ (i.e., $\omega_i > 0$ for all i), then $\check{y} \in X^p$;*
**(ii)** *if $\omega = (\omega_1, \omega_2, \ldots, \omega_\kappa) \geq 0$, but $\omega \neq 0$ (i.e., at least one component is nonzero and all others are nonnegative), then $\check{y} \in X^{\omega p}$.*

The following theorem is proved in the case of a convex feasible region and convex objective functions in [17].

**Theorem 10.2.** *[17] Suppose X is a convex set and all objective functions are convex. Then,*

**(i)** *if $\hat{y} \in X^p$, there exists some $\omega > 0$ such that $\hat{y}$ is an optimal solution of Eq.* (10.3);
**(ii)** *if $\hat{y} \in X^{\omega p}$, there exists some $\omega \geq 0$ such that $\hat{y}$ is an optimal solution of Eq.* (10.3).

In a MOLPP, all the objective functions are linear and convex, and the feasible set is polyhedral, a convex set. Thus all efficient solutions can be obtained by the weighted sum method.

In some real cases, we see that the coefficients of objective functions, constraints, and suitable hand vectors are crisp and uncertain. Depending on the nature and type of uncertainties, the MOLPPs are formed in a different fuzzy setup. The coefficients may be fuzzy numbers (triangular, trapezoidal, etc.), IFN, PFN, etc. The fuzzy MOLPP are available in the literature. If the coefficients are IFN, then the problem is referred to as an intuitionistic fuzzy MOLPP (IFMOLPP). If all the coefficients of problem (10.2) are IFN, then the problem can be written as:

$$\text{Minimize } \tilde{Z}(y) = [\tilde{Z}_1(y), \tilde{Z}_2(y), \ldots, \tilde{Z}_\kappa(y)]$$

$$\text{subject to} \sum_{j=1}^{n} \tilde{a}_{ij} y_j \geq \tilde{b}_i,\ for\ i = 1, 2, \ldots, m_1 \tag{10.4}$$
$$\sum_{j=1}^{n} \tilde{a}_{ij} y_j \leq \tilde{b}_i,\ for\ i = m_1 + 1, m_1 + 2, \ldots, m_2$$
$$\sum_{j=1}^{n} \tilde{a}_{ij} y_j = \tilde{b}_i,\ for\ i = m_2 + 1, m_2 + 2, \ldots, m$$
$$y_j \geq 0,\ j = 1, 2, \ldots, n$$

and

$$\tilde{Z}_s(y) = \sum_{j=1}^{n} \tilde{c}_{sj} y_j, \quad \text{for all } s = 1, 2, \ldots, \kappa$$

is the $s$th objective function.

$\sim$ notations represent all the IFNs. These IFNs may be linear, triangular, trapezoidal, hyperbolic, parabolic, etc.

The problem of Eq. (10.4) has been converted to a crisp one in many different ways. Most authors use score and accuracy functions to convert this IFMOLPP.

If $\mathcal{R}$ is the accuracy function for any IFN, the above problem is converted to the following crisp problem:

$$\text{Minimize } Z'(y) = [Z_1'(y), Z_2'(y), \ldots, Z_\kappa'(y)]$$
$$\text{subject to} \sum_{j=1}^{n} a_{ij}' y_j \geq b_i',\ i = 1, 2, \ldots, m_1 \tag{10.5}$$
$$\sum_{j=1}^{n} a_{ij}' y_j \leq b_i',\ i = m_1 + 1, m_1 + 2, \ldots, m_2$$
$$\sum_{j=1}^{n} a_{ij}' y_j = b_i',\ i = m_2 + 1, m_2 + 2, \ldots, m$$
$$y_j \geq 0,\ j = 1, 2, \ldots, n,$$

where each coefficient is converted as:

$$Z_s'(y) = \mathcal{R}(\tilde{Z}_s(y)) = \sum_{j=1}^{n} \mathcal{R}(\tilde{c}_{\kappa j}) y_j,\ s = 1, 2, \ldots, \kappa$$
$$a_{ij}' = \mathcal{R}(\tilde{a}_{ij}),\ i = 1, 2, \ldots, m;\ j = 1, 2, \ldots, n,$$
$$b_i' = \mathcal{R}(\tilde{b}_i),\ i = 1, 2, \ldots, m.$$

Note that all these coefficients are crisp. Now, using a suitable method for solving this MOLPP, the solution of the original IFMOLPP can be determined. The equivalence between the solutions of Eqs. (10.4) and (10.5) is established in the following theorem.

**Theorem 10.3.** *[1] An efficient solution of the MOLPP Eq.* (10.4) *is also an efficient solution of the IFMOLPP of Eq.* (10.5).

The weighted sum method can convert the MOLPP to a single-objective LPP. For this approach, [1] stated the following result:

**Theorem 10.4.** *[1] An optimal solution for the crisp single objective LPP is also an optimal solution for a single-objective IFLPP.*

## 10.4 Picture fuzzy multiobjective linear programming problem

We have created picture fuzzy multiobjective optimization techniques based on the PFS [12,13]. These techniques characterize three membership functions, such as maximization of membership degree and minimization of neutral and nonmembership degrees under a picture fuzzy environment. The picture fuzzy decision set also offers plenty of room for the decision maker(s) to express their neutral ideas, effectively representing indeterminacy behavior in the decision-making problem. The intuitionistic fuzzy optimization technique deals with the maximization of membership grades and minimizing of nonmembership grades under the relevant decision set, whereas fuzzy multiobjective programming solely takes into account the maximization of the membership function. Additionally, when membership and nonmembership degrees are not quite a good representation of the decision-maker(s) views, and an indeterminacy condition obtains in some situations, the problem arises. PFS could handle this circumstance effectively. Therefore when certain indeterminacy thoughts are present in the decision-making process, the picture fuzzy programming approach aids in decision making.

In order to depict the different membership functions for MOLPP under a picture fuzzy environment, the minimum and maximum values of each objective function have been represented by $L_s$ and $U_s$; and can be obtained as follows:

$$U_s = \max[Z_s(t)] \text{ and } L_s = \min[Z_s(t)] \text{ for all } s = 1, 2, \ldots, \kappa.$$

The bounds for the $s$th objective function under the picture fuzzy environment can be obtained as follows:

$$U_s^{\mathcal{Y}} = U_s,\ L_s^{\mathcal{Y}} = L_s \text{ for membership,}$$
$$U_s^{\mathcal{A}} = L_s^{\mathcal{A}} + \phi_s,\ L_s^{\mathcal{A}} = L_s^{\mathcal{Y}} \text{ for neutral membership,}$$
$$U_s^{N} = U_s^{\mathcal{Y}},\ L_s^{N} = L_s^{\mathcal{Y}} + \delta_s \text{ for nonmembership,}$$

where, $\delta_s$ and, $\phi_s \in (0, 1)$ are predetermined real numbers prescribed by decision makers.

The following can be offered as the picture fuzzy decision-making model for MOPP based on the Bellman and Zadeh [6] notion in Eq. (10.6):

$$\begin{aligned} &\max \quad \min_{s=1,2,\ldots,\kappa} \quad \mathcal{Y}_s(\tilde{Z}_s(y)) \\ &\min \quad \max_{s=1,2,\ldots,\kappa} \quad \mathcal{A}_s(\tilde{Z}_s(y)) \\ &\min \quad \max_{s=1,2,\ldots,\kappa} \quad \mathcal{N}_s(\tilde{Z}_s(y)) \end{aligned}$$

subject to

$$\sum_{j=1}^{n} \tilde{a}_{ij} y_j \geq \tilde{b}_i, \text{ for } i = 1, 2, \ldots, m_1 \tag{10.6}$$

$$\sum_{j=1}^{n} \tilde{a}_{ij} y_j \leq \tilde{b}_i, \text{ for } i = m_1 + 1, m_1 + 2, \ldots, m_2$$

$$\sum_{j=1}^{n} \tilde{a}_{ij} y_j = \tilde{b}_i, \text{ for } i = m_2 + 1, m_2 + 2, \ldots, m$$

$$y_j \geq 0, j = 1, 2, \ldots, n$$

$$\mathcal{Y}_s(\tilde{Z}_s(y)) \geq \mathcal{A}_s(\tilde{Z}_s(y)), \ \mathcal{Y}_s(\tilde{Z}_s(y)) \geq \mathcal{N}_s(\tilde{Z}_s(y))$$

$$0 \leq \mathcal{Y}_s(\tilde{Z}_s(y)) + \mathcal{A}_s(\tilde{Z}_s(y)) + \mathcal{N}_s(\tilde{Z}_s(y)) \leq 1. \tag{10.7}$$

The aforementioned mathematical model Eq. (10.6) can be rewritten as follows using the auxiliary variables:

$$\begin{aligned} &\max \ a \\ &\min \ b \\ &\min \ c \end{aligned}$$

subject to

$$\sum_{j=1}^{n} \tilde{a}_{ij} y_j \geq \tilde{b}_i, \text{ for } i = 1, 2, \ldots, m_1$$

$$\sum_{j=1}^{n} \tilde{a}_{ij} y_j \leq \tilde{b}_i, \text{ for } i = m_1 + 1, m_1 + 2, \ldots, m_2$$

$$\sum_{j=1}^{n} \tilde{a}_{ij} y_j = \tilde{b}_i, \text{ for } i = m_2 + 1, m_2 + 2, \ldots, m \tag{10.8}$$

$$y_j \geq 0, j = 1, 2, \ldots, n$$

$$\mathcal{Y}_s(\tilde{Z}_s(y)) \geq a,\ \mathcal{A}_s(\tilde{Z}_s(y)) \leq b, \mathcal{N}_s(\tilde{Z}_s(y)) \leq c$$
$$a \geq b, a \geq c,\ 0 \leq a+b+c \leq 1.$$

The following mathematical fuzzy multiobjective optimization model can be created by condensing the aforementioned model Eq. (10.8) as follows:

$$\max\ (a-b-c)$$

subject to

$$\begin{aligned}
&\sum_{j=1}^{n} \tilde{a}_{ij} y_j \geq \tilde{b}_i, \ \text{for } i=1,2,\ldots,m_1\\
&\sum_{j=1}^{n} \tilde{a}_{ij} y_j \leq \tilde{b}_i, \ \text{for } i=m_1+1,m_1+2,\ldots,m_2\\
&\sum_{j=1}^{n} \tilde{a}_{ij} y_j = \tilde{b}_i, \ \text{for } i=m_2+1,m_2+2,\ldots,m\\
&y_j \geq 0, j=1,2,\ldots,n\\
&\mathcal{Y}_s(\tilde{Z}_s(y)) \geq a,\ \mathcal{A}_s(\tilde{Z}_s(y)) \leq b, \mathcal{N}_s(\tilde{Z}_s(y)) \leq c\\
&a \geq b, a \geq c,\ 0 \leq a+b+c \leq 1.
\end{aligned} \tag{10.9}$$

The score and accuracy functions are used for defuzzification of PFNs. Such functions are defined below.

**Definition 10.6.** Let $\widetilde{D} = (([a_1,a_2,a_3],a),([a_1',a_2,a_3'],b),([a_1'',a_2,a_3''],c))$ be a TPFN. The **score function** for the membership $\mathcal{Y}_{\widetilde{D}}(z)$, neutral $\mathcal{A}_{\widetilde{D}}(z)$, and nonmembership $\mathcal{N}_{\widetilde{D}}(z)$ are denoted by $S_c(\mathcal{Y}_{\widetilde{D}})$, $S_c(\mathcal{A}_{\widetilde{D}})$, and $S_c(\mathcal{N}_{\widetilde{D}})$, respectively. These are defined as:

$$S_c(\mathcal{Y}_{\widetilde{D}}) = \left(\frac{a_1+2a_2+a_3}{4}\right)a \tag{10.10}$$

$$S_c(\mathcal{A}_{\widetilde{D}}) = \left(\frac{a_1'+2a_2+a_3'}{4}\right)b \tag{10.11}$$

$$S_c(\mathcal{N}_{\widetilde{D}}) = \left(\frac{a_1''+2a_2+a_3''}{4}\right)c. \tag{10.12}$$

The **accuracy function** of $\widetilde{D}$ is denoted by $\mathcal{R}(\widetilde{D})$ and is defined by:

$$\mathcal{R}(\widetilde{D}) = \frac{S_c(\mathcal{Y}_{\widetilde{D}})+S_c(\mathcal{A}_{\widetilde{D}})+S_c(\mathcal{N}_{\widetilde{D}})}{\mathcal{Y}_{\widetilde{D}}+\mathcal{A}_{\widetilde{D}}+\mathcal{N}_{\widetilde{D}}}. \tag{10.13}$$

It can be proved that the accuracy function $\mathcal{R}$ is linear.

## Steps to solve PFMOLPP

The following succinct statement summarizes the stepwise algorithm used to solve the PFMOLPP:

**Step 1:** Convert the given PFMOLPP stated in Eq. (10.4) to a MOLPP using the accuracy function defined in Eq. (10.13).
**Step 2:** Consider one objective function at a time among $\kappa$ objective functions and all constraints. Solve this LPP by any method. Suppose the optimal solution is $y^s$ and the optimal value of the $s$th objective function is $Z_s^*(y^s)$, starting from $s = 1$.
**Step 3:** Compute the values of the remaining $(\kappa - 1)$ objective functions at $y^s$.
**Step 4:** Repeat Steps 2 and 3 for all objective functions.
**Step 5:** Prepare the payoff table by considering all such solutions as shown in Table 10.1.

**Table 10.1** Payoff table.

| $y$ | $Z_1$ | $Z_2$ | ... | $Z_\kappa$ |
|---|---|---|---|---|
| $y^1$ | $Z_1^*(y^1)$ | $Z_2(y^1)$ | ... | $Z_\kappa(y^1)$ |
| $y^2$ | $Z_1(y^2)$ | $Z_2^*(y^2)$ | ... | $Z_\kappa(y^2)$ |
| $\vdots$ | $\vdots$ | $\vdots$ | $\vdots$ | $\vdots$ |
| $y^\kappa$ | $Z_1(y^\kappa)$ | $Z_2^*(y^\kappa)$ | ... | $Z_\kappa(y^\kappa)$ |

**Step 6:** Compute the upper and lower bounds for each objective function and each constraint for the membership, neutral, and nonmembership functions to the set of solutions determined in Step 5.
**Step 7:** Create a linear membership function under a picture fuzzy environment using the various membership functions.
**Step 8:** Construct the PFMOLPP of the form (10.8) and solve this problem by any available method or software.

Depending on the choice of the membership, neutral, and nonmembership functions, the modified problem (10.8) may be linear or nonlinear. Using available optimization methods or software, the problem (10.8) can be solved.

Different type of PFNs can be defined. In the following, the trapezoidal PFN (TrPFN) is defined.

**Definition 10.7.** A triplet $D = (([p_1, q, n, l_1], a), ([p_2, q, n, l_2], b), ([p_3, q, n, l_3], c))$ is a PFN of trapezoidal type if:

$$\mathcal{Y}_c(y) = \begin{cases} \frac{a(y-p_1)}{q-p_1} & \text{if } p_1 \le y < q \\ a & \text{if } q \le y < n \\ \frac{a(l_1-y)}{l_1-n} & \text{if } n \le y \le l_1 \\ 0 & \text{if } y < p_1 \text{ or } y > l_1, \end{cases}$$

$$\mathcal{A}_c(y) = \begin{cases} \frac{q-y+b(y-p_2)}{q-p_2} & \text{if} \quad p_2 \leq y < q \\ b & \text{if} \quad q \leq y < n \\ \frac{y-n+b(l_2-y)}{l_2-n} & \text{if} \quad n \leq y \leq l_2 \\ 0 & \text{if} \quad y < p_2 \text{ or } y > l_2, \end{cases}$$

$$\mathcal{N}_c(y) = \begin{cases} \frac{q-y+c(y-p_3)}{q-p_3} & \text{if} \quad p_3 \leq y < q \\ c & \text{if} \quad q \leq y < n \\ \frac{y-n+c(l_3-y)}{l_3-n} & \text{if} \quad n \leq y \leq l_3 \\ 0 & \text{if} \quad y < p_3 \text{ or } y > l_3, \end{cases}$$

are its membership, neutral, and nonmembership functions, such as $a, b, c \in [0, 1]$, and $0 \leq a + b + c \leq 1$.

If in all three cases of the above representation, $q = n$ instead of $q \leq y < n$, then the TrPFN becomes TPFN.

### 10.4.1 Linear-type membership functions

The linear-type membership $\mathcal{Y}_k^L(Z_k(y))$, neutral $\mathcal{A}_k^L(Z_k(y))$, and nonmembership $\mathcal{N}_k^L(Z_k(y))$ functions under a picture fuzzy environment can be designed as follows:

$$\mathcal{Y}_k^L(Z_k(y)) = \begin{cases} 1, & \text{if} \quad Z_k(y) \leq L_k^{\mathcal{Y}} \\ \frac{U_k^{\mathcal{Y}} - Z_k(y)}{U_k^{\mathcal{Y}} - L_k^{\mathcal{Y}}}, & \text{if} \quad L_k^{\mathcal{Y}} \leq Z_k(y) \leq U_k^{\mathcal{Y}} \\ 0, & \text{if} \quad Z_k(y) \geq U_k^{\mathcal{Y}} \end{cases},$$

$$\mathcal{A}_k^L(Z_k(y)) = \begin{cases} 0, & \text{if} \quad Z_k(y) \leq L_k^{\mathcal{A}} \\ \frac{Z_k(y) - L_k^{\mathcal{A}}}{U_k^{\mathcal{A}} - L_k^{\mathcal{A}}}, & \text{if} \quad L_k^{\mathcal{A}} \leq Z_k(y) \\ 1, & \text{if} \quad Z_k(y) \geq U_k^{\mathcal{A}} \end{cases},$$

$$\mathcal{N}_k^L(Z_k(y)) = \begin{cases} 0, & \text{if} \quad Z_k(y) \leq L_k^{\mathcal{N}} \\ \frac{Z_k(y) - L_k^{\mathcal{N}}}{U_k^{\mathcal{N}} - L_k^{\mathcal{N}}}, & \text{if} \quad L_k^{\mathcal{N}} \leq Z_k(y) \leq U_k^{\mathcal{N}} \\ 1, & \text{if} \quad Z_k(y) \geq U_k^{\mathcal{N}} \end{cases}.$$

## 10.5 Application of PFMOLPP

Any reputed IT shop has its expected requirements of branded laptops for its statistical experience from the customers. The number of customers for any particular brand is not fixed. Even the customers may be eager for a particular brand or prefer something other than the brand, or have no opinion on the brand name. Therefore we can tally the number of customers by a picture fuzzy number in our work. Here, we have used a triangular PFN to represent all fuzzy parameters. Also, the price is a picture fuzzy parameter because the price can be reduced, may be increased, or remain the same as the previous month, which are considered membership, nonmembership, and neutral membership values.

To construct a real-life problem, we have an assumption that there are two IT shops in Kolkata, namely 'Shree Kailash IT Service' and 'Infotech', with their statistical data of the demand and opinion of their customers per month for two particular brands of laptop like brand 'S' and 'A'.

Let the IT shop, 'Shree Kailash IT Service', have its statistical data of demand for customers per month as follows. Assuming the number of laptops of brand 'S' and brand 'A' are denoted by $t_1$ and $t_2$, respectively. Also, the cost for brand 'S' is approximately $\$(([50, 51.8, 52.9], 0.4), ([49, 51.8, 53.2], 0.3), ([50.8, 51.8, 52.5], 0.2))$ and $\$(([53, 55.7, 56], 0.6), ([52, 55.7, 57.9], 0.2), ([51, 55.7, 58], 0.1))$, respectively, for the first and second IT shops. Then, the same is true for the brand 'A' approximately $\$(([30, 33, 34.8], 0.5), ([32.6, 33, 35.2], 0.1), ([31.5, 33, 36.3], 0.3))$ and $\$(([38, 40.9, 42], 0.2), ([39, 40.9, 43], 0.5), ([37.5, 40.9, 43.3], 0.2))$, respectively. All the costs are given in units of thousands for approximated cases. On the other hand, the total cost for two brands of laptops of "Shree Kailash IT Service" is at least $\$(([249.7, 250, 252], 0.6), ([248, 250, 253.3], 0.2), ([239, 250, 256.7], 0.1))$, but not greater than $\$(([459, 500, 512], 0.1), ([492, 500, 509.2], 0.4), ([498.5, 500, 503.8], 0.5))$. Also, the least budget for 'Infotech' per month to order these two types of laptop brands is at least approximated $\$(([297, 300, 309], 0.6), ([290, 300, 308.2], 0.2), ([280, 300, 314.7], 0.1))$ in the case study. Therefore the problem is how many of each brand of laptops are to be bought to minimize the total cost in such a problem? Thus the mathematical formulation of the problem can be stated as follows:

$$
\begin{aligned}
&\text{Minimize } \widetilde{Z_1}(y) = \widetilde{51.8}^{PFN} y_1 + \widetilde{33}^{PFN} y_2\\
&\text{Minimize } \widetilde{Z_2}(y) = \widetilde{55.7}^{PFN} y_1 + \widetilde{40.9}^{PFN} y_2\\
&\text{subject to } \widetilde{51.8}^{PFN} y_1 + \widetilde{33}^{PFN} y_2 \geq \widetilde{250}^{PFN}\\
&\qquad \widetilde{51.8}^{PFN} y_1 + \widetilde{33}^{PFN} y_2 \leq \widetilde{500}^{PFN}\\
&\qquad \widetilde{55.7}^{PFN} y_1 + \widetilde{40.9}^{PFN} y_2 \geq \widetilde{300}^{PFN}\\
&\qquad \widetilde{111.4}^{PFN} y_1 + \widetilde{61.35}^{PFN} y_2 \geq \widetilde{408.9}^{PFN},
\end{aligned}
$$

where,

$$
\begin{aligned}
\widetilde{51.8}^{PFN} &= (([50, 51.8, 52.9], 0.4), ([49, 51.8, 53.2], 0.3), ([50.8, 51.8, 52.5], 0.2))\\
\widetilde{33}^{PFN} &= (([30, 33, 34.8], 0.5), ([32.6, 33, 35.2], 0.1), ([31.5, 33, 36.3], 0.3))\\
\widetilde{55.7}^{PFN} &= (([53, 55.7, 56], 0.6), ([52, 55.7, 57.9], 0.2), ([51, 55.7, 58], 0.1))\\
\widetilde{40.9}^{PFN} &= (([38, 40.9, 42], 0.2), ([39, 40.9, 43], 0.5), ([37.5, 40.9, 43.3], 0.2))\\
\widetilde{61.35}^{PFN} &= (([59.8, 61.35, 62], 0.6), ([58, 61.35, 64], 0.1), ([60, 61.35, 67], 0.2))\\
\widetilde{111.4}^{PFN} &= (([108, 111.4, 114], 0.5), ([106, 111.4, 112], 0.4), ([102, 111.4, 116], 0.1)),
\end{aligned}
$$

and $y_1, y_2 \geq 0$.

For solving this problem, let $y = (y_1, y_2)$ be the number vector for two types of brand laptops in the considered problem, i.e., $y_1$ is the number of 'S' brand laptops and $y_2$ is the number of 'A' brand laptops. Then, $y_1$ and $y_2$ are nonnegative parameters for this problem.

First, we have written the minimization problem in terms of a picture fuzzy programming problem. Then, we used the linear-type membership functions approach (LTMFA) to reduce it to crisp.

We have now applied the score and accuracy functions for the defuzzification method on all triangular picture fuzzy numbers:

$$\mathcal{R}(\widetilde{51.8}^{PFN}) = 51.63,$$

where $S_c(\mathcal{Y}_{\widetilde{51.8}}) = 20.65$, $S_c(\mathcal{A}_{\widetilde{51.8}}) = 15.435$, $S_c(\mathcal{N}_{\widetilde{51.8}}) = 10.38$.

Similarly for the others, we have obtained all the values:

$$\mathcal{R}(\widetilde{33}^{PFN}) = 33.03,$$

where $S_c(\mathcal{Y}_{\widetilde{33}}) = 16.35$, $S_c(\mathcal{A}_{\widetilde{33}}) = 3.345$, $S_c(\mathcal{N}_{\widetilde{33}}) = 10.035$,

$$\mathcal{R}(\widetilde{55.7}^{PFN}) = 55.15,$$

where $S_c(\mathcal{Y}_{\widetilde{55.7}}) = 33.06$, $S_c(\mathcal{A}_{\widetilde{55.7}}) = 11.065$, $S_c(\mathcal{N}_{\widetilde{55.7}}) = 5.51$

$$\mathcal{R}(\widetilde{40.9}^{PFN}) = 40.77,$$

where $S_c(\mathcal{Y}_{\widetilde{40.9}}) = 8.09$, $S_c(\mathcal{A}_{\widetilde{40.9}}) = 20.475$, $S_c(\mathcal{N}_{\widetilde{40.9}}) = 8.13$

$$\mathcal{R}(\widetilde{61.35}^{PFN}) = 61.41,$$

where $S_c(\mathcal{Y}_{\widetilde{61.35}}) = 36.675$, $S_c(\mathcal{A}_{\widetilde{61.35}}) = 6.1175$, $S_c(\mathcal{N}_{\widetilde{61.35}}) = 12.485$

$$\mathcal{R}(\widetilde{111.4}^{PFN}) = 110.7,$$

where $S_c(\mathcal{Y}_{\widetilde{111.4}}) = 55.6$, $S_c(\mathcal{A}_{\widetilde{111.4}}) = 44.08$, $S_c(\mathcal{N}_{\widetilde{111.4}}) = 11.02$

$$\mathcal{R}(\widetilde{250}^{PFN}) = 250.24,$$

where $S_c(\mathcal{Y}_{\widetilde{250}}) = 150.255$, $S_c(\mathcal{A}_{\widetilde{250}}) = 50.065$, $S_c(\mathcal{N}_{\widetilde{250}}) = 24.8925$

$$\mathcal{R}(\widetilde{500}^{PFN}) = 499.6825,$$

where $S_c(\mathcal{Y}_{\widetilde{500}}) = 49.275$, $S_c(\mathcal{A}_{\widetilde{500}}) = 200.12$, $S_c(\mathcal{N}_{\widetilde{500}}) = 250.2875$

$$\mathcal{R}(\widetilde{300}^{PFN}) = 300.753,$$

where $S_c(\mathcal{Y}_{\widetilde{300}}) = 180.9$, $S_c(\mathcal{A}_{\widetilde{300}}) = 59.91$, $S_c(\mathcal{N}_{\widetilde{300}}) = 29.8675$

$$\mathcal{R}(\widetilde{408.9}^{PFN}) = 408.06,$$

where $S_c(\mathcal{Y}_{\widetilde{500}}) = 203.975$, $S_c(\mathcal{A}_{\widetilde{500}}) = 122.55$, $S_c(\mathcal{N}_{\widetilde{500}}) = 40.73$.

Therefore the considered problem is reduced to:

$$\begin{aligned}
&\text{Minimize } Z_1(y) = 51.63y_1 + 33.03y_2 \\
&\text{Minimize } Z_2(y) = 55.15y_1 + 40.77y_2 \\
&\text{subject to } 51.63y_1 + 33.03y_2 \geq 250.24 \\
&\qquad 51.63y_1 + 33.03y_2 \leq 499.683 \\
&\qquad 55.15y_1 + 40.77y_2 \geq 300.753 \\
&\qquad 110.7y_1 + 61.41y_2 \geq 408.06, \\
&\qquad y_1, y_2 \geq 0.
\end{aligned} \tag{10.14}$$

Using the LTMFA, the problem will be equivalent to the following:

$$\begin{aligned}
&\text{Maximize } a - b - c \\
&\text{subject to } 51.63y_1 + 33.03y_2 + (U_1^{\mathcal{Y}} - L_1^{\mathcal{Y}})a \leq U_1^{\mathcal{Y}}, \\
&\qquad 51.63y_1 + 33.03y_2 - (U_1^{\mathcal{A}} - L_1^{\mathcal{A}})b \leq L_1^{\mathcal{A}}, \\
&\qquad 51.63y_1 + 33.03y_2 - (U_1^{\mathcal{N}} - L_1^{\mathcal{N}})c \leq L_1^{\mathcal{N}}, \\
&\qquad 55.15y_1 + 40.77y_2 + (U_2^{\mathcal{Y}} - L_2^{\mathcal{Y}})a \leq U_2^{\mathcal{Y}}, \\
&\qquad 55.15y_1 + 40.77y_2 - (U_2^{\mathcal{A}} - L_2^{\mathcal{A}})b \leq L_2^{\mathcal{A}}, \\
&\qquad 55.15y_1 + 40.77y_2 - (U_2^{\mathcal{N}} - L_2^{\mathcal{N}})c \leq L_2^{\mathcal{N}}, \\
&\qquad 51.63y_1 + 33.03y_2 \geq 250.24 \\
&\qquad 51.63y_1 + 33.03y_2 \leq 499.683 \\
&\qquad 55.15y_1 + 40.77y_2 \geq 300.753 \\
&\qquad 110.7y_1 + 61.41y_2 \geq 408.06 \\
&\qquad a \geq b, a \geq c, a + b + c \leq 1, a, b, c \in (0, 1) \\
&\qquad y_1, y_2 \geq 0.
\end{aligned} \tag{10.15}$$

For the programming problem (10.15), we have

$$\begin{aligned}
Z_1 &= \widetilde{250.24} \approx \widetilde{250}^{PFN} \\
&= (([248.9, 250, 252], 0.6), ([245, 250, 260], 0.2), ([242.8, 250, 258], 0.1)), \\
Z_2 &= \widetilde{308.8793} \\
&= (([302.3, 308.88, 309], 0.5), ([305, 308.88, 310.2], 0.3), ([301.9, 308.88, 312], 0.1)).
\end{aligned}$$

For the first stage and the second stage, we have the following:

$$Z_1 = \widetilde{281.56}^{PFN}$$
$$= (([279, 281.56, 284], 0.7), ([275, 281.56, 287], 0.1), ([276, 281.56, 284.5], 0.2)),$$
$$Z_2 = \widetilde{300.75}^{PFN}$$
$$= (([297.8, 300.75, 302], 0.6), ([295, 300.75, 305], 0.2), ([298.1, 300.75, 301.6], 0.1)).$$

Therefore we have to discuss the payoff matrix in this portion. Here, we have considered a MOLPP under a picture fuzzy environment. Then, we converted it into a crisp MOLPP using score and accuracy functions for triangular picture fuzzy numbers, then picked a single-objective problem subject to the given constraints. Then, we have found an optimal solution and optimal value of its objective function to form the payoff matrix, which is given in Table 10.2 related to the Pareto-optimal solution of the considered problem.

**Table 10.2** Payoff table for the considered problem.

| $y$ | $Z_1$ | $Z_2$ |
|---|---|---|
| (5.45, 0) | 281.56 | 300.75 |
| (0, 7.576) | 250.24 | 308.879 |

We have the following information by considering $\phi_k = \delta_k = 0.2$ for $k = 1, 2$ for the objective functions:

$$U_1^{\mathcal{Y}} = 309,\ L_1^{\mathcal{Y}} = 248.9,\ U_1^{\mathcal{A}} = 310.2,\ L_1^{\mathcal{A}} = 245,\ U_1^{N} = 312,\ L_1^{N} = 242.8$$
$$U_2^{\mathcal{Y}} = 302,\ L_2^{\mathcal{Y}} = 279,\ U_2^{\mathcal{A}} = 305,\ L_2^{\mathcal{A}} = 275,\ U_2^{N} = 301.6,\ L_2^{N} = 276.$$

Putting all the values of $U_i^{\mathcal{Y}}, L_i^{\mathcal{Y}}$ for $i = 1, 2$; $U_i^{\mathcal{A}}, L_i^{\mathcal{A}}$ for $i = 1, 2$; and, $U_i^{N}, L_i^{N}$ for $i = 1, 2$; then the obtained programming problem is:

$$\begin{aligned}
&\text{Maximize } a - b - c\\
&\text{subject to } 51.63y_1 + 33.03y_2 + 60.1a \le 309,\\
&\qquad 51.63y_1 + 33.03y_2 - 65.2b \le 245,\\
&\qquad 51.63y_1 + 33.03y_2 - 69.2c \le 242.8,\\
&\qquad 55.15y_1 + 40.77y_2 + 23a \le 302,\\
&\qquad 55.15y_1 + 40.77y_2 - 30b \le 275,\\
&\qquad 55.15y_1 + 40.77y_2 - 25.6c \le 276,\\
&\qquad 51.63y_1 + 33.03y_2 \ge 250.24\\
&\qquad 51.63y_1 + 33.03y_2 \le 499.683\\
&\qquad 55.15y_1 + 40.77y_2 \ge 300.753
\end{aligned}$$

$$110.7y_1 + 61.41y_2 \geq 408.06$$
$$a \geq b, a \geq c, a + b + c \leq 1, a, b, c \in (0, 1)$$
$$y_1, y_2 \geq 0. \tag{10.16}$$

Also, we have considered the solutions from multiobjective solutions of the two objective functions as:

$$y_1 = \widetilde{3.949}^{PFN}$$
$$= (([2.29, 3.949, 4.39], 0.7), ([1.78, 3.949, 5.23], 0.1), ([2.98, 3.949, 4.21], 0.13)),$$
$$y_2 = \widetilde{6.745}^{PFN}$$
$$= (([5.29, 6.745, 7, 99], 0.59), ([4.78, 6.745, 8.25], 0.2), ([3.72, 6.745, 8.19], 0.12)).$$

From the above-deduced values of $y_1$ and $y_2$ in the triangular picture fuzzy form, the highest possibility of attaining the approximate number of items to obtain a minimized budget for any laptop shop, is considered in our problem. Since the number of laptops is in fuzzy form, the number is always an integer. Hence, we have taken the most suitable values for the number of items in laptops of both brands 'S' and 'A'.

Putting in all these values, the deduced form is the required expression of MOLPP under a picture fuzzy environment to be solved.

The results are obtained on solving the programming problem for optimization under the picture fuzzy system of the model (10.16). The objective values are much better than the existing models, where intuitionistic numbers are used to handle fuzzy parameters concerning the linear membership function.

Therefore our proposed picture fuzzy MOLPP reflects the solutions by using the mathematical software LINGO software version 18.0.44; we have deduced the results:

$$y_1 = 3.949,\ y_2 = 6.745,\ a = 0.43,\ b = 0.15,\ \text{and}\ c = 0.27.$$

Hence, we can conclude that both IT shops "Shree Kailash IT Service" and "Infotech" have to buy 4 items of 'S' brand laptops and 7 numbers of 'A' brand laptops per month to minimize the cost in the uncertain environment.

## 10.6 Conclusion

The PFMOLPP is discussed and solved by considering the TPFN for membership, neutral, and nonmembership functions. These functions may be of other types, such as TrPFN, exponential, hyperbolic, parabolic, etc. Hence, for different types of such functions, different results may be obtained. The Pareto-optimal solution is determined for PFMOLPP. A general concept of the weighted sum method (where all objective functions are combined by multiplying variable weights by the objective functions) is discussed at the beginning but not illustrated in detail. The deviation degree measures and weighted max–min method may be applied to solve PFMOLPP. To our knowledge, this is the first work to solve PF-MOLPP.

## References


[1] A.A.H. Ahmadini, F. Ahmad, A novel intuitionistic fuzzy preference relations for multiobjective goal programming problems, J. Intell. Fuzzy Syst. 40 (3) (2021) 4761–4777.

[2] A.A.H. Ahmadini, F. Ahmad, Solving intuitionistic fuzzy multiobjective linear programming problem under neutrosophic environment, AIMS Math. 6 (5) (2021) 4556–4580, https://doi.org/10.3934/math.2021269.

[3] P.P. Angelov, Optimization in an intuitionistic fuzzy environment, Fuzzy Sets Syst. 86 (1997) 299–306.

[4] M.D.L. Asuncin, L. Castillo, J.F. Olivares, O.G. Prez, A. Gonzlez, F. Palao, Handling fuzzy temporal constraints in a planning environment, Ann. Oper. Res. 155 (2007) 391–415.

[5] K.T. Atanassov, Intuitionistic fuzzy sets, Fuzzy Sets Syst. 20 (1986) 87–96.

[6] R.E. Bellman, L.A. Zadeh, Decision-making in a fuzzy environment, Manag. Sci. 17 (4) (1970) 140–164.

[7] A.K. Bit, M.P. Biswal, S.S. Alam, Fuzzy programming approach to multicriteria decision making transportation problem, Fuzzy Sets Syst. 50 (1992) 135–141.

[8] A.K. Bit, M.P. Biswal, S.S. Alam, Fuzzy programming approach to multi-objective solid transportation problem, Fuzzy Sets Syst. 57 (1993) 183–194.

[9] E. Cascetta, M. Gallo, B. Montella, Models and algorithms for the optimization of signal settings on urban networks with stochastic assignment models, Ann. Oper. Res. 144 (2006) 301–328.

[10] D. Chanas, Fuzzy programming in multi-objective linear programming—a parametric approach, Fuzzy Sets Syst. 29 (1989) 303–313.

[11] H. Cheng, W. Huang, Q. Zhou, J. Cai, Solving fuzzy multi-objective linear programming problems using deviation degree measures and wrighted max-min method, Appl. Math. Model. 37 (2013) 6855–6869.

[12] B.C. Cuong, Picture fuzzy sets - first results. Part 1, Seminar "Neuro-fuzzy systems with applications", Tech. Rep., Institute of Mathematics, Hanoi, 2013.

[13] B.C. Cuong, Picture fuzzy sets - first results. Part 2, Seminar "Neuro-fuzzy systems with applications", Tech. Rep., Institute of Mathematics, Hanoi, 2013.

[14] S.K. Das, A. Goswami, S.S. Alam, Multi-objective transportation problem with interval cost, source and destination parameters, Eur. J. Oper. Res. 117 (1999) 100–112.

[15] S.K. De, S.S. Sana, Backlogging EOQ model for promotional effort and selling price sensitive demand-an intuitionistic fuzzy approach, Ann. Oper. Res. 233 (1) (2015) 57–76.

[16] D. Dubois, H. Prade, Fuzzy Sets and Systems, Theory and Applications, Academic Press, New York, 1980.

[17] M. Ehrgott, Multicriteria Optimization, Springer, Berlin, 2005.

[18] R.N. Gasimov, K. Yenilmez, Soving fuzzy linear programming with linear membership functions, Turk. J. Math. 26 (2002) 375–396.

[19] K. Ganesan, P. Veeramani, Fuzzy linear programs with trapezoidal fuzzy numbers, Ann. Oper. Res. 143 (2006) 305–315.

[20] H. Hassanpour, E. Hosseinzadeh, M. Moodi, Solving intuitionistic fuzzy multi-objective linear programming problem and its application in supply chain management, Appl. Math. 68 (3) (2023) 269–287.

[21] M. Inuiguchi, J. Ramik, Possibilistic linear programming: a brief review of fuzzy mathematical programming and comparison with stochastic programming portfolio selection problem, Fuzzy Sets Syst. 111 (2000) 3–28.

[22] B. Jana, T.K. Roy, Multi-objective fuzzy linear programming and its application in transportation model, Tamsui Oxf. J. Math. Sci. 21 (2) (2005) 243–268.

[23] B. Jana, T.K. Roy, Multi-objective intuitionistic fuzzy linear programming and its application in transportation model, Notes IFS 13 (1) (2007) 34–51.

[24] M. Jiménez, M. Rodri'guez, M. Arenas, A. Bilbao, Solving a possibilistic linear program through compromise programming, Mathw. Soft Comput. 7 (2000) 175–184.

[25] M. Jiménez, M. Arenas, M.V. Bilbao Rodri'guez, Linear programming with fuzzy parameters: an interactive method resolution, Eur. J. Oper. Res. 177 (1) (2007) 1599–1609.
[26] M.K. Luhandjula, Fuzzy mathematical programming: theory, applications and extensions, J. Uncertain Syst. 1 (2007) 123–135.
[27] D. Rani, T. Gulati, H. Garg, Multi-objective non-linear programming problem in intuitionistic fuzzy environment: optimistic and pessimistic view point, Expert Syst. Appl. 64 (2016) 228–238.
[28] H. Rommelfanger, Fuzzy linear programming and applications, Eur. J. Oper. Res. 92 (1996) 512–528.
[29] M. Sakawa, H. Yano, Interactive decision making for multi-objective linear fractional programming problems with fuzzy parameters, Cybern. Syst. 16 (1985) 377–394.
[30] M. Sakawa, Fuzzy Sets and Interactive Multiobjective Optimization, Springer, New York, 1993.
[31] S.K. Singh, S.P. Yadav, Intuitionistic fuzzy non linear programming problem: modeling and optimization in manufacturing systems, J. Intell. Fuzzy Syst. 28 (2015) 1421–1433.
[32] S.K. Singh, S.P. Yadav, Efficient approach for solving type-1 intuitionistic fuzzy transportation problem, Int. J. Syst. Assur. Eng. Manag. 6 (3) (2015) 259–267.
[33] S.K. Singh, S.P. Yadav, A new approach for solving intuitionistic fuzzy transportation problem of type-2, Ann. Oper. Res. 243 (2016) 349–363.
[34] S.K. Singh, S.P. Yadav, Intuitionistic fuzzy multi-objective linear programming problem with various membership functions, Ann. Oper. Res. 269 (2018) 693–707.
[35] V. Singh, S.P. Yadav, Modeling and optimization of multi-objective programming problems in intuitionistic fuzzy environment: optimistic, pessimistic and mixed approaches, Expert Syst. Appl. 102 (2018) 143–157.
[36] S.K. Singh, M. Goh, Multi-objective mixed integer programming and an application in a pharmaceutical supply chain, Int. J. Prod. Res. 57 (2019) 1214–1237.
[37] A.M.K. Tarabia, M.A.E. Kassem, N.M. El-Badry, A modified approach for solving a fuzzy multi-objective programming problem, Appl. Inform. 4 (2017) 1, https://doi.org/10.1186/s40535-016-0029-7.
[38] S. Tong, Interval number and fuzzy number linear programming, Fuzzy Sets Syst. 66 (1994) 301–306.
[39] L.D. Xu, A fuzzy multi-objective programming algorithm in decision support systems, Ann. Oper. Res. 12 (1988) 315–320.
[40] H.J. Zimmermann, Fuzzy programming and linear programming with several objective functions, Fuzzy Sets Syst. 1 (1978) 45–55.
[41] H.J. Zimmermann, Fuzzy mathematical programming, Comput. Oper. Res. 10 (1983) 291–298.
[42] H.J. Zimmermann, Fuzzy Set, Decision Making and Expert Systems, Kluwer Academic Publisher, Dordrecht, 1987.

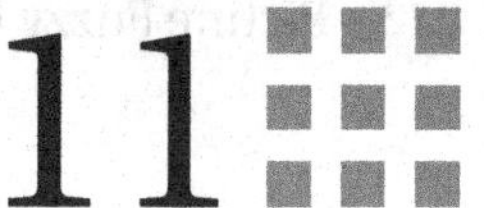

# 11 Picture fuzzy goal programming problem

## Abbreviations

| | |
|---|---|
| **FS** | Fuzzy set |
| **IFS** | Intuitionistic fuzzy set |
| **IFN** | Intuitionistic fuzzy number |
| **IF** | Intuitionistic fuzzy |
| **GP** | Goal programming |
| **FGP** | Fuzzy goal programming |
| **IFGP** | Intuitionistic fuzzy goal programming |
| **LPP** | Linear programming problem |
| **MOLPP** | Multiobjective LPP |
| **HTMFA** | Hyperbolic-type membership functions approach |
| **ETMFA** | Exponential-type membership functions approach |
| **PF** | Picture fuzzy |
| **PFS** | Picture fuzzy set |
| **PFN** | Picture fuzzy number |
| **PFGP** | Picture fuzzy goal programming |
| **GPFGP** | General picture fuzzy goal programming |

## 11.1 Introduction

Goal Programming (GP) is a very powerful and useful optimization technique that emerged as an extension of linear programming (LP) to handle single- and multiobjective (with conflicting types also) problems in decision-making processes. This method was developed in the mid-1960s and is used extensively in different areas such as project management, resource allocation, finance, engineering, etc. The essence of GP lies in its ability to address the complexities of real-world problems by accommodating various objectives while seeking an optimal compromise solution.

GP handles problems that contain multiple, often conflicting, objectives. In a conventional LP problem, the goal is to minimize or maximize only one objective function subject to certain constraints. In contrast, GP seeks to minimize the deviations from a set of predefined goals, while considering constraints for each objective.

In a GP, there are several components. The first component is decision variables. These variables represent the values that the decision maker seeks to determine to reach an optimal solution. Each objective function represents a specific goal or target to achieve, and the deviations from these goals are measured in terms of positive or negative slack variables. Another major component is constraints. Constraints restrict the values of the decision variable, which optimize the objective functions. Unlike LPP or multiobjective LPP (MOLPP), the GP introduces a new parameter called priority levels. To achieve the


https://doi.org/10.1016/B978-0-44-322024-1.00015-7

goal and address conflicting objectives, priority levels are assigned to each goal, indicating their relative significance. As optimization and decision science research advances, GP is likely to remain a prominent method for resolving intricate real-world challenges.

The advantages of GP are:

**(i)** It provides a systematic approach to handle situations where multiple objectives must be pursued simultaneously, allowing decision makers to consider diverse stakeholder interests.
**(ii)** It allows for easy modification of objective priorities and goals, enabling decision makers to adapt to changing circumstances and preferences.
**(iii)** By introducing slack variables, GP provides valuable insights into the tradeoffs between objectives, helping decision makers make informed choices.

Some important areas where GP are widely applied:

**(i)** Project management: It helps in scheduling projects by considering objectives like minimizing costs, maximizing efficiency, and meeting deadlines.
**(ii)** Environmental management: Goal programming is utilized to address environmental challenges by balancing ecological conservation, economic development, and social welfare.
**(iii)** Financial portfolio management: In investment decisions, goal programming aids in optimizing portfolios based on various financial objectives, such as risk minimization, profit maximization, and asset diversification.
**(iv)** Resource allocation: GP is used to allocate resources efficiently, considering multiple criteria, in fields like healthcare, education, and public services.

In 1955, Charnes et al. proposed GP for solving an LPP. After that, many authors proposed several methods in different directions based on GP. Charnes and Cooper [10], Ignizio [21], Ignizio [22], Romero [33], Romero [34], Chang [11], Caballero and Go [9], etc., explored GP both theoretically as well practically. GP works with the underpessimistic and overoptimistic accomplishment of the goals (objectives) such that each goal has deviated from their particular target as little as possible. In conventional GP, the aspiration level is simply the best achievement value in terms of the goal precisely mentioned, but in an uncertain world, the assignment of aspiration levels to a precise number is practically meaningless. Hence, in multiobjective decision-making problems, the aspiration level is not precise in most cases. During the last few years, the fuzzy set theory was successfully used in GP, and the new problem is known as fuzzy geometric programming (FGP). Bellman and Zadeh [5] elaborated the concept of fuzzy set theory in the decision-making process. Fuzzy decision-making theory is used in GP when some or all parameters are vague and ambiguous. The significance of the achievement within the optimizing goal is implicitly enhanced in the fuzzy optimization method, which will enable the decision maker (DM) to determine the best result. It becomes very difficult for the DM to take a decision when the problem contains vague preferences for the target goals and the corresponding aspiration level. Nowadays, several FGP models have been proposed by researchers, namely, Jana et al. [24], Pramanik and Roy [31], Cheng [13], Chen and Tsai [12], Jadidi et al. [23], Aköz and Petrovic [2], and Ghosh and Roy [19]. Among these models, the widely used FGP method proposed by Aköz and Petrovic [2] is used to solve multiobjective mathematical programming problems. In this approach, the goals are considered as linguistic terms such as Goal X is more important than Goal Y, or Goal X is slightly more important than Goal Y, etc. During the last few years, these types of preference relation were used in FGP to solve several real-world problems and were mathematically extended by many authors (for example, Torabi and Moghaddam [41], Petrovic and Aköz [30], Cheng [13], Díaz-Madroñero et al. [17], Bilbao-Terol et al. [7], Khalili-Damghani and Sadi-Nezhad [25], Khan et al. [26], Sheikhalishahi and Torabi [37], Arenas-Parra et al. [4], and Bilbao-Terol et al. [8]). All the available FGP models deal with imprecise preference relations and maximizing the degree of belongingness. However, in these models, the reasons for uncertainty due to vagueness, incomplete data, etc., are not considered. In uncertain systems, there is hesitation among the DMs due to a lack of information and the conflicting nature of the objective functions. Hence, the FGP methods that contain only the degree of belongingness for such cases are insufficient.

The more generalized versions of fuzzy sets, like an intuitionistic fuzzy set (IFS), a picture fuzzy set (PFS), etc., are used to overcome these situations. Angelov [3] introduced the optimization problem in an intuitionistic fuzzy (IF) environment. The IF optimization method considers simultaneously both aspects of satisfaction degree, such as maximization of belongingness and minimization of nonbelongingness for all objective functions. Thus GP in an IF environment (IFGP) deals with more imprecise preference relations compared to a fuzzy set. Many authors, namely, Seikh et al. [36], Bharati and Singh [6], Kour et al. [27], Ebrahimnejad and Verdegay [18], Zhao et al. [42], Mahajan and Gupta [28] Singh and Yadav [38], Roy et al. [35], Ghaffar [1], Ghosh and Roy [20], Tabrizi et al. [39], Dey and Roy [16], and Razmi et al. [32] have discussed the GP in an IF environment and applied the proposed methods to solve real-life problems that occur in an IF environment.

It is observed that most of the existing works on FGP and IFGP considered linear membership and nonmembership functions. The linear function is the simplest and most commonly used in the fuzzy setup. It is defined as fixing the lower- and upper-end points of the acceptability level. However, a linear function does not provide complete information about a problem; hence, other functions, such as exponential, hyperbolic, and parabolic membership, are defined. Also, the IFS does capture all information from an imprecise problem. Apart from the degree of acceptance and nonacceptance, there is another case called the degree of neutrality that occurs in many real-world problems. For example, for a specific proposition, say P, some people have accepted the proposition P, some others do not accept it, and some other people remain neutral (they do not give any response) regarding the proposition P. This type of situation cannot be represented by a fuzzy set or IFS, but is successfully represented by PFS.

To handle such a situation, we introduced GP on PFS, which contains membership, neutral, and nonmembership functions for each decision variable and objective function. Like FGP and IFGP, different types of picture fuzzy GPs (PFGPs) can be defined. In this chapter, we discuss the linear membership, neutral, and nonmembership functions in detail.

## 11.2 Preliminaries

The crisp multiobjective programming problem (MOPP) is stated below:

$$\left.\begin{array}{ll} \text{Maximize} & Z_k, k = 1, 2, \ldots, \kappa_1 \\ \text{Minimize} & Z_k, k = \kappa_1 + 1, \ \ldots, \kappa; \kappa \text{ is the total number of objective functions} \\ \text{Subject to} & \\ & Az \leq b \\ & z \geq 0 \end{array}\right\} . \tag{11.1}$$

When the coefficients of objective functions or constraints are fuzzy, the problem is called fuzzy MOPP. In addition, when the goals are uncertain aspiration levels, the problem becomes fuzzy goal programming (FGP). The FGP is a method that finds a solution of a fuzzy MOPP when the uncertain aspiration levels are assigned (generally, made by the decision maker or determined from the payoff table) to the objective functions. An objective function and an uncertain aspiration level are considered fuzzy goals. Let $Z = \{Z_k : k = 1, 2, \ldots, \kappa\}$ be a set of fuzzy objectives, some of them are maximization type, and others are minimization type. The FGP model proposed by Tiwari et al. [40] is defined as:

$$\left.\begin{array}{ll} \text{Optimize} & z \\ & Z_k(z) \succeq \varepsilon_k \text{ for a minimization objective function} \\ & Z_k(z) \preceq \varepsilon_k \text{ for a maximization objective function} \\ & k = 1, 2, \ldots, \kappa \\ \text{subject to} & \\ & Az \leq b \\ & z \geq 0 \end{array}\right\} . \tag{11.2}$$

The operators $\preceq$ and $\succeq$ indicate the fuzzy linguistic inequality meaning less than or equal to and greater than or equal to, respectively. The set of constraints is denoted by $Az \leq b$. Moreover, $\varepsilon_k$ represents the aspiration level for the $k$th goal $Z_k$.

The payoff table is obtained by solving the defuzzified version of the MOPP for each objective with all constraints and then determining other objective functions for each such solution.

Several methods are available to solve FGP and IFGP. Here, we propose a new method to solve GP in a picture fuzzy environment.

The PFS is an extension of the IFS. In PFS, one more piece of information is incorporated: the neutral function. Hence, the PFS is obtained from this third component in IFS as follows:

**Definition 11.1.** [14,15] A PFS $P$ over the fixed set $X$ is written as:

$$P = \{(\mathcal{Y}_P(t), \mathcal{A}_P(t), \mathcal{N}_P(t)) : t \in X\},$$

$\mathcal{Y}_P(t) : X \to [0, 1]$, $\mathcal{A}_P(t) : X \to [0, 1]$, and $\mathcal{N}_P(t) : X \to [0, 1]$ are, respectively, presented as a membership degree, a neutral membership degree, and a nonmembership degree in a fuzzy set $P$, where $0 \leq \mathcal{Y}_P(t) + \mathcal{A}_P(t) + \mathcal{N}_P(t) \leq 1$ for $t \in X$. Also, the term $\pi_P(t) = 1 - (\mathcal{Y}_P(t) + \mathcal{A}_P(t) + \mathcal{N}_P(t))$ is called a refusal membership degree for $t$.

By definition, a PFS carries more information than a FS and an IFS. Note that all three degrees/functions are not independent, as their sum is less than or equal to 1.

The generalized version of PFS can be defined as a generalized IFS [29]. In the above definition, all the functions membership, neutral, and nonmembership lie between 0 and 1, and their sum is less than or equal to 1. However, in a generalized PFS, the upper bounds are restricted by some numbers, which lie between 0 and 1, including 1, and their sum must be the sum of these upper bounds and less than or equal to 1.

**Definition 11.2.** [14] A generalized picture fuzzy set (GPFS) $\tilde{P^i}$ is given by the following equation:

$$\tilde{P^i} = \{z_i, \mathcal{Y}_{\tilde{pi}}(z_i), \mathcal{A}_{\tilde{pi}}(z_i), \mathcal{N}_{\tilde{pi}}(z_i) | z_i \in Z\},$$

where, $Z = \{z_1, z_2, ..., z_n\}$ is a finite universal set. Also, the following functions are known as membership, neutral membership, and nonmembership functions:

$$\mathcal{Y}_{\tilde{pi}}(z_i) : Z \to [0, w_1];\ z_i \in Z \to \mathcal{Y}_{\tilde{pi}}(z_i) \in [0, w_1],$$
$$\mathcal{A}_{\tilde{pi}}(z_i) : Z \to [0, w_2];\ z_i \in Z \to \mathcal{A}_{\tilde{pi}}(z_i) \in [0, w_2],$$
$$\mathcal{N}_{\tilde{pi}}(z_i) : Z \to [0, w_3];\ z_i \in Z \to \mathcal{N}_{\tilde{pi}}(z_i) \in [0, w_3].$$

The membership, neutral membership, and nonmembership functions satisfy the condition:

$$0 \leq \mathcal{Y}_{\tilde{pi}}(z_i) + \mathcal{A}_{\tilde{pi}}(z_i) + \mathcal{N}_{\tilde{pi}}(z_i) \leq w_1 + w_2 + w_3 \forall z_i \in Z,$$
$$0 \leq w_1 + w_2 + w_3 \leq 1;\ w_1, w_2, w_3 \in [0, 1].$$

Here, $w_1$, $w_2$, and $w_3$ are the gradations of the membership, neutral membership, and nonmembership functions, respectively.

**Definition 11.3.** A feasible solution $z^*$ is said to be a Pareto-optimal solution if and only if there does not exist another $z \in Z$ such that:

$$\mathcal{Y}_{\phi_a(z)}(z) \geq \mathcal{Y}_{\phi_a(z)}(z^*),$$
$$\mathcal{A}_{\phi_a(z)}(z) \leq \mathcal{A}_{\phi_a(z)}(z^*),$$
$$\mathcal{N}_{\phi_a(z)}(z) \leq \mathcal{N}_{\phi_a(z)}(z^*),$$

for all $a$, and strict inequality holds for at least one $a$.

## 11.3 Picture fuzzy goal programming problem

In this section, we have described the GP problem in a picture fuzzy environment. Throughout the chapter, we have considered $z$ as a vector variable and expressed it in the form $z = (z_1, z_2, ..., z_n)^T$. The problem in a formal way is represented as follows:

$$\left.\begin{array}{ll} \text{Find} & z = (z_1, z_2, ..., z_n)^T \\ \text{So as to} & \\ \text{Minimize}^{\,i} & \phi_a(z) \text{ with target value } G_a, \quad \text{for the } i\text{th objective} \\ & \quad \text{acceptance tolerance } t_a, \text{ neutral, and rejection tolerance } r_a; \\ & \quad \text{for } a = 1, 2, ..., n. \\ \text{subject to} & \\ & \theta_j(z) \le d_j,\ j = 1, 2, ..., m \\ & z > 0, \end{array}\right\}, \quad (11.3)$$

with certain membership, neutral membership, and nonmembership functions.

Several types of IFNs are available in the literature, but a very limited number of PFNs are defined. In this chapter, three types of PFNs, namely, linear, exponential, and hyperbolic, are defined. First, we discuss the linear-type membership, neutral, and nonmembership functions, and later the remaining two PFNs are discussed.

### Linear-type membership functions approach

Here, the PFGP for the linear case is considered first:

$$\mathcal{Y}_{\phi_a(z)}(\phi_a(z)) = \begin{cases} \delta_{\mathcal{Y}} & \text{if } \phi_a(z) \le G_a \\ \dfrac{(\phi_a(z) - G_a)\delta_{\mathcal{Y}}}{t_a} & \text{if } G_a \le \phi_a(z) \le G_a + t_a \\ 0 & \text{if } \phi_a(z) \ge G_a + t_a \end{cases},$$

$$\mathcal{A}_{\phi_a(z)}(\phi_a(z)) = \begin{cases} 0 & \text{if } \phi_a(z) \le G_a \\ \dfrac{(\phi_a(z) - G_a)\delta_{\mathcal{A}}}{r_a} & \text{if } G_a \le \phi_a(z) \le G_a + r_a \\ \delta_{\mathcal{A}} & \text{if } \phi_a(z) \ge G_a + r_a \end{cases}.$$

Two types of neutral functions can be defined. The first one is:

$$\mathcal{N}_{\phi_a(z)}(\phi_a(z)) = \begin{cases} 0 & \text{if } \phi_a(z) \le G_a \\ \dfrac{(\phi_a(z) - G_a)\delta_{\mathcal{N}}}{r_a} & \text{if } G_a \le \phi_a(z) \le G_a + r_a \\ \delta_{\mathcal{N}} & \text{if } \phi_a(z) \ge G_a + r_a \end{cases},$$

and the second one is:

$$\mathcal{N}_{\phi_a(z)}(\phi_a(z)) = \begin{cases} \delta_{\mathcal{N}} & \text{if } \phi_a(z) \leq G_a \\ \dfrac{(G_a + r_a - \phi_a(z))\delta_{\mathcal{N}}}{r_a} & \text{if } G_a \leq \phi_a(z) \leq G_a + r_a \,, \\ 0 & \text{if } \phi_a(z) \geq G_a + r_a \end{cases}$$

where, $\delta_{\mathcal{Y}} + \delta_{\mathcal{A}} + \delta_{\mathcal{N}} \leq 1$.

The first type of neutral function is monotonic increasing, while the second one is decreasing. The linear PFN is depicted in Fig. 11.1.

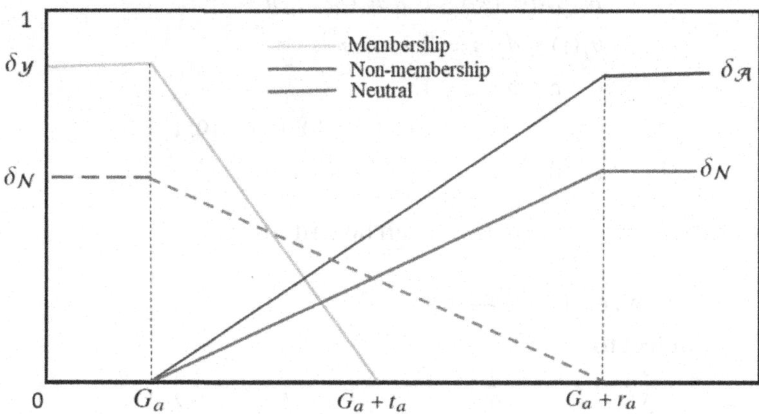


**FIGURE 11.1** Linear picture fuzzy number.

In Fig. 11.1, the membership and neutral functions are initiated from $G_a$ (a fixed point), and the membership function ends at the point $G_a + t_a$. However, this is not the general case. Any function can start from any point and end at any arbitrary point, but the condition is that at any point, the sum of membership, neutral, and nonmembership functions must not be more than 1.

The PFGP can be transformed into a crisp programming model using this linear-type membership, neutral membership, and nonmembership functions as:

$$\begin{aligned}
&\text{Maximize} \quad (\mathcal{Y}_{\phi_a(z)}(\phi_a(z))) \\
&\text{Minimize} \quad (\mathcal{A}_{\phi_a(z)}(\phi_a(z))) \\
&\text{Minimize} \quad (\mathcal{N}_{\phi_a(z)}(\phi_a(z))) \\
&\text{subject to} \\
&0 \leq \mathcal{Y}_{\phi_a(z)}(\phi_a(z)) + \mathcal{A}_{\phi_a(z)}(\phi_a(z)) + \mathcal{N}_{\phi_a(z)}(\phi_a(z)) \leq 1, a = 1, 2, ..., n \\
&\mathcal{Y}_{\phi_a(z)}(\phi_a(z)) \geq \mathcal{A}_{\phi_a(z)}(\phi_a(z)), a = 1, 2, ..., n \\
&\mathcal{Y}_{\phi_a(z)}(\phi_a(z)) \geq \mathcal{N}_{\phi_a(z)}(\phi_a(z)), a = 1, 2, ..., n \\
&\mathcal{N}_{\phi_a(z)}(\phi_a(z)) \geq 0 \quad \text{for } a = 1, 2, ..., n
\end{aligned}$$

$$\theta_j(z) \leq d_j, j = 1, 2, ..., m$$
$$z > 0.$$

This is equivalent to:

$$\begin{aligned}
&\text{Maximize } \ e, \text{ minimize } b, \text{ minimize } c \\
&\text{subject to} \\
&\mathcal{Y}_{\phi_a(z)}(\phi_a(z)) \geq e, \\
&\mathcal{A}_{\phi_a(z)}(\phi_a(z)) \leq b, \\
&\mathcal{N}_{\phi_a(z)}(\phi_a(z)) \leq c, a = 1, 2, ..., n \\
&\theta_j(z) \leq d_j, j = 1, 2, ..., m \\
&0 \leq e + b + c \leq 1 \\
&e \geq b, e \geq c, e \in [0, 1], b \in [0, 1], c \in [0, 1] \\
&z > 0.
\end{aligned} \tag{11.4}$$

Taking the arithmetic mean, Eq. (11.4) can be written as:

$$\begin{aligned}
&\text{Minimize } \ \left(\frac{e+2(1-c-b)}{3}\right)^{-1} \\
&\text{subject to} \\
&\frac{\phi_a(z)}{2}\left(\frac{1}{t_a} + \frac{1}{t_a+r_a} + \frac{1}{r_a}\right) \leq 1 + \frac{G_a}{2}\left(\frac{1}{t_a} + \frac{1}{t_a+r_a} + \frac{1}{r_a}\right) - \frac{e+2(1-c-b)}{3} \\
&\quad \text{for } \ a = 1, 2, ..., n \\
&\theta_j(z) \leq d_j \ \text{ for } \ j = 1, 2, ..., m \\
&0 \leq e + b + c \leq 1; e \in [0, 1], b \in [0, 1], c \in [0, 1] \\
&e \geq b, e \geq c \\
&z > 0.
\end{aligned} \tag{11.5}$$

Solve the above crisp model (11.5) by using an appropriate mathematical programming algorithm to obtain the optimal solution of objective functions.

**Theorem 11.1.** *$z^* \in Z$ is Pareto-optimal* (11.3) *if and only if $z^*$ is a Pareto-optimal solution of*

$$\begin{aligned}
&\textit{Minimize } \ \phi_a(z), a = 1, 2, ..., n \\
&\textit{subject to} \\
&\theta_j(z) \leq d_j, j = 1, 2, ..., m \\
&z > 0.
\end{aligned} \tag{11.6}$$

*Proof.* Let, $z^* \in Z$ be a Pareto-optimal solution of (11.3), then there does not exist any $z \in Z$ such that:

$$\mathcal{Y}_{\phi_a(z)}(z) \geq \mathcal{Y}_{\phi_a(z)}(z^*)$$

$$\mathcal{A}_{\phi_a(z)}(z) \leq \mathcal{A}_{\phi_a(z)}(z^*), \; and$$
$$\mathcal{N}_{\phi_a(z)}(z) \leq \mathcal{N}_{\phi_a(z)}(z^*)$$

for all $a$, and strictly inequality holds for at least one $a$.

From the expression of membership, neutral membership, and nonmembership functions, we have:

$$\delta y - \frac{\phi_a(z) - G_a}{t_a} \geq \delta y - \frac{\phi_a(z^*) - G_a}{t_a}$$
$$i.e., \; \phi_a(z) \leq \phi_a(z^*), \; and$$
$$\frac{\phi_a(z) - G_a}{r_a + t_a} \leq \frac{\phi_a(z^*) - G_a}{r_a + t_a}$$
$$i.e., \; \phi_a(z) \leq \phi_a(z^*),$$
$$\frac{\phi_a(z) - G_a}{r_a} \leq \frac{\phi_a(z^*) - G_a}{r_a}$$
$$i.e., \; \phi_a(z) \leq \phi_a(z^*),$$

with strict inequality holding for at least one $a$. Hence, $z^*$ is a Pareto-optimal solution of (11.5). On the other hand, $z^*$ is a Pareto-optimal solution of (11.4) such that $\phi_a(z) \leq \phi_a(z^*)$ with strict inequality holding for at least one $a$. Hence:

$$\phi_a(z) - G_a \leq \phi_a(z^*) - G_a$$
$$i.e., \; \frac{\phi_a(z) - G_a}{r_a} \leq \frac{\phi_a(z^*) - G_a}{r_a},$$

which tells us that $\mathcal{N}_{\phi_a(z)}(z) \leq \mathcal{N}_{\phi_a(z)}(z^*)$ and, $\delta y - \frac{\phi_a(z)-G_a}{t_a} \geq \delta y - \frac{\phi_a(z^*)-G_a}{t_a}$, which tells us that $\mathcal{Y}_{\phi_a(z)}(z) \geq \mathcal{Y}_{\phi_a(z)}(z^*)$. For optimal solution, it is clear that the acceptance tolerance $t_a$ is greater or less than the rejection tolerance $r_a$. Hence, we have $t_a + r_a > 0$. Therefore $\frac{\phi_a(z)-G_a}{r_a+t_a} \leq \frac{\phi_a(z^*)-G_a}{r_a+t_a}$, which reflects the following:

$$\mathcal{A}_{\phi_a(z)}(z) \leq \mathcal{A}_{\phi_a(z)}(z^*).$$

Hence, $z^*$ is a Pareto-optimal solution of (11.4). □

### 11.3.1 Generalized picture fuzzy goal programming problem

In the previous section, the PFGP is discussed for conventional PFSs. To increase flexibility, the generalized PFS is used to solve GP. The GP is defined by an overgeneralized PFS as:

$$\begin{aligned}
\text{Find} \quad & z = (z_1, z_2, \ldots, z_n)^T, \text{ so as to} \\
& \text{Minimize } \; \phi_a(z) \text{ with largest value } G_a, \text{ acceptance tolerance } t_a, \\
& \text{neutral, and rejection tolerance } r_a \\
& \text{subject to } \; \phi_j(z) \leq d_j, \; j = 1, 2, \ldots, m \\
& \qquad z > 0,
\end{aligned}$$

with membership, neutral membership, and nonmembership functions:

$$\mathcal{Y}^{w_1}_{\phi_a(z)}(\phi_a(z)) = \begin{cases} w_1, & \text{if } \phi_a(z) \le G_a \\ w_1\left(1-\frac{\phi_a(z)-G_a}{t_a}\right), & \text{if } G_a \le \phi_a(z) \le G_a + t_a \\ 0, & \text{if } \phi_a(z) \ge G_a + t_a \end{cases},$$

$$\mathcal{A}^{w_2}_{\phi_a(z)}(\phi_a(z)) = \begin{cases} 0, & \text{if } \phi_a(z) \le G_a \\ w_2\left(\frac{\phi_a(z)-G_a}{r_a}\right), & \text{if } G_a \le \phi_a(z) \le G_a + r_a \\ w_2, & \text{if } \phi_a(z) \ge G_a + r_a \end{cases},$$

$$\mathcal{N}^{w_3}_{\phi_a(z)}(\phi_a(z)) = \begin{cases} 0, & \text{if } \phi_a(z) \le G_a \\ w_3\left(\frac{\phi_a(z)-G_a}{r_a}\right), & \text{if } G_a \le \phi_a(z) \le G_a + r_a \\ w_3, & \text{if } \phi_a(z) \ge G_a + r_a \end{cases}.$$

The PFGP can be transformed into a crisp programming model using membership, neutral membership, and nonmembership functions as:

$$\begin{aligned}
&\text{Maximize} \quad (\mathcal{Y}^{w_1}_{\phi_a(z)}(\phi_a(z)))\\
&\text{Minimize} \quad (\mathcal{A}^{w_2}_{\phi_a(z)}(\phi_a(z)))\\
&\text{Minimize} \quad (\mathcal{N}^{w_3}_{\phi_a(z)}(\phi_a(z)))\\
&\text{subject to } 0 \le \mathcal{Y}^{w_1}_{\phi_a(z)}(\phi_a(z)) + \mathcal{A}^{w_2}_{\phi_a(z)}(\phi_a(z)) + \mathcal{N}^{w_3}_{\phi_a(z)}(\phi_a(z)) \le 1\\
&\qquad \mathcal{Y}^{w_1}_{\phi_a(z)}(\phi_a(z)) \ge \mathcal{A}^{w_2}_{\phi_a(z)}(\phi_a(z))\\
&\qquad \mathcal{Y}^{w_1}_{\phi_a(z)}(\phi_a(z)) \ge \mathcal{N}^{w_3}_{\phi_a(z)}(\phi_a(z))\\
&\qquad \mathcal{N}^{w_3}_{\phi_a(z)}(\phi_a(z)) \ge 0,\\
&\qquad Q_j(z) \le d_j,\ j = 1, 2, \ldots, m,\\
&\qquad z > 0.
\end{aligned}$$

This is equivalent to:

$$\begin{aligned}
&\text{Maximize} \quad e\\
&\text{Minimize} \quad b\\
&\text{Minimize} \quad c\\
&\text{subject to } \mathcal{Y}^{w_1}_{\phi_a(z)}(\phi_a(z)) \ge e,\\
&\qquad \mathcal{N}^{w_3}_{\phi_a(z)}(\phi_a(z)) \le c,\\
&\qquad \mathcal{A}^{w_2}_{\phi_a(z)}(\phi_a(z)) \le b,\\
&\qquad \theta_j(z) \le d_j,\ j = 1, 2, \ldots, m,\\
&\qquad 0 \le e + b + c \le 1, e \ge b, e \ge c, e \in [0, 1], b \in [0, 1], c \in [0, 1],\\
&\qquad z > 0, w_1, w_2, w_3 \in [0, 1].
\end{aligned} \tag{11.7}$$

Taking the arithmetic mean, Eq. (11.7) can be written as:

$$\begin{aligned}
&\text{Minimize} \quad \left(\tfrac{e+2(1-c-b)}{3}\right)^{-1} \\
&\text{subject to} \\
&\tfrac{\phi_a(z)}{2}\left(\tfrac{w_1}{t_a}+\tfrac{w_2}{t_a+r_a}+\tfrac{w_3}{r_a}\right) \le 1+\tfrac{G_a}{2}\left(\tfrac{w_1}{t_a}+\tfrac{w_2}{t_a+r_a}+\tfrac{w_3}{r_a}\right)-\left(\tfrac{e+2(1-c-b)}{3}\right) \qquad (11.8)\\
&\theta_j(z) \le d_j, \quad j=1,2,\ldots,m \\
&0 \le e+b+c \le 1, e \ge b, e \ge c, e \in [0,1], b \in [0,1], c \in [0,1], \\
&z > 0, 0 \le w_1+w_2+w_3 \le 1, w_1, w_2, w_3 \in [0,1].
\end{aligned}$$

Solve the above crisp model (11.8) by using an appropriate mathematical programming algorithm to obtain the optimal solution of objective functions.

**Theorem 11.2.** *For a GPFGP model, the sum of membership, neutral membership, and non-membership functions will lie between 0 and $w_1+w_2+w_3$.*

*Proof.* It is clear that $0 \le \mathcal{Y}^{w_1}_{\phi_a(z)}(\phi_a(z)) \le w_1$, $0 \le \mathcal{A}^{w_2}_{\phi_a(z)}(\phi_a(z)) \le w_2$ and $0 \le \mathcal{N}^{w_3}_{\phi_a(z)}(\phi_a(z)) \le w_3$. For $\phi_a(z) \le G_a$, $\mathcal{Y}^{w_1}_{\phi_a(z)}(\phi_a(z)) = w_1$, $\mathcal{A}^{w_2}_{\phi_a(z)}(\phi_a(z)) = \mathcal{N}^{w_3}_{\phi_a(z)}(\phi_a(z)) = 0$.

Therefore

$$\mathcal{Y}^{w_1}_{\phi_a(z)}(\phi_a(z)) + \mathcal{A}^{w_2}_{\phi_a(z)}(\phi_a(z)) + \mathcal{N}^{w_3}_{\phi_a(z)}(\phi_a(z)) = w_1 \le w_1+w_2+w_3.$$

Again, $w_1 \ge 0$ gives that:

$$\mathcal{Y}^{w_1}_{\phi_a(z)}(\phi_a(z)) + \mathcal{A}^{w_2}_{\phi_a(z)}(\phi_a(z)) + \mathcal{N}^{w_3}_{\phi_a(z)}(\phi_a(z)) \ge 0.$$

For $G_a < \phi_a(z) \le G_a + t_a$; when $\phi_a(z) > G_a$,

$$\begin{aligned}
\mathcal{Y}^{w_1}_{\phi_a(z)}(\phi_a(z)) + \mathcal{A}^{w_2}_{\phi_a(z)}(\phi_a(z)) + \mathcal{N}^{w_3}_{\phi_a(z)}(\phi_a(z)) &\le w_1 \frac{t_a}{r_a} \\
&< w_1, \; since \; \frac{t_a}{r_a} < 1 \\
&\le w_1+w_2+w_3.
\end{aligned}$$

For $G_a + t_a < \phi_a(z) \le G_a + r_a$; when $\phi_a(z) > G_a + t_a$;

$$\begin{aligned}
\mathcal{Y}^{w_1}_{\phi_a(z)}(\phi_a(z)) + \mathcal{A}^{w_2}_{\phi_a(z)}(\phi_a(z)) + \mathcal{N}^{w_3}_{\phi_a(z)}(\phi_a(z)) &> w_2 \frac{t_a}{r_a} > w_2 \ge 0, \\
&\text{since } \frac{t_a}{r_a} < 1.
\end{aligned}$$

When $\phi_a(z) \le G_a + r_a$;

$$\mathcal{Y}^{w_1}_{\phi_a(z)}(\phi_a(z)) + \mathcal{A}^{w_2}_{\phi_a(z)}(\phi_a(z)) + \mathcal{N}^{w_3}_{\phi_a(z)}(\phi_a(z)) \le w_1+w_3 \le w_1+w_2+w_3.$$

For $\phi_a(z) \leq G_a + r_a$; $\mathcal{Y}^{w_1}_{\phi_a(z)}(\phi_a(z)) = 0$, $\mathcal{A}^{w_2}_{\phi_a(z)}(\phi_a(z)) = w_2$, $N^{w_3}_{\phi_a(z)}(\phi_a(z)) = w_3$. Therefore

$$\mathcal{Y}^{w_1}_{\phi_a(z)}(\phi_a(z)) + \mathcal{A}^{w_2}_{\phi_a(z)}(\phi_a(z)) + N^{w_3}_{\phi_a(z)}(\phi_a(z)) = w_1 + w_3 \leq w_1 + w_2 + w_3.$$

Again, $w_2 \geq 0$, $w_3 \geq 0$ gives that $\mathcal{Y}^{w_1}_{\phi_a(z)} + \mathcal{A}^{w_2}_{\phi_a(z)} + N^{w_3}_{\phi_a(z)} \geq 0$.
Hence, in all cases, $0 \leq \mathcal{Y}^{w_1}_{\phi_a(z)} + \mathcal{A}^{w_2}_{\phi_a(z)} + N^{w_3}_{\phi_a(z)} \leq w_1 + w_2 + w_3$. □

Most of the authors considered linear membership, neutral membership, and non-membership functions for MOLPP and goal programming. However, several other such functions are reported in the literature, namely, exponential, hyperbolic, parabolic, etc.

Now, we define below the other two types of functions: exponential and hyperbolic.

## Exponential-type membership functions approach (ETMFA)

The exponential-type membership $\mathcal{Y}^{E}_{\phi_a(z)}(\phi_a(z))$, neutral membership $\mathcal{A}^{E}_{\phi_a(z)}(\phi_a(z))$, and nonmembership $N^{E}_{\phi_a(z)}(\phi_a(z))$ functions under a picture fuzzy environment can be stated as follows:

$$\mathcal{Y}^{E}_{\phi_a(z)}(\phi_a(z)) = \begin{cases} \delta^{E}_{\mathcal{Y}} & \text{if } \phi_a(z) \leq G_a \\ \dfrac{e^{-p\left(\delta_{\mathcal{Y}} - \dfrac{\phi_a(z) - G_a}{t_a}\right)} - e^{-p}}{1 - e^{-p}} & \text{if } G_a \leq \phi_a(z) \leq G_a + t_a \\ 0 & \text{if } \phi_a(z) \geq G_a + t_a \end{cases},$$

$$\mathcal{A}^{E}_{\phi_a(z)}(\phi_a(z)) = \begin{cases} 0 & \text{if } \phi_a(z) \leq G_a \\ \dfrac{e^{-p\left(\dfrac{\phi_a(z) - G_a}{r_a}\right)} - e^{-p}}{1 - e^{-p}} & \text{if } G_a \leq \phi_a(z) \leq G_a + r_a \\ \delta^{E}_{\mathcal{A}} & \text{if } \phi_a(z) \geq G_a + r_a \end{cases},$$

$$N^{E}_{\phi_a(z)}(\phi_a(z)) = \begin{cases} 0 & \text{if } \phi_a(z) \leq G_a \\ \dfrac{e^{-p\left(\dfrac{\phi_a(z) - G_a}{r_a}\right)} - e^{-p}}{1 - e^{-p}} & \text{if } G_a \leq \phi_a(z) \leq G_a + r_a \\ \delta^{E}_{N} & \text{if } \phi_a(z) \geq G_a + r_a \end{cases},$$

where, $\delta^{E}_{\mathcal{Y}} + \delta^{E}_{\mathcal{A}} + \delta^{E}_{N} \leq 1$.

Here, $p$ is the measure of vagueness degree, and it is assigned by the decision maker.

Therefore the PFGP problem can be modified for the exponential-type membership functions as follows:

$$\left.\begin{array}{l}
\text{Maximize} \quad (\mathcal{Y}^E_{\phi_a(z)}(\phi_a(z))) \\
\text{Minimize} \quad (\mathcal{A}^E_{\phi_a(z)}(\phi_a(z))) \\
\text{Minimize} \quad (\mathcal{N}^E_{\phi_a(z)}(\phi_a(z))) \\
\text{subject to} \\
0 \le \mathcal{Y}^E_{\phi_a(z)}(\phi_a(z)) + \mathcal{A}^E_{\phi_a(z)}(\phi_a(z)) + \mathcal{N}^E_{\phi_a(z)}(\phi_a(z)) \le 1, a = 1, 2, ..., n \\
\mathcal{Y}^E_{\phi_a(z)}(\phi_a(z)) \ge \mathcal{A}^E_{\phi_a(z)}(\phi_a(z)), a = 1, 2, ..., n \\
\mathcal{Y}^E_{\phi_a(z)}(\phi_a(z)) \ge \mathcal{N}^E_{\phi_a(z)}(\phi_a(z)), a = 1, 2, ..., n \\
\mathcal{N}^E_{\phi_a(z)}(\phi_a(z)) \ge 0 \quad \text{for } a = 1, 2, ..., n \\
\theta_j(z) \le d_j, j = 1, 2, ..., m \\
z > 0,
\end{array}\right\}. \tag{11.9}$$

Also, the picture fuzzy programming problem (11.9) can be transformed into the form of $e, b, c$ by using the arithmetic mean. Then, the objective function will be $\frac{e+2(1-c-b)}{3}$ and other conditions are similarly deduced. Additionally, a generalized version for exponential-type membership, neutral membership, and nonmembership functions can be modified by including respective weights as in the linear-type case. A more detailed study and corresponding application area can be explored by any researcher in the near future.

On the other hand, for exponential-type membership functions approach can be used in picture fuzzy multiobjective programming problem can be modeled. In that case, $\mathcal{Y}^E_{\phi_a(z)}(\phi_a(z)) \ge e$, $\mathcal{A}^E_{\phi_a(z)}(\phi_a(z)) \le b$, and $\mathcal{N}^E_{\phi_a(z)}(\phi_a(z)) \le c$, for all $a = 1, 2, ..., n$.

Then, using auxiliary parameters $e$, $b$, and $c$, the MOLPP with linear-type membership functions will be reduced to the following problem:

$$\begin{array}{ll}
\text{(ETMFA)} & \text{Maximize} \quad (e - b - c) \\
& \text{subject to} \\
& \dfrac{e^{-p\left(\delta_y - \frac{\phi_a(z) - G_a}{t_a}\right)} - e^{-p}}{1 - e^{-p}} \ge e \\
& \dfrac{e^{-p\left(\frac{\phi_a(z) - G_a}{r_a}\right)} - e^{-p}}{1 - e^{-p}} \le b \\
& \dfrac{e^{-p\left(\frac{\phi_a(z) - G_a}{r_a}\right)} - e^{-p}}{1 - e^{-p}} \le c \\
& 0 \le e + b + c \le 1, e \ge b, e \ge c, e \in [0, 1], b \in [0, 1], c \in [0, 1], \\
& \text{all other particular constraints for a considered MOLPP.}
\end{array} \tag{11.10}$$

Here, $p$ is the measures of vagueness degree and it is assigned by the decision maker.

**Remark.** If $p \longrightarrow 0$, then the exponential-type membership functions will be reduced to linear-type membership functions for MOLPP and goal programming problems under a picture fuzzy environment.

Also, it is notable that for MOLPP, the ETMFA will be the expressions of membership, neutral membership, and nonmembership functions as follows:

$$\mathcal{Y}^E_{\phi_a(z)}(\phi_a(z)) = \begin{cases} \delta^E_{\mathcal{Y}} & \text{if } \phi_a(z) \le L^{\mathcal{Y}}_{\phi_a(z)} \\ \dfrac{e^{-p\left(\frac{\phi_a(z) - L^{\mathcal{Y}}_{\phi_a(z)}}{U^{\mathcal{Y}}_{\phi_a(z)} - L^{\mathcal{Y}}_{\phi_a(z)}}\right)} - e^{-p}}{1-e^{-p}} & \text{if } L^{\mathcal{Y}}_{\phi_a(z)} \le \phi_a(z) \le U^{\mathcal{Y}}_{\phi_a(z)} \\ 0 & \text{if } \phi_a(z) \ge U^{\mathcal{Y}}_{\phi_a(z)} \end{cases},$$

$$\mathcal{A}^E_{\phi_a(z)}(\phi_a(z)) = \begin{cases} 0 & \text{if } \phi_a(z) \le L^{\mathcal{A}}_{\phi_a(z)} \\ \dfrac{e^{-p\left(\frac{\phi_a(z) - L^{\mathcal{A}}_{\phi_a(z)}}{U^{\mathcal{A}}_{\phi_a(z)} - L^{\mathcal{A}}_{\phi_a(z)}}\right)} - e^{-p}}{1-e^{-p}} & \text{if } L^{\mathcal{A}}_{\phi_a(z)} \le \phi_a(z) \le U^{\mathcal{A}}_{\phi_a(z)} \\ \delta^E_{\mathcal{A}} & \text{if } \phi_a(z) \ge U^{\mathcal{A}}_{\phi_a(z)} \end{cases},$$

$$N^E_{\phi_a(z)}(\phi_a(z)) = \begin{cases} 0 & \text{if } \phi_a(z) \le L^{N}_{\phi_a(z)} \\ \dfrac{e^{-p\left(\frac{U^{N}_{\phi_a(z)} - \phi_a(z)}{U^{N}_{\phi_a(z)} - L^{N}_{\phi_a(z)}}\right)} - e^{-p}}{1-e^{-p}} & \text{if } L^{N}_{\phi_a(z)} \le \phi_a(z) \le U^{N}_{\phi_a(z)} \\ \delta^E_{N} & \text{if } \phi_a(z) \ge U^{N}_{\phi_a(z)} \end{cases},$$

where, $\delta^E_{\mathcal{Y}} + \delta^E_{\mathcal{A}} + \delta^E_N \le 1$; and $U_{\phi_a(z)} = max[\phi_a(z)]$, $L_{\phi_a(z)} = min[\phi_a(z)]$ for all $a = 1, 2, ..., n$. Also,

$$U^{\mathcal{Y}}_{\phi_a(z)} = U_{\phi_a(z)}, L^{\mathcal{Y}}_{\phi_a(z)} = L_{\phi_a(z)} \text{ for a membership function,}$$
$$U^{\mathcal{A}}_{\phi_a(z)} = L^{\mathcal{Y}}_{\phi_a(z)} + \delta_{\phi_a(z)}, L^{\mathcal{A}}_{\phi_a(z)} = L^{\mathcal{Y}}_{\phi_a(z)} \text{ for a neutral membership function,}$$
$$U^{N}_{\phi_a(z)} = U^{\mathcal{Y}}_{\phi_a(z)}, L^{N}_{\phi_a(z)} = L^{\mathcal{Y}}_{\phi_a(z)} + \delta_{\phi_a(z)} \text{ for a nonmembership function.}$$

Here, $\delta_{\phi_a(z)} \in (0, 1)$ are predetermined real numbers prescribed by decision makers.

## Hyperbolic-type membership functions approach (HTMFA)

The hyperbolic-type membership, neutral membership, and nonmembership functions for picture fuzzy environment to formulate goal programming problems are as follows:

$$\mathcal{Y}^H_{\phi_a(z)}(\phi_a(z)) = \begin{cases} \delta^H_{\mathcal{Y}} & \text{if } \phi_a(z) \le G_a \\ \frac{1}{2}\left[1 + tanh\left(\theta_{\phi_a(z)}\left(\frac{U^{\mathcal{Y}}_{\phi_a(z)} + L^{\mathcal{Y}}_{\phi_a(z)}}{2} - \phi_a(z)\right)\right)\right] & \text{if } G_a \le \phi_a(z) \le G_a + t_a \\ 0 & \text{if } \phi_a(z) \ge G_a + t_a \end{cases},$$

$$\mathcal{A}^{H}_{\phi_a(z)}(\phi_a(z)) = \begin{cases} 0 & \text{if } \phi_a(z) \le G_a \\ \frac{1}{2}\left[1 + tanh\left(\theta_{\phi_a(z)}\left(\frac{U^{\mathcal{A}}_{\phi_a(z)} + L^{\mathcal{A}}_{\phi_a(z)}}{2} - \phi_a(z)\right)\right)\right] & \text{if } G_a \le \phi_a(z) \le G_a + r_a \\ \delta^{H}_{\mathcal{A}} & \text{if } \phi_a(z) \ge G_a + r_a \end{cases},$$

$$\mathcal{N}^{H}_{\phi_a(z)}(\phi_a(z)) = \begin{cases} 0 & \text{if } \phi_a(z) \le G_a \\ \frac{1}{2}\left[1 + tanh\left(\theta_{\phi_a(z)}\left(\phi_a(z) - \frac{U^{\mathcal{N}}_{\phi_a(z)} + L^{\mathcal{N}}_{\phi_a(z)}}{2}\right)\right)\right] & \text{if } G_a \le \phi_a(z) \le G_a + r_a \\ \delta^{H}_{\mathcal{N}} & \text{if } \phi_a(z) \ge G_a + r_a \end{cases},$$

for all $a = 1, 2, ..., n$. The meaning and notations for target value, acceptance, neutral, and rejection tolerances for a particular case study are to be modified as per requirements, and these are discussed in the linear-type situation.

Moving with the same procedure as in the exponential-type approach, we have obtained a PFGP problem as in (11.9) by replacing all $\mathcal{Y}^{E}_{\phi_a(z)}(\phi_a(z))$, $\mathcal{A}^{E}_{\phi_a(z)}(\phi_a(z))$, and $\mathcal{N}^{E}_{\phi_a(z)}(\phi_a(z))$ with $\mathcal{Y}^{H}_{\phi_a(z)}(\phi_a(z))$, $\mathcal{A}^{H}_{\phi_a(z)}(\phi_a(z))$, and $\mathcal{N}^{H}_{\phi_a(z)}(\phi_a(z))$, respectively:

$$\mathcal{Y}^{H}_{\phi_a(z)}(\phi_a(z)) = \begin{cases} \delta^{H}_{\mathcal{Y}} & \text{if } \phi_a(z) \le L^{\mathcal{Y}}_{\phi_a(z)} \\ \frac{1}{2}\left[1 + tanh\left(\theta_{\phi_a(z)}\left(\frac{U^{\mathcal{Y}}_{\phi_a(z)} + L^{\mathcal{Y}}_{\phi_a(z)}}{2} - \phi_a(z)\right)\right)\right] & \text{if } L^{\mathcal{Y}}_{\phi_a(z)} \le \phi_a(z) \le U^{\mathcal{Y}}_{\phi_a(z)} \\ 0 & \text{if } \phi_a(z) \ge U^{\mathcal{Y}}_{\phi_a(z)} \end{cases},$$

$$\mathcal{A}^{H}_{\phi_a(z)}(\phi_a(z)) = \begin{cases} 0 & \text{if } \phi_a(z) \le L^{\mathcal{A}}_{\phi_a(z)} \\ \frac{1}{2}\left[1 + tanh\left(\theta_{\phi_a(z)}\left(\frac{U^{\mathcal{A}}_{\phi_a(z)} + L^{\mathcal{A}}_{\phi_a(z)}}{2} - \phi_a(z)\right)\right)\right] & \text{if } L^{\mathcal{A}}_{\phi_a(z)} \le \phi_a(z) \le U^{\mathcal{A}}_{\phi_a(z)} \\ \delta^{H}_{\mathcal{A}} & \text{if } \phi_a(z) \ge U^{\mathcal{A}}_{\phi_a(z)} \end{cases},$$

$$\mathcal{N}^{H}_{\phi_a(z)}(\phi_a(z)) = \begin{cases} 0 & \text{if } \phi_a(z) \le L^{\mathcal{N}}_{\phi_a(z)} \\ \frac{1}{2}\left[1 + tanh\left(\theta_{\phi_a(z)}\left(\phi_a(z) - \frac{U^{\mathcal{N}}_{\phi_a(z)} + L^{\mathcal{N}}_{\phi_a(z)}}{2}\right)\right)\right] & \text{if } L^{\mathcal{N}}_{\phi_a(z)} \le \phi_a(z) \le U^{\mathcal{N}}_{\phi_a(z)} \\ \delta^{H}_{\mathcal{N}} & \text{if } \phi_a(z) \ge U^{\mathcal{N}}_{\phi_a(z)} \end{cases},$$

where, $\theta_{\phi_a(z)} = \frac{6}{U_{\phi_a(z)} - L_{\phi_a(z)}} \forall a = 1, 2, ..., n$.

Also, for MOLPP in a picture fuzzy environment by HTMFA will produce the following programming problem:

(HTMFA) Maximize $(e - b - c)$

subject to

$$\frac{1}{2}\left[1 + tanh\left(\theta_{\phi_a(z)}\left(\frac{U^{\mathcal{Y}}_{\phi_a(z)} + L^{\mathcal{Y}}_{\phi_a(z)}}{2} - \phi_a(z)\right)\right)\right] \ge e$$

$$\frac{1}{2}\left[1 + tanh\left(\theta_{\phi_a(z)}\left(\frac{U^{\mathcal{A}}_{\phi_a(z)} + L^{\mathcal{A}}_{\phi_a(z)}}{2} - \phi_a(z)\right)\right)\right] \le b$$

$$\frac{1}{2}\left[1+tanh\left(\theta_{\phi_a(z)}\left(\phi_a(z)-\frac{U^{\mathcal{N}}_{\phi_a(z)}+L^{\mathcal{N}}_{\phi_a(z)}}{2}\right)\right)\right]\leq c$$

$$0\leq e+b+c\leq 1, e\geq b, e\geq c, e\in[0,1], b\in[0,1], c\in[0,1],$$

all other particular constraints for a considered MOLPP. (11.11)

This problem (11.11) is equivalent to the following:

(HTMFA) Maximize $(e-b-c)$

subject to

$$\theta_{\phi_a(z)}\phi_a(z)+tanh^{-1}(2e-1)\leq\frac{\theta_{\phi_a(z)}}{2}\left(U^{\mathcal{Y}}_{\phi_a(z)}+L^{\mathcal{Y}}_{\phi_a(z)}\right),$$

$$\theta_{\phi_a(z)}\phi_a(z)-tanh^{-1}(2b-1)\leq\frac{\theta_{\phi_a(z)}}{2}\left(U^{\mathcal{A}}_{\phi_a(z)}+L^{\mathcal{A}}_{\phi_a(z)}\right),$$

$$\theta_{\phi_a(z)}\phi_a(z)-tanh^{-1}(2c-1)\leq\frac{\theta_{\phi_a(z)}}{2}\left(U^{\mathcal{N}}_{\phi_a(z)}+L^{\mathcal{N}}_{\phi_a(z)}\right)$$

$$0\leq e+b+c\leq 1, e\geq b, e\geq c, e\in[0,1], b\in[0,1], c\in[0,1],$$

all other particular constraints for a considered MOLPP, (11.12)

where, $\theta_{\phi_a(z)}=\frac{6}{U_{\phi_a(z)}-L_{\phi_a(z)}}\forall a=1,2,\ldots,n.$

All these MOLPP and goal programming problems can be generalized by including corresponding weights to the membership, neutral membership, and nonmembership functions, respectively.

## 11.4 An application of picture fuzzy goal programming in the recycling process of plastic

A plastic recycling plant is needed to be designed in such a way that the gathered plastics are to be recycled for use later in another way. The recycling industry units are eager to wash and resize the mass of plastic collected from different zones of the environment. The recycling process of plastic has many steps in any industry or plant.

This study takes three major consecutive steps of the plastic recycling process for the model design and decision-making purpose after collecting all the plastics.

Sorting ⟶ Washing ⟶ Resizing.

After each recycling process, the plastics are strictly washed and resized for fresh use. Let $z_i$ be the percentage of errors that are remaining in each step completion. Then, after completing the three major steps of the recycling procedure, the remaining percentage of error will be $z_1z_2z_3$. The decision maker's (DM) target is to minimize the remaining percentage of error with a minimum cost per year as much as possible. All the costs collected from a reputed plastic recycling plant for performing each step mentioned in this problem are supplied annually, see Table 11.1.

**Table 11.1** Charges for performing different steps of the recycling process per year.

| Steps | Cost in Rs. thousands |
|---|---|
| Sorting | $29.2z_1^{-1.23}$ |
| Washing | $23.4z_2^{-1.44}$ |
| Resizing | $98.3z_3^{-0.6}$ |

Now, let us note the series of steps of the recycling process with minimum annual cost and minimum percentage of error in each step. The DM has set some targets on total annual cost and the remaining percentage of error for the whole process. Also, he/she has given relaxation on his/her targets. Hence, the PFGP model is:

$$\widetilde{\text{Minimize}}^1 \; Cost\,(z_1, z_2, z_3) = 29.2z_1^{-1.23} + 23.4z_2^{-1.44} + 98.3z_3^{-0.6}$$

with target 296, acceptance tolerance 198,

neutral and rejection tolerance 297,

$$\widetilde{Minimize}^1 \; ERROR\,(z_1, z_2, z_3) = z_1 z_2 z_3$$

with target 0.014, acceptance tolerance 0.4,

neutral and rejection tolerance 0.6,

subject to $z_1, z_2, z_3 > 0.$

### 11.4.1 Modeling in picture fuzzy goal programming problems

All the membership, neutral membership, and nonmembership functions of objective functions are described as follows:

$$\mathcal{Y}_{cost}(z_1, z_2, z_3) = \begin{cases} \delta_{\mathcal{Y}}, & \text{if } cost(z_1, z_2, z_3) \le 296 \\ \delta_{\mathcal{Y}} - \frac{cost(z_1, z_2, z_3) - 296}{198}, & \text{if } 296 \le cost(z_1, z_2, z_3) \le 494 \\ 0, & \text{if } cost(z_1, z_2, z_3) \ge 494 \end{cases},$$

$$\mathcal{A}_{cost}(z_1, z_2, z_3) = \begin{cases} 0, & \text{if } cost(z_1, z_2, z_3) \le 296 \\ \frac{cost(z_1, z_2, z_3) - 296}{297}, & \text{if } 494 \le cost(z_1, z_2, z_3) \le 593 \\ \delta_{\mathcal{A}}, & \text{if } cost(z_1, z_2, z_3) \ge 593 \end{cases},$$

$$\mathcal{N}_{cost}(z_1, z_2, z_3) = \begin{cases} 0, & \text{if } cost(z_1, z_2, z_3) \le 296 \\ \frac{cost(z_1, z_2, z_3) - 296}{297}, & \text{if } 296 \le cost(z_1, z_2, z_3) \le 593 \\ \delta_{\mathcal{N}}, & \text{if } cost(z_1, z_2, z_3) \ge 593 \end{cases},$$

$$\mathcal{Y}_{ERROR}(z_1, z_2, z_3) = \begin{cases} \delta_{\mathcal{Y}}, & \text{if } ERROR(z_1, z_2, z_3) \le 296 \\ \delta_{\mathcal{Y}} - \frac{ERROR(z_1, z_2, z_3) - 0.014}{0.4}, & \text{if } 0.014 \le ERROR(z_1, z_2, z_3) \le 0.414 \\ 0, & \text{if } ERROR(z_1, z_2, z_3) \ge 0.414 \end{cases},$$

$$\mathcal{A}_{ERROR}(z_1, z_2, z_3) = \begin{cases} \delta_{\mathcal{Y}}, & \text{if} \quad ERROR(z_1, z_2, z_3) \leq 0.014 \\ \frac{ERROR(z_1,z_2,z_3)-0.014}{0.6}, & \text{if} \quad 0.414 \leq ERROR(z_1, z_2, z_3) \leq 0.614 \\ \delta_{\mathcal{A}}, & \text{if} \quad ERROR(z_1, z_2, z_3) \geq 0.614 \end{cases},$$

$$\mathcal{A}_{ERROR}(z_1, z_2, z_3) = \begin{cases} 0, & \text{if} \quad ERROR(z_1, z_2, z_3) \leq 0.014 \\ \frac{ERROR(z_1,z_2,z_3)-0.014}{0.6}, & \text{if} \quad 0.014 \leq ERROR(z_1, z_2, z_3) \leq 0.614 \\ \delta_{\mathcal{N}}, & \text{if} \quad ERROR(z_1, z_2, z_3) \geq 0.614 \end{cases}.$$

This PFGP model can be transformed into a crisp programming model with the help of membership, neutral membership, and nonmembership functions for both the objective functions represented in the above section:

$$\text{Minimize} \quad \left(\frac{e+2(1-c-b)}{3}\right)^{-1}$$

subject to

$$[29.2z_1^{-1.23} + 23.4z_2^{-1.44} + 98.3z_3^{-0.6}] + \frac{e+2(1-c-b)}{0.015} \leq 496$$

$$z_1z_2z_3 + \frac{e+2(1-c-b)}{7.749} \leq 0.401$$

$$z_1, z_2, z_3 > 0$$

$$0 \leq e+b+c \leq 1$$

$$0 \leq e \leq 1, 0 \leq b \leq 1, 0 \leq c \leq 1.$$

We have taken the values $\delta_{\mathcal{Y}} = 0.6$, $\delta_{\mathcal{A}} = 0.2$, $\delta_{\mathcal{N}} = 0.1$ as the choice of the DM for this particular case study to evaluate the membership, neutral membership, and nonmembership values in the decision-making process. Now, we have used the mathematical software LINGO with software version 18.0.44 to find the solution to the above programming problem, and all output data are provided in Table 11.2.

**Table 11.2** Optimized solutions for the case study.

| Weights | Variables | Optimal objective functions | $e^*, b^*, c^*$ | Membership value | Neutral membership value | Nonmembership value |
|---|---|---|---|---|---|---|
| $w_1 = 1$<br>$w_2 = 1$<br>$w_3 = 1$ | $z_1^* = 1.492$<br>$z_2^* = 0.17$<br>$z_3^* = 4.328$ | $Cost^* = 358.83$ | $e^* = 0.965$<br>$b^* = 3.32 \times 10^{-10}$ | $\mathcal{Y}_{Cost^*} = 0.382$ | $\mathcal{A}_{Cost^*} = 0.0424$ | $\mathcal{N}_{Cost^*} = 0.212$ |
| $w_1 = 1$<br>$w_2 = 1$<br>$w_3 = 1$ | $z_1^* = 1.492$<br>$z_2^* = 0.17$<br>$z_3^* = 4.328$ | $ERROR^* = 0.0189$ | $e^* = 0.965$<br>$b^* = 3.32 \times 10^{-10}$ | $\mathcal{Y}_{ERROR^*} = 0.587$ | $\mathcal{A}_{ERROR^*} = 0.0016$ | $\mathcal{N}_{ERROR^*} = 0.008$ |

Table 11.2 shows that membership, neutral membership, and nonmembership functions satisfy all the restrictions as in our constructed model and theorem. The minimized cost for the plastic recycling process is obtained as Rs. 358.83 thousand, and the minimized percentage of error in performing every step of the procedure is determined as a value of 0.0189, i.e., the error in the process in its minimum value is 1.89%.

It is obvious that the minimized error is very low. This is reflecting the sustainability and efficiency of our constructed model.

## 11.5 Conclusion

The GP is considered in a picture fuzzy environment. A detailed discussion on GP has been made over linear membership, neutral, and nonmembership functions. The exponential and hyperbolic membership, neutral, and nonmembership functions are also highlighted for GP and need more investigations. An application of GP in a picture fuzzy environment is also presented. In many research articles, the goal is considered as a linguistic term, and this approach is more realistic. For this case, more research is required for PFS.

## References

[1] A.R.A. Ghaffar, Md.G. Hasan, Z. Ashraf, M.F. Khan, Fuzzy goal programming with an imprecise intuitionistic fuzzy preference relations, Symmetry 12 (2020) 1548, https://doi.org/10.3390/sym12091548.

[2] O. Aköz, D. Petrovic, A fuzzy goal programming method with imprecise goal hierarchy, Eur. J. Oper. Res. 181 (2007) 1427–1433.

[3] P.P. Angelov, Optimization in an intuitionistic fuzzy environment, Fuzzy Sets Syst. 86 (1997) 299–306.

[4] M. Arenas-Parra, A. Bilbao-Terol, M. Jiménez, Standard goal programming with fuzzy hierarchies: a sequential approach, Soft Comput. 20 (2016) 2341–2352.

[5] R. Bellman, L. Zadeh, Decision making in a fuzzy environment, Manag. Sci. 17 (1970) 141–164.

[6] S.K. Bharati, S.R. Singh, Solving multi objective linear programming problems using intuitionistic fuzzy optimization method: a comparative study, Int. J. Model. Optim. 4 (2014) 10.

[7] A. Bilbao-Terol, M. Jiménez, M. Arenas-Parra, A group decision making model based on goal programming with fuzzy hierarchy: an application to regional forest planning, Ann. Oper. Res. 245 (2016) 137–162.

[8] A. Bilbao-Terol, M. Arenas-Parra, V. Cañal-Fernández, M. Jiménez, A sequential goal programming model with fuzzy hierarchies to sustainable and responsible portfolio selection problem, J. Oper. Res. Soc. 67 (2016) 1259–1273.

[9] R. Caballero, T. Go, Goal programming: realistic targets for the near future, J. Multi-Criteria Decis. Anal. 110 (2010) 79–110.

[10] A. Charnes, W.W. Cooper, Management models and industrial applications of linear programming, Manag. Sci. 4 (1957) 38–91.

[11] C.T. Chang, Efficient structures of achievement functions for goal programming models, Asia-Pac. J. Oper. Res. 24 (2007) 755–764.

[12] L.H. Chen, F.C. Tsai, Fuzzy goal programming with different importance and priorities, Eur. J. Oper. Res. 133 (2001) 548–556.

[13] H.W. Cheng, A satisficing method for fuzzy goal programming problems with different importance and priorities, Qual. Quant. 47 (2013) 485–498.

[14] B.C. Cuong, Picture fuzzy sets-first results, in: Seminar Neuro-Fuzzy Systems with Applications, Institute of Mathematics, Vietnam Academy of Science and Technology, Hanoi-Vietnam, 2013, Preprint 03/2013.

[15] B.C. Cuong, Picture fuzzy sets - first results. Part 2, Seminar "Neuro-fuzzy systems with applications", Tech. Rep., Institute of Mathematics, Hanoi, 2013.

[16] S. Dey, T.K. Roy, Intuitionistic fuzzy goal programming technique for solving non-linear multi-objective structural problem, J. Fuzzy Set Valued Anal. 2015 (3) (2015) 179–193.

[17] M. Díaz-Madroñero, J. Mula, M. Jiménez, Fuzzy goal programming for material requirements planning under uncertainty and integrity conditions, Int. J. Prod. Res. 52 (2014) 6971–6988.

[18] A. Ebrahimnejad, J.L. Verdegay, A new approach for solving fully intuitionistic fuzzy transportation problems, Fuzzy Optim. Decis. Mak. 17 (2017) 447–474.

[19] P. Ghosh, T.K. Roy, Goal geometric programming with crisp and imprecise targets, J. Glob. Res. Comput. Sci. 4 (8) (2013) 21–29.
[20] P. Ghosh, T.K. Roy, Intuitionistic fuzzy goal geometric programming problem, Notes IFS 20 (1) (2014) 63–78.
[21] J.P. Ignizio, A review of goal programming: a tool for multiobjective analysis, J. Oper. Res. Soc. 29 (1978) 1109–1119.
[22] J.P. Ignizio, Generalized goal programming: an overview, Comput. Oper. Res. 10 (1983) 277–289.
[23] O. Jadidi, S. Zolfaghari, S. Cavalieri, A new normalized goal programming model for multi-objective problems: a case of supplier selection and order allocation, Int. J. Prod. Econ. 148 (2014) 158–165.
[24] R.K. Jana, D.K. Sharma, B. Chakraborty, A hybrid probabilistic fuzzy goal programming approach for agricultural decision-making, Int. J. Prod. Econ. 173 (2016) 134–141.
[25] K. Khalili-Damghani, S. Sadi-Nezhad, A decision support system for fuzzy multi-objective multi-period sustainable project selection, Comput. Ind. Eng. 64 (2013) 1045–1060.
[26] M.F. Khan, M. Hasan, A. Quddoos, A. Fügenschuh, S.S. Hasan, Goal programming models with linear and exponential fuzzy preference relations, Symmetry 12 (2020) 934.
[27] D. Kour, S. Mukherjee, K. Basu, Solving intuitionistic fuzzy transportation problem using linear programming, Int. J. Syst. Assur. Eng. Manag. 8 (2017) 1090–1101.
[28] S. Mahajan, S. Gupta, On fully intuitionistic fuzzy multiobjective transportation problems using different membership functions, Ann. Oper. Res. (2019) 1–31.
[29] T.K. Mondal, S.K. Samanta, Generalised intuitionistic fuzzy sets, J. Fuzzy Math. 10 (2012) 839–861.
[30] D. Petrovic, O. Aköz, A fuzzy goal programming approach to integrated loading and scheduling of a batch processing machine, J. Oper. Res. Soc. 59 (2008) 1211–1219.
[31] S. Pramanik, T.K. Roy, Multiobjective transportation model with fuzzy parameters: priority based fuzzy goal programming approach, J. Transp. Syst. Eng. Inf. Technol. 8 (2008) 40–48.
[32] J. Razmi, E. Jafarian, S.H. Amin, An intuitionistic fuzzy goal programming approach for finding Pareto-optimal solutions to multi-objective programming problems, Expert Syst. Appl. 65 (2016) 181–193.
[33] C. Romero, Handbook of Critical Issues in Goal Programming, Pergamon Press, Oxford, UK, 1991, pp. 1–12.
[34] C. Romero, A general structure of achievement function for a goal programming model, Eur. J. Oper. Res. 153 (2003) 675–686.
[35] S.K. Roy, A. Ebrahimnejad, J.L. Verdegay, S. Das, New approach for solving intuitionistic fuzzy multi-objective transportation problem, Sadhana 43 (2018) 3.
[36] M.R. Seikh, P.K. Nayak, M. Pal, Application of intuitionistic fuzzy mathematical programming with exponential membership and quadratic non-membership functions in matrix games, Ann. Fuzzy Math. Inform. 9 (2015) 183–195.
[37] M. Sheikhalishahi, S. Torabi, Maintenance supplier selection considering life cycle costs and risks: a fuzzy goal programming approach, Int. J. Prod. Res. 52 (2014) 7084–7099.
[38] S. Singh, S. Yadav, Intuitionistic fuzzy multi-objective linear programming problem with various membership functions, Ann. Oper. Res. 269 (2017) 693–707.
[39] B.B. Tabrizi, K. Shahanaghi, M. Saeed Jabalameli, Fuzzy multi-choice goal programming, Appl. Math. Model. 36 (2012) 1415–1420.
[40] R. Tiwari, S. Dharmar, J. Rao, Fuzzy goal programming—an additive model, Fuzzy Sets Syst. 24 (1987) 27–34.
[41] S.A. Torabi, M. Moghaddam, Multi-site integrated production-distribution planning with transshipment: a fuzzy goal programming approach, Int. J. Prod. Res. 50 (2012) 1726–1748.
[42] X. Zhao, Y. Zheng, Z. Wan, Interactive intuitionistic fuzzy methods for multilevel programming problems, Expert Syst. Appl. 72 (2017) 258–268.

# 12

# Picture fuzzy linear assignment problem and its application in multicriteria group decision-making problems

## Abbreviations

| | |
|---|---|
| **PF** | Picture fuzzy |
| **PFS** | Picture fuzzy set |
| **PFN** | Picture fuzzy number |
| **LAM** | Linear assignment model |
| **AO** | Average operator |
| **DM** | Decision maker |
| **MADM** | Multiattributes decision making |
| **MCDM** | Multicriteria decision making |
| **MCGDM** | Multicriteria group decision making |
| **MABAC** | Multiattributive border approximation area comparison |

## 12.1 Introduction

In order to deal with ambiguous situations and inaccurate information, Zadeh's fuzzy sets theory [38] provides a practical and acceptable technique. Following its inception, fuzzy sets have gained enormous popularity across practically all scientific disciplines. The picture fuzzy sets (PFS) extension is one of the newest. Cuong [8,9] defined PFS, which are an extension of intuitionistic fuzzy sets (IFS) [6,7] that may independently model uncertainty using membership degree, nonmembership degree, and hesitant degree. Wei [34] defined weighted average operators (AOs) in this manner to aggregate PFS data. Wei [33] analyzed the entropy of PFS information and applied this idea to create a MADM model. Later, Wei employed the Hamacher AOs and built picture fuzzy Hamacher AOs using this concept [35]. Additionally, Peng and Dai [14] created an approach for creating distance-based MADM issues in a PFS. Singh [19] studied the coefficient of correlation measure using PFS arguments. New inference systems for PFS were proposed by Son et al. [21,23]. When developing decisions, Son [20] used distribution-based picture fuzzy clustering. In their research of the clustering approach to analyze complicated dates and weather forecasting in PFS environments, Thong and Son cited the following papers: [24–28]. In order to create customer requirements, Ping et al. [15] used a novel quality function deployment (QFD) environment. Projections-based MADM issues in a PFS are discussed by Wei et al. in [30]. A dice similarity measure approach was employed in PFSs by Wei and Gao to create

Picture Fuzzy Logic and Its Applications in Decision Making Problems. https://doi.org/10.1016/B978-0-44-322024-1.00016-9

MADM models [31]. Wei proposed a new similarity measure in a picture fuzzy environment in [32]. In the same setting, weighted Dombi AOs were proposed by Jana et al. [11] and utilized to build a MADM technique. Readers are referred to the following sources for more details on the decision-making process connected to the PFS [16,17,21–23,29,36].

The linear programming problem (LPP) is the most well-known, straightforward, and widely applied paradigm in mathematical programming. The LPP model's simplicity makes it easily adaptable to a wide range of practical applications, including supply-chain management, supplier selection, assignment issues, and challenges with transportation. Over several decades, the extension and development of standard LPP have been discussed. Examples of expanded LPP include goal linear programming problems, multiobjective linear programming problems, and bilevel or multilayer linear programming problems; for more information, we refer the reader to [1–5,18,39]. The assignment problem in linear programming for multiattribute decision making served as the inspiration for Bernardo and Blin's linear assignment method (LAM) [5]. The fundamental tenet of the LAM is that it combines the component ranking that results in an overall preference rating that results in the best possible compromise between the various component rankings. The objective of [12] was to provide an extended LAM to resolve multicriteria decision-making (MCDM) problems in a Pythagorean fuzzy environment. The LAM for interval-valued Pythagorean fuzzy sets was also developed by Liang et al. [13]. Chen [10] created an effective solution for resolving MCDM issues in the interval-valued intuitionistic fuzzy environment by expanding the conventional LAM.

As far as we are aware, no study has been done on the extension and use of the LAM in a picture fuzzy multicriteria group decision-making context. The purpose of this study is to create a novel multicriteria group decision-making method based on a linear assignment methodology using picture fuzzy sets, as well as to demonstrate how it can be applied to choosing a location for a sponge iron factory development. The following contributions are made by the suggested algorithm. First, judgement values are described using picture-language phrases that can take into account how hesitantly decision makers comment on alternatives and criteria. To reduce the impact of subjectivity, the ranking of options has been done using the LAM.

## 12.2 Preliminaries

In this section, we annotate some essential ideas of PFSs of the universe $Z$.

**Definition 12.1.** [8,9] A PFS $U$ over the fixed set $X$ is written as:

$$\mathcal{R}_f = \{\langle \mathcal{Y}_f(x), \eta_U(x), \mathcal{N}_f(x)\rangle | x \in X\},$$

$\mathcal{Y}_f(x) : X \to [0,1]$, $\eta_U(x) : X \to [0,1]$ and $\mathcal{N}_f(x) : X \to [0,1]$ are, respectively, the presented membership degree, neutral membership degree, and nonmembership degree in a fuzzy set $U$, where $0 \leq \mathcal{Y}_f(x) + \mathcal{A}_f(x) + \mathcal{N}_f(x) \leq 1$ for $x \in X$. Also, the refusal membership degree is denoted for $x$ as $\pi_U(x) = 1 - \mathcal{Y}_f(x) - \mathcal{A}_f(x) - \mathcal{N}_f(x)$. The group $(\mathcal{Y}_f, \mathcal{A}_f, \mathcal{N}_f)$ are named picture fuzzy numbers (PFNs).

Cuong et al. [8] introduced some basic operations on PFSs given as:

**Definition 12.2.** [8] Let $U = (\mathcal{Y}_f(x), \mathcal{A}_f(x), \mathcal{N}_f(x))$ and $V = (\mu_V(x), \eta_V(x), \nu_V(x))$ be any two PFNs over the set $X$. The operations between the two PFNs are:

**(i)** $U \subseteq V$, if $\mathcal{Y}_f(x) \leq mu_V(x)$, $\mathcal{A}_f(x) \leq \eta_V(x)$ and $\mathcal{N}_f(x) \geq \nu_V(x)$ for all $x \in X$;
**(ii)** $U = V$ iff $U \subseteq V$ and $V \subseteq U$;
**(iii)** $U \cup V = \{\langle x, \max\{\mathcal{Y}_f(x), \mu_V(x)\}, \min\{\mathcal{A}_f(x), \eta_V(x)\}, \min\{\mathcal{N}_f(x), \nu_V\}\rangle | x \in X\}$;
**(iv)** $U \cap V = \{\langle x, \min\{\mathcal{Y}_f(x), \mu_V(x)\}, \max\{\mathcal{A}_f(x), \eta_V(x)\}, \max\{\mathcal{N}_f(x), \nu_V\}\rangle | x \in X\}$;
**(v)** $\overline{U} = \{\langle x, \mathcal{N}_f(x), \mathcal{A}_f(x), \mathcal{Y}_f(x)\rangle | x \in X\}$ for all $x \in X$.

**Definition 12.3.** [34] Let $U = (\mathcal{Y}_f(x), \mathcal{A}_f(x), \mathcal{N}_f(x))$ and $V = (\mu_V(x), \eta_V(x), \nu_V(x))$ be any two PFNs over the set $X$, then more operations are as follows:

**(i)** $\overline{U} = \{\langle x, \mathcal{N}_f(x), \mathcal{A}_f(x), \mathcal{Y}_f(x)\rangle | x \in X\}$;
**(ii)** $U \wedge V = \{\langle x, \min\{\mathcal{Y}_f(x), \mu_V(x)\}, \max\{\mathcal{A}_f(x), \eta_V(x)\}, \max\{\mathcal{N}_f(x), \nu_V\}\rangle | x \in X\}$;
**(iii)** $U \vee V = \{\langle x, \max\{\mathcal{Y}_f(x), \mu_V(x)\}, \min\{\mathcal{A}_f(x), \eta_V(x)\}, \min\{\mathcal{N}_f(x), \nu_V\}\rangle | x \in X\}$;
**(iv)** $U \bigoplus V = (\langle \mathcal{Y}_f + \mu_V - \mathcal{Y}_f\mu_V, \mathcal{A}_f\eta_V, \mathcal{N}_f\nu_V\rangle)$;
**(v)** $U \bigotimes V = (\langle \mathcal{Y}_f\mu_V, \mathcal{A}_f + \eta_V - \mathcal{A}_f\eta_V, \mathcal{N}_f + \nu_V - \mathcal{N}_f\nu_V\rangle)$;
**(vi)** $\lambda U = \left(1 - (1 - \mathcal{Y}_f)^\lambda, \eta_U^\lambda, \nu_U^\lambda\right)$;
**(vii)** $U^\lambda = \left(\mu_U^\lambda, 1 - (1 - \eta_U)^\lambda, 1 - (1 - \nu_U)^\lambda\right)$.

Wei [34] derived the PF AOs provided in the next definitions.

**Definition 12.4.** [34] Let $\mathcal{R}_f = (\mathcal{Y}_f, \mathcal{A}_f, \mathcal{N}_f)$ $(f = 1, 2, \ldots \varpi)$ be a group of PFNs. Let PFWA be an operator of dimension $\varpi$ such that $\Theta^\varpi \to \Theta$ with weighting vector $\Xi = (\Xi_1, \Xi_2, \ldots, \Xi_\varpi)^T$, where $\Xi > 0$ and $\sum_{f=1}^{\varpi} \Xi_f = 1$, such as $PFWA_\Xi(\mathcal{R}_1, \tilde{\mathcal{R}}_2, \ldots, \mathcal{R}_\varpi) = \bigoplus_{f=1}^{\varpi}(\Xi_f\mathcal{R}_f) = \left(1 - \prod_{f=1}^{\varpi}(1 - \mathcal{Y}_f)^{\Xi_f}, \prod_{f=1}^{\varpi}\mathcal{A}_b{}^{\Xi_f}, \prod_{f=1}^{\varpi}\mathcal{N}_f{}^{\Xi_f}\right)$.

**Definition 12.5.** [11] Let $\mathcal{R}_1 = (\mathcal{Y}_1, \mathcal{A}_1, \mathcal{N}_1)$ be PFNs, then the score $\Lambda(\mathcal{R}_1)$ and accuracy $\Phi(\mathcal{R}_1)$ for PFN are defined as follows:

$$\begin{aligned} \Lambda(\mathcal{R}_1) &= \frac{1 + \mathcal{Y}_1 - \mathcal{N}_1}{2}, \quad \Lambda(\mathcal{R}_1) \in [0, 1], \\ \Phi(\mathcal{R}_1) &= \mathcal{Y}_1 - \mathcal{N}_1, \quad \Phi(\mathcal{R}_1) \in [-1, 1]. \end{aligned} \tag{12.1}$$

Based on Definition (12.1), prioritized relations between two PFNs $\mathcal{R}_1$ and $\mathcal{R}_2$ are defined in the following ways.

**Definition 12.6.** **(i)** If $\Lambda(\mathcal{R}_1) < \Lambda(\mathcal{R}_2)$, indicates $\mathcal{R}_1 \prec \mathcal{R}_2$:
**(ii)** If $\Lambda(\mathcal{R}_1) > \Lambda(\mathcal{R}_2)$, indicates $\mathcal{R}_1 \succ \mathcal{R}_2$;
**(iii)** If $\Lambda(\mathcal{R}_1) = \Lambda(\mathcal{R}_2)$, then
  **(1)** If $\Phi(\mathcal{R}_1) < \Phi(\mathcal{R}_2)$, indicates $\mathcal{R}_1 \prec \mathcal{R}_2$;

**(2)** If $\Phi(\mathcal{R}_1) > \Phi(\mathcal{R}_2)$, indicates $\mathcal{R}_1 \succ \mathcal{R}_2$;
**(3)** If $\Phi(\mathcal{R}_1) = \Phi(\mathcal{R}_2)$, indicates $\mathcal{R}_1 \sim \mathcal{R}_2$.

## 12.3 Linear assignment method on picture fuzzy set

This segment builds a PFS-based decision environment initially. Then, we create a new LAM using an MCDM issue in which the criterion values and importance are represented as PF sets. The proposed extended LAM can be utilized to assign criteria significance in the scalar form in addition to PFS important values. As a result, we discuss the topic of ordinary importance weights. We obtain the initial ranking of the maintenance strategies using the picture fuzzy LAM by transforming the linguistic assessment into picture fuzzy numbers and aggregating the rating and the weights. This model is a modified and a combined version of the LAM that ranks the maintenance solutions in a PF environment using both quantitative and qualitative data.

### 12.3.1 Decision environment defined on PFS

The PF linear assignment model is an extension of the traditional LAM. The PF-LAM is divided into the following phases. The language terms and their matching image fuzzy numbers are shown in Table 12.1. A decision matrix whose components display the values assigned to each alternative according to each criterion in a fuzzy environment. Think about a $k$-person decision-making team, $\gamma = (\gamma_1, \gamma_2, \ldots, \gamma_k)$ participating in a problem requiring group decision making. Let $\aleph = \{\aleph_1, \aleph_2, \ldots, \aleph_\xi\}$ be a set of decision alternatives, here $\xi$ represents the number of alternatives set and that $A_f$ $(f = 1, 2, \ldots, \varpi)$ is a finite set of $\varpi$ criteria, where $\varpi$ is the number of criteria that is used to evaluate the alternative performances. The set $A$ is partitioned into two sets $A_b$ (benefit criteria) and $A_c$ (cost criteria), where $A_b \cap A_c = \varnothing$ and $A_b \cup A_c = A$.

Let $\aleph^b_{gf}$ and $\aleph^c_{gf}$ represent the rating of the alternative $\aleph_g \in \aleph$ $(g = 1, 2, \ldots, \varpi)$ for the criteria $A_f \in A_b$ and $A_c$ $(f = 1, 2, \ldots, \xi)$. Then, $\aleph^b_{gf}$ and $\aleph^c_{gf}$ can be expressed as

$$\aleph^b_{gf} = (\mathcal{Y}_{gf}, \mathcal{A}_{gf}, N_{gf}) \text{ for } A_f \in A_b \qquad \aleph^c_{gf} = (N_{gf}, \mathcal{A}_{gf}, \mathcal{Y}_{gf}) \text{ for } A_f \in A_c. \tag{12.2}$$

We must achieve comparable scales of all ratings since the criterion set $A$ is partitioned into $A_b$ and $A_c$, which prevents computational issues brought on by the varied directions of the benefit and cost criteria. Be aware that PF membership degrees imply expectations for successful performance evaluations. The PF nonmembership degree, on the other hand, implies unfavorable performance rating outcome predictions. The PF neutral membership degree, on the other hand, implies passive performance rating outcome predictions. In order to create a comparable rating system for the benefit and cost criteria, we use the idea of complements of PF sets.

In this study, we make the assumption that a rating's preference increases with size. The cost criteria are regarded as benefit criteria by taking the complement of $\aleph^c_{gf}$ the benefit criteria in order to obtain a single scale system with a consistent orientation. Let $\aleph gf$

represent the assessment of option $\aleph_g \in \aleph$ based on criteria $A_f \in A$, and let

$$\aleph_{gf} = \begin{cases} \aleph^b_{gf}(= (\mathcal{Y}^b_{gf}, \mathcal{A}^b_{gf}, N^b_{gf})), & \text{if } A_f \in A_b \\ \aleph^c_{gf}(= (N^c_{gf}, \mathcal{A}_{gf}, \mathcal{Y}^c_{gf})), & \text{if } A_f \in A_c \end{cases}. \tag{12.3}$$

The characteristics of the alternative $\aleph_g$ are represented by the PF set as follows:

$$\begin{aligned} \aleph_g &= \left\{ \langle A_1, (\mathcal{Y}_{g1}, \mathcal{A}_{g1}, N_{g1}\rangle, \langle A_2, (\mathcal{Y}_{g2}, \mathcal{A}_{g2}, N_{g2}\rangle, \ldots, \langle A_\xi, (\mathcal{Y}_{g\xi}, \mathcal{A}_{g\xi}, N_{g\xi}\rangle \right\} \\ &= \left\{ \langle A_f, (\mathcal{Y}_{gf}, \mathcal{A}_{gf}, N_{gf}\rangle | A_f \in A,\ f = 1, 2, \ldots, \xi,\ g = 1, 2, \ldots, \varpi \right\}. \end{aligned} \tag{12.4}$$

A matrix format is used to clearly explain a PF multicriteria group decision-making (MCGDM) problem. Let $D$ be a PF decision matrix with the following definitions:

$$D = \begin{array}{c} \\ \aleph_1 \\ \aleph_2 \\ \vdots \\ \aleph_\varpi \end{array} \begin{array}{c} \begin{array}{cccc} A_1 & A_2 & \cdots & A_\xi \end{array} \\ \begin{bmatrix} \aleph^{(h)}_{11} & \aleph^{(h)}_{12} & \cdots & \aleph^{(h)}_{1\xi} \\ \aleph^{(h)}_{21} & \aleph^{(h)}_{22} & \cdots & \aleph^{(h)}_{2\xi} \\ \vdots & \vdots & \ddots & \vdots \\ \aleph^{(h)}_{\varpi 1} & \aleph^{(h)}_{\varpi 2} & \cdots & \aleph^{(h)}_{\varpi \xi} \end{bmatrix} \end{array}. \tag{12.5}$$

Similar to this, the decision maker's preference judgements regarding the significance of the various evaluative criteria can also be expressed using the PF set. Criteria $A_f$'s relevance $\Xi_f$ can be summed up as follows:

$$\Xi_f = (\Xi_1, \Xi_2, \ldots, \Xi_\varpi). \tag{12.6}$$

The PFS $\Xi$ is described as an item of the following form in the discourse $A$ universe:

$$\begin{aligned} \Xi &= \left\{ \langle A_1, (\Psi_1, \Omega_1, \Theta_1)\rangle, \langle A_2, (\Psi_2, \Omega_2, \Theta_2)\rangle, \ldots, \langle A_\xi, (\Psi_\xi, \Omega_\xi, \Theta_\xi)\rangle \right\} \\ &= \{\langle A_f, (\Psi_f, \Omega_f, \Theta_f)\rangle | A_f \in A,\ f = 1, 2, \ldots, \xi\}. \end{aligned} \tag{12.7}$$

## 12.3.2 Extended linear assignment model

We have created a PF LAM to handle judgements based on various criteria inspired by the LAM's basic structure. The priority ranking system-based alternate ordering is related to the MCDA problem. By contrasting the score function and accuracy function of PF sets in the presented approach, we proposed a "rank frequency matrix" based on the idea of the product-attribute matrix introduced by Bernardo and Blin [5].

We contrast the PF numbers' scoring functions and accuracy functions indices. In relation to the MCDM issue with the decision matrix D, let $\aleph_{g_1 f}$ and $\aleph_{g_2 f}$ be any two PFNs in each column of D. We evaluate the score functions $\Lambda(\aleph_{g_1 f})$ and $\Lambda(\aleph_{g_2 f})$, and accuracy functions $\Phi(\aleph_{g_1 f})$ and $\Phi(\aleph_{g_2 f})$. The ranking method can be applied to compare any two

PFNs for each column of D. In accordance with the ranking method, the $\varpi$ alternatives can be rated in terms of each criterion $A_f \in A$.

The frequency with which $\aleph_g$ is ranked as the $k$th criterion-wise ranking is represented by the elements of the rank frequency matrix, $\pi^0$, which is a square $(\varpi \times \varpi)$ nonnegative matrix. The matrix produced by this definition is shown as:

$$\pi^0 = \begin{array}{c} \\ \aleph_1 \\ \aleph_2 \\ \vdots \\ \aleph_\varpi \end{array} \begin{array}{c} \begin{array}{cccc} 1 & 2 & \cdots & \varpi \end{array} \\ \begin{bmatrix} \pi^0_{11} & \pi^0_{12} & \cdots & \pi^0_{1\varpi} \\ \pi^0_{21} & \pi^0_{22} & \cdots & \pi^0_{2\varpi} \\ \vdots & \vdots & \ddots & \vdots \\ \pi^0_{\varpi 1} & \pi^0_{\varpi 2} & \cdots & \pi^0_{\varpi\varpi} \end{bmatrix} \end{array}. \tag{12.8}$$

The rank frequency matrix $\pi^0$ can be viewed as a performance matrix where each criterion is given equal weight. For practical applications, giving each criterion a different level of importance is acceptable. This study implied that the $f$th ranking be replicated by $\Xi_f$ times. Let $A_{f_1}, A_{f_2}, \ldots, A_{f_{\pi^0_g}}$ indicate the criteria for which the alternative $\aleph_g$ is ranked $k$th. Thus a "weighted-ranked frequency matrix" $\Pi^0$ can be computed as follows:

$$\Pi^0 = \begin{array}{c} \\ \aleph_1 \\ \aleph_2 \\ \vdots \\ \aleph_\varpi \end{array} \begin{array}{c} \begin{array}{cccc} 1 & 2 & \cdots & \varpi \end{array} \\ \begin{bmatrix} \Pi^0_{11} & \Pi^0_{12} & \cdots & \Pi^0_{1\varpi} \\ \Pi^0_{21} & \Pi^0_{22} & \cdots & \Pi^0_{2\varpi} \\ \vdots & \vdots & \ddots & \vdots \\ \Pi^0_{\varpi 1} & \Pi^0_{\varpi 2} & \cdots & \Pi^0_{\varpi\varpi} \end{bmatrix} \end{array}, \tag{12.9}$$

where,

$$\Pi^0_{gk} = \Xi_{f_1} + \Xi_{f_2} + \ldots, + \Xi_{f_{\pi^0_{gk}}} = \left(1 - \prod_{f=f_1}^{f_{\pi^0_{gk}}} (1 - \Psi_f), \prod_{f=f_1}^{f_{\pi^0_{gk}}} \Omega_f, \prod_{f=f_1}^{f_{\pi^0_{gk}}} \Theta_f\right). \tag{12.10}$$

Keep in mind that the significance of each element $\Pi^0_{ik}$ of $\Pi^0$ is a measurement of the agreement between all of the factors used to rank the $g$th alternative $k$th.

When the $q$ options are tied for a given criterion, the initial ranking must be divided into $\rho!$ equalized ranks. Each of these equalized rankings is also given a weight of $1/\rho!$. Consider the five choices $\aleph_1, \aleph_2, \ldots, \aleph_5$, for example, in relation to criterion $A_f$. Let us consider five alternative $\aleph_{gf}$ $(g = 1, 2, 3, 4, 5)$ and comparison of them that yield $(\aleph_2 \sim \aleph_4 \sim \aleph_5) \succ \aleph_3 \succ \aleph_1$. Hence, we obtain the following $\aleph_2 \succ \aleph_4 \succ \aleph_5 \succ \aleph_3 \succ \aleph_1$, $\aleph_2 \succ \aleph_5 \sim \aleph_4 \succ \aleph_3 \succ \aleph_1$, $\aleph_4 \succ \aleph_2 \sim \aleph_5 \succ \aleph_3 \succ \aleph_1$, $\aleph_4 \succ \aleph_5 \sim \aleph_2 \succ \aleph_3 \succ \aleph_1$, $\aleph_5 \succ \aleph_2 \sim \aleph_4 \succ \aleph_3 \succ \aleph_1$, and $\aleph_5 \succ \aleph_4 \sim \aleph_2 \succ \aleph_3 \succ \aleph_1$. Additionally, each of these six ranks is given an equal weight at 1/6(3!) Let $\rho_f$ be the total number of tied options for a given criterion $A_f$. After dividing the first criterion-wise ranking into $\rho_f$, let $\pi_{gk}$ represent the frequency with which $\aleph_g$ is ranked as the $k$th criterion-wise ranking after the initial criterion-wise ranking has been separated into $\rho_f$ equalized rankings for all $A_f \in A$. Let $A_{f_1}, A_{f_2}, \ldots, A_{f_{\pi_{gk}}}$ be the criterion for which the alternative $\aleph_g$ is ranked $k$th. Thus the following is the expression for a modified weighted-rank frequency matrix P:

$$\Pi = \begin{array}{c} \\ \aleph_1 \\ \aleph_2 \\ \vdots \\ \aleph_\varpi \end{array} \begin{array}{c} \begin{array}{cccc} 1 & 2 & \cdots & \varpi \end{array} \\ \begin{bmatrix} \Pi_{11} & \Pi_{12} & \cdots & \Pi_{1\varpi} \\ \Pi_{21} & \Pi_{22} & \cdots & \Pi_{2\varpi} \\ \vdots & \vdots & \ddots & \vdots \\ \Pi_{\varpi 1} & \Pi_{\varpi 2} & \cdots & \Pi_{\varpi\varpi} \end{bmatrix} \end{array}, \tag{12.11}$$

where,

$$\Pi_{gk} = \frac{1}{\rho_{f_1}!}\Xi_{f_1} + \frac{1}{\rho_{f_2}!}\Xi_{f_2} + \ldots, + \frac{1}{\rho_{f_{\pi_{gk}}}!}\Xi_{f_{\pi_{gk}}} = \left(1 - \prod_{f=f_1}^{f_{\pi_{gk}}} (1 - \Psi_f), \prod_{f=f_1}^{f_{\pi_{gk}}} \Omega_f, \prod_{f=f_1}^{f_{\pi_{gk}}} \Theta_f\right). \tag{12.12}$$

The adjusted weighted-rank frequency matrix $\Pi$ can then be used to determine an overall ranking. The overall ranking closely agrees with the criterion-wise rankings, and the aggregate ranking would effectively incorporate the relative performance of each alternative for each criterion. Therefore we can use $\Pi_{gk}$ to gauge how much the alternative $\aleph_g$ contributed to the overall ranking when it was given to the $k$th overall rank. It is evident that the greater concordance that results from placing $\aleph_g$ in the $k$th overall rank, the larger the contribution indicated by $\Pi_{gk}$. As a result, we must compare the $\Pi_{gk}$ values for each of $g, k = 1, 2, \ldots, \varpi$. This study systematically applied the idea of a fresh accuracy function $M$ suggested by Ye [37] to produce a comparable value of $\Pi_{gk}$ since the values of $\Pi_{gk}$ are PFNs. By using existing score and accuracy functions in some circumstances, this innovative accuracy function helps circumvent the challenges that incomplete information can pose in a decision-making context. We can then determine the $\aleph_g$ for each $k$ $(g, k = 1, 2, \ldots, \varpi)$ that maximizes $\sum_{k=1}^{\varpi} M(\Pi_{gk})$ by computing the unique accuracy function $M(\Pi_{gk})$ of $\Pi_{gk}$. This is an $\varpi!$ comparison difficulty. To resolve the $\varpi!$ comparison problem in the event of a large $\varpi$, we can create a linear programming model.

Let the permutation matrix $P$ represent a $(\varpi \times \varpi)$ square matrix with an entry $\Pi_{gk}$ that is either $P_{gk} = 0$ or $P_{gk} = 1$ depending on whether $\aleph_g$ is assigned to the overall rank $k$. It goes without saying that we favor the overall ranking in which $M(\Pi_{gk})$, $P_{gk}$ is at its highest value. The explanation is that, among all of the criterion-wise rankings, the overall ranking with the highest $M(\Pi_{gk})$. $P_{gk}$ yields the optimal compromise. The ranking produces the highest value of $M(\Pi_{gk})$. Then, $P_{gk}$ will be picked after considering every possible ranking. We know that $\sum_{k=1}^{\varpi} P_{gk}) = 1$ since alternative $\aleph_g$ can only be given to one rank in the total ranking. Similar to this, there can only be one alternative allocated to a particular rank $k$, and thus $\sum_{k=1}^{\varpi} P_{gk}) = 1$. The following linear programming format can therefore be used to define the expanded linear assignment model:

$$\max \sum_{g=1}^{\varpi} \sum_{k=1}^{\varpi} M(\Pi_{gk}).P_{gk} = \max \sum_{g=1}^{\varpi} \sum_{k=1}^{\varpi} M\left(\left(1 - \prod_{f=f_1}^{f_{\pi_{gk}}} (1 - \Psi_f), \prod_{f=f_1}^{f_{\pi_{gk}}} \Omega_f, \prod_{f=f_1}^{f_{\pi_{gk}}} \Theta_f\right)\right).P_{gk} \tag{12.13}$$

subject to

$$\sum_{k=1}^{m} P_{gk} = 1, \quad \forall g = 1, 2, \ldots, \varpi$$

$$\sum_{g=1}^{m} P_{gk} = 1, \quad \forall k = 1, 2, \ldots, \varpi$$

$$P_{gk} = 0 \text{ or } 1 \text{ for all } g \text{ and } k.$$

Be aware that the extended linear assignment model has not yet been able to determine $P_{gk}$, although the innovative accuracy function $M(\Pi_{gk})$ of $\Pi_{gk}$ is derived from the criterion-wise rankings. We may resolve the aforementioned linear programming issue using the Simplex approach to obtain the ideal permutation matrix $P^*$. By multiplying $\aleph$ by $P^*$, one can sequentially arrive at the ideal arrangement.

**Table 12.1** Picture fuzzy linguistic terms.

| Linguistic terms | Rating |
|---|---|
| Very High Importance (VHI) | (0.85, 0.0, 0.05) |
| High Importance (HI) | (0.65, 0.05, 0.2) |
| Slightly More Importance (SMI) | (0.5, 0.0, 0.4) |
| Equal Importance (EI) | (0.4, 0.1, 0.5) |
| Slightly Low Importance (SLI) | (0.3, 0.0, 0.6) |
| Low Importance (LI) | (0.26, 0.06, 0.6) |
| Very Low Importance (VLI) | (0.1, 0.0, 0.85) |

### 12.3.3 The proposed algorithm

The following sequence of stages summarizes the methodology of the proposed extended LAM that employs picture fuzzy data for MCDM:

**Step 1:** Create the possible alternative ($\aleph = \{\aleph_1, \aleph_2, \ldots, \aleph_\varpi\}$ and identify the criteria ($A = \{A_1, A_2, \ldots, A_\xi\}$. In the meantime, we determine the criterion's importance value $\Xi = (\Psi_f, \Omega_f, \Theta_f)$. Suppose, $D = \{D_1, D_2, \ldots, D_k)$ is the set of experts with corresponding weight $w = (w_1, w_2, \ldots, w_k)^T$.

**Step 2:** Establish the appropriate rating values of the alternatives given in linguistics proposed by the decision makers in relation to each criterion. Next, create the decision matrix $D$ by aggregating the individual decision matrices based on the PFWA operator defined in Definition 12.4 and the importance $\Xi$ for each criterion using these evaluation data as PFS.

**Step 3:** Calculating the score function $\lambda(\aleph)$ and accuracy function $\Phi(\aleph)$ are given Eq. (12.1) for all $\aleph_h \in \aleph$ and $A_f \in A$.

**Step 4:** Use the ranking process to order the $\varpi$ choices according to each criterion. When the $\rho$ options are tied for a criterion, then $\rho!$ equalized rankings should be listed separately. We also created a criteria-wise frequency rank matrix.

**Step 5:** Create the adjusted weighted-rank frequency matrix $P$ after computing $\Pi_{gk}$. A weight of $1/\rho!$ should be applied to each of the $\rho!$ equalized rankings as well. Next, determine $\Pi_{gk}$'s score function $\lambda(\Pi_{gk})$.

**Step 6:** Create a linear assignment model based on the $\lambda(\Pi_{gk})$ values and define the permutation matrix $P$ as a square $(\varpi \times \varpi)$ matrix.

**Step 7:** Obtain the ideal permutation matrix $P^*$ by solving the linear assignment model using the 0–1 integer programming method. The best order for the $\varpi$ options can then be determined by applying the permutation matrix $P^*$ to $\aleph$. The linear programming format can be used to express the linear assignment model as follows:

$$\max \sum_{g=1}^{\varpi} \sum_{k=1}^{\varpi} \lambda(\Pi_{gk}).P_{gk} \tag{12.14}$$

subject to

$$\sum_{k=1}^{m} P_{gk} = 1, \quad \text{for all } g = 1, 2, \ldots, \varpi$$

$$\sum_{g=1}^{m} P_{gk} = 1, \quad \text{for all } k = 1, 2, \ldots, \varpi$$

$$P_{gk} = 0 \text{ or } 1 \text{ for all } g \text{ and } k.$$

## 12.4 An application to a sponge iron factory location selection

Sponge iron plants fall under the "red category" of industries, which means they pose serious environmental and health risks. Extreme heat and smoke, as well as silica and unburned carbon particles, are released during manufacturing. Plants that use an electrostatic precipitator (ESPS), which are required in the majority of states to check emissions, produce slightly less dust, though how they do so is debatable. Other pollutants that require safe disposal include coal char, iron dust, and carbon dust gathered from ESPS. A business chooses the best location for that use and appoints a committee of three experts to assess five potential sites. Based on their rankings, the expert team selected the most suitable location. The DMs provide interim ratings for the PFNs, and the three experts' combined weight vector is $(0.5, 0.23, 0.27)^T$. Assume the business has considered five potential sites as alternatives for potential sponge iron factories: $\aleph_1$, $\aleph_2$, $\aleph_3$, $\aleph_4$, and $\aleph_5$. The following primary criteria are used by the expert team of engineers to determine the best location:

$A_1$: Infrastructures;
$A_2$: Environmental conditions;
$A_3$: Social impact;
$A_4$: Governmental policies.

**Step 1:** Here, we considered five locations $\aleph_1$, $\aleph_2$, $\aleph_3$, $\aleph_4$, and $\aleph_5$ by the experts under the four criteria $A_1$, $A_2$, $A_3$, and $A_4$. Here, invite the three decision makers $D = (D_1, D_2, D_3)$, whose weight vector is $w = (w_1, w_2, w_3)^T = (0.5, 0.23, 0.27)^T$ to select the best location for the sponge iron factory. Assuming that the criteria's significance levels are represented as $\Xi = (\Xi_1, \Xi_2, \Xi_3, \Xi_4)^T$ as $((0.5, 0.1, 0.2), (0.4, 0.02, 0.3), (0.6, 0.1, 0.1), (0.7, 0.01, 0.2))^T$. As a result, PFNs are used to represent the expert assessments of the four locations in relation to each criterion.
**Step 2:** Decision-makers' assessments for the criteria and alternatives based on Table 12.1 are given in Table 12.2, Table 12.3, and Table 12.4, respectively. Aggregating the individual decision matrices based on the PFWA operator defined in Definition 12.4 by three decision makers with significance levels of 0.5, 0.23, and 0.27 based on the criteria listed in Table 12.5.

**Table 12.2** Decision matrix 1.

| $DM1$ | $A_1$ | $A_2$ | $A_3$ | $A_4$ |
|---|---|---|---|---|
| $\aleph_1$ | VHI | HI | VHI | VLI |
| $\aleph_2$ | SLI | VHI | SLI | HI |
| $\aleph_3$ | HI | VLI | SMI | EI |
| $\aleph_4$ | SLI | HI | HI | LI |
| $\aleph_5$ | HI | SLI | SMI | HI |

**Table 12.3** Decision matrix 2.

| $DM2$ | $A_1$ | $A_2$ | $A_3$ | $A_4$ |
|---|---|---|---|---|
| $\aleph_1$ | HI | SHI | HI | EI |
| $\aleph_2$ | SHI | LI | HI | EI |
| $\aleph_3$ | HI | VLI | VHI | SMI |
| $\aleph_4$ | EI | HI | EI | LI |
| $\aleph_5$ | SLI | HI | EI | EI |

**Step 3:** Utilizing the score function Eq. (12.1), compute the components of the scored $\lambda(\mathcal{R}_{gk})$ and accuracy function $\Phi(\mathcal{R}_{gk})$ that are given in Table 12.6.

**Table 12.4** Decision matrix 3.

| *DM*3 | $A_1$ | $A_2$ | $A_3$ | $A_4$ |
|---|---|---|---|---|
| $\aleph_1$ | SLI | SHI | SLI | VHI |
| $\aleph_2$ | LI | HI | VLI | HI |
| $\aleph_3$ | SHI | HI | VHI | HI |
| $\aleph_4$ | VHI | SMI | VHI | VLI |
| $\aleph_5$ | HI | SMI | VLI | LI |

**Table 12.5** Aggregated decision matrix.

| | $A_1$ | $A_2$ | $A_3$ | $A_3$ |
|---|---|---|---|---|
| $\aleph_1$ | (0.7237,0.00,0.1345) | (0.5817,0.00,0.2828) | (0.7237,0.00,0.1345) | (0.4946,0.00,0.3501) |
| $\aleph_2$ | (0.3423,0.00,0.5465) | (0.7278,0.00,0.1287) | (0.3613,0.00,0.5119) | (0.6038,0.00,0.2469) |
| $\aleph_3$ | (0.7216,0.00,0.1376) | (0.3026,0.00,0.5751) | (0.7261,0.00,0.1414) | (0.5026,0.00,0.3708) |
| $\aleph_4$ | (0.5543,0.00,0.2941) | (0.6146,0.00,0.1562) | (0.6038,0.00,0.2469) | (0.2198,0.00,0.6516) |
| $\aleph_5$ | (0.4406,0.00,0.2574) | (0.4549,0.00,0.1913) | (0.3889,0.00,0.5161) | (0.5150,0.00,0.3322) |

**Table 12.6** Score matrix.

| *DM*3 | $\lambda(\mathcal{R}_{gf})$ | $\Phi(\mathcal{R}_{gf})$ |
|---|---|---|
| $\mathcal{R}_{11}$ | 0.7946 | 0.5892 |
| $\mathcal{R}_{12}$ | 0.3979 | -0.2042 |
| $\mathcal{R}_{13}$ | 0.7920 | 0.5840 |
| $\mathcal{R}_{14}$ | 0.6301 | 0.2602 |
| $\mathcal{R}_{15}$ | 0.5916 | 0.1832 |
| $\mathcal{R}_{21}$ | 0.6495 | 0.2989 |
| $\mathcal{R}_{22}$ | 0.7996 | 0.5991 |
| $\mathcal{R}_{23}$ | 0.3638 | -0.2725 |
| $\mathcal{R}_{24}$ | 0.7292 | 0.4584 |
| $\mathcal{R}_{25}$ | 0.6318 | 0.2636 |
| $\mathcal{R}_{31}$ | 0.7946 | 0.5892 |
| $\mathcal{R}_{32}$ | 0.4247 | -0.1506 |
| $\mathcal{R}_{33}$ | 0.7924 | 0.5847 |
| $\mathcal{R}_{34}$ | 0.6785 | 0.3569 |
| $\mathcal{R}_{35}$ | 0.4364 | -0.1272 |
| $\mathcal{R}_{41}$ | 0.5723 | 0.1445 |
| $\mathcal{R}_{42}$ | 0.6785 | 0.3569 |
| $\mathcal{R}_{43}$ | 0.5659 | 0.1318 |
| $\mathcal{R}_{44}$ | 0.2841 | -0.4318 |
| $\mathcal{R}_{45}$ | 0.5914 | 0.1828 |

**Step 4:** Utilizing the result of Definition (12.1) and Eq. (12.5), Table 12.7 represents the criteria-wise rankings. Based on the results of Table 12.7 and Eq. (12.8), we obtained a frequency rank matrix. Table 12.8 represents the rank frequency matrix.

**Table 12.7** Criteria-wise ranking of the alternatives.

| Rank | $A_1$ | $A_2$ | $A_3$ | $A_4$ |
|---|---|---|---|---|
| $1st$ | $\aleph_1$ | $\aleph_2$ | $\aleph_1$ | $\aleph_2$ |
| $2nd$ | $\aleph_3$ | $\aleph_4$ | $\aleph_3$ | $\aleph_5$ |
| $3rd$ | $\aleph_4$ | $\aleph_1$ | $\aleph_4$ | $\aleph_1$ |
| $4th$ | $\aleph_5$ | $\aleph_5$ | $\aleph_5$ | $\aleph_3$ |
| $5th$ | $\aleph_2$ | $\aleph_3$ | $\aleph_2$ | $\aleph_4$ |

**Table 12.8** Rank frequency matrix $\Pi^0$.

| $\Pi^0$ | 1st | 2nd | 3rd | 4th | 5th |
|---|---|---|---|---|---|
| $\aleph_1$ | 2 | 0 | 2 | 0 | 0 |
| $\aleph_2$ | 2 | 0 | 0 | 0 | 2 |
| $\aleph_3$ | 0 | 2 | 0 | 1 | 1 |
| $\aleph_4$ | 0 | 1 | 2 | 1 | 1 |
| $\aleph_5$ | 0 | 1 | 0 | 3 | 0 |

**Step 5:** The results of Table 12.8 and (12.11) were used to construct a weighted-rank frequency matrix that is given in Table 12.9. For Table 12.9, we imposed the importance degrees of the criteria $\Xi = (\Xi_1, \Xi_2, \Xi_3, \Xi_4)^T$. We consider $\Pi_{13}$ as an example. Based on the ranking results of Table 12.8, the frequency that the alternative $\aleph_1$ is ranked the $3rd$ criteria-wise ranking is 2, i.e., $\Pi^0_{13} = 2$. In light of the results of Table 12.7, $A_2$ and $A_4$ are the criteria for which $\aleph_1$ is ranked $3rd$. In this case, $\rho_{f_2} = 1$ and $\rho_{f_{\pi^0_{13}}} = \rho_{f_4} = 1$. $\Pi_{13}$ is computed as follows: $\Pi_{13} = \left(1 - (1-0.4)^{\frac{1}{1}}(1-0.7)^{\frac{1}{1}}, (0.2)^{\frac{1}{1}}(0.01)^{\frac{1}{1}}, (0.3)^{\frac{1}{1}}(0.2)^{\frac{1}{1}}\right) = (0.82, 0.002, 0.06)$.

**Table 12.9** Weighted-Rank frequency matrix.

| $\Pi$ | 1st | 2nd | 3rd | 4th | 5th |
|---|---|---|---|---|---|
| $\aleph_1$ | (0.80,0.003,0.002) | (0.00,0.00,0.00) | (0.82,0.002,0.06) | (0.00,0.00,0.00) | (0.00,0.00,0.00) |
| $\aleph_2$ | (0.82,0.002,0.06) | (0.00,0.00,0.00) | (0.00,0.00,0.00) | (0.00,0.00,0.00) | (0.80,0.003,0.02) |
| $\aleph_3$ | (0.00,0.00,0.00) | (0.80,0.003,0.02) | (0.00,0.00,0.00) | (0.70,0.01,0.20) | (0.60,0.03,0.10) |
| $\aleph_4$ | (0.00,0.00,0.00) | (0.40,0.02,0.30) | (0.80,0.003,0.02) | (0.00,0.00,0.00) | (0.70,0.01,0.20) |
| $\aleph_5$ | (0.00,0.00,0.00) | (0.70,0.01,0.20) | (0.00,0.00,0.00) | (0.88,0.001,0.01) | (0.00,0.00,0.00) |

**Step 6:** We calculated the scores of the components shown in the weighted-rank frequency matrix $\Pi$, which are listed in Table 12.9, using the scoring function of Definition (12.1), and the results are shown in Table 12.10.

**Step 7:** We may create an extended linear assignment model for the PF MCGDM based on the scoring results of Table 12.10 and Eq. (12.14). The linear assignment model is built as follows:

$$\text{Maximize } 0.88P_{11} + 0.9P_{15} + 0.89P_{22} + 0.75P_{24} + 0.75P_{25} + 0.89P_{34} + 0.75P_{35}$$
$$+0.55P_{42} + 0.89P_{43} + 0.75P_{45} + 0.75P_{52} + 0.94P_{54}$$

**Table 12.10** The score of $\lambda(\Pi_{gf})$.

| $\lambda(\Pi_{gf}$ | 1st | 2nd | 3rd | 4th | 5th |
|---|---|---|---|---|---|
| $\aleph_1$ | 0.90 | 0.00 | 0.88 | 0.00 | 0.00 |
| $\aleph_2$ | 0.88 | 0.00 | 0.00 | 0.00 | 0.90 |
| $\aleph_3$ | 0.00 | 0.89 | 0.00 | 0.75 | 0.75 |
| $\aleph_4$ | 0.00 | 0.55 | 0.89 | 0.00 | 0.75 |
| $\aleph_5$ | 0.00 | 0.75 | 0.00 | 0.94 | 0.00 |

subject to

$$\begin{aligned}
P_{11} + P_{12} + P_{13} + P_{14} + P_{15} &= 1\\
P_{21} + P_{22} + P_{23} + P_{24} + P_{25} &= 1\\
P_{31} + P_{32} + P_{33} + P_{34} + P_{35} &= 1\\
P_{41} + P_{42} + P_{43} + P_{44} + P_{45} &= 1\\
P_{51} + P_{52} + P_{53} + P_{54} + P_{55} &= 1\\
P_{11} + P_{21} + P_{31} + P_{41} + P_{51} &= 1\\
P_{12} + P_{22} + P_{32} + P_{42} + P_{52} &= 1\\
P_{13} + P_{23} + P_{33} + P_{43} + P_{53} &= 1\\
P_{14} + P_{24} + P_{34} + P_{44} + P_{54} &= 1\\
P_{15} + P_{25} + P_{35} + P_{45} + P_{55} &= 1.
\end{aligned}$$

All the variables

$$P_{11}, P_{12}, P_{13}, P_{14}, P_{15}, P_{21}, P_{22}, P_{23}, P_{24}, P_{25}, P_{31}, P_{32}, P_{33}, P_{34}, P_{35}, P_{41},$$
$$P_{42}, P_{43}, P_{44}, P_{45}, P_{51}, P_{52}, P_{53}, P_{54}, P_{55}$$

are either 0 or 1.

We used the 0–1 integer programming method to solve the aforementioned linear programming problem and obtained 4.35 as the optimal value. Furthermore, we observed that the ideal permutation matrix $P^*$ is as follows:

$$P^* = \begin{matrix} \\ \aleph_1 \\ \aleph_2 \\ \aleph_3 \\ \aleph_4 \\ \aleph_5 \end{matrix} \begin{matrix} \begin{matrix} 1 & 2 & 3 & 4 & 5 \end{matrix} \\ \begin{bmatrix} 1 & 0 & 0 & 0 & 0 \\ 0 & 1 & 0 & 0 & 0 \\ 0 & 0 & 0 & 0 & 1 \\ 0 & 0 & 1 & 0 & 0 \\ 0 & 0 & 0 & 1 & 0 \end{bmatrix} \end{matrix}. \tag{12.15}$$

The ordering $(\aleph_1, \aleph_2, \aleph_3, \aleph_4, \aleph_5)$ can then be obtained by multiplying by $P^*$ to $\aleph$. The resulting matrix is as follows:

$$(\aleph_1, \aleph_2, \aleph_3, \aleph_4, \aleph_5).\begin{bmatrix} 1 & 0 & 0 & 0 & 0 \\ 0 & 1 & 0 & 0 & 0 \\ 0 & 0 & 0 & 0 & 1 \\ 0 & 0 & 1 & 0 & 0 \\ 0 & 0 & 0 & 1 & 0 \end{bmatrix} = (\aleph_1, \aleph_2, \aleph_4, \aleph_5, \aleph_3). \tag{12.16}$$

The optimal ranking order of the alternatives is $\aleph_1 \succ \aleph_2 \succ \aleph_4 \succ \aleph_5 \succ \aleph_3$. The desirable alternative is therefore $\aleph_1$.

## 12.5 Conclusions

This work presented an expanded LAM to obtain the optimal priority ranking of the options for the MCGDM issue in the picture fuzzy environment. Using the picture fuzzy set theory, the proposed method can capture ambiguous and imprecise information in the MCGDM process In order to obtain the criterion-wise preference of alternatives in the PF context, the proposed extended LAM makes use of the notions of score functions and accuracy functions. In order to create a linear programming model for the criteria-wise rankings and the PF values of criterion importance, we used the modified weighted-rank frequency matrix to construct a linear programming model for determining the optimal preference ranking. The proposed extended LAM can be used to address the MCGDM problem with nonfuzzy weights in addition to the PF significance values. The development of new generalized aggregation operators based on the varied relationships between input arguments in interval-valued picture fuzzy conditions may be the main topic of future research.

## References

[1] F. Ahmad, A.Y. Adhami, F. Smarandache, Neutrosophic optimization model and computational algorithm for optimal shale gas water management under uncertainty, Symmetry 11 (4) (2019) 544.
[2] F. Ahmad, A.Y. Adhami, F. Smarandache, Modified neutrosophic fuzzy optimization model for optimal closed-loop supply chain management under uncertainty, in: Optimization Theory Based on Neutrosophic and Plithogenic Sets, Academic Press, 2020, pp. 343–403.
[3] P.P. Angelov, Optimization in an intuitionistic fuzzy environment, Fuzzy Sets Syst. 86 (3) (1997) 299–306.
[4] R.E. Bellman, L.A. Zadeh, Decision-making in a fuzzy environment, Manag. Sci. 17 (4) (1970) 140–164.
[5] J.J. Bernardo, J.M. Blin, A programming model of consumer choice among multi-attributed brands, J. Consum. Res. 4 (2) (1977) 111.
[6] K. Atanassov, Intuitionistic Fuzzy Sets: Theory and Applications, Studies in Fuzziness and Soft Computing, vol. 35, Physica-Verlag, Heidelberg, 1999.
[7] K.T. Atanassov, G. Gargov, Interval-valued intuitionistic fuzzy sets, Fuzzy Sets Syst. 31 (1989) 343–349.
[8] B.C. Cuong, Picture fuzzy sets - first results. Part 1, Seminar "Neuro-fuzzy systems with applications", Tech. Rep., Institute of Mathematics, Hanoi, 2013.
[9] B.C. Cuong, Picture fuzzy sets - first results. Part 2, Seminar "Neuro-fuzzy systems with applications", Tech. Rep., Institute of Mathematics, Hanoi, 2013.
[10] T.Y. Chen, The extended linear assignment method for multiple criteria decision analysis based on interval-valued intuitionistic fuzzy sets, Appl. Math. Model. 38 (7–8) (2014) 2101–2117.

[11] C. Jana, T. Senapati, M. Pal, R.R. Yager, Picture fuzzy Dombi aggregation operators: application to MADM process, Appl. Soft Comput. 74 (1) (2019) 99–109.
[12] D. Liang, A.P. Darko, Z. Xu, Y. Zhang, Partitioned fuzzy measure-based linear assignment method for Pythagorean fuzzy multi-criteria decision-making with a new likelihood, J. Oper. Res. Soc. 15 (2019) 1–15.
[13] D. Liang, A.P. Darko, Z. Xu, W. Quan, The linear assignment method for multicriteria group decision making based on interval-valued Pythagorean fuzzy Bonferroni mean, Int. J. Intell. Syst. 33 (11) (2018) 2101–2138.
[14] X. Peng, J. Dai, Algorithm for picture fuzzy multiple attribute decision making based on new distance measure, Int. J. Uncertain. Quantificat. 7 (2017) 177–187.
[15] Y.J. Ping, R. Liu, W. Lin, H.C. Liu, A new integrated approach for engineering characteristic prioritization in quality function deployment, Adv. Eng. Inform. (ISSN 1474-0346) 45 (2020).
[16] P.H. Phong, D.T. Hieu, R.T.H. Ngan, P.T. Them, Some compositions of picture fuzzy relations, in: Proceedings of the 7th National Conference on Fundamental and Applied Information Technology Research, FAIR'7, Thai Nguyen, 2014, pp. 19–20.
[17] P.T.M. Phuong, P.H. Thong, L.H. Son, Theoretical analysis of picture fuzzy clustering, J. Comput. Sci. Cybern. 34 (1) (2018) 17–31.
[18] R.M. Rizk-Allah, A.E. Hassanien, M. Elhoseny, A multi-objective transportation model under neutrosophic environment, Comput. Electr. Eng. 69 (2018) 705–719.
[19] P. Singh, Correlation coefficients for picture fuzzy sets, J. Intell. Fuzzy Syst. 27 (2014) 2857–2868.
[20] L.H. Son, DPFCM: a novel distributed picture fuzzy clustering method on picture fuzzy sets, Expert Syst. Appl. 2 (2015) 51–66.
[21] L.H. Son, Generalized picture distance measure and applications to picture fuzzy clustering, Appl. Soft Comput. 46 (2016) 284–295.
[22] L.H. Son, Measuring analogousness in picture fuzzy sets: from picture distance measures to picture association measures, Fuzzy Optim. Decis. Mak. 16 (3) (2017) 1–20.
[23] L.H. Son, P. Viet, P. Hai, Picture inference system: a new fuzzy inference system on picture fuzzy set, Appl. Intell. 46 (3) (2017) 652–669.
[24] P.H. Thong, L.H. Son, Picture fuzzy clustering for complex data, Eng. Appl. Artif. Intell. 56 (2016) 121–130.
[25] P.H. Thong, L.H. Son, A novel automatic picture fuzzy clustering method based on particle swarm optimization and picture composite cardinality, Knowl.-Based Syst. 109 (2016) 48–60.
[26] P.H. Thong, L.H. Son, A new approach to multi-variables fuzzy forecasting using picture fuzzy clustering and picture fuzzy rules interpolation method, in: 6th International Conference on *Knowledge and Systems Engineering*, Hanoi, Vietnam, 2015, pp. 679–690.
[27] P.H. Thong, L.H. Son, H. Fujita, Interpolative picture fuzzy rules: a novel forecast method for weather nowcasting, in: Fuzzy Systems (FUZZ-IEEE), 2016 IEEE International Conference on IEEE, Vancouver, Canada, July 24-29, 2016, pp. 86–93.
[28] N.T. Thong, L.H. Son, HIFCF: an effective hybrid model between picture fuzzy clustering and intuitionistic fuzzy recommender systems for medical diagnosis, Expert Syst. Appl. 42 (7) (2015) 3682–3701.
[29] P.V. Viet, H.T.M. Chau, L.H. Son, P.V. Hai, Some extensions of membership graphs for picture inference systems, in: Knowledge and Systems Engineering (KSE), 2015 Seventh International Conference on, Ho Chi Minh City, Vietnam, IEEE, October 8-10, 2015, pp. 192–197.
[30] G.W. Wei, F.E. Alsaadi, T. Hayat, A. Alsaedi, Projection models for multiple attribute decision making with picture fuzzy information, Int. J. Mach. Learn. Cybern. 9 (4) (2018) 713–719.
[31] G.W. Wei, H. Gao, The generalized Dice similarity measures for picture fuzzy sets and their applications, Informatica 29 (1) (2018) 1–18.
[32] G.W. Wei, Some similarity measures for picture fuzzy sets and their applications, Iran. J. Fuzzy Syst. 15 (1) (2018) 77–89.
[33] G.W. Wei, Picture fuzzy cross-entropy for multiple attribute decision making problems, J. Bus. Econ. Manag. 17 (4) (2016) 491–502.
[34] G.W. Wei, Picture fuzzy aggregation operators and their application to multiple attribute decision making, J. Intell. Fuzzy Syst. 33 (2017) 713–724.

[35] G.W. Wei, Picture fuzzy Hamacher aggregation operators and their application to multiple attribute decision making, Fundam. Inform. 157 (3) (2018) 271–320.
[36] Y. Yang, C. Liang, S. Ji, T. Liu, Adjustable soft discernibility matrix based on picture fuzzy soft sets and its applications in decision making, J. Intell. Fuzzy Syst. 29 (4) (2015) 1711–1722.
[37] J. Ye, Intuitionistic fuzzy hybrid arithmetic and geometric aggregation operators for the decision-making of mechanical design schemes, Appl. Intell. 47 (2017) 743–751.
[38] L.A. Zadeh, Fuzzy sets, Inf. Control 8 (1965) 338–353.
[39] H.J. Zimmermann, Fuzzy programming and linear programming with several objective functions, Fuzzy Sets Syst. 1 (1) (1978) 45–55.

# Index

## Q



## R



## S



## T

Printed in the United States
by Baker & Taylor Publisher Services

Printed in the United States
by Baker & Taylor Publisher Services